St. Louis Community College

Forest Park
Florissant Valley
Meramec

Insructional Resources
St. Louis, Missouri

DIGITAL
ELECTRONICS
WITH
MICROPROCESSOR
APPLICATIONS

NEW TITLES IN ELECTRONIC TECHNOLOGY

Alan Dixon and James Antonakos
DIGITAL ELECTRONICS WITH MICROPROCESSOR APPLICATIONS
(1987)

Rodney B. Faber
ESSENTIALS OF SOLID-STATE ELECTRONICS (1985)

Joseph D. Greenfield
PRACTICAL DIGITAL DESIGN USING ICs, 2nd edition (1983)

Larry Jones and A. Foster Chin
ELECTRONIC INSTRUMENTS AND MEASUREMENTS (1983)

Adi J. Khambata
MICROPROCESSORS/MICROCOMPUTERS: ARCHITECTURE, SOFT-
WARE AND SYSTEMS, 2nd edition (1987)

Sol Lapatine
ELECTRONICS IN COMMUNICATIONS, 2nd edition (1986)

Donald P. Leach
BASIC ELECTRIC CIRCUITS, 3rd edition (1984)

Martin Newman
INDUSTRIAL ELECTRONICS AND CONTROLS (1986)

Dan Porat and Arpad Barna
INTRODUCTION TO DIGITAL TECHNIQUES, 2nd edition (1987)

Mohamed Rafiquzzaman
MICROCOMPUTER THEORY AND APPLICATIONS WITH INTEL
SDK-85, 2nd edition (1987)

James Reynolds
APPLIED TRANSFORMED CIRCUIT THEORY FOR TECHNOLOGY (1985)

Theodore Wildi
BASIC ELECTRICITY FOR INDUSTRY: CIRCUITS AND MACHINES (1985)

Henry Zanger
SEMICONDUCTOR DEVICES AND CIRCUITS (1984)

DIGITAL ELECTRONICS WITH MICROPROCESSOR APPLICATIONS

ALAN C. DIXON
Broome Community College

JAMES ANTONAKOS
Broome Community College

JOHN WILEY & SONS
NEW YORK   CHICHESTER   BRISBANE   TORONTO   SINGAPORE

Library of Congress Cataloging in Publication Data:

Dixon, Alan C.
  Digital electronics with microprocessor applications.

  Includes index.
  1. Digital electronics.   2. Microprocessors.
3. Intel 8080A (Microprocessor)   4. Intel 8085
(Microprocessor)   I. Antonakos, James.   II. Title.
TK7868.D5D59   1987   621.391′6   86-1350
ISBN 0-471-89018-9

Printed in the United States of America

10 9 8 7 6 5 4 3 2 1

TO OUR PARENTS

PREFACE

We are in the age of the microprocessor. Even the simplest tasks are being dedicated to microprocessor-based circuitry. Automatic telephone dialers, burglar alarms, VCRs, and digital clocks have been equipped with microprocessors, and these are but a few of the low-level applications. Microprocessors are also being used by the thousands in industry, in both commercial and military applications. Most of us, in fact, come in contact with microprocessors many times each day without realizing it. Cash registers, automobile dashboards, telephone equipment, automatic bank tellers, and other devices have dedicated microprocessors working away to make our lives easier. With microprocessors appearing in ever-increasing numbers, one fact becomes clear: microprocessors are here to stay.

The technician, engineer, or programmer who is preparing to use microprocessors should study both their software and hardware in depth. This book is designed to do just that. A student with basic math skills, an understanding of electrical circuit theory, and an earnest desire to learn, will profit immensely from the information contained in the following 16 chapters, both in direct and immediate application and also as a stepping-stone to larger understanding.

Before the potential user learns microprocessor architecture and programming, a strong understanding of digital electronics must be developed, since the microprocessor after all is a digital machine. For this reason, this text may be separated into two parts. The first half (Chapters 1 to 8) deals exclusively with digital electronics and the information related to microprocessors in these chapters is plentiful.

The second half of the book offers instruction on the two most popular microprocessors now on the market: the Intel 8080A and the Intel 8085. Whereas the first of these, the 8080A, has been around since the early 1970s and may be considered by some as an "old" micro, the 8080A still enjoys widespread use. From an educational viewpoint, learning an 8-bit instruction set like the 8080As is faster and easier than tackling the instruction set of a 16- or 32-bit microprocessor, especially for a student with no previous knowledge of computers.

Chapter 1 explains the binary, octal, and hexadecimal number systems, and the methods involved in converting between any of these bases. A brief explanation of how binary numbers are used in computers is also included.

Chapter 2 covers the rules necessary for adding, subtracting, and multiplying binary numbers. Division is covered later in Chapter 12, due to its complexity.

Chapter 3 introduces logic gates, simple Boolean algebra, truth tables, and a few of the most popular logic families such as TTL, CMOS, and ECL. Gate applications include switch debouncers, oscillators, and adder circuits.

Chapter 4 covers the design of custom digital circuitry from the "black box" approach. Methods of gate reduction by use of the Karnaugh map are covered in detail. Displays and PAL devices are introduced.

Chapter 5 presents more complex logic functions such as flip-flops, counters, and shift registers. Applications include custom counters and dividers, and the use of the shift register to produce time delays.

Chapter 6 continues with complex logic functions, with an emphasis on encoders/multiplexers and decoders/demultiplexers. The use of logic gates to control digital signals is covered. Gates used in microprocessor based systems are included.

Chapter 7 covers various types of semiconductor memories. RAMs, ROMs, PROMs, static and dynamic memories, and mass storage devices, such as disks and bubble memories, are included, core memories and character generator ROMs are also covered. Applications include psuedorandom generators, diode-matrix ROMs, and video display logic.

Chapter 8 details the ASCII character set, its use in computers, and forms of serial and parallel data transmission. Baud rate, DMA, UARTs, and the RS232 standard are also covered.

Chapter 9 deals with the computer from a block-diagram approach. The four main elements of any computer (CPU, memory, timing and control, and I/O) are covered, as well as interrupts, mnemonics, and the basic instruction groups common to all computers.

Chapter 10 introduces the 8080A and 8085 microprocessor instruction sets, with examples for each instruction type.

Chapter 11 employs the assembly language instructions from the previous chapter in programming applications including delay loops, stack control, subroutines, bit manipulation, serial I/O, data tables, command recognizers, IC testing, traffic light controllers, interrupts, and memory mapped video displays.

Chapter 12 continues with programming applications of an advanced nature, such as a software controlled security system, a real-time clock, an 8080A/8085 disassembler, a telephone dialer, a BCD math package, D/A and A/D converters, and the efficient use of code.

Chapter 13 covers all hardware aspects of the 8080A and 8085 microprocessors, and discusses the single board computer. Common industrial busses such as the S-100, Multibus, GPIB, and STD are discussed.

Chapter 14 gives an overview of a few of the many peripheral controllers available for microprocessor use. These include DMA devices, high-speed binary math chips, and disk controllers.

Chapter 15 details the instruction sets of many current microprocessors. Eight, 16-, and 32-bit machines are included, such as the 6800, Z80, 6502, 8086, and 68000.

Chapter 16 introduces the student to digital troubleshooting, with many helpful hints. Logic analyzers and microcomputer development systems are also discussed.

The appendix contains a complete set of labs for the digital section of the course and a set of working 8080A/8085 assembly language programs, which can be used with the text.

The book is based on material used in a digital logic and microprocessors course that we have taught over an 8-year period. Through an accelerated teaching process involving simultaneous lectures, digital labs, and micro labs, the students are able to cover many of the book's topics in 15 weeks, although the book is detailed enough to allow for its use over two semesters.

We thank our reviewers for their helpful criticism and suggestions. They are: Dave Terrell, ITT Technical Institute; Leo D. Martin, Iowa Western Community College; John L. Morgan, De Vry Institute of Technology; John Debo, University of Central Florida; Vernon Hartshorn, Mount Hood Community College; Art Seidman, Pratt Institute; George Alexander, New Mexico State University; V. S. Anandu, East Texas State University; Micheal Mason, Cabrillo College; Jack Carter, Memphis State University; and Daniel Merkel, Milwaukee Area Technical School.

We also thank the many students who offered their ideas, original programs, and numerous corrective comments during the initial phases of the manuscript. We are grateful to Professor Michael Coppola for his valued proofreading of many sections and to Donovan McCarty for his technical advice. Most of all, we are indebted to Maureen Gillette for typing the original manuscript, thus making the book possible.

Alan C. Dixon

James Antonakos

CONTENTS

CHAPTER 1
NUMBER SYSTEMS   1

   1   Instructional Objectives   1
   2   Self-Evaluation Questions   1
   3   Introduction   2
   4   Current Number Systems   2
   5   Decimal, Binary, Octal, and Hexadecimal   5
   6   Similarities Between Bases   11
   7   Counting in Different Bases   12
   8   Converting Between Bases   13
   9   Use of Hex and Binary in Computers   19
  10   *Summary*   *21*
  11   *Glossary*   *22*

CHAPTER 2
BINARY ARITHMETIC   25

   1   Instructional Objectives   25
   2   Self-Evaluation Questions   25
   3   Introduction   26
   4   Simple Addition in Binary   26
   5   Complex Addition in Binary   26
   6   Simple Binary Subtraction   28
   7   Multiplication and Division   35
   8   Binary Coded Decimal (BCD) Arithmetic   42
   9   *Conclusion*   *47*
  10   *Summary*   *47*
  11   *Glossary*   *47*

CHAPTER 3
LOGIC GATES   51

1   Instructional Objectives   51
2   Self-Evaluation Questions   51
3   Introduction   52
4   Logic Gates   59
5   Gate Applications   76
6   Software Logic Gates   87
7   *Summary*   *88*
8   *Glossary*   *89*

CHAPTER 4
LOGIC CIRCUIT DESIGN   97

1   Instructional Objectives   97
2   Self-Evaluation Questions   97
3   Introduction   98
4   Three-Input Digital Black Boxes   98
5   Reducing Logic Functions   99
6   The Cost of Digital Design   102
7   Karnaugh Mapping   108
8   Larger Karnaugh Maps   114
9   Design of the BCD to Seven-Segment Readout Decoder/Driver   116
10 Gate Array Logic   121
11 Programmable Array Logic: The PAL   121
12 *Summary*   *126*
13 *Glossary*   *126*

CHAPTER 5
FLIP-FLOPS, COUNTERS, AND SHIFT REGISTERS   129

1   Instructional Objectives   129
2   Self-Evaluation Questions   129
3   Introduction   129
4   Flip-Flops   130
5   Other Flip-Flops   138
6   Enables   139
7   Counters   139
8   Complex Counters   147
9   Shift Registers   148
10 Shift Register Operation   148
11 Other Shift Registers   153
12 Race Conditions   153
13 *Summary*   *154*
14 *Glossary*   *154*

CHAPTER 6
DATA CONTROL DEVICES   157

1   Instructional Objectives   157
2   Self-Evaluation Questions   157
3   Introduction   157
4   Selectors/Multiplexers   158
5   Decoders/Demultiplexers   168
6   Microcomputer Interface Circuits   176
7   Transceivers and Latches   176
8   Bringing It All Together   180
9   Timing Diagram Fundamentals   180
10   *Summary*   *184*
11   *Glossary*   *184*

CHAPTER 7
MEMORIES   189

1   Instructional Objectives   189
2   Self-Evaluation Questions   189
3   Introduction   189
4   Shift Register Memories   190
5   Static Versus Dynamic   194
6   RAM and ROM   195
7   Random Access Memory   195
8   Other Memories   214
9   *Summary*   *226*
10   *Glossary*   *226*

CHAPTER 8
DATA TRANSMISSION   229

1   Instructional Objectives   229
2   Self-Evaluation Questions   229
3   Introduction   229
4   The American Standard Code for Information Interchange
     (ASCII)   230
5   The Parity Bit   233
6   Transmission Methods   234
7   Eleven-Bit Transmission Code   235
8   RS232C: The Electronic Industries Association Standard   238
9   Handshaking and Line Protocol   239
10   Cable Length   240
11   Simplex, Half Duplex, and Full Duplex   241
12   Converting Between TTL and RS232C   242
13   Current Loop Transmission   242

14   Decoding ASCII Control Characters   243
15   Generating an RS232C Waveform Using a Digital Black Box   244
16   Large-Scale Integrated Receiver/Transmitter Chips   246
17   The Extended Binary Coded Decimal Interchange Code
     (EBCDIC)   248
18   Communication Over Telephone Lines   249
19   The Electronic Industries Association RS422 Standard   252
20   *Summary*   *253*
21   *Glossary*   *253*

CHAPTER 9
ORGANIZATION OF COMPUTERS   257
1   Instructional Objectives   257
2   Self-Evaluation Questions   257
3   Introduction   257
4   The Operation of a Computer   263
5   Computer Instructions   264
6   Direct Memory Access   272
7   The Interrupt   272
8   *Summary*   *273*
9   *Glossary*   *273*

CHAPTER 10
A SOFTWARE LOOK AT THE 8080A/85 MICROPROCESSOR   277
1   Instructional Objectives   277
2   Self-Evaluation Questions   277
3   Introduction   278
4   Internal Organization   278
5   The Instruction Set   280
6   The Complete 8080A/85 Instruction Set   318
7   Simple 8080A/85 Programming Examples   321
8   *Summary*   *329*
9   *Glossary*   *330*

CHAPTER 11
8080A/85 PROGRAMMING APPLICATIONS   335
1   Instructional Objectives   335
2   Self-Evaluation Questions   335
3   Introduction   336
4   Double-Register Delay   336
5   Improved Double-Register Delay   340
6   Using the Stack   343
7   Subroutines   345

8   The Caterpillar   347
9   Bit Flipping: A Software Flip-Flop   348
10   Serial Input/Output Techniques   350
11   Parallel Data Handling   364
12   A Traffic Signal Controller   372
13   Scanning a Keypad   374
14   The Hardware Interrupt   377
15   The Software Interrupts   379
16   Memory Mapping   382
17   Data Handling   386
18   Base Conversions   388
19   Binary Coded Decimal Arithmetic   392
20   A Disco Light Show   395
21   Using One-Shots   397
22   *Summary*   *399*
23   *Glossary*   *400*

CHAPTER 12
ADVANCED PROGRAMMING   405

1   Instructional Objectives   405
2   Self-Evaluation Questions   405
3   Introduction   406
4   Advanced Topics   406
5   The Software UART   406
6   Security Systems: Polling Many Lines   411
7   A Software Clock   412
8   HEXDUMP: A Program for Memory Dumps   419
9   An 8085 Disassembler   424
10   The Automatic Telephone Dialer   440
11   Binary Coded Decimal Arithmetic   442
12   WAND: A Universal Product Code Label Reader   459
13   A Digital Voltmeter   474
14   A Large Multiplex Display System   485
15   Downline Loading   497
16   Sorting   504
17   Saving Bytes   505
18   *Summary*   *506*
19   *Glossary*   *509*

CHAPTER 13
A HARDWARE LOOK AT THE 8080A/85   513

1   Instructional Objectives   513
2   Self-Evaluation Questions   513

3   Introduction   513
4   The 8080A Central Processing Unit   514
5   The 8085 Central Processing Unit   529
6   Using the 8085   535
7   Internal and External Busses   539
8   *Summary*   *554*
9   *Glossary*   *554*

CHAPTER 14
PERIPHERAL CONTROLLERS   557

1   Instructional Objectives   557
2   Self-Evaluation Questions   557
3   Introduction   558
4   The 8741A Universal Peripheral Interface   558
5   The 8202 Dynamic RAM Controller   561
6   The 8232 Floating-Point Processor   563
7   The 8231 Arithmetic Processor   567
8   The 8253 Programmable Interval Timer   571
9   Floppy Disk Controllers   572
10 The 8275 Programmable CRT Controller   574
11 The 8279 Programmable Keyboard/Display Interface   575
12 *Conclusion*   *579*
13 *Glossary*   *580*

CHAPTER 15
OTHER MICROPROCESSORS   583

1   Instructional Objectives   583
2   Self-Evaluation Questions   583
3   Introduction   584
4   The 8080A, the 8085, and the Z80   584
5   The Z80 Microprocessor   586
6   The 6800 Microprocessor   597
7   The 6502 Microprocessor   609
8   Sixteen-Bit Microprocessors   613
9   The 8086/8088   617
10 The 68000 Microprocessor   633
11 The 8048 Single-Chip Computer   643
12 The 8051 Single-Chip Microcomputer   645
13 The 8096 Sixteen-Bit Microcontroller   649
14 The Three-Chip 432 Set   653
15 The 2901 Four-Bit Processor Slice   654

16   The LSI-11 Microcomputer   658
17   Array Processors   662
18   *Summary*   662
19   *Glossary*   663

CHAPTER 16
TROUBLESHOOTING TECHNIQUES   667
1   Instructional Objectives   667
2   Self-Evaluation Questions   667
3   Introduction   667
4   Development Systems   668
5   Logic Analyzers   673
6   Testing Serial Data Communications   680
7   Signature Analysis   684
8   Troubleshooting Microprocessor-Based Systems   684
9   *Summary*   691
10   *Glossary*   692

APPENDIX A
Hexadecimal/Decimal Conversion; Powers of 2 and 16   A-1

APPENDIX B
8085 Microcomputer Programs for Laboratory Use   B-1

APPENDIX C
555 Timer Details   C-1

APPENDIX D
Digital and Microprocessor Integrated Circuits   D-1

APPENDIX E
Digital Laboratory Exercises for the 7400 Series   E-1

Answers to Selected Problems   P-1

INDEX   I-1

DIGITAL ELECTRONICS WITH MICROPROCESSOR APPLICATIONS

CHAPTER 1
NUMBER SYSTEMS

1 INSTRUCTIONAL OBJECTIVES

All chapters begin with chapter objectives and self-evaluation questions, to help you get more from the book. Read this material as you start a chapter and then reread it when you have finished. If you cannot answer any of the questions, reread the section that applies. Your reading of this chapter should enable you to:

1. Count in all four bases described.
2. Use the standard and shortcut methods to convert a decimal number to other bases.
3. Convert hexadecimal pairs into bytes for use in microcomputer programming.
4. Be familiar with powers of 2, 8, and 16.
5. Tell the differences between nibbles, bytes, words, and long words.
6. See how numbers compare from one base or radix to another.
7. Understand the difference between analog and digital signals.
8. Find the least significant bit (LSB) and the most significant bit (MSB) in a binary number.

2 SELF-EVALUATION QUESTIONS

Keep the following questions in mind and try to answer them when you have completed the chapter:

1. What are the four bases described, and what are the legal characters used in each?
2. How can all number systems be represented by the radix method?
3. What is the difference between a digital and an analog signal?
4. How do hexadecimal pairs relate to a microprocessor program?
5. What is a register used for?
6. What is the largest number that can be stored in a byte, a word, or a long word?

7. How does counting in hexadecimal differ from counting in decimal?

8. Can a number like 0.7 (7/10) be accurately represented in binary, octal, or hexadecimal? (Answer: No. Why not?)

3 INTRODUCTION

Before we begin to learn the four number systems important to digital logic and microcomputers, let us take a brief look at the history of computers.

In 1642 the French mathematician-philosopher Blaise Pascal invented the first mechanical adding machine, which consisted of eight 10-digit wheels connected to advance each other by the rules of addition. In 1670 the German mathematician Gottfried Wilhelm von Leibniz improved Pascal's design by adding multiplication gears.

In the 19th century in Great Britain Charles Babbage designed complicated machines that could have solved complex mathematical problems. The machines were never built, however, because they were too complex for the technology of the times.

Alan M. Turing, a 20th-century mathematician, introduced the idea of a general-purpose computer in 1937 when he proposed a simple hypothetical machine, since named the universal Turing machine. A general-purpose computer is one that can be programmed to handle a variety of diverse problems. The computers in this text are all general purpose in nature.

The first electronic computer appeared in 1946 at the University of Pennsylvania. The computer was called ENIAC, which is the acronym for electronic numerical integrator and computer. It contained 18,000 vacuum tubes and occupied 15,000 square feet of floor space. ENIAC was able to perform 5000 additions or between 360 and 500 multiplications per minute.

With the invention of the transistor and later on the integrated circuit, computers shrunk in size, and now powerful devices can sit on the top of a desk. Some calculators are even small enough to fit inside a wristwatch.

The one link from Pascal's adding machine through ENIAC to today's microchip computers is found not in the machine's internal circuitry, but in its external communication. After all, computers are number-oriented machines, and in the next section we will see exactly why numbers are so important.

4 CURRENT NUMBER SYSTEMS

Today's microcomputers deal in four common number systems: decimal, binary, octal, and hexadecimal. Although most people are very much at ease with decimals, the other three number systems can seem awkward at first. In this chapter we examine all four systems in detail and learn how they operate in digital computers. After you have had some practice, these number systems will become powerful tools for designing and using all types of digital circuits.

The circuitry on the inside of any microcomputer must deal with the binary

FIGURE 1-1 Microcomputer Chip

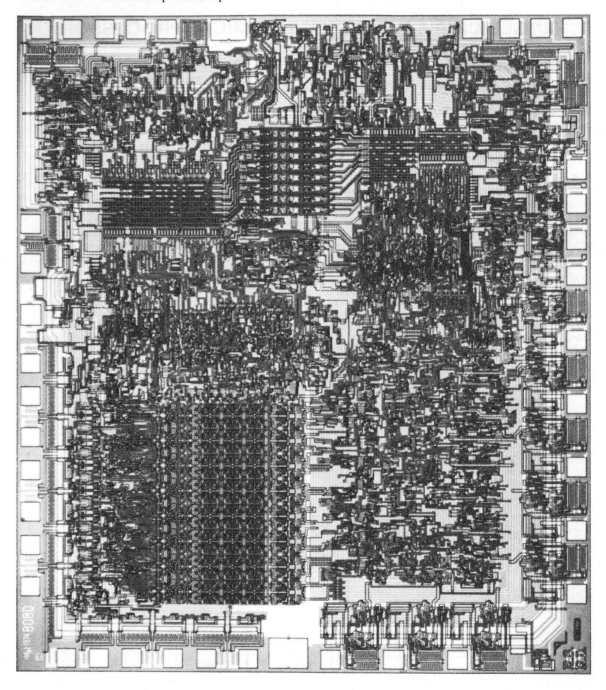

TABLE 1-1   Summary of Number Systems

Number Systems	Base or Radix[a]
Decimal	10
Binary	2
Octal	8
Hexadecimal	16

[a] The words base and radix are used interchangeably to refer to the base of the number system.

number system only. On the outside, however, the users of microcomputers deal with all four number systems or bases. Table 1-1 shows a summary of the four systems.

It can be easily shown why a microcomputer must deal in binary on the inside. Modern microcomputer chips have thousands of transistors packed into an area the size of your thumbnail (see Fig. 1-1). Each transistor can be thought of as a switch that is either completely turned on or completely turned off. These two states, on and off, are used to represent the two characters of the binary number system, one (1) and zero (0). These two states may also be thought of as high/low, true/false, yes/no—or on/off. A circuit that has only two states and can be represented by a pair of voltage levels is said to be a *digital* circuit. All microcomputer chips and integrated circuits that make up a microcomputer are digital circuits. Electronic circuits that consist of many voltage levels and are not limited to two different voltages are considered to be nondigital. These circuits are used to amplify or produce music or voice or complex waveforms, and they have an infinite number of voltage levels. Such circuits are called *analog* circuits. Simple digital and analog signals are shown in Fig. 1-2.

There are two worlds to consider then, the digital and the analog. This text is mainly concerned with digital signals and circuitry. To combine the two forms of circuitry, a converter or conversion process is necessary. To connect digital circuitry to analog circuitry, a digital-to-analog converter is used. Likewise, to

FIGURE 1-2 Simple digital (*a*) and analog (*b*) signal

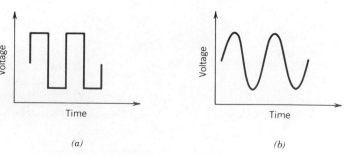

(a)   (b)

connect an analog circuit to a digital one, an analog-to-digital converter is used. The subject of D-to-A and A-to-D converters is discussed at length in a later chapter. First we must study the four number systems that we are going to use in our work with digital circuitry and computers.

5 DECIMAL, BINARY, OCTAL, AND HEXADECIMAL

5.1 Decimal

The *decimal* number system has 10 legal characters or symbols. We are accustomed to calling them digits and are aware that they begin with zero and end with nine. Decimal numbers are made up of the characters zero to nine. We are also familiar with the weight or importance that we ascribe to the position each digit has within a number. The units, tens, hundreds, and thousands have the weights of 1, 10, 100, and 1000, respectively.

Let's consider the number 491.68. This decimal quantity contains 4 hundreds, 9 tens, 1 unit, 6 tenths, and 8 one-hundredths. Where do these weights come from? A weight can be easily shown to be the base of a number system raised to a power. Here is a simple table detailing the weights involved in our example number.

$$\begin{array}{ccccccc} 4 & 9 & 1 & . & 6 & 8 \\ 10^2 & 10^1 & 10^0 & . & 10^{-1} & 10^{-2} \end{array}$$

This is the basis for all our number systems. The weight of a position is related to the base you are operating in, raised to a power. This also holds when we deal with the binary number system.

5.2 Binary Number System

The *binary* or base 2 number system operates in a similar manner. In a binary number there are only two legal characters or symbols: zero (0) and one (1). Hence, binary numbers consist entirely of zeros and ones. Each binary character (i.e., each "0" or "1") is called a *bit*.

A sample binary number might look like this: 10101100.1101 B, where the B denotes the binary number system. This serves to avoid confusing binary numbers with decimal or hexadecimal ones, which may be written 491.68, 491.68 D or 491.68 T (D means decimal, T means ten). We will use this method, following a number with a letter when dealing with all four bases. We evaluate the binary number by first determining the weight of each binary position or bit. Once again, the weight is a function of the base raised to a power.

EXAMPLE 1-1

Find the decimal equivalent of the binary number 10101100.1101 B.

SOLUTION
The number is converted to base 10 by using the following method:

Row 1	1	0	1	0	1	1	0	0	·	1	1	0	1
	2^7	2^6	2^5	2^4	2^3	2^2	2^1	2^0	·	2^{-1}	2^{-2}	2^{-3}	2^{-4}
Row 3	128	64	32	16	8	4	2	1	·	$\frac{1}{2}$	$\frac{1}{4}$	$\frac{1}{8}$	$\frac{1}{16}$

The left-most bit of a number is called the *most significant bit* (MSB), since it has the most weight. The left-most bit of our example number has a total weight of 128. The right-most bit is called the *least significant bit* (LSB) because it has the least weight. To calculate the binary number, you now simply multiply the number in row 1 by the number directly underneath it in row 3. Thus we have

$$
\begin{array}{rcl}
1 \times 128 &=& 128 \\
+\ 0 \times 64 &=& 0 \\
+\ 1 \times 32 &=& 32 \\
+\ 0 \times 16 &=& 0 \\
+\ 1 \times 8 &=& 8 \\
+\ 1 \times 4 &=& 4 \\
+\ 0 \times 2 &=& 0 \\
+\ 0 \times 1 &=& 0 \\
+\ 1 \times 0.5 &=& 0.5 \\
+\ 1 \times 0.25 &=& 0.25 \\
+\ 0 \times 0.125 &=& 0 \\
+\ 1 \times 0.0625 &=& \underline{0.0625} \\
\text{Total} &=& 172.8125
\end{array}
$$

TABLE 1-2   Powers of 2

n	2^n	2^{-n}
0	1	1
1	2	0.5
2	4	0.25
3	8	0.125
4	16	0.0625
5	32	0.03125
6	64	0.015625
7	128	0.0078125
8	256	0.00390625
9	512	0.001953125
10	1,024	0.0009765625
11	2,048	0.00048828125
12	4,096	0.000244140625
13	8,192	0.0001220703125
14	16,384	0.00006103515625
15	32,768	0.000030517578125
16	65,536	0.0000152587890625
17	131,072	0.00000762939453125
18	262,144	0.000003814697265625
19	524,288	0.0000019073486328125
20	1,048,576	0.00000095367431640625

TABLE 1-3   Basic Data Sizes

$$1 \text{ Nibble} = 4 \text{ bits}$$
$$1 \text{ Byte} = 2 \text{ nibbles} = 8 \text{ bits}$$
$$1 \text{ Word} = 2 \text{ bytes} = 16 \text{ bits}$$
$$1 \text{ Long word} = 4 \text{ bytes} = 32 \text{ bits}$$

Nibble	1101
Byte	11011101
Word	1001101001101010
Long word	10011010011111011011100110101100

The powers of 2 are very important, and you are encouraged to learn them as soon as possible. You will need to know the range from 2^0 through 2^{20} (very important in microcomputer work) and also the numbers 2^{-1} through 2^{-8}. Use Tables 1-2 and 1-3 to familiarize yourself with these numbers, which will appear in your work. To help acquaint you further with the binary number system, let's evaluate another sample number.

EXAMPLE 1-2

Convert the number 1001011011.101101 B into base 10.

SOLUTION
As in Example 1-1, we sum only the weights of the positions that have the value "1"; thus we have

$$
\begin{array}{r}
512 \\
+ \quad 64 \\
+ \quad 16 \\
+ \quad 8 \\
+ \quad 2 \\
+ \quad 1 \\
+ \quad 0.5 \\
+ \quad 0.125 \\
+ \quad 0.0625 \\
+ \quad 0.015625 \\
\hline
603.703125
\end{array}
$$

We have seen that in a binary number each character is called a bit. Other data sizes (nibble, byte, etc.) are listed, with examples, in Table 1-3. The number of bits in a binary number determines the maximum size of the quantity to be represented. For instance, the largest number that can be represented with four bits is 1111 B. This represents the quantity 15 in decimal. Table 1-4 shows the largest number for some representative numbers of bits. This largest number can be computed from the expression $2^n - 1$, where n is the number of bits.

TABLE 1-4   Calculation of $2^n - 1$

Binary Bits	Largest Number
5	31 D = 11111 B
8	255 D = 11111111 B
12	4,095 D = 111111111111 B
16	65,535 D = 1111111111111111 B
18	262,143 D = 111111111111111111 B
36	6.8719×10^{10} D = 111111111111111111111111111111111111 B

We will observe later that the size of a computer's memory is limited by its ability to handle large numbers of bits in representing numbers.

5.3 Octal

Another number system that finds limited use in computer systems is *octal* or base 8. Octal numbers require eight legal characters, and we use the digits 0 through 7, so an octal number would never contain the digits 8 or 9. An octal number will be followed by the letter O or Q to distinguish it from other bases. A Q is preferred, since an O might be mistaken for a zero and the number interpreted as a base 10 number. A sample octal number is 1423.41 Q. The weights of the positions are again related to the base raised to a power.

EXAMPLE 1-3

Evaluate the following octal number: 1423.41 Q.

SOLUTION
Unlike binary, where we used powers of 2 to calculate the weights, we will now use powers of eight.

$$
\begin{array}{cccccccc}
1 & 4 & 2 & 3 & . & 4 & 1 & Q \\
8^3 & 8^2 & 8^1 & 8^0 & . & 8^{-1} & 8^{-2} & \\
512 & 64 & 8 & 1 & . & .125 & .015625 &
\end{array}
$$

Now, by multiplying the first and third rows and taking the sum, we have

$$
\begin{array}{lcl}
1 \times 512 & = & 512 \\
+\ 4 \times 64 & = & 256 \\
+\ 2 \times 8 & = & 16 \\
+\ 3 \times 1 & = & 3 \\
+\ 4 \times 0.125 & = & 0.5 \\
+\ 1 \times 0.015625 & = & \underline{0.015625} \\
\text{Total} & = & 787.515625
\end{array}
$$

EXAMPLE 1-4

Convert the binary number 11011101.11011 B to decimal form.

$$
\begin{array}{rcl}
1 \times 128 &=& 128 \\
+\ 1 \times \ \ 64 &=& 64 \\
+\ 1 \times \ \ 16 &=& 16 \\
+\ 1 \times \ \ \ 8 &=& 8 \\
+\ 1 \times \ \ \ 4 &=& 4 \\
+\ 1 \times \ \ \ 1 &=& 1 \\
+\ 1 \times \frac{1}{2} &=& 0.5 \\
+\ 1 \times \frac{1}{4} &=& 0.25 \\
+\ 1 \times \frac{1}{16} &=& 0.0625 \\
+\ 1 \times \frac{1}{32} &=& \underline{0.03125} \\
\text{Total} &=& 221.84375
\end{array}
$$

EXAMPLE 1-5

Convert the octal number 3407.521 Q to decimal form.

$$
\begin{array}{rcl}
3 \times 512 &=& 1536 \\
4 \times 64 &=& 256 \\
7 \times 1 &=& 7 \\
5 \times \frac{1}{8} &=& 0.625 \\
2 \times \frac{1}{64} &=& 0.03125 \\
1 \times \frac{1}{512} &=& \underline{0.00195313} \\
\text{Total} &=& 1799.6582
\end{array}
$$

5.4 Hexadecimal

The *hexadecimal* or base 16 number is very important and enjoys widespread use in computers. Hexadecimal (sometimes called just "hex") requires 16 legal characters. Since we only have 10 digits, we use not only zero to nine, but also the first six letters of the alphabet, A through F. In hexadecimal we write the letter A to represent 10 decimal. Table 1-5 shows the conversion.

A typical hexadecimal number *may* contain both digits and letters as follows: 2AC6.4B H. Notice the H to indicate that we are dealing with a hexadecimal number. (We need the H to ensure that a decimal number, which is followed by a D, will not be mistaken for a hexadecimal number, where D is a legal character.) As in the three other number systems, the weight of each position is related to the base.

TABLE 1-5   Conversion from Hex to Decimal

Decimal	0	1	2	3	4	5	6	7	8	9	10	11	12	13	14	15
Hex	0	1	2	3	4	5	6	7	8	9	A	B	C	D	E	F

EXAMPLE 1-6

Calculate the value of the hexadecimal number 2AC6.4B H.

SOLUTION
As in Example 1-1, we make a small table of weights and take the sum of the products in the columns.

2	A	C	6	.	4	B	H
16^3	16^2	16^1	16^0	.	16^{-1}	16^{-2}	
4096	256	16	1	.	0.0625	0.00390625	

$$
\begin{aligned}
2 \times 4096 &= 8{,}192 \\
+\ 10 \times 256 &= 2{,}560 \\
+\ 12 \times 16 &= 192 \\
+\ 6 \times 1 &= 6 \\
+\ 4 \times 0.0625 &= 0.25 \\
+\ 11 \times 0.00390625 &= 0.04296875 \\
\hline
\text{Total} &= 10{,}950.29297
\end{aligned}
$$

Notice that in the conversion, A becomes 10, C becomes 12 and B becomes 11.

We frequently use hex numbers to four places, and it is useful to know that the largest number in hex to four places left of the decimal point is 0FFFF H (0FFFF H = 65535 D = $2^{16} - 1$). Notice the zero (0) in the "hex" number. This is common when dealing with hex numbers. If the hexadecimal number begins with a letter, we customarily precede it with a zero. Both the following are correct: 0A412.4B H or 426.1B H. (When we deal with assemblers in a later chapter, you will see why the zero becomes important.)

Let us look at another example of a hexadecimal-to-decimal conversion.

EXAMPLE 1-7

Convert the following hexadecimal number to decimal:

$$4A29.C3\ H$$

$$
\begin{aligned}
4 \times 4{,}096 &= 16{,}384 \\
10 \times 256 &= 2{,}560 \\
2 \times 16 &= 32 \\
9 \times 1 &= 9 \\
12 \times \tfrac{1}{16} &= 0.75 \\
3 \times \tfrac{1}{256} &= 0.01171875 \\
\hline
\text{Total} &= 18{,}985.762
\end{aligned}
$$

An alternate method of representing different numbers by number systems, illustrated in Table 1-6, consists of placing the base of the number at the end of the number being considered.

TABLE 1-6	Different Ways of Representing Numbers
Hexadecimal	3F7C.44B3 H $= (3F7C.44B3)_{16}$
Decimal	489.24 D $= (489.24)_{10} = 489.24$
Binary	1101.11 B $= (1101.11)_2$
Octal	4531.45 Q $= (4531.45)_8$

For instance, 47.9 D will become $(47.9)_{10}$ and 3F7 H will become $(3F7)_{16}$. Both methods work to convey the idea of a different number system, but we will use the first because of its ready and immediate application to assembly language programming.

A final word about number systems. Every computer has many locations in memory for storing data, and the value of a computer is frequently related to the number of data storage locations it has. We can count the locations in decimal, binary, and hexadecimal. A typical microcomputer might be able to have a memory with 2^{16} (or 65,536) locations. We begin numbering these locations at zero; therefore, the locations are numbered from 0 through 65,535, or in mathematical terms, 0 to $2^{16} - 1$. They can be numbered in binary from 0 B to 1111 1111 1111 1111 B, or in hexadecimal from 0 H to 0FFFF H. It is left to you to prove that 65,535 = 0FFFF H = 1111 1111 1111 1111 B. Notice that we use four hex characters or 16 binary bits to represent the same number. The newest microcomputers are capable of working with 2^{20} (1,048,576) memory locations! They are numbered from 0 to $2^{20} - 1$, or 0FFFFF H, or 1111 1111 1111 1111 1111 B. Once again, you may prove that these numbers are all equal. In this case the same number uses 5 hex digits, or 20 binary bits.

6 SIMILARITIES BETWEEN BASES

Many bases are possible, and we are discussing only those used most commonly in computer systems. Regardless of the base (or radix), the weight assigned to a position is always the radix raised to a power as follows:

$$R^n \ldots R^3\ R^2\ R^1\ R^0.\ R^{-1}\ R^{-2}\ R^{-3} \ldots R^{-m}$$

To review then, in base N the weights of the positions are as follows:

Base (N)	R^4	R^3	R^2	R^1	R^0	.	R^{-1}	R^{-2}	R^{-3}
$N = 10$	10,000	1,000	100	10	1	.	0.1	0.01	0.001
$N = 2$	16	8	4	2	1	.	$\frac{1}{2}$	$\frac{1}{4}$	$\frac{1}{8}$
$N = 8$	4,096	512	64	8	1	.	$\frac{1}{8}$	$\frac{1}{64}$	$\frac{1}{512}$
$N = 16$	65,536	4,096	256	16	1	.	$\frac{1}{16}$	$\frac{1}{256}$	$\frac{1}{4096}$

These known weighting factors make it easy (with practice) to convert any binary, octal, or hexadecimal number to our familiar decimal number system. Section 8 shows how to convert in the other direction—that is, from decimal to another base.

7 COUNTING IN DIFFERENT BASES

We know that computers require many memory locations, which are numbered from zero up. We can count or number these locations in any of the bases we are studying. Therefore, we need a good understanding of how to count in the various number systems. Table 1-7 offers a comparison of a continuous count for you to study. Memory locations are frequently numbered only in hexadecimal to four hex positions. A full memory may range from 0000 H to 0FFFF H. Some sequential locations are as follows:

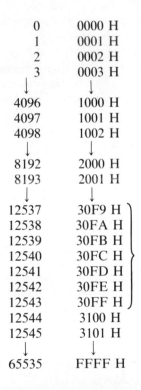

0	0000 H
1	0001 H
2	0002 H
3	0003 H
↓	↓
4096	1000 H
4097	1001 H
4098	1002 H
↓	↓
8192	2000 H
8193	2001 H
↓	↓
12537	30F9 H
12538	30FA H
12539	30FB H
12540	30FC H
12541	30FD H
12542	30FE H
12543	30FF H
12544	3100 H
12545	3101 H
↓	↓
65535	FFFF H

This sequence is provided because you may feel uncomfortable counting in hex at first. You *will* get used to it!

It is useful to refer to the number of a memory location as its *address*. The address of each memory location is unique and is a sequential number beginning with zero and usually given in hexadecimal. Since we begin counting memory locations at address zero, the address of the fifth location is 4 decimal, or 0004 H. Similarly, the address of the 12,540th location is 12539 decimal, or 30FB H. It is always one less because our count starts at zero and not one. In other words, location 1 is address zero. You see now why it is important to be able to count in the different bases.

TABLE 1-7   Comparison of a Continuous Count

Decimal	Binary	Octal	Hexadecimal
0	00000 B	0 Q	0 H
1	00001 B	1 Q	1 H
2	00010 B	2 Q	2 H
3	00011 B	3 Q	3 H
4	00100 B	4 Q	4 H
5	00101 B	5 Q	5 H
6	00110 B	6 Q	6 H
7	00111 B	7 Q	7 H
8	01000 B	10 Q	8 H
9	01001 B	11 Q	9 H
10	01010 B	12 Q	A H
11	01011 B	13 Q	B H
12	01100 B	14 Q	C H
13	01101 B	15 Q	D H
14	01110 B	16 Q	E H
15	01111 B	17 Q	F H
16	10000 B	20 Q	10 H
17	10001 B	21 Q	11 H
18	10010 B	22 Q	12 H
19	10011 B	23 Q	13 H
20	10100 B	24 Q	14 H

8 CONVERTING BETWEEN BASES

We will now learn how to convert decimal numbers to any of the other three bases in use today. There are two separate methods, one for converting the integer part of a number and one for converting the fractional part. The integer part uses a divide-down procedure while the fractional method uses a multiply procedure. These two methods convert a decimal number to any base.

$$293 \cdot 625$$

Integer part   Fractional part

8.1 Decimal Integers to Another Base

To convert a decimal number (integer part only) to another base, we divide by the base we are converting to, and find the number of times that this division is possible. Any remainders will be the converted result in the new base.

EXAMPLE 1-8

Convert the decimal number 293 to base 2.

SOLUTION

To convert, we simply divide by 2 and write the remainders in a separate column, which will be the converted answer.

$$
\begin{array}{ll}
2 \underline{\; 293 \;} & \cdot \text{ binary point} \\
146 & 1 \;(\text{i.e., remainder of } 1) \\
73 & 0 \;(\text{i.e., remainder of } 0) \\
36 & 1 \\
18 & 0 \\
9 & 0 \\
4 & 1 \\
2 & 0 \\
1 & 0 \\
0 & 1
\end{array}
$$

When we get to zero, we are finished with the dividing, so now we read the remainder digits from the bottom up to the decimal point. Doing this we have 100100101 B. The result is always carefully checked by the method we know from Section 5.2: 256 + 32 + 4 + 1 = 293. It checks!

This method works when converting decimal integers to any base. Let us now convert the same number to octal:

$$
\begin{array}{ll}
8 \underline{\; 293 \;} & \cdot \\
36 & 5 \\
4 & 4 \\
0 & 4
\end{array}
$$

The result is 445 Q and we also check the result:

$$
\begin{array}{rl}
5 \times 1 = & 5 \\
+\; 4 \times 8 = & 32 \\
+\; 4 \times 64 = & \underline{256} \\
& 293 \qquad \text{Check!}
\end{array}
$$

And, let's convert 293 into hexadecimal:

$$
\begin{array}{ll}
16 \underline{\; 293 \;} & \cdot \\
18 & 5 \\
1 & 2 \\
0 & 1
\end{array}
$$

The result is 125 H, which we check quickly (by the same method):

$$
\begin{array}{rl}
5 \times 1 = & 5 \\
+\; 2 \times 16 = & 32 \\
+\; 1 \times 256 = & \underline{256} \\
& 293 \qquad \text{Once again!}
\end{array}
$$

Let us now learn how to convert the fractional part of a decimal number to another base.

8.2 Decimal Nonintegers to Another Base

To convert decimal numbers to any other base, a multiply procedure is used. We simply multiply the number to be converted by the new base, with one limit: Only multiply what is to the right of the decimal point! Numbers appearing to the left of the decimal point are part of the result.

EXAMPLE 1-9

Convert 0.625 decimal to base 2.

SOLUTION
To solve, we simply multiply all numbers to the right of the decimal point by 2 until the number immediately to the right of the decimal is zero. Then we read the digits on the *left* of the decimal point from the top down. Doing this, we have

$$
\begin{array}{r}
0.625 \\
\times\ 2 \\
\hline
\underline{1}.250
\end{array}
$$

$$
\begin{array}{r}
0.250 \\
\times\ 2 \\
\hline
\underline{0}.500
\end{array}
$$

$$
\begin{array}{r}
0.500 \\
\times\ 2 \\
\hline
\underline{1}.000
\end{array}
$$

Now we are done because only zeros are on the right of the decimal point. Reading from the top down (opposite of the integer routine), we get .101 B = 0.625 decimal. We also check our result to be sure:

$$
\begin{array}{rl}
1 \times 0.5 & = 0.5 \\
+\ 0 \times 0.25 & = 0.0 \\
+\ 1 \times 0.125 & = \underline{0.125} \\
& \ \ 0.625 \quad \text{Check!}
\end{array}
$$

The conversion of fractions is not always exact, so it will be useful to do another example.

EXAMPLE 1-10

Convert the number 0.2 to binary.

SOLUTION
Some numbers, like 0.2, will not convert exactly. Rather, they continue forever with a particular binary pattern of ones and zeros.

$$
\begin{array}{r}
0.2 \\
\times\ 2 \\
\hline
0.4 \\
0.8 \\
1.6 \\
1.2 \\
0.4 \\
0.8 \\
1.6 \\
1.2 \\
0.4
\end{array}
$$

The result is .00110011$\overline{0011}$ B: The sequence 0011 will continue forever. (*Note*: The repeating portion of such figures is indicated with an overline.)

$$
\begin{array}{r}
1 \times .125 \\
+ 1 \times .0625 \\
+ 1 \times .0078125 \\
+ 1 \times .00390625 \\
+ 1 \times .00048828125 \\
+ 1 \times \underline{.000244140625} \\
\hline
.199951171875
\end{array}
$$

This is not quite 0.2, nor will it ever be. To simplify matters, if we take only the first six 1s (we selected six 1s for convenience here, but this choice does not represent any standard and you must determine your own requirements for accuracy) we will find that we are closely approximating 0.2.

In Chapter 2 we discuss how decimal numbers may be more accurately represented in binary codes. The error between 0.2 and 0.199951171 can be lessened, but not eliminated, by going to more places. This is one problem with binary numbers that must be overcome. It is usually solved by adding a slight "fuzz," or rounding many digits down the line. In other words, once the result is 0.19999999, a small part is added to this (0.00000001) to force the result to 0.2, which is the correct answer. This is how very large computer systems handle the situation. Microcomputers use binary coded decimal (BCD) codes, which are covered in Chapter 2.

The fractional conversion from decimal to another base can also work for octal, so let us convert both example numbers 0.625 and 0.2.

EXAMPLE 1-11

$$
\begin{array}{cc}
0.625 & 0.2 \\
\underline{\times\ 8} & \underline{\times\ 8} \\
5.00 & 1.6 \\
0.00 & 4.8 \\
& 6.4 \\
& 3.2 \\
& 1.6 \\
& 4.8
\end{array}
$$

By this example you can see that .625 decimal = .5 Q and that .2 decimal once again does not convert exactly, but repeats: .1463$\overline{1463}$ Q. If we use our familiar method of checking our results, we will get

$$5 \times 0.125 = .625 \quad \text{Check!}$$

$$
\begin{array}{rll}
1 \times .125 & = .125 \\
+ 4 \times .015625 & = .0625 \\
+ 6 \times .001953125 & = .01171875 \\
+ 3 \times .00024414 & = \underline{.000732421} \\
& \quad\ \ .199951171 & \text{Close!}
\end{array}
$$

In doing fractional conversions that never stop, we will find that the answer always closes in on the expected result. It does this from below; that is, the result gets closer and closer to .2 (in this case), but never exceeds .2.

Let us try another example.

EXAMPLE 1-12

Convert 0.625 to binary.

SOLUTION

$$
\begin{array}{r}
0.625 \\
\times\ \ 2 \\
\hline
1.25 \\
0.50 \\
1.00 \\
0.00
\end{array}
$$

The result is .101 B. Therefore, .625 D = .101 B = .5 Q.

$$
\begin{array}{rll}
1 \times .5 & = .5 \\
+\ 0 \times .25 & = .0 \\
+\ 1 \times .125 & = \underline{.125} \\
& \ \ \ .625 & \text{Correct!}
\end{array}
$$

EXAMPLE 1-13

Try a hexadecimal conversion with the two numbers used in Example 1-11.

SOLUTION

$$
\begin{array}{cc}
0.625 & 0.2 \\
\times\ 16 & \times\ 16 \\
\hline
10.00 & 3.2 \\
0.00 & 3.2 \\
 & 3.2 \\
 & 3.2
\end{array}
$$

We convert the 10 to an A, obtaining our hexadecimal answer .625 decimal = .A H. Once again we see, too, that .2 converts into a repeating number, namely .333\overline{3} H. To check:

$$
A \times .0625 = 10 \times .0625 = .625!
$$

$$
\begin{array}{rll}
3 \times .0625 & = .1875 \\
+\ 3 \times .00390625 & = .01171875 \\
+\ 3 \times .00024414 & = \underline{.000732421} \\
& \ \ \ .199951171 & \text{Close!}
\end{array}
$$

8.3 Shortcut Between Bases

A relationship exists between bases 2, 8, and 16 that is very useful in making conversions from binary to or from octal and hexadecimal. This relationship comes from the fact that $2^3 = 8$ and $2^4 = 16$. To convert 46 Q to binary, we allow three binary places for each character. We then use a 4-2-1 weighting scheme—the 421 code, as it is called.

EXAMPLE 1-14

Convert 46 Q into binary.

SOLUTION
By making use of the 421 code, we have, simply:

```
4 2 1   4 2 1
1 0 0   1 1 0 . B

  4       6          Therefore 46 Q = 100110 B.
```

The 4 and 6 are each encoded using the 421 code.

EXAMPLE 1-15

Convert 763.421 Q into binary.

SOLUTION
Making use of the 421 code again, we have

```
4 2 1   4 2 1   4 2 1  .  4 2 1   4 2 1   4 2 1
1 1 1   1 1 0   0 1 1  .  1 0 0   0 1 0   0 0 1 B

  7       6       3    .   4        2       1
```

There is also an easy way to go between binary and hexadecimal numbers. To convert 4AC3 H to binary, an 8421 code or weighting scheme is used. In this case, four binary places (or bits) are allowed for each character. Knowing this, let us look at another example.

EXAMPLE 1-16

Convert 4AC3 H and 2864.D2C3 H into binary.

SOLUTION
The characters are encoded using the 8421 code as follows:

```
8 4 2 1   8 4 2 1   8 4 2 1   8 4 2 1  .  8 4 2 1   8 4 2 1   8 4 2 1   8 4 2 1
0 1 0 0   1 0 1 0   1 1 0 0   0 0 1 1 B
   4         A         C         3 H
0 0 1 0   1 0 0 0   0 1 1 0   0 1 0 0  .  1 1 0 1   0 0 1 0   1 1 0 0   0 0 1 1 B
   2         8         6         4     .    D         2         C         3 H
```

These shortcuts are very useful in working with numbers in various computer systems. Let us try one more example, converting first to base 16 and using the shortcut to obtain the binary and octal results.

EXAMPLE 1-17

Convert 293.41 decimal to binary, octal, and hexadecimal.

SOLUTION
First do the integer part:

$$
\begin{array}{r|rl}
16 & 293 & \quad\cdot \\
& 18 & 5 \\
& 1 & 2 \\
& 0 & 1 \\
\end{array}
$$

Then the fractional part:

$$
\begin{array}{r}
.41 \\
\times\ \ 16 \\
\hline
6.56 \\
8.96 \\
15.36 \\
5.76 \\
\end{array}
$$

The result is then 125.68F5 H. Let's check the result:

$$
\begin{array}{rcl}
5 \times 1 & = & 5 \\
+\ 2 \times 16 & = & 32 \\
+\ 1 \times 256 & = & \underline{256} \\
& & 293 \\
\end{array}
$$

$$
\begin{array}{rll}
6 \times .0625 & = .375 \\
+\ 8 \times .00390625 & = .03125 \\
+\ F \times .00024414 & = .0036621 \\
+\ 5 \times .000015258 & = \underline{.000076293} \\
& \quad\ .409988393 & \text{Close!} \\
\end{array}
$$

We can now use the shortcut to convert over to binary.

125.68F5 H

1 0010 0101 . 0110 1000 1111 0101 B

And now we convert into octal by grouping the binary bits above into triplets.

100 100 101 . 011 010 001 111 010 B

445.32172 Q

9 USE OF HEX AND BINARY IN COMPUTERS

The circuitry inside a computer must deal in the binary number system because transistors operating either fully on or fully off can be used to represent a binary "1" or "0." While bases 2, 8, and 16 are used in computers, the predominant mode is binary (inside the machine) and hexadecimal (on the outside, whereby the user specifies the binary). A person could easily slip an extra 0 or 1 into a string of binary bits, but one is less likely to drop a hexadecimal character; hence the use of both numbering systems. Octal is used in some applications, but it has been largely supplanted by hexadecimal. Frequently when programming a micro-computer, it is necessary to rapidly convert from hex pairs to 8 bits of binary. For example, a programmer will need to convert 0C3 H to 11000011 B. Several examples are now shown.

EXAMPLE 1-18

Convert the following hex pairs into binary:
(a) 03E H   (b) 2F H   (c) 76 H   (d) 0CD H

SOLUTION
We simply use the 8421 code to convert separate characters, and then group them into 8-bit numbers.

(a) 03E H = 0011 1110 B

(b) 2F H = 0010 1111 B

(c) 76 H = 0111 0110 B

(d) 0CD H = 1100 1101 B

Frequently a program appears as a hexadecimal string, and it is left to the programmer to divide the string into hex pairs and then convert quickly into binary.

EXAMPLE 1-19

Convert the following hexadecimal string into binary numbers (8 bits each):
3EFFD300C20010AFCD102176.

SOLUTION
To solve, we take two characters at a time and convert into binary.

$$
\begin{array}{l}
3E - 0\,0\,1\,1\quad 1\,1\,1\,0 \\
FF - 1\,1\,1\,1\quad 1\,1\,1\,1 \\
D3 - 1\,1\,0\,1\quad 0\,0\,1\,1 \\
00 - 0\,0\,0\,0\quad 0\,0\,0\,0 \\
C2 - 1\,1\,0\,0\quad 0\,0\,1\,0 \\
00 - 0\,0\,0\,0\quad 0\,0\,0\,0 \\
10 - 0\,0\,0\,1\quad 0\,0\,0\,0 \\
AF - 1\,0\,1\,0\quad 1\,1\,1\,1 \\
CD - 1\,1\,0\,0\quad 1\,1\,0\,1 \\
10 - 0\,0\,0\,1\quad 0\,0\,0\,0 \\
21 - 0\,0\,1\,0\quad 0\,0\,0\,1 \\
76 - 0\,1\,1\,1\quad 0\,1\,1\,0
\end{array}
$$

Practice using hexadecimal pair–binary conversion and become expert at it. This is a common requirement for microcomputer programming.

While a computer's large memory must be able to store whole programs, it is also necessary to have a separate area in which to store smaller numbers—the bits, bytes, words, and so on listed in Table 1-3. These numbers are usually byte sized (8 bits) and are stored in one of several 8-bit locations called *registers*. A register in a microcomputer is capable of storing either an 8-bit number (a byte) or a word (double byte). We will be making good use of registers in the future. A large general-purpose computer has even larger registers, capable of storing 36 bits and more at one time. Registers of this size are not common in microcomputers . . . yet.

Before concluding, a few words on entering a microcomputer program. As we have seen in this chapter, a program may consist of hexadecimal pairs. These pairs must be *loaded* or *deposited* into the computer's memory locations beginning at a specific location or address. Each additional instruction must be loaded in the next sequential location. We may read the program in hexadecimal, but the program must be in binary in the memory location. Using the program from before, we deposit or load the program in sequential locations beginning, say, at address 1000 H. The program in hexadecimal is: 3EFFD300C20010AFCD102176. The program loaded or deposited into memory is:

Address	Memory Byte	Hex Form
1000	0 0 1 1 1 1 1 0	3E
1001	1 1 1 1 1 1 1 1	FF
1002	1 1 0 1 0 0 1 1	D3
1003	0 0 0 0 0 0 0 0	00
1004	1 1 0 0 0 0 1 0	C2
1005	0 0 0 0 0 0 0 0	00
1006	0 0 0 1 0 0 0 0	10
1007	1 0 1 0 1 1 1 1	AF
1008	1 1 0 0 1 1 0 1	CD
1009	0 0 0 1 0 0 0 0	10
100A	0 0 1 0 0 0 0 1	21
100B	0 1 1 1 0 1 1 0	76

This preview of our concerns in succeeding chapters of this text illustrates the importance of both binary and hexadecimal number systems. You should be able to quickly (and mentally) convert hexadecimal pairs to binary to deposit them into memory.

10 SUMMARY

In this chapter we learned about the four most common bases used in microcomputers: decimal (base 10), binary (2), octal (8), and hexadecimal (16). To convert the integer part of a number to a different base, a divide-down procedure is used. To convert the fractional part of a number, a multiply procedure is used. Numbers are easily converted between the bases 2, 8, and 16 by using the 421 and 8421 codes.

11 GLOSSARY

Address.   A number identifying a location in memory where data are stored.

Analog.   A continuously changing signal (i.e., having an infinite number of possible levels).

Binary.   A numbering system based on the number 2.

Bit.   The smallest element of a binary number.

Byte.   A group of 8 bits.

Digital.   A signal with only two possible states.

Hexadecimal.   A numbering system based on the number 16.

Legal character.   Any character belonging to a particular number system.

LSB.   The least significant bit: the character in the right-most position of a number. It has the smallest weight.

MSB.   The most significant bit: the character in the left-most position of a number. It has the largest weight.

Octal.   A numbering system based on the number 8.

Radix.   The base of the number system.

Register.   A temporary storage device used for one or more bytes.

Weight.   The importance of a position in a number.

Word.   Two bytes (16 bits) of information.

PROBLEMS

1-1 Convert each binary number to a decimal number.
- (a) 1101.101 B
- (b) 110111.110 B
- (c) 1101111010.11 B
- (d) .0001110101 B
- (e) 110111010011 B
- (f) 10110111.111 B
- (g) 11111111.00101 B
- (h) 1101101111.100101 B

1-2 Convert each octal number to a decimal number.
- (a) 26.32 Q
- (b) .001634 Q
- (c) 46721.43 Q
- (d) 5472.623 Q

1-3 Convert each hexadecimal number to a decimal number.
- (a) 6AC.4B H
- (b) .46ACB H
- (c) 4CFA.B2 H
- (d) 2631.42 H

1-4 In a microcomputer memory with addresses beginning at address 0, what is the hexadecimal address of each of the following locations?
- (a) The 2467 Dth location
- (b) The 65,000 Dth location

    (c) The 4000 Dth location
    (d) The 8191 Dth location

1-5 In the same microcomputer, what is the decimal location of each of the following addresses (location zero is the "first" location)?
    (a) 2000 H
    (b) 200 H
    (c) 200D H
    (d) 0CC00 H

1-6 Convert the following integers to bases 2, 8, and 16.
    (a) 75 D
    (b) 289 D
    (c) 4071 D
    (d) 28431 D

1-7 Convert the following decimals to bases 2, 8, and 16.
    (a) .2 D
    (b) .1 D
    (c) .0461 D
    (d) .879 D

1-8 Convert the following "hex" pairs into 8-bit binary.
    (a) 0FE H
    (b) 55 H
    (c) 0AA H
    (d) 0DB H
    (e) 0C3 H
    (f) 0CD H
    (g) 94 H

1-9 Convert the following microcomputer program to byte form (8-bit binary):
210040DB40FE02C2034076.

1-10 Convert the following octal numbers to binary, hexadecimal, and decimal values.
    (a) 263.41 Q   (e) 777.77 Q
    (b) 46.235 Q   (f) .1 Q
    (c) 20634 Q   (g) 22.22 Q
    (d) .00431 Q   (h) 234.4 Q

1-11 Convert the following decimal numbers to bases 2, 8, and 16.
    (a) 268.49   (d) 999.99
    (b) 34019.2   (e) 2.414
    (c) 222.41   (f) .000141

1-12 What is the largest number (decimal) that can be represented if a 20-bit binary number is all 1s? What is the hex equivalent?

1-13 What is the largest number (decimal) that can be represented if a 24-bit binary number is all 1s? What is the hex equivalent?

1-14 Convert the following numbers to binary.
    (a) 4768219.4083   (c) 372.43 Q
    (b) 20A.4CD H

1-15 Convert the following numbers to decimal.
    (a) 11011010111011.10011 B   (c) 333.47 Q
    (b) 473F.4C2 H

1-16 Convert the following numbers to octal.
    (a) 2A3.41 H   (c) 3476.004
    (b) 1100010011010.11111101 B

1-17 Fill in the following table:

N	2^N	8^N	16^N
0			
1			
2			
3			
4			
5		****	****
6		****	****
7		****	****
8		****	****

1-18 Compute the following:

$2^{20} =$ $8^3 =$
$2^{16} =$ $8^4 =$
$2^8 =$ $16^2 =$
$16^3 =$ $16^4 =$

1-19 Convert to 8-bit binary:

A7 H =   21 H =
D3 H =   E7 H =
C9 H =   DB H =
76 H =   F8 H =

CHAPTER 2
BINARY ARITHMETIC

1 INSTRUCTIONAL OBJECTIVES

When finished with this chapter, you should be able to:

1. Directly add or subtract complex binary numbers.
2. Multiply binary numbers by using the shift and accumulate method.
3. Add and subtract BCD numbers.
4. Determine the sign of a binary number.
5. Represent negative numbers using binary.
6. Correct the result of a BCD add operation.
7. Have an understanding of adding multiple binary numbers.
8. Use the carry bit and the auxiliary carry bit in math operations.

2 SELF-EVALUATION QUESTIONS

Keep the following questions in mind and try to answer them when you have completed the chapter.

1. What are the simple rules for binary addition?
2. How does the "control word" control the result in the accumulator in a binary multiplication?
3. Why do BCD numbers have to be adjusted after addition or subtraction?
4. What are the carry bit and the auxiliary carry bit?
5. When performing BCD subtraction, how is the number corrected when the result will be negative?
6. In the addition of hexadecimal numbers, how much does a carry represent?
7. Why is 2's complement arithmetic done?
8. What are the four variations of binary addition?

3 INTRODUCTION

In Chapter 1 we became familiar with binary numbers and their importance in microcomputers. Now we will see how a computer performs addition, subtraction, and multiplication with binary numbers. Since binary division is not a simple task on any level, we will postpone working with this function until we have more knowledge of microcomputer programming.

4 SIMPLE ADDITION IN BINARY

As we learn to add in binary it is helpful to remember how we add numbers in decimal.

$$
\begin{array}{r}
1 \quad \leftarrow \text{Carry into tens column} \\
8 \\
+\ 9 \\
\hline
17
\end{array}
$$

In the addition of 8 and 9, the total is 7 for the units column with a 1 to carry into the tens column.

The carry into the next column is common in binary addition as well. A few rules of binary addition are now in order.

$$
\begin{array}{ccccc}
 & & & & 1 \\
 & & & 1 & 1 \\
 & & 1 & 1 & 1 \\
0 & 1 & 1 & 1 & 1 \\
+\ 1 & +\ 1 & +\ 1 & +\ 1 & +\ 1 \\
\hline
1 & 10 & 11 & 100 & 101
\end{array}
$$

In such additions as 2 + 3, a carry occurs in the second and third columns:

$$
\begin{array}{ll}
1 \quad\quad \leftarrow \text{Carry bits} \\
010\ \text{B} & (2) \\
+\ 011\ \text{B} & +\ (3) \\
\hline
101\ \text{B} & (5)
\end{array}
$$

5 COMPLEX ADDITION IN BINARY

The larger the binary number, the more complex the addition can become. It is important to watch any carries and place them in the correct column or columns. Let's add 47 and 29 in binary:

$$
\begin{array}{rl}
 & 111111 \quad \leftarrow \text{Carry bits} \\
47 & 101111\ \text{B} \\
+\ 29 & +\ 011101\ \text{B} \\
\hline
76 & 1001100\ \text{B}
\end{array}
$$

Multinumber addition is a bit trickier.

EXAMPLE 2-1

Add 47 + 29 + 37.

SOLUTION

$$
\begin{array}{rr}
47 & 101111 \text{ B} \\
+\ 29 & 011101 \text{ B} \\
+\ 37 & +\ 100101 \text{ B} \\
\hline
113 & 1110001 \text{ B}
\end{array}
$$

To check our result:

$$
\begin{array}{rcl}
1 \times 1 & = & 1 \\
+\ 1 \times 16 & = & 16 \\
+\ 1 \times 32 & = & 32 \\
+\ 1 \times 64 & = & 64 \\
\hline
& & 113 \quad \text{Check!}
\end{array}
$$

In a computer, the addition of binary numbers is performed in a special area called the arithmetic logic unit (ALU). The ALU is capable of performing additions, subtractions, and many other operations. The results are almost always placed in a general-purpose register called the *accumulator*. In many microcomputers the accumulator is an 8-bit register. For this reason it is common to add 8-bit numbers together. Example 2-2 shows the addition of hex pairs.

EXAMPLE 2-2

Add 2C H and 48 H in both binary and hexadecimal.

SOLUTION

$$
\begin{array}{rl}
1 & 1 \qquad \leftarrow \text{Carry bits} \\
2C\ H & 00101100 \text{ B} \\
+\ 48\ \ H & 01001000 \text{ B} \\
\hline
74\ \ H & 01110100 \text{ B}
\end{array}
$$

In the hex addition of C + 8 we must convert C to 12 and call this the addition of 12 + 8 = 20. Then, since this is base 16, we subtract 16 from 20 and find that C + 8 is 4 with a 1 to carry to the next column. Let's try another example.

EXAMPLE 2-3

Add C3 H and B7 H in both binary and hexadecimal.

SOLUTION

$$
\begin{array}{rl}
1 & 1 \quad\ \ 111 \quad \leftarrow \text{Carry bits} \\
C3\ H & 11000011 \text{ B} \\
+\ B7\ H & 10110111 \text{ B} \\
\hline
17A\ H & 101111010 \text{ B}
\end{array}
$$

In the hex addition, $3 + 7 = 10$ or A, $C + B$ becomes $12 + 11 = 23$, and $23 - 16 = 7$. Therefore, $C + B$ is 7 with 1 to carry.

The final result of adding C3 H and B7 H is 17A H. The binary equivalent, which is really what appears in a computer, is also shown.

In Example 2-3 the result ends up in a carry in a ninth binary position, which is outside the 8-bit range of many microcomputer accumulators. This can be handled easily with clever programming, as covered in a later chapter.

6 SIMPLE BINARY SUBTRACTION

In subtraction with decimal numbers we have become accustomed to "borrowing" from another column to find a result. This is not easily done on a computer, and so computer subtraction is not done in the way that is intuitively familiar to us. The computer subtracts by using the *complement* of a number. The complement of a binary number (1's complement) is found by inverting each bit (changing each zero to a one and each one to a zero).

6.1 One's Complement

Consider the binary number for 28: 28 D = 11100 B. The 1's complement of 11100 B is 00011 B. In working with complements, it is very important to hold an equal number of places for numbers being subtracted. That is:

$$
\begin{array}{r}
101101 \\
-\ 000110 \\
\end{array}
$$

EXAMPLE 2-4

Find the 1's complement in binary of 47 H, 0B3 H, and 76 H.

SOLUTION

$$47 \text{ H} = 01000111 \text{ B}, \quad 1\text{'s complement} = 10111000 \text{ B}$$
$$0\text{B3 H} = 10110011 \text{ B}, \quad 1\text{'s complement} = 01001100 \text{ B}$$
$$76 \text{ H} = 01110110 \text{ B}, \quad 1\text{'s complement} = 10001001 \text{ B}$$

Subtraction in binary is accomplished by adding the 1's complement of the number that is to be subtracted to the second number.

EXAMPLE 2-5

Subtract 23 H from 81 H in binary.

SOLUTION

$$
\begin{array}{cc}
81 \text{ H} & 10000001 \text{ B} \\
-\ 23 \text{ H} & -00100011 \text{ B} \\
\hline
5\text{E H} & ? \\
\end{array}
$$

Rewrite the problem so that the − 00100011 B is replaced with its 1's complement; then add.

$$
\begin{array}{r}
1 \qquad\qquad\qquad \leftarrow \text{Carry bits} \\
10000001 \text{ B} \\
+ \quad 11011100 \text{ B} \\
\hline
\text{End-around carry} \;\ulcorner 101011101 \text{ B} \quad \rightarrow \text{15D H} \\
\llcorner\!\!\longrightarrow 1 \\
\hline
01011110 \text{ B} \;\; \rightarrow \text{5E H}
\end{array}
$$

The correct answer must be 5E H. So far we have 15D H. The "1" at the left-most end of the binary result indicates a positive result and is a reminder to add 1 to 5D H to obtain the correct result, namely, 5E H.

Actually, the "1" can be added to the least significant bit of the result as shown. Adding the "1" from the carry into the LSB position is called an "end-around carry." Let's try another example.

EXAMPLE 2-6

Subtract 27 D from 72 D.

SOLUTION

$$
\begin{array}{r}
72 \qquad\quad 1001000 \text{ B} \\
- \;27 \qquad - \;0011011 \text{ B} \\
\hline
45
\end{array}
$$

First, be sure both binary numbers are of the same length. We rewrite the problem using the 1's complement of 27 in binary.

$$
\begin{array}{r}
1 \qquad\qquad\qquad \leftarrow \text{Carry bits} \\
1001000 \text{ B} \\
+ \; 1100100 \text{ B} \\
\hline
\text{End-around carry} \;\ulcorner 10101100 \text{ B} \\
\llcorner\!\!\longrightarrow 1 \\
\hline
0101101 \text{ B}
\end{array}
$$

Check it for yourself. It's 45! The "1" in the eighth position indicates two things: first, that the result is positive and, second, that an end-around carry is necessary.

6.2 Two's Complement

It has been found that most results on a computer are positive and the end-around carry is required; therefore, to save computer hardware, the "1" can be added to the complement before the add operation in the accumulator. In this case, the complement is called the 2's complement.

EXAMPLE 2-7

Find the 2's complement of 28 in binary.

SOLUTION

$$28 \text{ D} = 11100 \text{ B}, \qquad 1\text{'s complement} = 00011 \text{ B}$$

Add 1 to find the 2's complement.

$$
\begin{array}{r}
00011 \text{ B} \\
+ \qquad 1 \\
\hline
00100 \text{ B}
\end{array}
$$

EXAMPLE 2-8

Subtract 47 D from 123 D using 2's complement binary arithmetic.

$$
\begin{array}{rr}
123 & 1111011 \text{ B} \\
-\ 47 & -0101111 \text{ B} \\
\hline
76 & ?
\end{array}
$$

SOLUTION

First find the 1's complement of 47 D, then add 1.

$$
\begin{array}{l}
-\ 0101111 \text{ B} \\
\underline{\quad 1010000 \text{ B}} \leftarrow 1\text{'s complement} \\
\quad 1010001 \text{ B} \leftarrow 2\text{'s complement}
\end{array}
$$

Now perform the addition.

$$
\begin{array}{r}
11 \quad 11 \ \leftarrow \text{Carry bits} \\
1111011 \text{ B} \qquad (123) \\
+\ 1010001 \text{ B} \\
\hline
\text{Ignored} \rightarrow 1\underline{1001100}\,\text{B} \\
\text{Answer}
\end{array}
$$

$$4 + 8 + 64 = 76!$$

When using 2's complement arithmetic, the resultant carry has a meaning different from its meaning in 1's complement arithmetic. A "1" means that the answer is positive and a "0" means that it is negative. In the event of a zero, the answer is recomplemented to obtain the correct result. The zero in the carry position then also means that the result is negative. In a computer, the use of 2's complement arithmetic generally saves a step in the addition process.

   Let us reverse the numbers of Example 2-8 to see how this may work.

EXAMPLE 2-9

Subtract 123 D from 47 D using 2's complement arithmetic.

$$
\begin{array}{rr}
47 & 0101111 \text{ B} \\
-123 & -\ 1111011 \text{ B} \\
\hline
-76 & ?
\end{array}
$$

SOLUTION

$$0101111 \text{ B}$$
$$+ \ 0000101 \text{ B} \qquad \text{(2's complement)}$$
$$\overline{00110100 \text{ B}} \qquad Note: \text{ The carry is 0.}$$

Since the carry is zero, the result is negative, and recomplementing causes the number to appear in proper form:

$$- \ 1001100 \text{ B}$$
$$(47 \ - \ 123 \ = \ - \ 76)$$

6.3 Signed Two's Complement Arithmetic

In a microcomputer, math operations are frequently saved in an accumulator of fixed length—either 8 or 16 bits for today's microprocessor chips. In the case of an 8-bit accumulator, the eighth bit (MSB)* is used for the sign bit. Arithmetic subtraction is done using 2's complement arithmetic. If the eighth bit is used to indicate a minus (i.e., if MSB = 1), that leaves only 7 bits for a positive number. Therefore, the largest positive number we could represent is 127 D.

$$01111111 \text{ B} = 127 \text{ D}$$

When 2's complement arithmetic is coupled with using the MSB as a sign bit, numbers are said to be in signed 2's complement form.

To represent a negative number in signed 2's complement form, the following rules are applied:

1. Complement the number to form the 1's complement.

2. Add 1 to the result to form the 2's complement.

Ignore any carry into a ninth position.

EXAMPLE 2-10

Find the signed 2's complement of -26 D.

SOLUTION

$$+ \ 26 \text{ D} = 00011010 \text{ B}$$

$$\begin{array}{lr} \text{1's complement} = & 11100101 \text{ B} \\ \text{Add 1} & + \qquad 1 \\ \hline \text{Signed 2's complement} & 11100110 \text{ B} \end{array}$$

Therefore, the signed 2's complement of -26 D is 0E6 H. (Note that the sign bit is a "1," indicating a negative number.)

---

*We call this bit 7 because we start counting at zero.

The range of positive numbers that can be represented in 8 bits using signed 2's complement notation is 0–127.

$$0 = 00000000 \text{ B} = 0 \text{ H}$$
$$1 = 00000001 \text{ B} = 1 \text{ H}$$
$$\downarrow$$
$$126 \text{ D} = 01111110 \text{ B} = 7E \text{ H}$$
$$127 \text{ D} = 01111111 \text{ B} = 7F \text{ H}$$

The range of negative numbers that can be represented in 8 bits using signed 2's complement notation is -1 to -128.*

$$-1 = 11111111 \text{ B} = 0FF \text{ H}$$
$$-2 = 11111110 \text{ B} = 0FE \text{ H}$$
$$\downarrow$$
$$-127 \text{ D} = 10000001 \text{ B} = 81 \text{ H}$$
$$-128 \text{ D} = 10000000 \text{ B} = 80 \text{ H}$$

Let's perform a subtraction using signed 2's complement notation.

EXAMPLE 2-11

Use signed 2's complement notation to find the result of 68 D $-$ 49 D.

SOLUTION

$$68 \text{ D} = 01000100 \text{ B}$$
$$- 49 \text{ D} = 00110001 \text{ B}$$
$$19 \text{ D} \qquad ?$$

Find the 1's complement of -49.

$$11001110 \text{ B}$$
$$\text{Add 1} \qquad + \qquad 1$$
$$\text{2's complement} \quad 11001111 \text{ B}$$

Now perform the addition.

$$01000100 \text{ B}$$
$$+ 11001111 \text{ B}$$
$$100010011 \text{ B}$$

The correct answer is 00010011 B or 19 D. The carry bit indicates a positive result.

*We get an extra negative number because we do not need to represent zero twice (since $+0 = -0$, one 0 is enough).

6.4 Unsigned Two's Complement Arithmetic

A number that ignores the MSB as a sign bit is called unsigned. Positive numbers may then be in the range 0–255.

$$0 = 00000000 \text{ B} = 0 \text{ H}$$
$$1 = 00000001 \text{ B} = 1 \text{ H}$$
$$\downarrow$$
$$127 \text{ D} = 01111111 \text{ B} = 7\text{F H}$$
$$128 \text{ D} = 10000000 \text{ B} = 80 \text{ H}$$
$$\downarrow$$
$$255 \text{ D} = 11111111 \text{ B} = 0\text{FF H}$$

Subtraction is still performed using 2's complement notation, but all 8 bits are available for use.

EXAMPLE 2-12

Using unsigned 2's complement arithmetic, find the result of 112 D − 49 D.

SOLUTION

$$
\begin{array}{rll}
112 \text{ D} = & 01110000 \text{ B} & = 70 \text{ H} \\
- \ 49 \text{ D} = & - \ 00110001 \text{ B} & = 31 \text{ H} \\
\hline
63 \text{ D} & ? & = 3\text{F H}
\end{array}
$$

$$
\begin{array}{lr}
\text{1's complement of } -49 = & 11001110 \text{ B} \\
\text{Add 1} & + \quad\quad\quad 1 \\
\hline
\text{2's complement} & 11001111 \text{ B}
\end{array}
$$

$$
\begin{array}{r}
01110000 \text{ B} \\
+ \ 11001111 \text{ B} \\
\hline
100111111 \text{ B} \\
\uparrow \\
\text{Carry bit}
\end{array}
$$

In a microcomputer the carry out of the eighth bit is called the *carry bit*. A "1" indicates a positive result. The answer is +63 D or +3F H or + 00111111 B. If the result is negative, the carry bit is a zero and the result must be complemented to obtain the correct answer.

EXAMPLE 2-13

Subtract 112 D from 49 D.

SOLUTION

$$
\begin{array}{rll}
49 \text{ D} = & 00110001 \text{ B} & = 31 \text{ H} \\
- 112 \text{ D} = & - \ 01110000 \text{ B} & = 70 \text{ H} \\
\hline
- \ 63 \text{ D} & ? &
\end{array}
$$

Find the 2's complement of − 112 D.

$$
\begin{array}{lr}
\text{1's complement} & \text{10001111 B} \\
\text{Add 1} & +\quad\quad 1 \\
\hline
\text{2's complement} & \text{10010000 B}
\end{array}
$$

$$
\begin{array}{lr}
\text{49 D} & \text{00110001 B} \\
\text{2's complement of } -112\text{ D} & +\ \text{10010000 B} \\
\hline
\text{Carry} = 0 \longrightarrow & \text{011000001 B}
\end{array}
$$

To obtain the correct answer, we note that the carry bit is a zero. The result is a negative number. Take the 2's complement and you have the correct result!

$$
\begin{array}{lr}
& \text{011000001 B} \\
\text{1's complement} & \text{00111110 B} \\
\text{Add 1} & +\quad\quad 1 \\
\hline
(-)\ & \text{00111111 B} = -\text{3F H}
\end{array}
$$

The correct answer is − 00111111 B or − 3F H or − 63 D.

If this seems complicated, remember that all arithmetic on a computer must be done using binary numbers and must follow a routine easily adhered to by computer circuitry. The programmer of a microcomputer frequently must perform such operations and understand how they are accomplished. We have discussed three types of subtraction:

1. 1's complement.

2. Signed 2's complement.

3. Unsigned 2's complement.

The use of method 3, unsigned 2's complement, is preferred and is often extended to be used in a 16-bit accumulator. An 8-bit accumulator can operate on numbers ranging from + 255 to − 255. A 16-bit accumulator can operate on numbers ranging from + 65,535 to − 65,535. Chapter 12 on advanced programming will deal with 16-bit subtractions in more detail. Let us work one more example.

EXAMPLE 2-14

Subtract 255 D from 47 D.

SOLUTION
Convert both numbers to binary.

$$
\begin{array}{lll}
47\text{ D} = & 00101111\text{ B} = 2\text{F H} \\
-255\text{ D} = - & 11111111\text{ B} = -0\text{FF H} \\
\hline
-208\text{ D}
\end{array}
$$

Find the 1's complement of − 255 D.

$$
\begin{array}{lr}
\text{1's complement} & \text{00000000 B} \\
\text{Add 1} & +\quad\quad 1 \\
\hline
\text{2's complement} & \text{00000001 B}
\end{array}
$$

Now perform the addition.

$$
\begin{array}{llr}
47\ \text{D} & & 00101111\ \text{B} \\
\text{2's complement of } -255\ \text{D} & & +\ \underline{00000001\ \text{B}} \\
\text{Carry bit } = 0 & & 000110000\ \text{B}
\end{array}
$$

The answer is negative, and we must find the 2's complement to get the correct result.

$$
\begin{array}{lr}
 & 000110000\ \text{B} \\
\text{1's complement} & 11001111\ \text{B} \\
\text{Add 1} & +\ \underline{\qquad\quad 1} \\
\text{2's complement} & 11010000\ \text{B}
\end{array}
$$

The answer is -11010000 B $= -0$D0 H $= -208$ D.

7 MULTIPLICATION AND DIVISION

There are several methods to accomplish multiplication and division on a computer. Some are less accurate than others as a result of rounding errors and the nature of representing decimal numbers in binary.

One way to perform a multiplication that gives a satisfactory result is to do successive additions the correct number of times. A relatively simple microcomputer program is necessary to accomplish this. For instance, to multiply 5 × 17, we add 17 to itself a total of five times. This method works well for integers that are also small numbers.

EXAMPLE 2-15

Find 5 × 17 using 8-bit binary.

SOLUTION
5 × 17 = 85.

$$
\begin{array}{r}
1\quad 1 \\
00010001\ \text{B} \\
00010001\ \text{B} \\
00010001\ \text{B} \\
00010001\ \text{B} \\
+\ \underline{00010001\ \text{B}} \\
01010101\ \text{B}
\end{array}
$$

A similar method to divide that also works involves successive subtractions. In this case, the answer may be only approximate, but in many cases that is sufficient. Such a method is used with the bar code reader described in Chapter 12. To divide 62 by 19, we subtract 19 and count the number of times we can subtract without going negative.

$$
\begin{array}{r}
62 \\
-\ 19 \\
\hline
43 \\
-\ 19 \\
\hline
24 \\
-\ 19 \\
\hline
5 \\
-\ 19 \\
\hline
-\ 14
\end{array}
$$

Once

Twice

Three times

Negative

The answer is that 62 divided by 19 is 3 (an approximate answer).

Computers usually require a method that allows much greater accuracy, particularly when mathematics is the essential function being performed. The next section describes a very accurate technique for multiplication.

7.1 Accurate Multiplication

We have already described the concept of a register as storage for binary numbers. Even though a register may not be long enough in reality, it can be considered to have storage capacity for a very large number of bits if a few programming tricks are used.

Such a register can be made to shift or rotate the bits to the left or to the right. This is usually done by the application of a clock pulse, a fast binary signal used to sequence or control digital circuitry. It takes one clock pulse to shift the entire binary string stored in the register. Figure 2-1 shows such a register with the binary number 11011.011 B stored in it. This is the equivalent of 27.375 D.

If we cause the register to perform a shift left by the application of a single clock pulse, the bits in the register move one place left with respect to the binary point. The new number looks like this:

$$
\boxed{1\ 1\ 0\ 1\ 1\ 0\ .\ 1\ 1\ 0}
$$

Checking the new result, we find a larger number, 54.75 D. This number is exactly twice the size of the original number. From this we can immediately and correctly conclude that each time a string of bits (binary number) is shifted to the left, it doubles. Furthermore, each shift to the right will halve the number. Shifting to

---

FIGURE 2-1 Shift register.

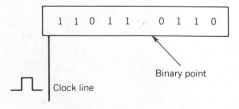

Binary point

Clock line

the left again will give us a second doubling of the number, or four times the original.

The highly accurate multiplication by computer of binary numbers is based on the use of a shift register, which enables us to shift bits left and right, and an accumulator.

7.2 Integer Multiplication

We first learn how to multiply a number by an integer using a shift register and an accumulator.

We will multiply 18.75 by the integer 12 for a result of 225. We enter 18.75 into the shift register and use 12 as the *control word*, to tell us exactly what operations to perform. We use the accumulator, initialized to zero, as a place in which to add numbers. First let's convert the problem to binary.

$$\begin{array}{r} 18.75 \\ \times \quad 12 \\ \hline 225. \end{array} \qquad \begin{array}{r} 10010.11 \text{ B} \leftarrow \text{Shift register number} \\ \times \quad 1100. \text{ B} \quad \leftarrow \text{Control word} \\ \hline \end{array}$$

We enter 10010.11 B into the shift register. We will interpret the control word (1100 B) in the following way: A "0" will mean to shift left; a "1" will mean to add what is in the shift register to the accumulator, then shift left again.

Shift Register

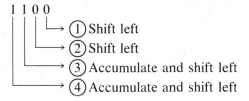

10010.110000 B

We have entered 10010.11 B into the shift register.

The control word is actually a set of instructions to be read, beginning with the least significant bit. The instructions are as follows:

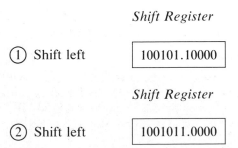

1 1 0 0

① Shift left
② Shift left
③ Accumulate and shift left
④ Accumulate and shift left

We will follow the instructions in that order.

Shift Register

① Shift left   100101.10000

Shift Register

② Shift left   1001011.0000

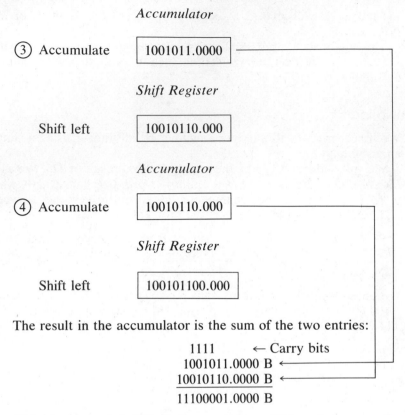

Accumulator

③ Accumulate   1001011.0000

Shift Register

Shift left   10010110.000

Accumulator

④ Accumulate   10010110.000

Shift Register

Shift left   100101100.000

The result in the accumulator is the sum of the two entries:

$$\begin{array}{r} 1111 \qquad \leftarrow \text{Carry bits} \\ 1001011.0000 \text{ B} \leftarrow \\ \underline{10010110.0000 \text{ B}} \leftarrow \\ 11100001.0000 \text{ B} \end{array}$$

This binary result is 225, the correct answer. This technique works well for integers and is similar to the method we use to multiply by a decimal number.

7.3 Multiplication by a Decimal

Multiplication by a decimal means that the result will be a smaller number than we start with. Let us multiply 26.75 by .2 D. We place 26.75 converted to binary in the shift register and use .2 D converted to binary as the control word. In this case we are expecting a smaller result, so we will be shifting right.

Let us first convert each number to binary.

$$\begin{array}{cc} 26.75 & 11010.11 \text{ B} \\ \times \quad .2 & \times .\,00110011001\overline{10011} \text{ B} \\ \hline 5.35 & \end{array}$$

.2 D is a repeating binary sequence. It goes on forever. We will work down to five 1s (five accumulator entries) in the control word and see how close we are to the answer. So, we will consider that .2 D is .00110011001 B. This is rounded off, but it will illustrate the method. Greater accuracy can be obtained only by using more places.

The rules for using the control word are as follows: A ''0'' means shift right, and a "1" means shift right and then make an entry into the accumulator. Thus the control word is a set of instructions that runs like this:

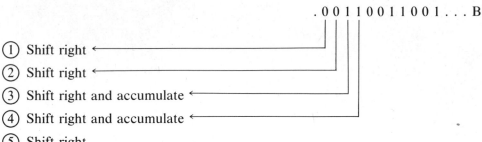

.0 0 1 1 0 0 1 1 0 0 1 . . . B

① Shift right
② Shift right
③ Shift right and accumulate
④ Shift right and accumulate
⑤ Shift right
⑥ Shift right
⑦ Shift right and accumulate
⑧ Shift right and accumulate
⑨ Shift right
⑩ Shift right
⑪ Shift right and accumulate

Notice that there will be five entries to the accumulator, one for each 1 in the control word.

Shift Register (contains 26.75)

| . 0 . 0 . 0 . 0 . 0 . 0 . 1 . 1 . 0 . 1 . 0 . 1 1 0 0 |
| 11 10 9   8   7   6   5   4   3   2   1 |

You will move the point to see how the five entries are made in the accumulator. Remember, the bits shift right and the point moves the other way.

The accumulator contains each entry and the totaled result. It is up to you to check the entire operation!

Accumulator

1	Carry Bits
11 1 1 1 1 1 1 1	
11.01011000000B	Step ③
1.10101100000B	Step ④
.00110101100B	Step ⑦
.00011010110B	Step ⑧
.0000001101011B	Step ⑪
101.0101011110011B	

The answer is checked as follows:

$$
\begin{array}{r}
5. \\
+\ .25 \\
.0625 \\
.015625 \\
.0078125 \\
.00390625 \\
.001953125 \\
.00024414 \\
\underline{.00012207} \\
5.342163085
\end{array}
$$

The result is close to but does not exceed 5.35. More accuracy is obtained by further use of the control word to additional significant places.

This is not only a good exercise but also great practice in adding binary numbers. However, you will agree, the sooner we make a computer do this work, the better! And remember that we are adapting the circuitry of a machine operating in binary to a decimal world. The results are not always going to be easy to get, and they may not always be exact.

7.4 Multiplication of Decimal Numbers

The multiplication of 8.5×7.25 requires the use of both methods outlined previously. First, 8.5×7 is performed using the integer rules. Then, $8.5 \times .25$ is performed using the decimal rules. The results are added together. Simple numbers are chosen here to illustrate the point. A more complex example follows later.

EXAMPLE 2-16

Multiply 8.5×7.25.

$$
\begin{array}{ll}
\quad\ 8.5 & \quad 1000.1\ \text{B} \qquad \leftarrow \text{Shift register} \\
\underline{\times\ 7.25} & \underline{\times\ 0111.01\ \text{B}} \quad \leftarrow \text{Control word(s)} \\
\quad 61.625 &
\end{array}
$$

SOLUTION
First,

$$
\begin{array}{l}
\quad\ 1000.1\ \text{B} \\
\underline{\times\ \ 111.\ \ \text{B}} \quad \leftarrow 0 = \text{Shift left} \\
\qquad\qquad\qquad\ \ 1 = \text{Accumulate and shift left}
\end{array}
$$

Shift Register

1	0 . 0 . 0 . 1 . 0 . 0 . 0
	5 4 1 2 3

111.Control word

① Accumulate and shift left

② Accumulate and shift left

③ Accumulate and shift left

Now we do the fractional part:

$$
\begin{array}{r}
1000.1 \ \ B \\
\times \quad .01 \ B \\
\end{array}
$$

④ Shift right

⑤ Shift right and accumulate

Accumulator

1 0 0 0 . 1 0 0 0 B	Step ①
1 0 0 0 1 . 0 0 0 0 B	Step ②
1 0 0 0 1 0 . 0 0 0 0 B	Step ③
1 0 . 0 0 1 0 B	Step ⑤

1 1 1 1 0 1 . 1 0 1 0 B

$$
\begin{array}{r}
0.5 \\
0.125 \\
1 \\
4 \\
8 \\
16 \\
\underline{32} \\
61.625 \quad \text{Check!}
\end{array}
$$

And now, a more complex example.

EXAMPLE 2-17

Multiply 63.7 × 13.8.

SOLUTION

We will show only the accumulator entries. It is left to you to work the problem.

$$
\begin{array}{r}
63.7 \\
\times \ 13.8 \\
\hline
879.06
\end{array}
$$

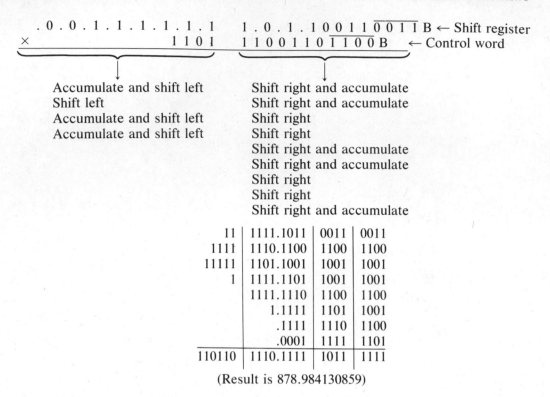

$$. 0 . 0 . 1 . 1 . 1 . 1 . 1 . 1 \qquad 1 . 0 . 1 . 1 0 0 1 1 \overline{0 0 1 1} \text{ B} \leftarrow \text{Shift register}$$
$$\times \underline{\hspace{4cm} 1 1 0 1} \qquad 1 1 0 0 1 1 0 \overline{1 1 0 0} \text{ B} \quad \leftarrow \text{Control word}$$

Accumulate and shift left	Shift right and accumulate
Shift left	Shift right and accumulate
Accumulate and shift left	Shift right
Accumulate and shift left	Shift right
	Shift right and accumulate
	Shift right and accumulate
	Shift right
	Shift right
	Shift right and accumulate

	1111.1011	0011	0011
11	1111.1011	0011	0011
111↑	1110.1100	1100	1100
11111	1101.1001	1001	1001
1	1111.1101	1001	1001
	1111.1110	1100	1100
	1.1111	1101	1001
	.1111	1110	1100
	.0001	1111	1101
110110	1110.1111	1011	1111

(Result is 878.984130859)

In this problem, the carry bits are not shown for purposes of clarity. To insert them would make for confusion. However, you can check the accumulator entries against your own. The result, which is for the entries shown, is close to the actual answer of 879.06 and does not exceed it. A closer result may be obtained by using more significant bits.

8 BINARY CODED DECIMAL (BCD) ARITHMETIC

We have seen how computers can use binary numbers to perform math operations. All these methods use numbers in pure binary and frequently are subject to errors due to the imprecision of representing decimal numbers with a limited number of binary bits. For example, .2 D converts to the repeating binary sequence .0011001100110$\overline{0011}$ B. No computer can afford an infinite number of places to represent such a number. This causes certain rounding errors in large computers that must be "covered up." Additional programming or circuitry must be used to obtain answers that are apparently correct to the user.

Many microcomputers use languages that avoid these problems by making use of a 4-bit binary code called binary coded decimal (BCD). Each digit to be used in an operation is encoded using 4 bits in an 8421 code or 8421 weighting scheme. In this way, decimal numbers are stored in a computer using 0s and 1s but without the necessity of converting to pure binary. All the inaccuracies due to rounding disappear when using BCD arithmetic.

For example,

$$48 \text{ D} = 01001000 \text{ in BCD}$$

The 4 becomes 0100 and the 8 becomes 1000. So any decimal number is easily represented using BCD. Each digit requires 4 bits, making the size (number of bits) of a BCD number easily predictable. For instance, an 8-digit decimal number requires precisely 32 bits (8 × 4). Such estimations are not so easy when using pure binary.

Doing arithmetic with BCD numbers requires a few tricks, but they are easily learned and implemented on a computer. Let us see how this is done.

8.1 Addition Using BCD Numbers

In performing addition of BCD numbers, we must remember that the addition will actually be addition in hexadecimal. Example 2-18 illustrates this point.

EXAMPLE 2-18

Case I: Add 46 D + 23 D.

SOLUTION

$$
\begin{array}{r}
46 \text{ D} \\
+ 23 \text{ D} \\
\hline
69 \text{ D}
\end{array}
\qquad
\text{In BCD:}
\quad
\begin{array}{r}
01000110 \\
00100011 \\
\hline
01101001
\end{array}
$$

6   9   The correct answer!

However, when a carry must occur, things are different.

EXAMPLE 2-19

Case II: Add 46 D + 29 D.

SOLUTION

$$
\begin{array}{r}
46 \text{ D} \\
+ 29 \text{ D} \\
\hline
75 \text{ D}
\end{array}
\qquad
\begin{array}{r}
01000110 \\
+ 00101001 \\
\hline
01101111
\end{array}
\qquad
\begin{array}{r}
46 \text{ H} \\
+ 29 \text{ H} \\
\hline
6\text{F H}
\end{array}
$$

6   F

In this case, 6 F is *not* an acceptable answer. The 6 and 9 added to result in a hex F. Since we are trying to represent decimal numbers, only the codes for 0–9 are acceptable. When an illegal code (A–F) appears, we must perform a correction.

The correction is to add 06 to the result:

$$
\begin{array}{cc}
01101111 & \text{6F H} \\
+\ 00000110 & +\ 06\ \text{H} \\
\hline
01110101 & 75\ \text{H}
\end{array}
$$

7   5

Let us try several more examples to determine other necessary corrections.

EXAMPLE 2-20

Case III: Add 57 + 92.

SOLUTION

$$
\begin{array}{ccc}
57 & 01010111 & 57\ \text{H} \\
+\ 92 & +\ 10010010 & +\ 92\ \text{H} \\
\hline
149 & 11101001 & \text{E9 H}
\end{array}
$$

E   9

The E appears and is an illegal code. To obtain the correct result we add 60 H to the result.

$$
\begin{array}{cc}
11101001 & \text{E9 H} \\
+\ 01100000 & +\ \ 60\ \text{H} \\
\hline
101001001 & 149\ \text{H}
\end{array}
$$

1   4   9

The correction causes a carry into the third hex position or the ninth binary bit. This is called the carry bit. The *auxiliary carry* is at the halfway point (i.e., meaning after the fourth binary bit—bit 3). Another example illustrates this effect.

EXAMPLE 2-21

Case IV: Add 75 D + 49 D.

SOLUTION

$$
\begin{array}{ccc}
75\ \text{D} & 01110101 & 75\ \text{H} \\
+\ 49\ \text{D} & +\ 01001001 & +\ 49\ \text{H} \\
\hline
124\ \text{D} & 10111110 & \text{BE H}
\end{array}
$$

In this case, both parts of the result are over the code for 9 and are illegal. The correction is to add 66 H to the result. We see that the first "digit" (E) carries in auxiliary fashion into the second "digit" (B). We also see the second "digit" *carry* into the third digit.

$$C \quad AC$$
$$
\begin{array}{ll}
10111110 & BE\ H \\
+\ 01100110 & +\ \ 66\ H \\
\hline
100100100 & 124\ H \\
\end{array}
$$

$$\underbrace{}_{1}\quad\underbrace{}_{2}\quad\underbrace{}_{4}$$

The correction then involves observing the result and deciding whether to add the binary representation of 00 H, 06 H, 60 H, or 66 H to obtain the correct result. In a microcomputer with an 8-bit accumulator, the digits of a very long number would be added two at a time until the entire result became available. In a microcomputer with a 16-bit accumulator, the addition could be done four digits at a time. It becomes a simple function of the microcomputer program to keep track of the pairs of digits.

We are also able to handle decimals with this method without the problem of rounding errors.

EXAMPLE 2-22

Add 26.21 D + 29.46 D.

SOLUTION

$$
\begin{array}{ll}
26.21\ D & 00100110.00100001 \\
29.46\ D & +\ 00101001.01000110 \\
\hline
55.67\ D & 01001111.01100111 \\
\text{6 added as a correction} \longrightarrow & +\ \ \ \ \ \ \ \ \ 0110. \\
\hline
& 01010101.01100111 \\
\end{array}
$$

$$\underbrace{}_{5}\ \underbrace{}_{5}\ .\ \underbrace{}_{6}\ \underbrace{}_{7}$$

Chapter 11 will show how such additions are performed a pair of digits at a time and also how such numbers are entered into a computer. Also, we will find that most microcomputers incorporate an instruction to perform the appropriate corrections for BCD additions.

8.2 BCD Subtraction

Subtracting using BCD numbers is a bit trickier. A variation of complementing must be used, but in decimal. The *9's complement* of a decimal number is found by subtracting from the largest number in the base using the same number of places. The complement of -475 is found by subtracting from 999. The complement of 42 is found by subtracting from 99. A 6-digit number would be subtracted from 999999. The *10's complement* is found by adding 1 to the 9's complement.

To perform the subtraction 46 D − 29 D, we do the following:

$$\begin{array}{r} 46\text{ D} \\ -\ 29\text{ D} \\ \hline 17\text{ D} \end{array}$$

First, find the 9's complement of − 29.

$$\begin{array}{r} 99 \\ -\ 29 \\ \hline 70 \end{array}$$

Add 1 $\begin{array}{r} +\ \ 1 \\ \hline \end{array}$

10's complement   71

Now add 46 to 71.

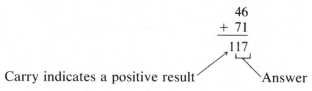

$$\begin{array}{r} 46 \\ +\ 71 \\ \hline 117 \end{array}$$

Carry indicates a positive result   Answer

Accomplishing this in a microcomputer required to do BCD subtraction involves a similar route. Add 1 to 99 to obtain 9A. Subtract 29 from 9A (similar to 2's complement arithmetic).

$$\begin{array}{r} 9\text{A H} \\ -\ 29\text{ H} \\ \hline 71\text{ H} \end{array}$$

Add 46 D to 71 D.

$$\begin{array}{r} 46\text{ D} \\ +\ 71\text{ D} \\ \hline 117\text{ D} \end{array}$$

Answer   (46 − 29 = 17)

Let's try another example.

EXAMPLE 2-23

Subtract 43 D from 72 D using binary.

$$\begin{array}{r} 72\text{ D} \\ -\ 43\text{ D} \\ \hline 29\text{ D} \end{array} \qquad \begin{array}{r} 01110010\text{ B} \\ -\ 01000011\text{ B} \\ \hline ? \end{array} \qquad \begin{array}{r} 72\text{ H} \\ -\ 43\text{ H} \\ \hline ? \end{array}$$

SOLUTION
First subtract 43 H from 9A H.

$$\begin{array}{r} 9\text{A H} \\ -\ 43\text{ H} \\ \hline 57\text{ H} \end{array}$$

Now add 57 H to 72 H.

$$\begin{array}{r} 72\ H \\ +\ 57\ H \\ \hline 129\ H \end{array}$$

Answer   (72 − 43 = 29)

This will serve as an introduction to the later programming examples that implement BCD addition and subtraction.

In cases like the examples for subtraction of 43 from 9A, a subtract instruction will be available on your microcomputer.

9 CONCLUSION

There are many ways of doing arithmetic on computers. All deal in binary bits (0s or 1s). Large computers may use pure binary, but many microcomputers use binary coded decimal numbers. A further choice exists—the selection of programming or electronic circuitry (hardware) to perform the mathematics. Computers use different combinations of these approaches. In your applications of microcomputers to the control of things around you, you will probably run into all these techniques.

10 SUMMARY

In this chapter we learned the simple rules for binary addition. We also learned how to multiply in binary by the use of a control word and an accumulator. Finally, we covered BCD addition and subtraction and saw how BCD requires that some adjustments be made after an arithmetic operation.

11 GLOSSARY

Accumulator.   Register in a computer that contains the result of math and logical operations.

ALU.   Arithmetic logic unit; the portion of a computer that performs math and logical operations.

Auxiliary carry.   The auxiliary carry is a carry from bit 3 to bit 4 and pertains to adding BCD numbers.

BCD.   Binary coded decimal (also 8421 coded binary).

Carry bit.   A bit located at MSB + 1 that is set or reset depending on the result of adding the two MSBs.

Complement, 1's.   Representation of a binary word in which the value of each bit is exchanged.

Complement, 2's.   1's complement plus 1; most commonly used in binary subtraction.

Complement, 9's.   Representation of a BCD word in which the original word was subtracted from 9's.

Complement, 10's.   9's complement plus 1; used most commonly in BCD subtraction.

Control word.   Group of bits that controls the additions to the accumulator during binary multiplication.

PROBLEMS

2-1   Add the following numbers both in binary and in hexadecimal.
   (a) 37 + 29
   (b) 75 + 94
   (c) 26.125 + 18.675
   (d) 57.4 + 27.9

2-2   Convert the following numbers to binary and subtract using 1's complement arithmetic.
   (a) 88 − 26
   (b) 94 − 138
   (c) 18.7 − 14.8

2-3   Subtract the following numbers using 2's complement arithmetic.
   (a) 68 − 49
   (b) 57 − 83
   (c) 26.41 − 18.27

2-4   Subtract the following numbers using signed 2's complement arithmetic.
   (a) 83 − 47
   (b) 242 − 126
   (c) 148 − 206

2-5   Multiply the following numbers using the shift register and accumulator technique.
   (a) 206.4 × 36
   (b) 141.3 × .6
   (c) 27.5 × 14.3

2-6   Add the following numbers using BCD arithmetic. (Show the operations in binary.)
   (a) 26 + 49
   (b) 16.3 + 18.9
   (c) 93.2 + 46.8

2-7   Subtract the following numbers using BCD arithmetic. (Show the operations in binary.)
   (a) 27 − 13
   (b) 286 − 431
   (c) 14.3 − 17.9

2-8   Add the following numbers in binary.
   (a) 111010.1101 B   (b) 11010.1101 B
       100101.1001 B   1100.1011 B
                                    11.1001 B

2-9 Find the difference between the following pairs of numbers using signed 2's complement arithmetic.

(a)   69 D
    − 47 D

(b)   28 H
    − 79 H

2-10 Find the binary product of the following pairs of numbers using a shift register and accumulator. Show all accumulator entries.

(a) 11101.101 B
    × 110.01 B

(b)   10111.111 B
    × 1110.101 B

2-11 Add the following numbers using BCD arithmetic.

(a)   47.12
    + 29.41

(b)   69.99
    + 47.21

2-12 Subtract the following numbers using BCD arithmetic.

(a)   27.4
    − 32.9

(b)   19.61
    − 12.08

2-13 Convert the number 6598.457 D to hexadecimal, binary, and octal.

2-14 Multiply 546.85 × 453.2 using the shift register and accumulator technique discussed in this chapter.

2-15 Explain why the 10's complement of zero (in BCD) is always zero.

2-16 Show that the 2's complement of any number can be found by subtracting the number from the next highest binary multiple. For example, the 2's complement of 5 can be found by subtracting 5 from 8, or the 2's complement of 1000 can be found by subtracting 1000 from 1024.

CHAPTER 3
LOGIC GATES

1 INSTRUCTIONAL OBJECTIVES

When finished with this chapter you should be able to:

1. Know the different logic families and their characteristics.
2. Understand the characteristics of the TTL logic family.
3. Understand the function and operation of the simplest TTL gates: inverter, buffer, AND, OR, NAND, NOR, XOR.
4. Apply Boolean algebra rules and truth tables when using TTL gates.
5. Use truth tables to determine whether a gate is good or bad.
6. Substitute one type of gate for another.
7. Use DeMorgan's theorem.
8. Understand open-collector and tri-state output gates.
9. Understand pulse rise time, fall time, and pulse width.
10. Understand the binary adder circuitry.

2 SELF-EVALUATION QUESTIONS

Keep the following questions in mind and try to answer them when you have completed the chapter.

1. What are a few of the most common logic families?
2. Why is a noise margin important?
3. What is an open-collector output, and what is it used for?
4. What is the difference between 5400 and 7400 series logic gates?
5. What is meant by the expression "tri-state logic"?
6. Why is a bounceless switch circuit necessary?
7. What is a decoder circuit (data selector)?
8. What is the difference between a half adder, a full adder, and a parallel adder?

3 INTRODUCTION

This chapter will familiarize you with a number of the basic building blocks of digital electronics. These building blocks are logic gates and include the following types:

1. Inverter.
2. Buffer.
3. AND gate.
4. OR gate.
5. Exclusive OR gate.
6. NAND gate.
7. NOR gate.
8. Combinations of these gates.

All the logic gates use transistors to represent and store binary numbers or bits. Through the discussion, ways of representing a ''0'' or a ''1'' are considered. These gates are built in integrated circuit (IC) form and sold for use in digital circuitry.

3.1 Logic Families

Digital logic gates and their associated circuitry have been in a constant state of change and improvement. Thus it is appropriate to present a brief history and comparison of the series of families of logic that evolved. In the 1960s a form of logic that included all the aforementioned logic gates (AND, OR, NAND, etc.) was developed. It used diodes and resistors and was referred to as diode-resistor logic (DRL). This family of logic was soon replaced by resistor-transistor logic (RTL), which was replaced by a family that is very popular today called transistor-transistor logic or TTL (or T^2L). Two other basic types that are popular and important now are emitter-coupled logic (ECL) and CMOS (complementary symmetry metal oxide semiconductor). Each has advantages and disadvantages.

TTL logic offers good speed at a reasonable price. ECL offers high speed (> 500 MHz) at a hefty price, while CMOS, which represents field effect transistor (FET) technology, offers low power consumption. Highly complex circuitry using

TABLE 3-1 Comparison of Major Characteristics of Various Logic Families

Logic	Binary Element Clock Frequency, Typical (MHz)	Delay Time (nsec)	Power/ Gate (mW)	Operating Voltage (V)
DRL	40	30	8	5.0 ±10%
RTL	8	24	2.5	3.6 ±10%
TTL	50	10	15	5.0 ±10%
CMOS	25	25	Low (varies with f)	+3 to +14 ±10%
ECL	400	4	40	−5.2 ±20%

tens of thousands of field effect transistors can perform digital miracles at relatively low power in the CMOS family. CMOS keeps undergoing improvements, and PMOS, NMOS, HMOS, and VMOS are also now available.

Regardless of the family of logic, you can find the logic gates you will require in any of the available families. Table 3-1 compares time and electrical characteristics of five logic families; Appendix A lists many of the popular logic gates and their numbers in the various families.

Microcomputers use a working combination of logic gates from both the TTL and CMOS families. In our next few chapters we will be thinking in terms of the TTL logic family. It is reasonable in cost and features a wide range of applications at a relatively high operating speed (> 50 MHz).

3.2 The 5400/7400 Series of TTL Logic

A very popular series of TTL logic is the 54/74 series. The 54 series is intended for military use and operates over a wide temperature range ($-55°C$ to $+125°C$) to accommodate a variety of environments. The 74 series is a consumer-oriented version of the TTL family and is restricted to a narrower range of operating temperatures (0–70°C). For many applications this range is adequate, since television receivers and other consumer items operate in environments in which we all feel comfortable. It is strongly recommended that you have a TTL data book for a reference. The 7404 is a hex inverter, as is the 5404, with one difference being its operating temperature range (see *The TTL Data Book* of Texas Instruments Company).

The familiar acronym DIP stands for dual in-line package, and a suffix (J, N, W, T) is affixed to the number to indicate the type and style of package: J for ceramic DIP, N for plastic DIP, W for ceramic flat pack, and T for flat pack. Figure 3-1 illustrates TTL logic in each of these types. The 54/74 series comes in a variety of available subfamilies, designated by letter(s) after the 54 or 74 (e.g., 7400, 74S00, 74LS00, 74H00). The ''plain'' 7400 is becoming obsolete and the 74LS00 is gradually becoming the industry standard.* Table 3-2 summarizes the capability of these improvements to the 54/74 series.

TABLE 3-2   Summary of Capabilities of the 54/74 Series

Family	Typical Gate Propagation Delay (nsec)	Typical Power Dissipation Per Gate (mW)	Comments
74xxx	10	10	
74Lxxx	33	1	Low power
74Hxxx	6	22	High level
74Sxxx	3	19	Schottky
74LSxxx	10	2	Low-power Schottky
74ASxxx	1.5	22	Advanced Schottky
74ALSxxx	5	1	Advanced LS

*A newer form of TTL called ALS (advanced low-power Schottky) offers faster speed and lower power.

FIGURE 3-1 Types and styles of TTL logic. (*a*) 14-Pin J ceramic,
(*b*) 14-pin N plastic, (*c*) 14-pin W ceramic, and (*d*) T flat package.
(Courtesy of Texas Instruments Incorporated.)

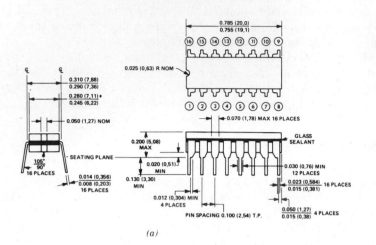

(*a*)

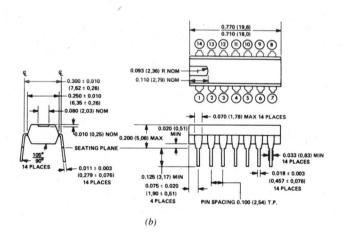

(*b*)

FIGURE 3-1 (continued)

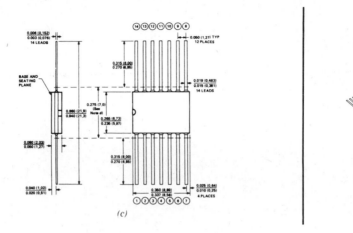

(c)

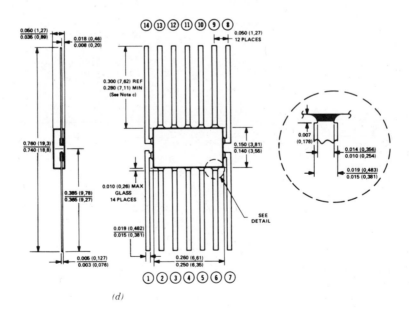

(d)

Figure 3-2 shows a typical dual in-line package. Two numbers appear on the top: SN7400N is the part number, and 7847 is the manufacturer's date code. The date code indicates that this particular integrated circuit was manufactured in the 47th week of 1978. This information is useful if it ever becomes necessary to identify a bad batch.

FIGURE 3-2 Typical dual in-line packages. (Courtesy of Texas Instruments Incorporated.)

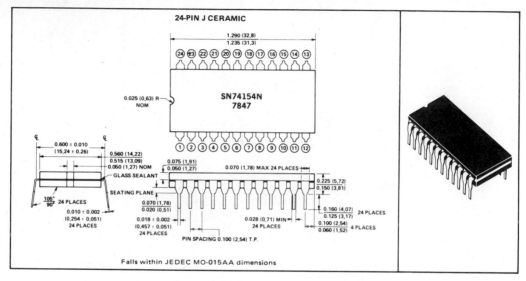

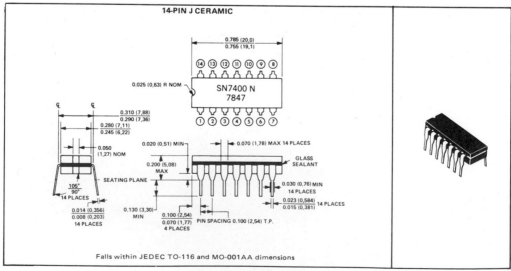

3.3 CMOS Logic

A great deal of logic constructed of field effect transistors is in use in the computer field. This logic uses insulated-gate FETs and is very susceptible to damage caused by static discharge. Such circuitry must be handled very carefully and protected against this very destructive influence. Circuit boards containing CMOS devices are shipped in special static-protective bags. Integrated circuits containing CMOS devices are shipped in a conductive foam material. One problem with this type of damage is that the results cannot be seen, only suspected. The only symptom is gate failure.

At least two great CMOS families have emerged: the 4000 series and the 8000 series. Standard logic functions are available in the 4000 series, and many micro-computer circuits are a part of the 8000 series. It should be said that any micro-computer chip or its associated support chips contain MOS devices as a means of reducing the required power to large numbers of logic gates. While many of these chips contain MOS devices inside that operate on + 12 V, they are made TTL compatible before leaving the chip. That is, the connections are converted to TTL levels for use on the printed circuit board outside the CMOS chip. Several varieties of MOS devices are available (e.g., PMOS, NMOS, VMOS, and HMOS). All are sensitive to static discharge and must be handled carefully. The primary advantage of such circuitry is in its low power consumption per gate.

When a circuit entails the use of more than 100 logic gates, it is said to use large-scale integration (LSI). Popular microcomputer chips involving thousands of logic gates and tens of thousands of transistors are definitely LSI and may be classified as VLSI (very-large-scale integration).

3.4 Logic Levels

In a digital circuit, the binary values 0 and 1 must be represented using a pair of voltage levels. Each logic family has a different standard, and great care must be used when attempting to mix logic families. Usually one family is selected for design based on its desirable characteristics. Table 3-3 shows the voltages used by several families.

Logic families that use positive voltages are called *positive level* families. Logic families that use the most positive value to represent a binary "1" are called *positive logic*. TTL is both *positive level* and *positive logic*. ECL is *negative level* and *positive logic*.

TABLE 3-3   Voltages Used by Several Logic Families

Family	Voltage (V)	
	Logic 0	Logic 1
TTL	0 V	+5 V
ECL	−1.7 V	−.9 V[a]
CMOS	0 V	3−14 V

[a]ECL uses −5.2 V supply voltage.

FIGURE 3-3 Range of allowable TTL levels. (*a*) TTL family output limits and (*b*) TTL family input limits.

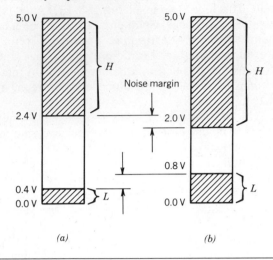

(*a*)   (*b*)

FIGURE 3-4 Noise induced in wire connected from TTL output to TTL input.

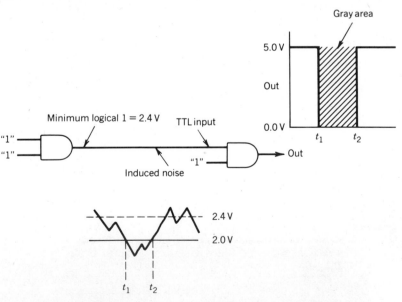

Between t_1 and t_2 the noise drops below 2.0 V, which is the minimum input level for a TTL input. During this time the output is not at the 5-V level. Since it is in a gray area, the output cannot be predicted. Such drops should be avoided.

TABLE 3-4 Logic Levels for a "0" or "1"

"0"	"1"	
Low	High	
Down	Up	
Off	On	
False	True	
0	+5	← TTL
0	3 to 14	← CMOS
−1.7	−.9	← ECL
0	3.6	← RTL (obsolete)

While it is understood that a family like TTL uses a +5-V power supply (+5 ±10%) and that a "0" is represented with 0 V and a "1" is represented with 5 V, a certain range is acceptable. A logic zero on the input of a TTL gate can be a level from 0 to 0.8 V. A logic "1" on the input can be from 2 to 5 V.

The range of allowable TTL levels (Fig. 3-3) is defined differently for the output of a TTL gate. The allowable range for a logic zero at the output is 0 to 0.4 V. The range for a one at the output is 2.4 to 5 V.

The difference between the input maximum "zero" and the output maximum "zero" is called the noise margin of the family. It is also the difference between the output minimum "one" and the input minimum "one." For TTL, the noise margin is 400 mV. This allows the gate to function correctly with up to 400 mV of noise (peak voltage) existing at an input. This noise may be induced in a wire connected from a TTL output to a TTL input by some external influence. Figure 3-4 illustrates this.

Table 3-4 summarizes various ways of representing and referring to the logic levels for a "0" or "1."

Within a digital circuit, then, logic levels will be found that represent a binary 0 or a binary 1. In TTL circuitry these levels will generally be 0 or +5 V or within the specified limits from 0 to 0.8 and 2 to 5 V.

4 LOGIC GATES

The first logic gate typically used in digital circuitry is the *inverter*. Figure 3-5 shows the logic symbol for an inverter and the corresponding RTL schematic. The inverter's sole function is to change a zero to a one, or a one to a zero. The RTL gate acts as a simple transistor switch. When the input is pulled low (to ground for a zero) the transistor turns off and the output floats up to a 1 level. On the other hand, when a "1" is applied to the input, the transistor turns on and the output is pulled low. This type of circuit always uses more current than necessary to turn the transistor on and is referred to as a *saturated* form of logic. Saturated logic is slower than nonsaturated logic because of the added time needed to remove excess electrical charge from the base of the transistor. Figure 3-6 shows modern TTL, CMOS, and ECL inverters.

FIGURE 3-5 The logic inverter. (*a*) Inverter symbol. (*b*) Resistor–transistor logic gate. (*c*) 7404 TTL inverter. (Courtesy of Texas Instruments Incorporated.)

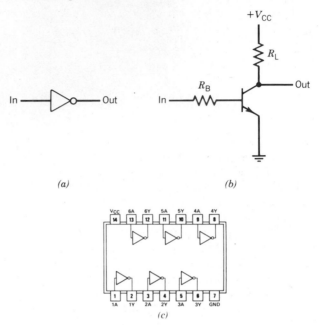

(a)   (b)

(c)

The T²L family features a number of integrated circuits with *open-collector* (OC) outputs. In such circuitry (OC outputs), the outputs usually require an external load resistor to pull the output high. Such a resistor is called a pullup resistor. Figure 3-7 shows a 7405 inverter stage with an OC output.

The operation of an inverter is described in a table called a truth table or function table. Such a chart exists for most logic gates and describes how the gate behaves for all different input combinations. With a one-input gate like the inverter, only two possible input conditions exist. The output is labeled *f* because each logic gate produces some type of logic function. Figure 3-8 shows the inverter and an expression to describe its behavior, and the associated truth table.

The circle on the output of the symbol indicates the inverting or complement operation. The same function is performed if the symbol incorporates the circle on the input side. Only one circle is proper, since two circles (or two inversions) would get you back where you started. The logic expression accompanying Fig. 3-8*a* describes in algebraic terms the behavior of the gate. This type of algebra, called *Boolean algebra*, is named for the work done by George Boole in the mid-19th century and is a form of symbolic logic used in the design of digital circuitry. The line over the $\overline{A}$ is the inverting or complement function and means NOT A. The expression $f = \overline{A}$ is read, "*f* will be a '1' when *A* is not a '1.' "

FIGURE 3-6 (*a*) TTL inverter. (*b*) CMOS inverter. (*c*) ECL inverter.

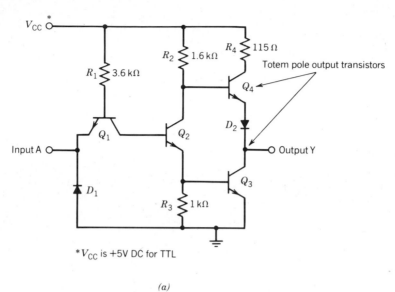

*V_{CC} is +5V DC for TTL

(*a*)

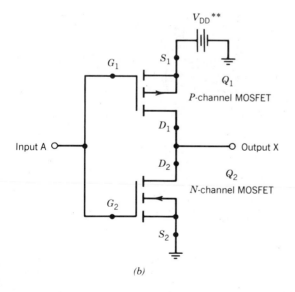

(*b*)

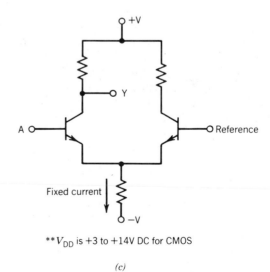

**V_{DD} is +3 to +14V DC for CMOS

(*c*)

FIGURE 3-7 (*a*) Open-collector output. (*b*) 7405 inverter (open-collector) stage.

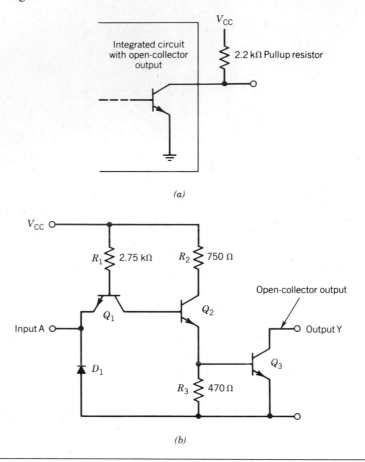

(a)

(b)

FIGURE 3-8 Logic inverter (*a*) and truth table (*b*).

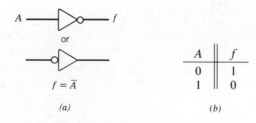

$f = \overline{A}$

(a)

A	f
0	1
1	0

(b)

TABLE 3-5   Boolean Expressions and Identities

Expressions	
Terms	Meaning
A	A is a 1
$\overline{A}$	A is a 0
=	when
x, ·, Multiply	AND
+, Sum	OR

Identities
1. $A \cdot 0 = 0$
2. $A \cdot 1 = A$
3. $A + 0 = A$
4. $A + 1 = 1$
5. $A + A = A$
6. $A + \underline{A} + A = A$
7. $A + \overline{A} = 1$
8. $1 \cdot 1 = 1$
9. $1 \cdot 0 = 0$
10. $1 + 0 = 1$
11. $1 + 1 = 1$
12. $A\,(B + \overline{B}) = A$

A term or symbol alone, (e.g., B) is read "B is a '1' "; an inverted or complemented symbol is read "$\overline{B}$ is a '0.' " Table 3-5 summarizes some common Boolean operators and Boolean identities.

4.1 The Schmitt Trigger Gate

Often it is necessary to square up a signal for use by a digital circuit. For instance, a 60-Hz a-c sine wave is often used to synchronize digital circuitry and video display circuitry. The 16-msec period of the waveform is very slow in comparison to a microcomputer's abilities. A 60-Hz pulse waveform is needed to accurately time events and operate logic circuitry. A Schmitt trigger switches at one specific voltage, not a range of voltages. Most TTL gates expect a logic zero to be from 0 to 0.8 V and also expect a logic one to be from 2 to 5 V. The region from 0.8 to 2 V is a forbidden zone or gray area. A signal in this area will not produce predictable results on a standard TTL gate. The Schmitt trigger, however, switches on at a specific voltage (the lower threshold voltage) and switches off at another voltage (the upper threshold voltage). Figure 3-9 shows a sine wave input to a Schmitt inverter.

The Schmitt trigger is not concerned with voltages in the gray area, since it has already switched. The Schmitt trigger will switch back when the input signal passes the upper threshold voltage. Figure 3-10 shows a simple circuit to square up a sine wave input for use by logic circuitry.

FIGURE 3-9 (*a*) Schmitt trigger and (*b*) sine wave input. (Courtesy of Texas Instruments Incorporated.)

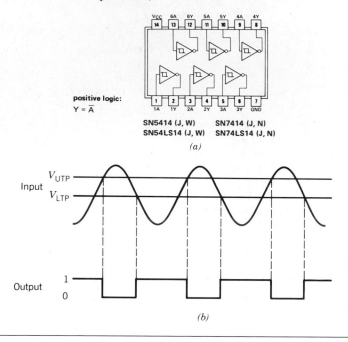

(a)

(b)

FIGURE 3-10 Schmitt trigger used to clean up a sine wave.

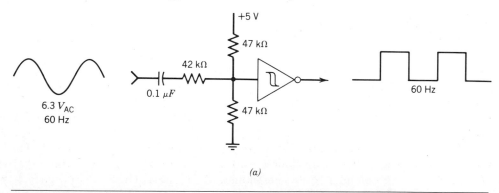

(a)

4.2 The Buffer

The buffer is a very simple logic gate. At first glance it appears to perform no function at all. It produces the same level that is applied to it. Figure 3-11 shows this gate, its Boolean expression, and its truth table. Such a gate, however, is used to buffer a signal. It removes noise from a line by acting as an amplifier. If

FIGURE 3-11 Logic buffer. (*a*) 7407 TTL gate (open collector) and (*b*) truth table. (Courtesy of Texas Instruments Incorporated.)

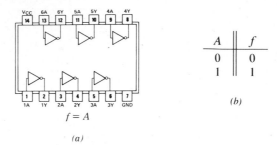

$f = A$

(a)

A	f
0	0
1	1

(b)

a signal has become lower in amplitude because it has had to drive a number of other TTL inputs, a buffer brings it back up. For example, a TTL "1" can be as low as 2 V and still be considered a "1." It is, however, marginal. Passing through a buffer gives the signal new life. A "1" is again a 5-V level.

The number of TTL gate inputs that a TTL output can safely drive is referred to as the *fanout* of the gate. The fanout for a TTL gate may be as high as 20. Figure 3-11 shows a TTL buffer, the 7407.

4.3 The AND Gate

The function of an AND gate is to produce a "1" at the output only when all inputs are at this level. If even one input is low, the output will be low. Figure 3-12 shows the symbol for a two-input AND gate and associated Boolean expres-

FIGURE 3-12 (*a*) Two-input AND gate and (*b*) truth table.

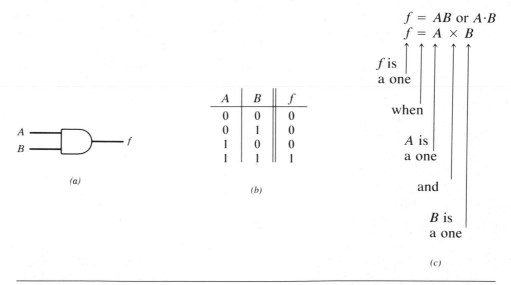

(a)

A	B	f
0	0	0
0	1	0
1	0	0
1	1	1

(b)

$$f = AB \text{ or } A \cdot B$$
$$f = A \times B$$

f is a one

when

A is a one

and

B is a one

(c)

FIGURE 3-13 7408 quad two-input AND gate. (Courtesy of Texas Instruments Incorporated.)

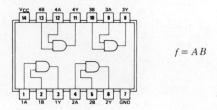

$f = AB$

FIGURE 3-14 (*a*) Three-input AND gate and (*b*) truth table. (*c*) Schematic for 7411 TTL triple three-input AND gate. (Courtesy of Texas Instruments Incorporated.)

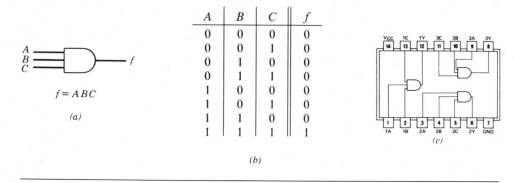

A	B	C	f
0	0	0	0
0	0	1	0
0	1	0	0
0	1	1	0
1	0	0	0
1	0	1	0
1	1	0	0
1	1	1	1

$f = ABC$

(*a*)

(*b*)

(*c*)

sion and truth table. (AND gates come with many more inputs if necessary.) The truth table lists the four possible inputs that can be applied to the gate. Both inputs could be low (0), or one or the other could be high (1), or both could be high. The truth table lists every possible input combination. There are *no more* with only two inputs. The number of entries on a truth table is always 2^N, where N is the number of inputs on the gate.

The Boolean equation reads, "*f* is a '1' when A is a '1' AND B is a '1.' " Figure 3-13 shows a 7408 T²L AND gate. The Boolean equation for the AND operation can be written in more than one way: $f = A \times B$ or $f = A \cdot B$ or $f = AB$. In any case, a multiply is taken as the word AND to specify that particular logic function.

Figure 3-14 shows a three-input AND gate and a truth table that is $2^N = 2^3 = 8$ lines long.

4.4 OR Gate and Exclusive OR Gate (XOR)

The OR gate will produce an output if any input is high. A plus sign (+) is used to represent the OR operation in Boolean algebra. Figure 3-15 shows a three-input

FIGURE 3-15 (*a*) Three-input OR gate and (*b*) truth table.

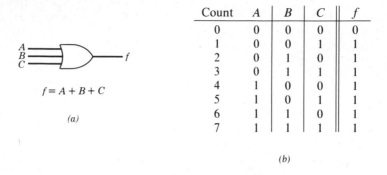

$$f = A + B + C$$

(a)

Count	A	B	C	f
0	0	0	0	0
1	0	0	1	1
2	0	1	0	1
3	0	1	1	1
4	1	0	0	1
5	1	0	1	1
6	1	1	0	1
7	1	1	1	1

(b)

FIGURE 3-16 (*a*) Two-input exclusive *OR* gate (XOR) and (*b*) truth table. (*c*) Schematic of 7486 quad two-input XOR gate. (Courtesy of Texas Instruments Incorporated.)

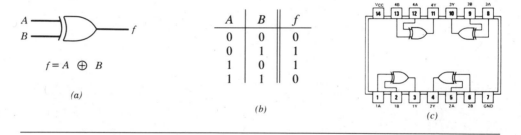

$$f = A \oplus B$$

(a)

A	B	f
0	0	0
0	1	1
1	0	1
1	1	0

(b)

(c)

OR gate and truth table. The equation reads, "*f* is a '1' when *A* is a '1' $\overline{OR}$ '*B*' is a '1' $\overline{OR}$ C is a '1.'" The truth table has eight lines and represents a $\overline{\text{binary}}$ count from 0 to 7. By counting, each truth table will always look the same; thus we will not miss any input sequences.

The exclusive OR gate is a special OR gate. The output will be high if either input is high, but *not* both. Figure 3-16 shows such a gate and its logic symbol. The new symbol $\oplus$ represents the exclusive OR function. This equation can be written in two ways, $f = A \oplus B$ and $f = A\overline{B} + \overline{A}B$. The second expression is read: "*f* is a '1' when *A* is a '1' and *B* is a '0' OR when *A* is a '0' and *B* is a '1.'"

4.5 The NAND Gate

The NAND gate is an inverted AND gate. The inversion is at the output of the AND. Figure 3-17 shows two- and three-input NAND gates and then truth tables. A long line indicates that the inversion is at the output on the Boolean expression.

FIGURE 3-17 (*a*) and (*b*) Three-input NAND gate and truth table; (*c*) 7410 triple three-input NAND. (*d*) and (*e*) Two-input NAND gate and truth table; (*f*) 7400 quad two-input NAND. (Courtesy of Texas Instruments Incorporated.)

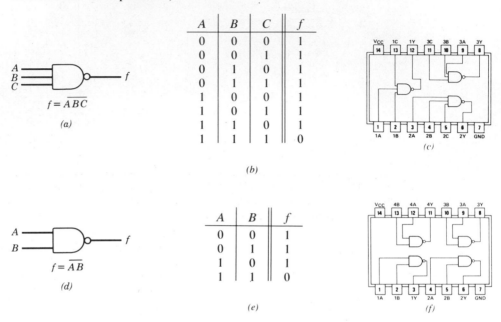

A	B	C	f
0	0	0	1
0	0	1	1
0	1	0	1
0	1	1	1
1	0	0	1
1	0	1	1
1	1	0	1
1	1	1	0

$f = \overline{ABC}$

(*a*)

(*b*)

(*c*)

$f = \overline{AB}$

(*d*)

A	B	f
0	0	1
0	1	1
1	0	1
1	1	0

(*e*)

(*f*)

4.6 The NOR Gate

The NOR gate is an inverted OR gate. The inversion takes place at the output of the OR gate. Figure 3-18 shows the NOR gate in two forms. Care must be used in writing the Boolean expressions when showing the inversions. Short lines may *never* be replaced by a single long line.

Caution: Two shorts do not make a long! When writing Boolean expressions, great care must be taken to prevent lines (inversions) from running together. An inverted inversion cancels itself ($\overline{\overline{A}} = A$), but $\overline{A}\ \overline{B} \neq \overline{AB}$! This can be shown by comparing the truth tables for $\overline{A}\ \overline{B}$ and $\overline{AB}$ as in Fig. 3-19. Since the truth tables are different, we conclude that $\overline{A}\ \overline{B} \neq \overline{AB}$. So, use care in drawing the lines!

4.7 Troubleshooting Logic Gates

Learning the behavior of the various logic gates is important and can lead to the identification of faulty gates in digital circuitry. For instance, if a NAND gate has two inputs that are both low and an output that is also low, is this a possible faulty gate? The answer is yes! A NAND gate should have a "1" at its output when both inputs are low.

FIGURE 3-18 (*a*) and (*b*) Two-input NOR gate and truth table; (*c*)
7402 NOR. (*d*) and (*e*) Three-input NOR gate and truth table; (*f*)
7427 NOR. (Courtesy of Texas Instruments Incorporated.)

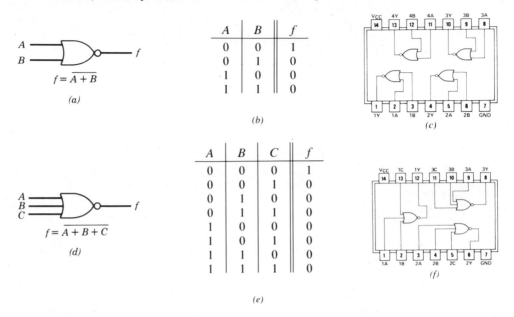

A	B	f
0	0	1
0	1	0
1	0	0
1	1	0

$f = \overline{A + B}$

(*a*)

(*b*)

(*c*)

A	B	C	f
0	0	0	1
0	0	1	0
0	1	0	0
0	1	1	0
1	0	0	0
1	0	1	0
1	1	0	0
1	1	1	0

$f = \overline{A + B + C}$

(*d*)

(*e*)

(*f*)

FIGURE 3-19 Comparison of $\overline{A}\ \overline{B}$ and $\overline{AB}$.

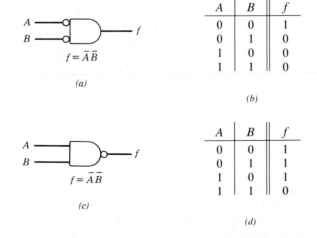

$f = \overline{A}\,\overline{B}$

(*a*)

A	B	f
0	0	1
0	1	0
1	0	0
1	1	0

(*b*)

$f = \overline{A}\,\overline{B}$

(*c*)

A	B	f
0	0	1
0	1	1
1	0	1
1	1	0

(*d*)

EXAMPLE 3-1

Identify each of the following logic gates as either good or possibly bad.

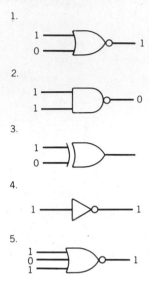

SOLUTION

1. Bad gate—output should be low.   4. Bad gate—no inversion.

2. Good gate.   5. Bad gate—output should be low.

3. Bad gate—output should be high.

EXAMPLE 3-2

In the circuit shown, are there any bad gates for the levels indicated?

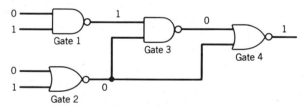

SOLUTION
Yes, gate 3 is faulty.

Logic gates usually do not have d-c or static signals applied to them. More often the signals are constantly changing, and we refer to the inputs as dynamic. Such a signal is a series of pulses. If a series of pulses is applied to an inverting type of gate, the pulses become inverted as they pass through the gate. This is normal because each "0" is complemented to a "1" and each "1" is complemented to a "0." Consider a two-input AND gate as shown in Fig. 3-20. If the

FIGURE 3-20 Enabling a set of pulses.

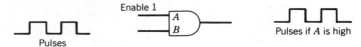

FIGURE 3-21 NAND gate enable.

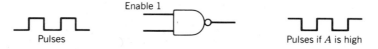

A input is high, the pulses pass through the AND gate. However, if *A* is low, no pulses pass, since the AND operation allows the output to be high only when all inputs are high.

If a NAND gate is used, the effect is similar, but the pulses are inverted. The *A* input is used to switch the pulses on or off in both cases and such an operation is referred to as an *ENABLE* (Fig. 3-21). This technique allows a logic gate to be used to turn signals on and off in digital computers.

Consider the application of dissimilar pulse trains to the inputs to an AND gate. The output will be high only when all inputs are high (Fig. 3-22). If a NAND gate is used, the output is inverted, as in Fig. 3-23.

FIGURE 3-22 Using an AND gate as a switch.

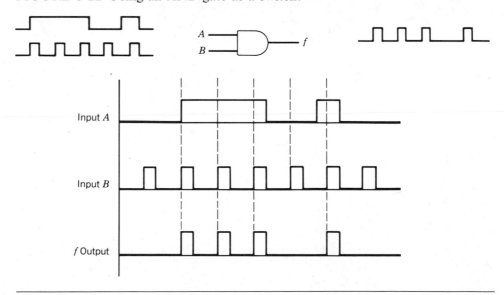

FIGURE 3-23 NAND gate enable.

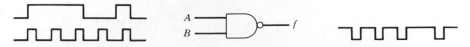

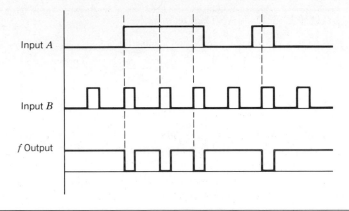

FIGURE 3-24 OR gate enable.

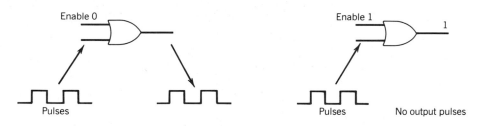

FIGURE 3-25 NOR gate enable.

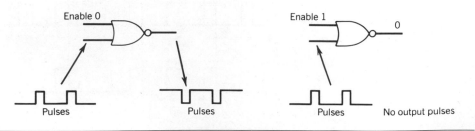

FIGURE 3-26 Complex waveform on a NAND gate.

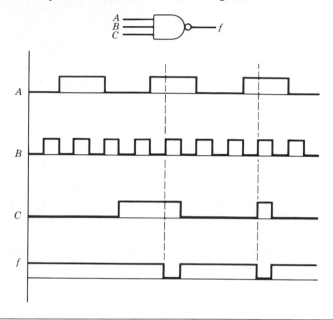

An OR gate operates in a different way. The output of an OR gate is low only when all inputs are low. For pulses to pass through an OR gate, the enable line must be low. If it is high, the output stays high. Figure 3-24 shows these two conditions. The output is inverted for the NOR gate, as shown in Fig. 3-25. Figure 3-26 shows three complex waveforms applied to a three-input NAND gate and the resulting output. The output can be low only when all inputs are high.

4.8 Gate Substitutions and DeMorgan's Theorem

Frequently in designing logic circuits, a particular type of gate is needed and we are reluctant to use another integrated circuit. If an inverter is needed and a spare NOR gate is available, then an answer is at hand. This section shows how one gate may substitute for another without changing the logic. Figure 3-27 illustrates several ways to build an inverter. Each circuit will replace an inverter as shown. Tying inputs together, which frequently increases the load on a gate, is not a good idea. The better approach is shown. In the case of the NAND gate, unused inputs are tied high through a 2.2-KΩ resistor. This is done to prevent a faulty input from disabling the gate.

It is a relatively simple matter to construct OR gates from NOR gates by adding an inverter. The same may be done to construct AND gates from NAND gates. These substitutions may be seen in Fig. 3-28.

FIGURE 3-27 Inverters.

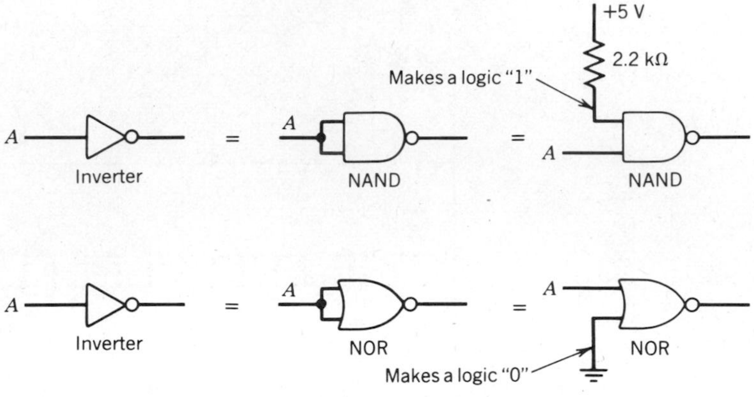

FIGURE 3-28 Substitutions of OR and AND gates.

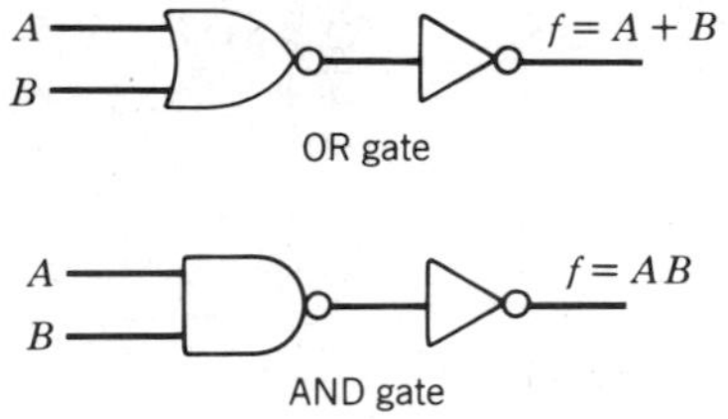

Another 19th century mathematician, Augustus DeMorgan, a friend of George Boole, is known for two theorems that find extensive use in digital design:

1. $\overline{A}\,\overline{B} = \overline{A + B}$

2. $\overline{AB} = \overline{A} + \overline{B}$

Theorem 1 shows that the AND gate may be replaced by a NOR gate, provided the inputs are complemented. Theorem 2 shows that a NAND gate may replace an OR gate if the inputs are complemented. Figure 3-29 shows these gate equivalents.

Frequently in the logic drawing for a piece of equipment we find the symbol presented in Fig. 3-30. It should be now recognized as an AND gate performing the NOR function.

Table 3-6 lists some interesting conversions using DeMorgan's theorem. Study it to become more familiar with these gate substitutions. In the next chapter we will be using all the logic gates and the gate substitution ideas in more detail. These form the basis for all digital circuit design.

FIGURE 3-29 DeMorgan's theorems illustrating gate equivalence.

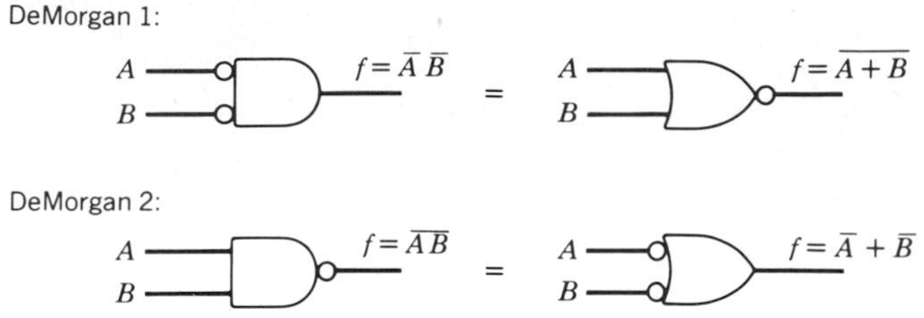

DeMorgan 1:

DeMorgan 2:

FIGURE 3-30 Inverted input AND gate performing the NOR function.

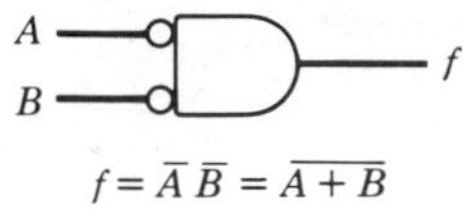

$$f = \overline{A}\,\overline{B} = \overline{A + B}$$

TABLE 3-6   Gate Conversions Using De Morgan's Theorem

$\overline{A + B}$	$=$	$\overline{A}\ \overline{B}$
$\overline{\overline{A} + B}$	$=$	$A\ \overline{B}$
$A + B$	$=$	$\overline{\overline{A} \cdot \overline{B}}$
$A + \overline{B}$	$=$	$\overline{\overline{A} \cdot B}$
$\overline{AB + BC}$	$=$	$\overline{AB} \cdot \overline{BC}$
$\overline{AB + BC}$	$=$	$\overline{A} + \overline{B} + \overline{B} + \overline{C} = \overline{A} + \overline{B} + \overline{C}$

Figure 3-31 shows the relations between various logic gates using DeMorgan's theorem. For instance, in Fig. 3-31a we can see that a NAND gate can convert to an OR gate if the inputs are inverted. We also see that an AND gate can convert to a NOR gate if the inputs are inverted. Clearly, the difference between an AND gate and a NAND gate is that the outputs are inverted. Figure 3-31b shows the same relationships in equation form.

You may be interested to see the original form of DeMorgan's theorem as he developed it in the early 1800s. It reads as follows:

> The negative or contradictory of an alternative proposition is a conjunction in which the conjuncts are the contradictions of the corresponding alternates, and that the negative of a conjunctive is an alternative proposition in which the alternates are the contradictories of the corresponding conjuncts.

Then again, you may prefer $\overline{A} \cdot \overline{B} = \overline{A + B}$ and also $\overline{A} + \overline{B} = \overline{A \cdot B}$. We include this material for its historical significance.

FIGURE 3-31 Gate relationships.

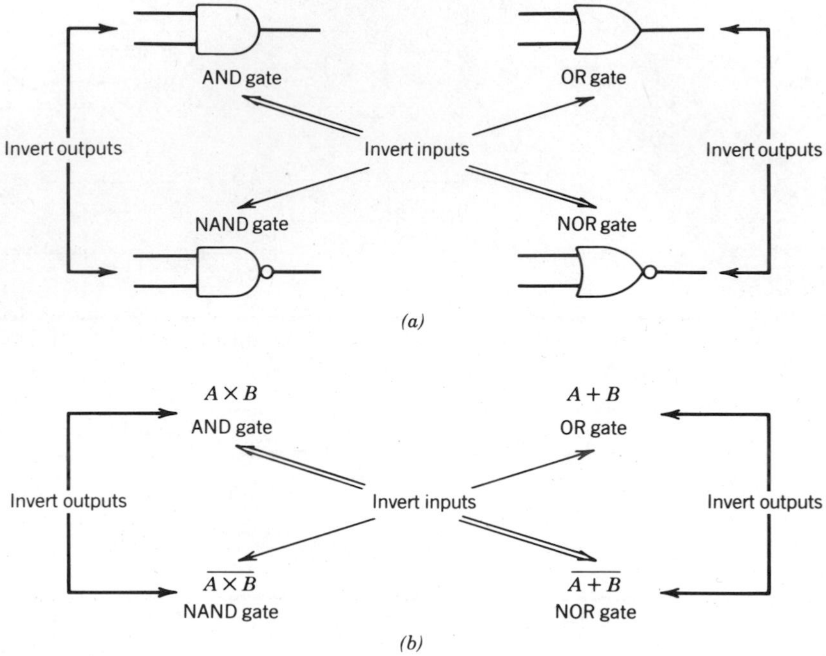

(a)

(b)

4.9 Propagation Delay

One very important aspect of digital logic that is often overlooked (but becomes very important in high-speed applications) is the effect of gate propagation delay time. Figure 3-32 demonstrates that the output of a 74LS00 NAND gate does not go low the instant the input goes high. This actually occurs 10 nsec later. This 10-nsec period is the propagation delay time of the 74LS00 and is a function of the time needed to turn transistors on and off inside the gate. Propagation delay can cause disastrous results if a circuit is not designed correctly. For example, the designer of the circuit in Fig. 3-33 ignored the propagation delays (use 10 nsec per gate to figure it out for yourself) and got an extra pulse, or glitch, of 20 nsec. It is important to keep track of the gate delays when designing a digital circuit.

5 GATE APPLICATIONS

A number of functional circuits can be easily constructed to show the usefulness of logic gates. One example is a digital oscillator that produces a free-running pulse train. Figure 3-34 shows a capacitor-controlled ring oscillator and a crystal oscillator.

FIGURE 3-32 Propagation delay.

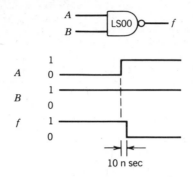

FIGURE 3-33 Errors due to propagation delay. (*a*) Simple digital circuit. (*b*) Desired timing. (*c*) Actual timing due to gate delays (10 nsec each).

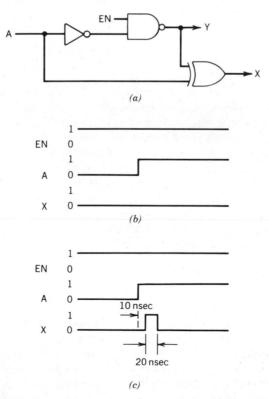

FIGURE 3-34 Digital oscillators. (*a*) Ring oscillator using 7404 inverters and (*b*) crystal oscillator using 74LS04 inverter.

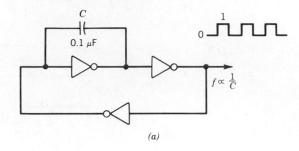

(a)

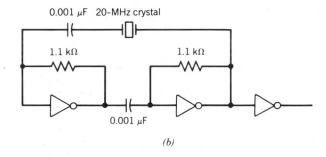

(b)

Another useful circuit is an LED (light-emitting diode) driver. Either a buffer or an inverter may be used to turn on the LED. This circuit is very useful as an indicator of the state of a logic gate's output. Figure 3-35 shows these circuits. The circuit is called a driver circuit because it is able to provide the drive current required by the LED. The resistor limits the current to protect both the LED and the logic gate. Such driver circuits are used in quantity to operate indicator lights.

One of the main reasons for producing open-collector outputs is to be able to turn high-current indicators on and off. While TTL operates on +5 V d-c, it is possible to operate a higher voltage indicator. The 7406 is an open-collector gate that can switch up to 30 V d-c. Figure 3-36 shows the circuit. A logic "1" on the input drives the output low, providing a ground for the lamp to turn it on.

5.1 Wired OR Gate

An open-collector output also can be used when a wire is common to many circuits. The wire turns a single indicator lamp on. At any point on the line, an open collector may pull the line down, thus lighting the lamp. For example, a warning indicator might be set off from several different sources. Figure 3-37 illustrates this important principle. Any open-collector gate can turn on the lamp. This function is called a *(WIRED OR)* function. A logic "1" on *A* or *B* or *C* or *D* will turn on the light.

FIGURE 3-35 LED drivers. (*a*) Buffer type and (*b*) inverter type.

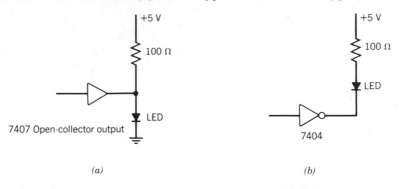

(*a*)   (*b*)

FIGURE 3-36 High-voltage lamp driver.

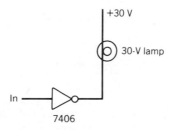

In a later chapter we will make use of a very special type of logic gate related to this discussion of the wired OR. This special gate is referred to as a *tri-stated* logic gate. This special-purpose gate has the ability to do one of three things to a digital signal line:

FIGURE 3-37 Wired OR function.

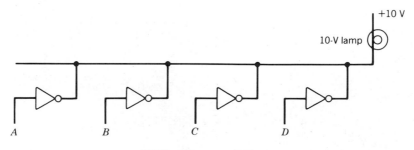

1. Apply a "0" to the line.
2. Apply a "1" to the line.
3. Disconnect from the line.

   Tri-state gates have a control line to connect or disconnect. If disconnected, the gate is said to be tri-stated. A tri-state inverter is shown in Fig. 3-38.
   Figure 3-39 shows two versions of a TTL 54/74 series tri-state buffer—the 74125

FIGURE 3-38 (*a*) Tri-state logic gate. (*b*) Equivalent three-state output.

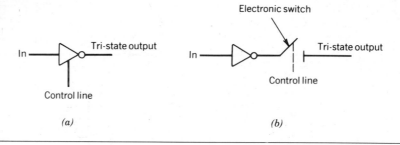

FIGURE 3-39 Tri-state buffers. (*a*) 74125 and (*b*) 74126. (Courtesy of Texas Instruments Incorporated.)

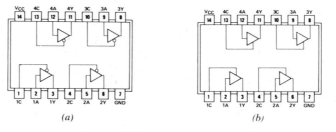

FIGURE 3-40 Tri-state NAND gate. 74134. (Courtesy of Texas Instruments Incorporated.)

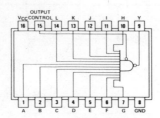

and 74126. The difference is in the use of a "0" or a "1" to tri-state the gate.

Figure 3-40 shows a 74134 12-input NAND gate with tri-state output. We will find that many microcomputer circuits are able to connect or disconnect themselves to digital circuit paths. Their ability to be tri-stated is very necessary to the sharing of signal paths by many digital devices.

5.2 Bounceless Switch

Logic circuitry demands that a pulse have a very clean edge. In most cases, this edge must rise in less than 100 nsec. Unfortunately, every mechanical switch is slow and subject to contact bounce. Figure 3-41 shows the effect of a common mechanical switch on closing and the desired waveform. Figure 3-41c shows the standard measurement for rise time. As Fig. 3-42 shows, the rise time for a pulse is the time to travel from 10% of maximum to 90% of maximum or from 0.5 to 4.5 V. Similarly, the fall time (t_f) is the time to go from 90% to 10% of the maximum. The pulse width (t_{pw}) is the time between 50% points.

FIGURE 3-41 Contact bounce. (*a*) Dirty switch. (*b*) Resulting waveform. (*c*) Desired waveform.

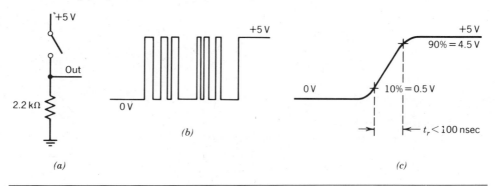

FIGURE 3-42 Rise time t_r and fall time t_f of a pulse.

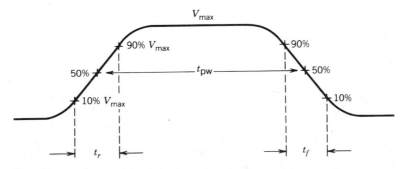

FIGURE 3-43 Bounceless electronic switches. (*a*) 7400 NAND switch and (*b*) 7402 NOR switch.

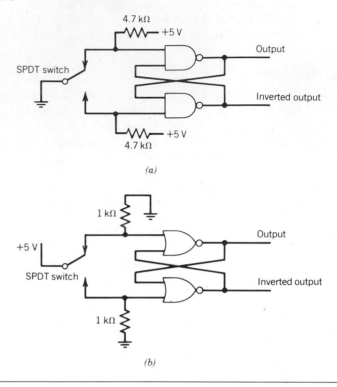

(a)

(b)

For a TTL gate the rise time should be less than 100 nsec, and it must be clean. Figure 3-43 shows two ways to completely clean up the waveform from a mechanical switch. The circuits both use a single-pole, double-throw switch (SPDT). One advantage of these circuits is that each has two separate outputs that are mutual complements. This circuit is also referred to as a d-c latch because of its locking action.

5.3 Alarm Circuit

A variation of the bounceless switch circuit can produce a latching circuit to set off an alarm. A separate switch is required to perform the reset operation. Figure 3-44 shows the alarm circuit complete with LED indicator.

5.4 Decoder Circuit

Two wires containing binary logic levels can have four states. The two wires can be at the following levels: 00, 01, 10, or 11. A circuit sometimes called a one-of-four selector circuit can be used to determine which of the four cases exists on the two wires. Indicators are provided as shown in Fig. 3-45.

FIGURE 3-44 Alarm circuit with indicator.

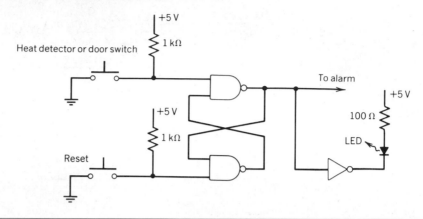

FIGURE 3-45 (*a*) One-of-four selector and (*b*) truth table.

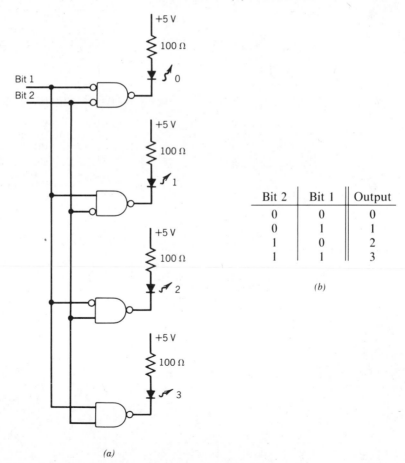

Bit 2	Bit 1	Output
0	0	0
0	1	1
1	0	2
1	1	3

(*b*)

(*a*)

FIGURE 3-46 Data selector.

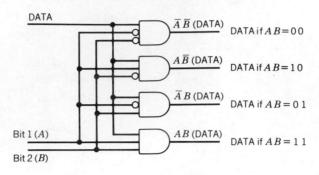

FIGURE 3-47 Decoder/multiplexers. (*a*) 74138 and (*b*) 74139.
(Courtesy of Texas Instruments Incorporated.)

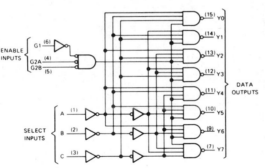

| INPUTS | | | | OUTPUTS | | | | | | | |
| ENABLE | | SELECT | | | | | | | | | |
G1	G2*	C	B	A	Y0	Y1	Y2	Y3	Y4	Y5	Y6	Y7
X	H	X	X	X	H	H	H	H	H	H	H	H
L	X	X	X	X	H	H	H	H	H	H	H	H
H	L	L	L	L	L	H	H	H	H	H	H	H
H	L	L	L	H	H	L	H	H	H	H	H	H
H	L	L	H	L	H	H	L	H	H	H	H	H
H	L	L	H	H	H	H	H	L	H	H	H	H
H	L	H	L	L	H	H	H	H	L	H	H	H
H	L	H	L	H	H	H	H	H	H	L	H	H
H	L	H	H	L	H	H	H	H	H	H	L	H
H	L	H	H	H	H	H	H	H	H	H	H	L

*G2 = G2A + G2B

H = high level, L = low level, X = irrelevant

(a)

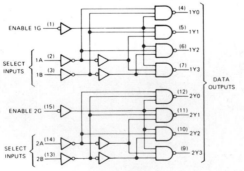

| INPUTS | | | OUTPUTS | | | |
| ENABLE | SELECT | | | | | |
G	B	A	Y0	Y1	Y2	Y3
H	X	X	H	H	H	H
L	L	L	L	H	H	H
L	L	H	H	L	H	H
L	H	L	H	H	L	H
L	H	H	H	H	H	L

H = high level, L = low level, X = irrelevant

(b)

The one-of-four selector is frequently used to "select" one of four signals. A variation of this circuit can be used to route a digital signal to one of four places. This requires an extra input that is common to each gate. We will refer to this signal as DATA (see Fig. 3-46). DATA can be high, low, or a series of pulses (high/low transitions). In any case, DATA will be routed to output 1, 2, 3, or 4, depending on which output was selected. The 54/74 series of TTL logic offers several types of data selector. The 74139 is a dual one-of-four selector. The 74138 is a one-of-eight selector. Figure 3-47 shows both these prepackaged circuits, which are called decoder/multiplexers.

5.5 Binary Adder

To add two binary bits together, we must be able to have both a sum bit and a carry bit. For instance, to add $0 + 1$, the result is 1 with 0 to carry. To add $1 + 1$, the result is 0 with 1 to carry (i.e., 10 B). A circuit that does this is relatively simple and must follow a truth table (Fig. 3-48). We can see from the truth table that the sum requires an exclusive OR gate and the carry requires an AND gate. The XOR function is written in two ways: sum $= A \oplus B$ or sum $= \overline{A}B + A\overline{B}$. The other implementation of the sum and carry functions are shown in Fig. 3-49. If either of these is installed in a box, we have a packaged binary adder called a *half adder circuit*. A full adder includes a carry input as well and can be made

FIGURE 3-48 Binary adder circuit. (*a*) Half adder and (*b*) truth table.

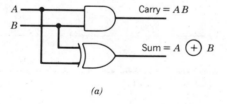

(a)

A	B	Sum	Carry
0	0	0	0
0	1	1	0
1	0	1	0
1	1	0	1

(b)

FIGURE 3-49 Binary adder circuit.

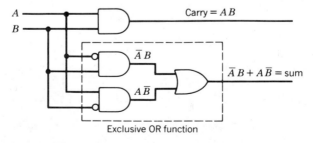

Exclusive OR function

from a pair of half adders (HA) as shown in Fig. 3-50. The full adder (FA) can be used with other full adders to build a parallel binary full adder. Such an adder can add large binary numbers together quickly. To illustrate the usefulness, in Fig. 3-51 we add two numbers: 9 D and 14 D. Pure binary (four bits) are used. This type of adder is called a hardware adder, since no programming is involved. Hardware adders are used in large computer systems and are finding increasing use in smaller computers. The 54/74 series of TTL integrated circuits offers a 4-bit parallel binary adder complete and ready for use. Figure 3-52 shows the 74283 4-bit binary full adder.

We are now ready to design some digital circuitry using the basic logic gates. Boolean algebra and the truth table will be important aspects of this design.

FIGURE 3-50 Full adder circuit. (*a*) Two half adders and (*b*) full adder truth table.

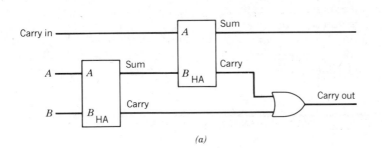

A	B	Carry in	Sum	Carry out
0	0	0	0	0
0	0	1	1	0
0	1	0	1	0
0	1	1	0	1
1	0	0	1	0
1	0	1	0	1
1	1	0	0	1
1	1	1	1	1

(*b*)

FIGURE 3-51 Four-bit parallel adder.

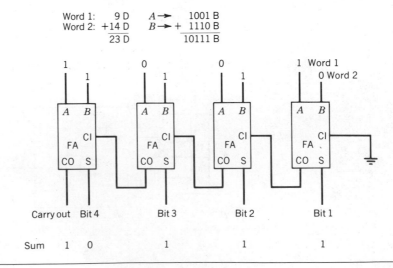

FIGURE 3-52 Parallel full adder. (Courtesy of Texas Instruments Incorporated.)

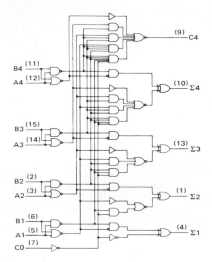

6 SOFTWARE LOGIC GATES

Later, when we implement the basic logic gates in a microcomputer program, we will use software rather than hardware to produce the logic functions AND, OR, XOR, Invert, and so on. At this stage we will be able to work with a pair of bytes at a time. For instance, if we desire to logically AND the following pair of bytes, we AND bits in identical positions between the two bytes (e.g., b_1 with b_1, b_2 with b_2, b_3 with b_3, etc.).

EXAMPLE 3-3

AND of 3E H and 46 H

SOLUTION

$$\begin{array}{rl}
3E\ H = & 00111110 \\
\text{AND}\quad \underline{46\ H =} & \underline{01000110} \\
06\ H \longleftarrow & 00000110
\end{array}$$

In the AND operation, a "1" is produced only when both matching bits are high. The hexadecimal result of this operation then is 06 H. We take the numbers to binary, AND them, and return to hex to express the result. Here are a few more examples.

EXAMPLE 3-4

OR of A6 H and 32 H

SOLUTION

$$
\begin{array}{rl}
& \text{A6 H} = \quad 10100110 \\
\text{OR} & \underline{\text{32 H}} = \quad \underline{00110010} \\
& \text{B6 H} \longleftarrow 10110110
\end{array}
$$

EXAMPLE 3-5

XOR of 47 H and 22 H

SOLUTION

$$
\begin{array}{rl}
& \text{47 H} = \quad 01000111 \\
\text{XOR} & \underline{\text{22 H}} = \quad \underline{00100010} \\
& \text{65 H} \longleftarrow 01100101
\end{array}
$$

EXAMPLE 3-6

Complement of 27 H (1's complement)

SOLUTION

$$
\begin{array}{l}
\text{27 H} = \quad 00100111 \\
\text{D8 H} \longleftarrow 11011000
\end{array}
$$

It is important to understand the basic logic gates in terms of both hardware and software. These examples are a hint of what is to come in our discussions of the software aspect of microcomputers. We will see that these operations are very valuable tools for the microprocessor engineer.

7 SUMMARY

The logic gates are the building blocks of the digital electronics industry, and an understanding of each gate and its various uses is essential to later work with microprocessors. Figure 3-53, which summarizes the logic gates and shows them in their different forms when DeMorgan's theorem is applied, constitutes a review of information presented in this chapter. Study these logic gates well, as they are used over and over again in this text. Also in this chapter, we learned about logic families: RTL, ECL, CMOS, but primarily TTL. TTL is relatively high-speed logic and easy to use (simple positive supply). The basic TTL building blocks are the inverter, the AND gate, and the OR gate. Specialized gates such as the NAND, NOR, and XOR can be made from combinations of these gates. With an understanding of truth tables and simple Boolean algebra, it is possible to design digital logic circuits to perform important tasks or functions such as oscillation, latching, switch debouncing, and adding. Finally, it is important to keep track of propagation delay when designing a digital circuit.

FIGURE 3-53 Summary of logic gates.

Logic Symbols	Truth Tables		
	Input A	Input B	Output f

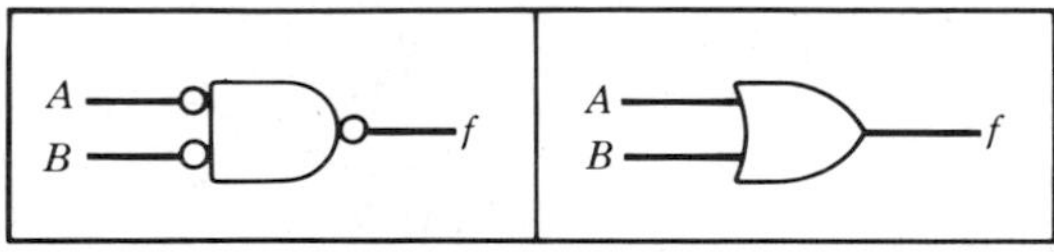

OR

	Input A	Input B	Output f
	0	0	0
	0	1	1
	1	0	1
	1	1	1

AND

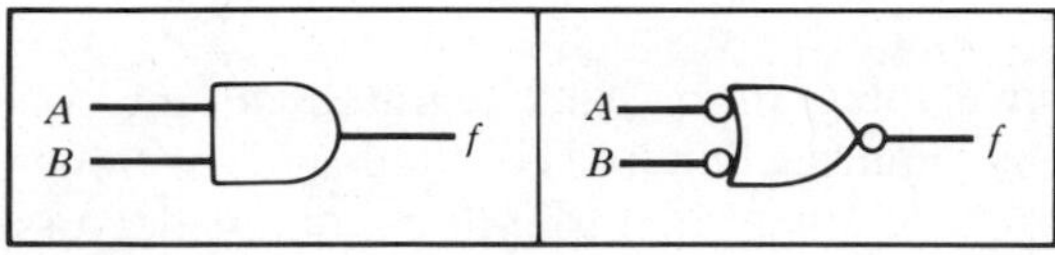

	Input A	Input B	Output f
	0	0	0
	0	1	0
	1	0	0
	1	1	1

NAND

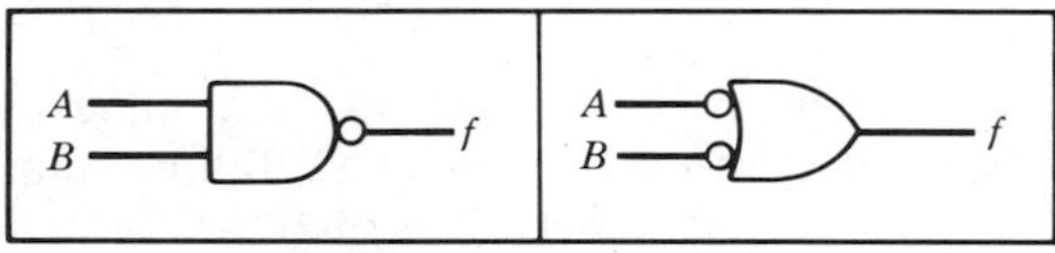

	Input A	Input B	Output f
	0	0	1
	0	1	1
	1	0	1
	1	1	0

NOR

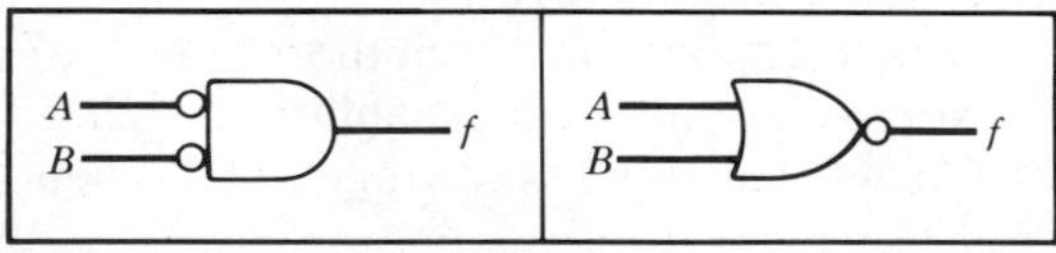

	Input A	Input B	Output f
	0	0	1
	0	1	0
	1	0	0
	1	1	0

Exclusive OR

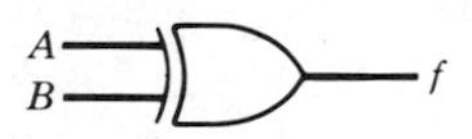

	Input A	Input B	Output f
	0	0	0
	0	1	1
	1	0	1
	1	1	0

8 GLOSSARY

AND gate.   Logic gate that produces a logic "1" only when all inputs are logic 1s.

Boolean algebra.   Symbolic logic (algebra) used in the design of digital circuits.

Buffer.   A logic gate with a high fanout used to drive many other gates.

Contact bounce.   Noise resulting from the action of closing a mechanical switch.

Decoder/multiplexer.   Logic gate used to select one of two (one of four, one of eight, etc.) inputs and transfer them to an output. Decoders are very useful when looking at many logic signals.

DeMorgan's theorem.   Rules for gate substitution that allow changes in the way a circuit is constructed: $\overline{A}\ \overline{B} = \overline{A + B}$ and $\overline{AB} = \overline{A} + \overline{B}$.

Exclusive OR.   A modified OR gate that produces a logic "1" only when one and only one of its inputs is at a logic "1" level.

Fanout.   The number of inputs a single TTL ouput can drive.

Full adder.   A modified half adder. A full adder adds the carry signal from a previous stage.

Gate delay.   See Propagation delay.

Half adder.   A binary adder used to add two logic levels together and produce a carry and a sum.

Hardware.   Actual electronic devices (transistors, resistors, capacitors, integrated circuits, LEDs, etc.).

Inverter.   A device used to switch logic levels. An inverter converts a logic "0" to a logic "1" and vice versa.

LSI.   Large-scale integration. An integrated circuit containing many complex logic functions requiring hundreds or thousands of transistors (ALUs, UARTs).

MSI.   Medium-scale integration (flip-flops, data selectors).

NAND gate.   AND gate with an inverted output.

Negative level.   Describing a family of integrated circuits that operate on negative voltages and signals.

Negative logic.   Logic in which the least positive value is a logical "1."

Noise margin.   Voltage difference between a TTL minimum output "1" and minimum input "1" ($2.4 - 2.0 = 0.4$). It is also the difference between an output minimum "1" and an input minimum "1."

NOR gate.   An OR gate with an inverted output.

Open collector.   Output circuitry in which there is no internal load. The collector is brought out to provide pulldown capability.

OR gate.   A logic gate that produces a logic "1" when any input is at a logic "1" level.

Positive level.   Describing a family of integrated circuits that operates on positive voltages and signals.

Positive logic.   Most positive voltage signal represents a logic "1."

Propagation delay.   The time needed for a gate's output to change state in response to an input.

Rise time.   In pulse analysis, the amount of time a signal takes to rise from 10% of a high level to 90% of a high level.

Saturated logic.   Form of logic in which the transistors are overdriven, causing reduced speed.

SSI.   Small-scale integration (AND, OR).

Tri-state.   To electronically disconnect the output of a gate from a wire.

Tri-state logic.   Logic gates that are capable of being electronically disconnected from a wire.

Truth table.   Table containing the key to the logical operation of an integrated circuit.

VLSI.   Very-large-scale integration. Advanced form of LSI (8080, Z80, and other large microprocessor chips).

Wired OR.   Function in which many gates are tied together at the output to commonly affect an input or produce a result.

PROBLEMS

3-1 Reword each of the following Boolean expressions to read in English describing the function. For example, $f = AB$ becomes "f is a '1' when A is a '1' and B is a '1.' "

(a) $f = A \overline{B} C$

(c) $f = A B \overline{C} + A \overline{B} C$

(b) $f = A + \overline{B} + C$

(d) $f = \overline{\overline{AB} + \overline{A} C}$

3-2 Draw truth tables for a four-input NOR gate and for a four-input NAND gate.

3-3 Show by comparing truth tables that $\overline{\overline{A} \overline{B}} \neq A B$.

3-4 Identify each of the logic gates shown in Fig. 3-54 as either good or bad, given the logic levels indicated.

3-5 Use DeMorgan's theorem to change each equation to NAND logic only.

(a) $f = \overline{A} \overline{B} + \overline{C} D$

(b) $f = A \overline{B} \overline{C} + \overline{A} B C$

(c) $f = \overline{A B + A \overline{B} + \overline{A} B C}$

3-6 Use DeMorgan's theorem to change each equation to NOR logic only.

(a) $f = \overline{A B + \overline{B} C}$

(b) $f = A C D + \overline{A} \overline{B} C$

(c) $f = (A + \overline{B}) (B + \overline{C})$

3-7 Determine the Boolean equation for the circuit shown in Fig. 3-55.

FIGURE 3-54 Problem 3-4.

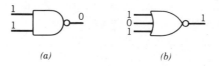

(a)   (b)

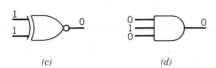

(c)   (d)

FIGURE 3-55 Problem 3-7.

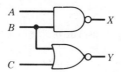

FIGURE 3-56 Problem 3-8.

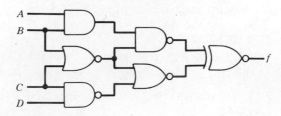

FIGURE 3-57 Problem 3-9.

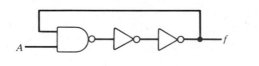

FIGURE 3-58 Problem 3-10.

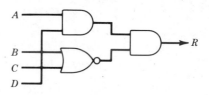

3-8 Determine the Boolean equation for the circuit shown in Fig. 3-56. Can the circuit be simplified?

3-9 Draw the timing diagram for the circuit shown in Fig. 3-57. Assume that all gates have a delay of 10 nsec.

3-10 In the circuit shown in Fig. 3-58, it is desired that the R output go high 40 nsec after the inputs go to 1001 ($ABCD$). Will this happen? Assume that all gates have a delay of 15 nsec.

3-11 The circuitry of Fig. 3-59 has five inputs and two outputs. Given the inputs on the truth table, follow the signals through the circuitry to determine the values of f_1 and f_2. If the output is PULSES or $\overline{PULSES}$, indicate using the symbols P and $\overline{P}$. Otherwise, just label the output 0 or 1.

3-12 Given the waveforms A and B shown in Fig. 3-60, sketch the resulting waveforms from the following 2-input gates: AND, OR, NAND, NOR, XOR, and from a NAND gate with one input inverted.

3-13 Determine the logical AND of the following hexadecimal values.

AE	45	C3	21
14	22	19	A7

4E	27	DB	F2
FE	02	22	0F

3-14 Calculate the complement of each hexadecimal value.

AA	55	44	E0
03	CD	9A	2C

FIGURE 3-59 Problem 3-11. For each truth table entry, fill in f_1 and f_2.

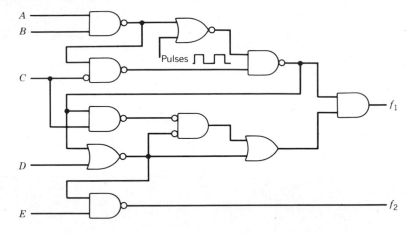

A	B	C	D	E	f_1	f_2
0	0	0	0	0		
1	1	1	1	1		
1	0	1	0	1		
0	1	0	1	0		
1	1	0	0	0		
0	0	1	0	1		
0	1	1	0	0		
1	1	1	0	1		
0	1	0	1	1		
0	0	0	1	1		

For each truth table entry, fill in f_1 and f_2.

3-15 Determine the logical OR of each pair of hexadecimal numbers.

26	5E	AB	07
45	24	44	CE

55	49	AB	CD
08	44	AB	DB

3-16 What is the XOR of each hexadecimal pair?

27	4E	CD	DB
D3	01	80	10

58	9A	21	3D
35	21	FF	00

FIGURE 3-60 Problem 3-12. Sketch each waveform.

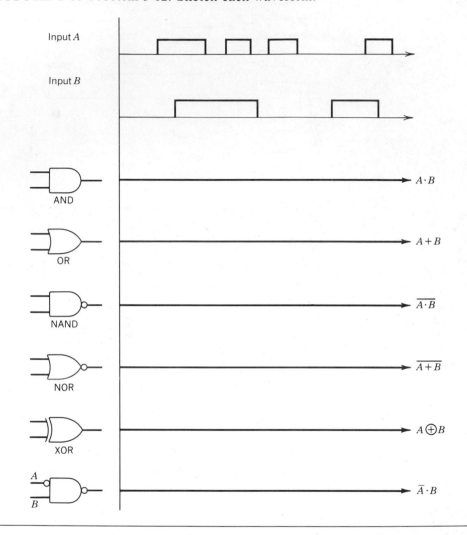

3-17 Calculate the 1's complement of the AND of each operation.

25	43	A7	14
01	02	22	41

3-18 Calculate the 2's complement of each hexadecimal number.

49	8E	22	C9
44	AB	FF	00

3-19 It has been said that the fanout of a TTL gate is the number of inputs that a single output can drive. Fanout is computed by dividing the driving gate output current by the input gate input current. For example, if a single gate's output current is 2 mA and the gates being driven by this output require 0.4 mA, then the fanout is 2/0.4 = 5, which means that five inputs can be driven from the output. Use a TTL data book to find out how many 74LS04s can be driven from a 74LS00 output. How many 74LS02s can be driven from a 74LS04?

3-20 A TTL gate is being used to drive four other TTL gate inputs. If the output current is 1.5 mA and each gate requires 0.4 mA, can the driving gate drive the four gates? If not, explain.

CHAPTER 4
LOGIC CIRCUIT DESIGN

1 INSTRUCTIONAL OBJECTIVES

When finished with this chapter, you should be able to:

1. Understand the function of a digital black box (DBB) and logic circuit design.
2. Use truth tables, symbolic logic, Boolean algebra, and Boolean reduction to reduce the gate cost and package count of a DBB.
3. Use Karnaugh mapping as a second method of reducing DBB logic (most useful with four or more variables).
4. Use DeMorgan's theorem for converting DBB logic to all-NAND or all-NOR logic.
5. Understand the operation of seven-segment LED displays.
6. Understand the purpose of gate array logic.
7. Show familiarity with multidigit displays.
8. Determine the cost of building a logic circuit and sketch the circuitry from logic equations.

2 SELF-EVALUATION QUESTIONS

Keep the following questions in mind and try to answer them when you have completed the chapter.

1. What is meant by "sum of products" and by "product of sums"?
2. What are the rules for reducing a Boolean equation?
3. How are the ones (or zeros) from a truth table output column used in a Karnaugh map to reduce a Boolean equation?
4. Why is it important not to overuse terms in reading from a Karnaugh map?
5. Why is it important to try to reduce a Boolean equation?
6. Why is Boolean reduction very difficult on problems involving more than four variables?
7. Do a sum-of-products result and a product-of-sums result produce the same logic function?

8. What is the advantage of strobing the digits of a multidigit display?

9. What is the purpose of gate array logic? Which type is available for circuits using fewer than 100 logic gates?

3   INTRODUCTION

This chapter explores several methods of digital design to produce circuitry that will behave in certain desirable ways. In each case, the design begins with the examination of a truth table outlining the desired logic function. The truth table yields a Boolean equation that is simplified and reduced to a logic drawing, which produces the logic function. We then practice the techniques learned by designing and observing logic circuitry. The subject of gate array logic introduces a fast way of building logic circuitry that is compact and easy to use.

4   THREE-INPUT DIGITAL BLACK BOXES

The construction of a DBB begins with the problem of designing the logic circuitry to make a particular logic function work. When the design is complete, we will know what circuitry to put into the box. The basic box consists of three inputs and one output. We begin by defining when the output is to be a "1" ($f = 1$). The following expression defines our first design problem: $f = m(0, 1, 6, 7)$.

This expression is read "f is to be a '1' for lines number 0, 1, 6, and 7" (on the truth table). This is a neat way of defining the project. The m numbers* are

---

FIGURE 4-1 (*a*) Three-input digital black box and (*b*) truth table.

m	A	B	C	f	
0	0	0	0	1	$\leftarrow \overline{A}\,\overline{B}\,\overline{C}$
1	0	0	1	1	$\leftarrow \overline{A}\,\overline{B}\,C$
2	0	1	0	0	
3	0	1	1	0	
4	1	0	0	0	
5	1	0	1	0	
6	1	1	0	1	$\leftarrow A\,B\,\overline{C}$
7	1	1	1	1	$\leftarrow A\,B\,C$

$$f = \overline{A}\,\overline{B}\,\overline{C} + \overline{A}\,\overline{B}\,C + A\,B\,\overline{C} + A\,B\,C$$

(a)   *(b)*

---

*The use of the letter m (as in m numbers) comes from an obsolete form of truth table notation. Terms of the form $A\,B\,C$, $A\,\overline{B}\,C$, and $\overline{A}\,\overline{B}\,C$ are called minimum terms (or min terms). Terms of the form $A. + B + C$, $A + \overline{B} + C$, $\overline{A} + \overline{B} + \overline{C}$ are called maximum terms (or max terms). For a min term to be a "1," each variable must be active. This is the case of the AND operation. For the max term to be a "1," only one variable needs to be active. This is the OR operation. This text avoids any further discussion of min terms or max terms and concentrates on the design of logic circuitry, referring occasionally to "m" numbers.

FIGURE 4-2 Digital black box circuitry.

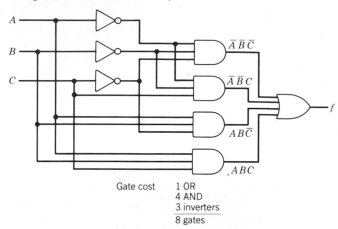

Gate cost   1 OR
               4 AND
               3 inverters
               _____
               8 gates

the binary count each line on the truth table represents. Figure 4-1 shows the
truth table and resulting logic function.

Each line on the table represents a binary count. For m_0 (the count of line 0),
$A = 0$, $B = 0$, and $C = 0$. These occur at the same time, and we write this as
an AND function: $\overline{A}\,\overline{B}\,\overline{C}$. This is one of four times that f is to be a "1." Each of
the four occurrences causes f to be a "1" and the four occurrences are ORed
together. The logic function is then: $f = \overline{A}\,\overline{B}\,\overline{C} + \overline{A}\,\overline{B}\,C + A\,B\,\overline{C} + A\,B\,C$.
Figure 4-2 shows the circuitry that is required.

The circuit is constructed to have three inputs, A, B, and C, and a single output.
It will fit neatly into the box, but what a lot of wiring! And there are a lot of logic
gates. We will soon see that there is a much better way to build the same box.
The improved method is called Boolean reduction.

5 REDUCING LOGIC FUNCTIONS

If you have a function $WX + W\overline{X}$, it may be factored to $W(X + \overline{X})$. Since $X + \overline{X} = 1$, the expression reduces to just W. Let's do that again (Fig. 4-3). You can
see from the figure that as factoring and simplifying proceed, the number of
required logic gates dwindles:

Step ① requires one OR, one inverter, and two AND gates for a total of four
       gates.

Step ② requires one OR, one inverter, and one AND gate for a total of three
       gates.

Step ③ requires *no* gates, just a piece of wire! We will make *very* good use of
       Boolean reduction.

FIGURE 4-3 Boolean reduction.

Steps in Boolean reduction   Circuit

(1) $f = WX + W\overline{X}$

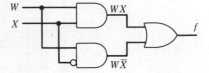

Factoring,

(2) $f = W(X + \overline{X})$

( ) $X + \overline{X} = 1$ always!

(3) $f = W$

(a cheap circuit!)

This method of reduction tells us to factor out terms whenever possible. It also allows for the simplification of digital circuitry.

EXAMPLE 4-1

Reduce $f = A B + A \overline{B}$.

SOLUTION
$f = A(B + \overline{B}) = A(1) = A$ (since $B + \overline{B} = 1$).

EXAMPLE 4-2

Reduce $f = A B C + A B \overline{C}$.

SOLUTION
$f = A B (C + \overline{C}) = A B (1) = A B$.

EXAMPLE 4-3

Reduce $f = A \overline{B} \overline{C} + \overline{A} \overline{B} \overline{C}$.

SOLUTION
$\overline{B} \overline{C}$ is common to both terms, so it is factored out: $f = (A + \overline{A}) \overline{B} \overline{C} = (1) \overline{B} \overline{C} = \overline{B} \overline{C}$.

Let us formalize this method of reduction by saying that we always compare two similar terms looking for *one* term that is different. To compare two terms,

FIGURE 4-4 Saving a gate by factoring.

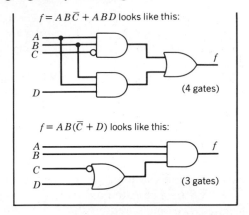

all the variables (letters) must be present. For instance, in $f = A B \overline{C} + A B D$, there is no hope of comparing because the letters are different.

Note: $f = A B \overline{C} + A B D$ can be factored, but no terms drop out: $f = A B (\overline{C} + D)$. However, we *do* save one logic gate by factoring, a useful thing to remember (see Fig. 4-4).

To continue with Boolean reduction, we can compare two terms that have *one* element different. The different element is eliminated. A few examples appear in Table 4-1.

We have discussed two methods of reducing the number of logic gates required:

① Boolean reduction to eliminate a different term from a pair.

② Factoring to drop one gate.

TABLE 4-1 · Some Simple Reductions

Logic Function	Reduced Function
$f = A \overline{B} C + A B C$	$f = A C (B + \overline{B}) = A C$
$f = A \overline{B} \overline{C} D + A \overline{B} C \overline{D}$	$f = A \overline{B} \overline{C}$
$f = A \overline{B} C + \overline{A} \overline{B} C$	$f = \overline{B} C$
$f = \overline{A} \overline{B} C + \overline{A} \overline{B} \overline{C}$	$f = \overline{A} \overline{B}$
$f = \overline{A} \overline{B} C + A B \overline{C}$	$f = \overline{A} \overline{B} C + A B \overline{C}^a$
$f = \overline{A} \overline{B} C \overline{D} + \overline{A} \overline{B} \overline{C} \overline{D}$	$f = \overline{A} \overline{B} \overline{D}$
$f = \overline{A} B \overline{C} + \overline{A} B C$	$f = B \overline{C}$
$f = B \overline{C} D + \overline{B} \overline{C} D$	$f = \overline{C} D$
$f = \overline{A} B D + \overline{A} \overline{B} D + A B C$	$f = \overline{A} D + A B C$
$f = A \overline{B} C + A \overline{B} \overline{C} + A B D$	$f = A \overline{B} + A B D$

[a]No reduction—can compare only terms that have one term different.

6 THE COST OF DIGITAL DESIGN

Our concern with the number of gates required to build a circuit is well justified. If 10,000 black boxes are to be built by a manufacturer, a reduction of one gate per black box will produce significant savings not only in logic gates but in wiring, sockets, and work hours.

We speak of two types of cost in the design of logic circuitry:

① Gate cost—the number of gates required.

② Package cost—the number of ICs required.

Let us reexamine the logic circuit of Fig. 4-2, where $f = \overline{A}\ \overline{B}\ \overline{C} + \overline{A}\ \overline{B}\ C + A\ B\ \overline{C} + A\ B\ C$. The gate cost was eight gates. To build such a circuit from the 54/74 series, we might require the following: two 7411 triple three-input AND gates, one 7404 hex inverter, and one 7425 dual four-input NOR (we'll have to make an OR by inverting a NOR). The cost of that circuit is eight logic gates and four T²L packages. If we apply Boolean reduction to that circuit, we should find considerable savings in both types of cost.

Continuing our example, we will compare the terms a pair at a time, trying all combinations. To be sure that no comparison is missed, we compare the first with second through last, then the second with third through last, and so on. Here is the function again:

$$f = \overline{A}\ \overline{B}\,\overline{C} + \overline{A}\ \overline{B}\ C + A\ B\,\overline{C} + A\ B\ C$$

$$\overline{A}\ \overline{B} \qquad\qquad A\ B$$

The first two terms compare and reduce, so do the last two. The new function is $f = \overline{A}\ \overline{B} + A\ B$.

This is *greatly* reduced from the original. We use a term twice only if there is a lone term that does not compare in any other way. We do *not* always use every comparison, lest the function get bigger. Remember, the object is to build a working circuit at the lowest possible cost. The new term requires one OR gate, two AND gates, and two inverters—only five gates and a lot less wire, since there are two input gates instead of three.

Figure 4-5 shows the new logic diagram. This new function reads as follows: "f is a '1' when $A = 0$ and $B = 0$ or when $A = 1$ and $B = 1$." Checking the truth table (Figure 4-6) reveals that this is just what we asked for! When $A = 0$

FIGURE 4-5 DBB for $f = m\ (0, 1, 6, 7)$.

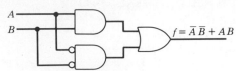

FIGURE 4-6 Truth table.

m	A	B	C	f
0	0	0	0	1
1	0	0	1	1
2	0	1	0	0
3	0	1	1	0
4	1	0	0	0
5	1	0	1	0
6	1	1	0	1
7	1	1	1	1

$f = \overline{A}\,\overline{B}$

$$f = \overline{A}\,\overline{B} + A\,B$$

$f = A\,B$

and $B = 0$, f is to be a "1"; when $A = 1$ and $B = 1$, f is to be a "1." When reading the 1s from the truth table we are using a method called the *sum of the products* (S/P).

An alternate method of logic design is the *product of the sums* (P/S) method. In P/S we read the zeros from the table. In P/S we want to find when the output f is at a "0." The problem is specified as follows: $\bar{f} = m(2, 3, 4, 5)$. The expression reads, "f is a '0' for counts 2, 3, 4, and 5." We reduce the function using Boolean reduction. Reading from the same truth table:

$$\bar{f} = \overline{A}\,B\,\overline{C} + \overline{A}\,B\,C + A\,\overline{B}\,\overline{C} + A\,\overline{B}\,C$$

$$\overline{A}\,B \qquad\qquad A\,\overline{B}$$

$\bar{f} = \overline{A}\,B + A\,\overline{B}$, and inverting both sides, $f = \overline{\overline{A}\,B + A\,\overline{B}}$. The product of sums result is *very* different from the sum of product result, but both are valid and satisfy the design requirement for our digital black box. The result is inverted to "f is a '1'" form by complementing each side of the equation. The result requires one NOR gate, two AND gates, and two inverters, as shown in Fig. 4-7. Both methods are used in design, the simpler of the two being the winner.

If we apply DeMorgan's theorem to the product-of-sums result such that we convert the output NOR gate to an AND gate, the equation becomes

$$\text{P/S} \qquad f = \overline{\overline{A}\,B + A\,\overline{B}} = \overline{\overline{A}\,B} \cdot \overline{A\,\overline{B}}$$

If we now change the NAND gates to OR gates using DeMorgan's theorem, we have

FIGURE 4-7 Product of sums result.

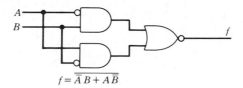

$f = \overline{\overline{A}\,B + A\,\overline{B}}$

$$\text{P/S} \qquad f = \overline{A\,B} \cdot \overline{A}\,\overline{B} = (A + \overline{B})\,(\overline{A} + B)$$

This form appears to be a product of sums and explains the name of the P/S method. Here are four of the many ways that we might build this circuit.

S/P	$f = \overline{A}\,\overline{B}\,\overline{C} + \overline{A}\,\overline{B}\,C + A\,B\,\overline{C} + A\,B\,C$	cost = 8
Reduced S/P	$f = \overline{A}\,\overline{B} + A\,B$	cost = 5
P/S	$f = \overline{\overline{A}\,B\,\overline{C} + \overline{A}\,B\,C + A\,\overline{B}\,\overline{C} + A\,\overline{B}\,C}$	cost = 8
Reduced P/S	$f = \overline{\overline{A}\,B + A\,\overline{B}} = (A + \overline{B})\,(\overline{A} + B)$	cost = 5 (for either)

All the equations and the resulting circuitry will perform the original task of design, which was $f = m(0, 1, 6, 7)$. We would be likely to implement one of the equations that costs only five gates.

Let us try another example and a new twist by using DeMorgan's theorem on the result to use *all*-NAND or *all*-NOR logic.

EXAMPLE 4-4

Design a three-input DBB for $f = m(0, 2, 3, 6)$.

SOLUTION

The sum-of-products information is given and we derive the P/S information from what is left (Fig. 4-8). Draw the truth table (Fig. 4-9) and read both ones and

---

FIGURE 4-8 Three-input DBB.

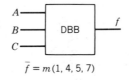

$$\bar{f} = m\,(1, 4, 5, 7)$$

FIGURE 4-9 Truth table.

m	A	B	C	f
0	0	0	0	1
1	0	0	1	0
2	0	1	0	1
3	0	1	1	1
4	1	0	0	0
5	1	0	1	0
6	1	1	0	1
7	1	1	1	0

S/P $\qquad f = \overline{A}\,\overline{B}\,\overline{C} + \overline{A}\,B\,\overline{C} + \overline{A}\,B\,C + A\,B\,\overline{C}$

$\qquad\qquad\qquad\quad \overline{A}\,\overline{C} \qquad\quad \overline{A}\,B \qquad\quad B\,\overline{C}$

P/S $\qquad \bar{f} = \overline{A}\,\overline{B}\,C + A\,\overline{B}\,\overline{C} + A\,\overline{B}\,C + A\,B\,C$

$\qquad\qquad\qquad\quad \overline{B}\,C \qquad\quad A\,\overline{B} \qquad\quad A\,C$

zeros. The reductions are trickier this time. In both S/P and P/S one term is used three times to our advantage. This common term allows the others to be reduced.

$$\text{S/P} \qquad f = \overline{A}\,\overline{C} + \overline{A}\,B + B\,\overline{C}$$
$$\text{P/S} \qquad f = \overline{B}\,C + A\,\overline{B} + A\,C$$

Let us change the S/P result to all NAND logic using DeMorgan's theorem (OR to NAND—invert the inputs).

$$\text{S/P} \qquad f = \overline{\overline{}\cdot\overline{}\cdot\overline{}}$$
$$f = \overline{\overline{A\,C} \cdot \overline{A\,B} \cdot \overline{B\,C}} \qquad \text{all NAND gates}$$

Let us change the P/S result to all-NOR logic (AND to NOR—invert the inputs).

$$\text{P/S} \qquad f = \overline{\overline{} + \overline{} + \overline{}}$$
$$f = \overline{(+) + (+) + (+)}$$
$$f = \overline{(B + \overline{C}) + (\overline{A} + B) + (\overline{A} + \overline{C})}$$

Study the *solution* carefully to see how the conversions are made, a step at a time. Let's try another example.

EXAMPLE 4-5

Design a three-input DBB for $f = m(0, 2, 3)$.

SOLUTION

The other half of the problem must be: $\bar{f} = m(1, 4, 5, 6, 7)$.

$$\text{S/P} \qquad f = \overset{0}{\overline{A}\,\overline{B}\,\overline{C}} + \overset{2}{\overline{A}\,B\,\overline{C}} + \overset{3}{\overline{A}\,B\,C}$$
$$\underset{\overline{A}\,\overline{C}}{} \qquad \underset{\overline{A}\,B}{}$$

To NAND:

$$f = \overline{\overline{A}\,\overline{C} \cdot \overline{A}\,B}$$

$$\text{P/S} \qquad \bar{f} = \overset{1}{\overline{A}\,\overline{B}\,C} + \overset{4}{A\,\overline{B}\,\overline{C}} + \overset{5}{A\,\overline{B}\,C} + \overset{6}{A\,B\,\overline{C}} + \overset{7}{A\,B\,C}$$
$$\underset{\overline{B}\,C}{} \qquad \underset{A\,\overline{C}}{} \quad \underset{A\,B}{}$$

$$\bar{f} = \overline{B}\,C + A\,\overline{C} + A\,B$$
$$f = \overline{\overline{B}\,C + A\,\overline{C} + A\,B}$$

To NOR:

$$f = \overline{(B + \overline{C}) + (\overline{A} + C) + (\overline{A} + \overline{B})}$$

In the P/S reduction, two extra comparisons are ignored. These extra terms are not needed to reduce the function. Using them *all* is a mistake; the extra terms would defeat the purpose of reducing circuitry.

In making comparisons, we try to use each term once. This causes the term to be reduced in size. If a term must be used again, it is only done to help another term become smaller. In Example 4-5, the S/P result reduced as follows: The m_0

and m_2 terms are replaced with $\overline{A}\ \overline{C}$. To shorten the m_3 term, it is necessary to use the m_2 term again. This reduces the m_2 and m_3 terms to $\overline{A}\ B$. Although m_2 already had been "covered" once, it was used again to reduce m_3.

In the P/S result, the possibility of overcomparing exists. Since m_1 and m_5 reduce to $\overline{B}\ C$ and m_6 and m_7 reduce to $A\ B$, this leaves m_4 to be reduced if possible. Here it is used with m_6 to reduce to $A\ \overline{C}$. This completes the reduction, since each term has been used at least once. There are two other possible comparisons: m_4 with m_5 and m_5 with m_7. We call these "redundant" terms because using them is unnecessary (we've already reduced all five elements) and would bring the cost back up. Once an original term has been "covered," it is not necessary to cover it again.

The terms m_4 and m_5 (not used) *could* have been used *instead* of m_4 and m_6 (again, not both). We just need to "cover" m_4.

An alternate result then could be

$$\bar{f} = \overline{B}\ C + A\ \overline{B} + A\ B$$

In fact, this is a better result, since it is reduced even more. The last two terms combine, and the new function is

$$f = \overline{B}\ C + A$$

This shows that Boolean reduction may lead to the choice of wrong pairs, producing a result that is not the shortest one possible. This example is a clue that a better method is desirable. The next section introduces the more desirable approach called Karnaugh mapping, but first let's try Boolean reduction once again.

EXAMPLE 4-6

Design a three-input DBB for $f = m(1, 2, 5)$.

SOLUTION

We deduce that $\bar{f} = m(0, 3, 4, 6, 7)$ for P/S. Figure 4-10 gives the truth table.

$$\text{S/P} \quad f = \overset{1}{\overline{A}\ \overline{B}\ C} + \overset{2}{\overline{A}\ B\ \overline{C}} + \overset{5}{A\ \overline{B}\ C}$$

$$\overline{B}\ C$$

$$f = \overline{B}\ C + \overline{A}\ B\ \overline{C}$$

FIGURE 4-10 Truth table.

A	B	C	f
0	0	0	0
0	0	1	1
0	1	0	1
0	1	1	0
1	0	0	0
1	0	1	1
1	1	0	0
1	1	1	0

FIGURE 4-11 Boolean Algebra Reduction.

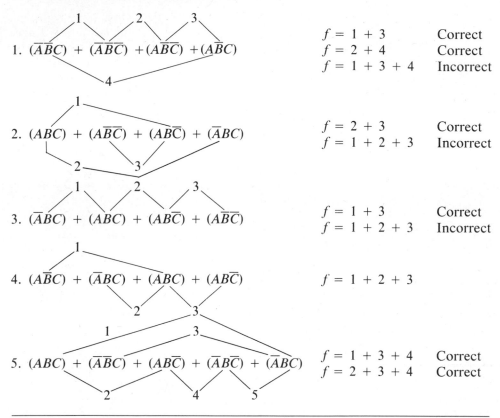

$$f = 1 + 3 \qquad \text{Correct}$$
$$f = 2 + 4 \qquad \text{Correct}$$
$$f = 1 + 3 + 4 \qquad \text{Incorrect}$$

$$f = 2 + 3 \qquad \text{Correct}$$
$$f = 1 + 2 + 3 \qquad \text{Incorrect}$$

$$f = 1 + 3 \qquad \text{Correct}$$
$$f = 1 + 2 + 3 \qquad \text{Incorrect}$$

$$f = 1 + 2 + 3$$

$$f = 1 + 3 + 4 \qquad \text{Correct}$$
$$f = 2 + 3 + 4 \qquad \text{Correct}$$

To NAND:

$$f = \overline{\overline{\overline{B}\,C} \cdot \overline{\overline{A}\,B\,\overline{C}}}$$

P/S $\bar{f} = \overset{0}{\overline{A}\ \overline{B}\ \overline{C}} + \overset{3}{\overline{A}\ B\ C} + \overset{4}{A\ \overline{B}\ \overline{C}} + \overset{6}{A\ B\ \overline{C}} + \overset{7}{A\ B\ C}$

$\overline{B}\,\overline{C} \qquad B\,C \qquad A\,\overline{C} \quad$ Redundant

$$\bar{f} = \overline{B}\,\overline{C} + B\,C + A\,\overline{C}$$

So, $f = \overline{\overline{B}\,\overline{C} + B\,C + A\,\overline{C}}$.

To NOR:

$$f = \overline{(B + C) + (\overline{B} + \overline{C}) + (\overline{A} + C)}$$

The ability to select the proper terms in Boolean reduction improves with practice. The use of redundant terms will increase the cost of a logic circuit, and this is not what we wish to do. Figure 4-11 presents some more examples for you to study to better learn which terms should be selected and which ones should be left alone.

7 KARNAUGH MAPPING

It is inevitable in black box design that another input will be required. With a fourth input the equations become more complex and we need help in finding terms that will be mutually comparable. The Karnaugh map is a visual method for spotting terms that are similar and yet offer a difference of one variable. Figure 4-12 shows a Karnaugh map for the three-input DBB. The three-input box has eight levels on the truth table ($2^N = 2^3 = 8$). There is a place for each m number on the Karnaugh map. Each of these eight boxes represents a possible input combination. The labeling of the Karnaugh map is important. The AB terms are across the top of the map (00, 01, 11, 10). Terms that are next to each other are different by only one bit. In this way, and because of the careful labeling, terms that are side by side differ by only one variable. Similarly, the C terms (0, 1) are down the side. Two boxes vertically adjacent also differ by one term. Let's try an example.

EXAMPLE 4-7

Simplify $f = m(0, 1, 4, 6)$.

SOLUTIONS

Boolean reduction:

$$f = \overline{A}\,\overline{B}\,\overline{C} + \overline{A}\,\overline{B}\,C + A\,\overline{B}\,\overline{C} + A\,B\,\overline{C}$$

$$\overline{A}\,\overline{B} \qquad\qquad A\,\overline{C}$$

$$f = \overline{A}\,\overline{B} + A\,\overline{C}$$

Karnaugh map method: See Fig. 4-13.

The Karnaugh map shows that m_0 and m_1, representing a vertical pair, are to be compared. Since $A = 0$ and $B = 0$ are the same, the C term is eliminated. A filled column points to $\overline{A}\,\overline{B}$ as the result. The map also shows that m_4 and m_6 differ by only one term. The B term is eliminated. Terms m_0 and m_4 also compare, but are not needed. This shows that the map can compare from one side to the other. For example, m_0 and m_4 compare, also m_1 and m_5. Since m_0 was used with m_1, it is "covered." It need not be used again. The same is true for m_4. Since m_4

FIGURE 4-12 Karnaugh map.

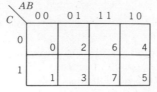

FIGURE 4-13 Example solution.

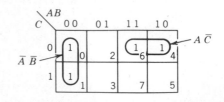

was compared with m_6, it need not be reused. To do so would bring the cost of the circuit up again. We use the Karnaugh map to direct us to terms that can be compared:

1. We look for pairs, horizontal or vertical, but not diagonal. Diagonal terms have a difference of two terms and are of no use.

2. Groups of four are also important. A group of four can be a full row or a square.

Figure 4-14 shows three similar ways to label a Karnaugh map. Each way of showing the map has advantages. The object is to select one and become adept at map reading. Let's read some sample Karnaugh maps by doing a few examples.

EXAMPLE 4-8

Design a three-input DBB in simplest form for $f = m(0, 2, 4, 5)$.

SOLUTION

See Fig. 4-15. Terms m_0 and m_2 are a pair and reduce to $\overline{A}\,\overline{C}$; m_4 and m_5 are a pair and reduce to $A\,\overline{B}$; m_0 and m_4 are a pair, but are not needed, since m_0 and m_4 are already "covered." Thus $f = \overline{A}\,\overline{C} + A\,\overline{B}$ is the answer.

FIGURE 4-14 Three ways to label a Karnaugh map.

FIGURE 4-15 Karnaugh map.

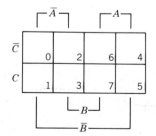

EXAMPLE 4-9

Design a three-input DBB in simplest form for $f = m(1, 3, 4, 7)$.

SOLUTION

See Fig. 4-16. Terms m_1 and m_3 are a pair and reduce to $\overline{A}\ C$; m_3 and m_7 are a pair and reduce to $B\ C$; m_4 does not compare with any other term. Thus the S/P answer is: $f = \overline{A}\ C + B\ C + A\ \overline{B}\ \overline{C}$. An alternate result is obtained from reading the zeros from the map: m_0 and m_2 compare to $\overline{A}\ \overline{C}$; m_2 and m_6 compare to $B\ \overline{C}$; m_5 is alone.

$$\text{P/S} \qquad \overline{f} = \overline{A}\ \overline{C} + B\ \overline{C} + A\ \overline{B}\ C$$
$$f = \overline{\overline{A}\ \overline{C} + B\ \overline{C} + A\ \overline{B}\ C}$$

FIGURE 4-16 Karnaugh map.

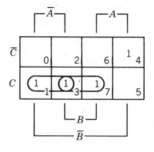

EXAMPLE 4-10

Design a three-input DBB in simplest form for $f = m(1, 2, 3, 6, 7)$.

SOLUTION

See Fig. 4-17. Terms m_1 and m_3 reduce to $\overline{A}\ C$; m_2 and m_6 and m_3 and m_7 are two pairs side by side and reduce to B. Thus $f = \overline{A}\ C + B$.

FIGURE 4-17 Karnaugh map.

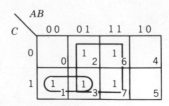

These three examples are done using Karnaugh maps labeled in different ways. By experimenting, you will develop your own preference. Figure 4-18 shows three maps and the appropriate solutions. Read each map. It becomes easy with practice!

A four-input digital black box has $2^4 = 16$ entries on a truth table. This means the Karnaugh map is twice as large. Figure 4-19 shows a four-input Karnaugh map in the different labeling schemes. The labeling is also important on the four-variable map. The m numbers are shown in each box, and adjacent boxes differ by one variable. To read the map, look for:

1. Vertical or horizontal pairs.

2. Rows or columns of four.

3. Squares of four.

4. Groups of eight.

The left and right columns of the map are next to each other. The top and bottom rows are next to each other. Diagonal terms do not compare. Here are some examples.

FIGURE 4-18 Sample Karnaugh maps.

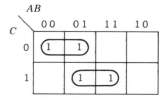

S/P $f = \overline{A}\ \overline{C} + B\ C$

P/S $\overline{f} = A\ \overline{C} + \overline{B}\ C$

$f = \overline{A\ \overline{C} + \overline{B}\ C}$

S/P $f = \overline{B}$

P/S $\overline{f} = B$

$f = \overline{B}$

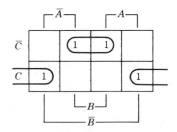

S/P $f = B\ \overline{C} + \overline{B}\ C$

P/S $\overline{f} = \overline{B}\ \overline{C} + B\ C$

$f = \overline{\overline{B}\ \overline{C} + B\ C}$

FIGURE 4-19 Four-variable Karnaugh maps.

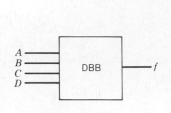

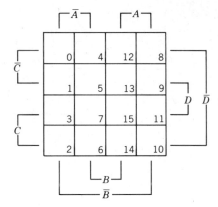

EXAMPLE 4-11

Design a four-input DBB in simplest form to produce $f = m(1, 5, 6, 7, 9, 13, 14, 15)$.

SOLUTION
Filling in the Karnaugh map, we obtain Fig. 4-20.

Sum of Products: Terms m_1, m_5, m_{13}, and m_9 are a group of four $= \overline{C} D$; m_7, m_6, m_{15}, and m_{14} are a group of four $= B C$. Therefore, $f = B C + \overline{C} D$ (S/P). The result $\overline{C} D$ is obtained by observing that m terms 1, 5, 13, and 9 fill a row; $\overline{C} D$ is the common term. The common term for m's 6, 7, 14, and 15 is $B C$. This can be done in parts: $m_6 + m_7 = \overline{A} B C$, $m_{14} + m_{15} = A B C$. The two pairs reduce to $B C$.

Product of Sums: Reading zeros from the map: m_0, m_4, m_8, and m_{12} form a row of four $= \overline{C} \overline{D}$; m_2, m_3, m_{10}, and m_{11} form a square $= \overline{B} C$. Therefore $\overline{f} = \overline{C} \overline{D} + \overline{B} C$ and $f = \overline{\overline{C} \overline{D} + \overline{B} C}$.

FIGURE 4-20 Karnaugh map.

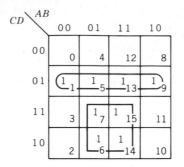

EXAMPLE 4-12

Design a four-input DBB in simplest form to produce $f = m(0, 2, 5, 8, 9, 10, 13, 15)$.

SOLUTION
Using an alternate form of the map, we obtain Fig. 4-21.

Sum of Products: Square: m_0, m_2, m_8, $m_{10} = \overline{B}\,\overline{D}$. Pairs: m_5, $m_{13} = B\,\overline{C}\,D$, m_{13}, $m_{15} = A\,B\,D$, m_{13}, $m_9 = A\,\overline{C}\,D$. Thus $f = \overline{B}\,\overline{D} + B\,\overline{C}\,D + A\,B\,D + A\,\overline{C}\,D$.

Product of Sums: Square: m_4, m_{12}, m_6, $m_{14} = B\,\overline{D}$. Pairs: m_1, $m_3 = \overline{A}\,\overline{B}\,D$, m_3, $m_7 = \overline{A}\,C\,D$, m_3, $m_{11} = \overline{B}\,C\,D$. Thus $\overline{f} = B\,\overline{D} + \overline{A}\,\overline{B}\,D + \overline{A}\,C\,D + \overline{B}\,C\,D$ and $f = \overline{B\,\overline{D} + \overline{A}\,\overline{B}\,D + \overline{A}\,C\,D + \overline{B}\,C\,D}$.

FIGURE 4-21 Karnaugh map.

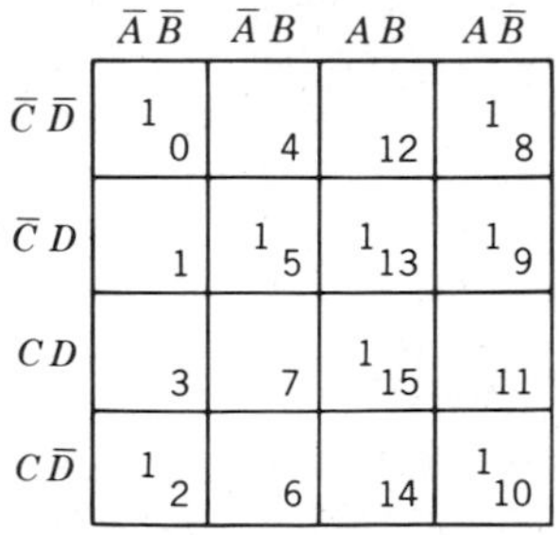

	$\overline{A}\,\overline{B}$	$\overline{A}\,B$	$A\,B$	$A\,\overline{B}$
$\overline{C}\,\overline{D}$	1 0	4	12	1 8
$\overline{C}\,D$	1	1 5	1 13	1 9
$C\,D$	3	7	1 15	11
$C\,\overline{D}$	1 2	6	14	1 10

EXAMPLE 4-13

Design a four-input DBB in simplest form to produce $f = m(1, 3, 9, 11, 12, 14)$.

SOLUTION
See Fig. 4-22.

Sum of Products: Square: $m_1, m_3, m_9, m_{11} = \overline{B} D$. Pair: $m_{12}, m_{14} = A B \overline{D}$.
Thus $f = \overline{B} D + A B \overline{D}$.

Product of Sums: Square: $m_5, m_7, m_{13}, m_{15} = B D$. Column: $m_4, m_5, m_6, m_7 = \overline{A} B$. Square: $m_0, m_2, m_8, m_{10} = \overline{B} \overline{D}$. Thus $\overline{f} = B D + \overline{A} B + \overline{B} \overline{D}$ and $f = \overline{B D + \overline{A} B + \overline{B} \overline{D}}$.

FIGURE 4-22 Karnaugh map.

	$\overline{A}\,\overline{B}$	$\overline{A}\,B$	$A\,B$	$A\,\overline{B}$
$\overline{C}\,\overline{D}$	0	4	^{1}12	8
$\overline{C}\,D$	^{1}1	5	13	^{1}9
$C\,D$	^{1}3	7	15	^{1}11
$C\,\overline{D}$	2	6	^{1}14	10

8 LARGER KARNAUGH MAPS

We have seen that in the design of a four-input digital black box Karnaugh maps are very useful in determining which terms differ by only one element. This approach is a great improvement over Boolean reduction and the labor of figuring out which terms to compare. The map shows directly which terms compare by their positions relative to one another. The object is always to produce a simplified logic circuit. Overcomparing is always a danger. If any term is used more than once, it must always be with a lone term. Once a term has been used once, it need not be used again, unless a lone term would miss comparison otherwise.

Designing five- and six-input digital black boxes is also done using Karnaugh maps. To do this, the 4×4 map just studied is used two or four at a time. After six inputs it is best to find a computer to do the simplification. Many inputs and complex logic equations can be solved using computer programs written for that purpose.

Consider the design of a five-variable DBB. Such a box would require a long truth table and would have $2^N = 2^5 = 32$ different input combinations. Using a pair of 4×4 Karnaugh maps solves this problem with one 4×4 for variable A and the other 4×4 map for variable $\overline{A}$. The two maps are envisioned as being one on top of the other to give depth.

EXAMPLE 4-14

Design a five-input DBB for $f = m(1, 5, 8, 9, 13, 22, 23, 24, 30, 31)$.

SOLUTION
First a look at the truth table, then the Karnaugh map, as shown in Fig. 4-23.

Sum of Products: Fours: $m_1, m_5, m_9, m_{13} = \overline{A}\,\overline{D}\,E$, $m_{22}, m_{23}, m_{30}, m_{31} = A\,C\,D$. Pair: $m_8, m_{24} = B\,\overline{C}\,\overline{D}\,\overline{E}$. The result is $f = \overline{A}\,\overline{D}\,E + A\,C\,D + B\,\overline{C}\,\overline{D}\,\overline{E}$.

   The next example gives a problem and the answer. It is left to you to draw the Karnaugh maps and verify the result.

FIGURE 4-23 Black box design: five inputs.

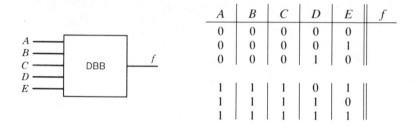

A	B	C	D	E	f
0	0	0	0	0	
0	0	0	0	1	
0	0	0	1	0	
1	1	1	0	1	
1	1	1	1	0	
1	1	1	1	1	

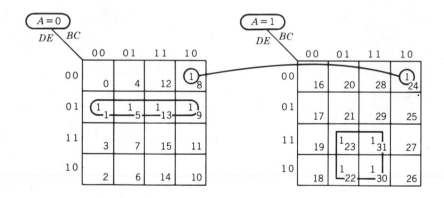

EXAMPLE 4-15

Design a five-variable DBB for $f = m(0, 2, 3, 4, 8, 12, 14, 15, 16, 20, 24, 28, 30, 31)$.

SOLUTION
$f = \overline{D}\,\overline{E} + B\,C\,D + \overline{A}\,\overline{B}\,\overline{C}\,D$.

9 DESIGN OF THE BCD TO SEVEN-SEGMENT READOUT DECODER/DRIVER

The seven-segment readout (Fig. 4-24) uses light-emitting diodes arranged in a figure eight to represent digits. Such displays come with the LED anodes all tied together (common anode-type display) or with all cathodes tied together (common cathode-type display).

FIGURE 4-24 (*a*) The seven-segment display. (*b*) Common anode. (*c*) Common cathode.

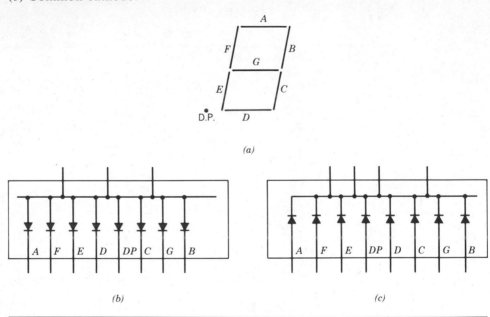

FIGURE 4-25 Decade counter driving a decoder.

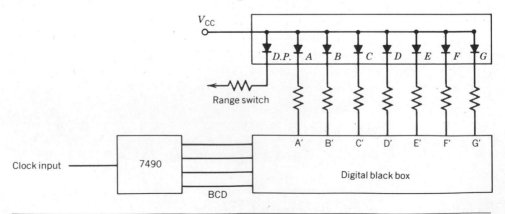

The segments are neatly identified with letters A-G. Often there is also a decimal point or colon as an additional indicator. Typically, such a display will require current-limiting resistors in series with each LED to protect the device from too much current. These are usually in the 150–220Ω range for use with a 5-V power supply. A more precise value would be determined from the data sheet for the readout device.

Since the information in a digital circuit is frequently BCD encoded, it is necessary to design a decoder that converts 4-bit BCD into the form the display requires. For instance, if the BCD vvalue is 0110 (six), then segments C, D, E, F, and G must be illuminated on the seven-segment readout. Figure 4-25 shows a 7490 decade counter driving a digital black box, which in turn drives the LED display. The DBB must provide a ground to the LED cathodes for the proper illumination of the display. Table 4-2 shows the truth table that the DBB must follow for correct operation of the display. We will suggest a DBB design for your consideration and study.

From Table 4-2 we observe that a seven-output DBB is required. The purpose of each output is to ground the appropriate segment at the proper BCD input. It is necessary then to use Karnaugh mapping to reduce each of the seven logic functions to build the circuitry for the digital black box. For practice, you should manually complete this prodigious task. However, we now present the 7447 BCD to seven-segment decoder driver that is available commercially to do the job, thus saving us the design of such a circuit. Figure 4-26 shows the 7447 driving an LED seven-segment readout.

Application of a BCD 4-bit count to the BCD input lines causes the proper digit to light on the display. The lamp test input is used to test the segments of the display. When lamp test is grounded, all segments are illuminated. The ripple blanking input (RBI) causes the figure zero to be blanked (not illuminated) when a BCD zero (0000) is applied to the input. This is useful in multidigit displays in suppressing leading zeros of a number (e.g., 456, not 000000456). The ripple blanking output (RBO) is used to drive the RBI line of an adjacent display. Figure 4-27 shows a typical brightness characteristic for a display as a function of current.

TABLE 4-2   Decoder Truth Table and Logic Equations

Input Code				Output State							Display	Logic Equations[a]
d	c	b	a	A'	B'	C'	D'	E'	F'	G'		
0	0	0	0	0	0	0	0	0	0	1	0	$A' = m(1, 4, 6)$
0	0	0	1	1	0	0	1	1	1	1	1	$B' = m(5, 6)$
0	0	1	0	0	0	1	0	0	1	0	2	$C' = m(2)$
0	0	1	1	0	0	0	0	1	1	0	3	$D' = m(1, 4, 7, 9)$
0	1	0	0	1	0	0	1	1	0	0	4	$E' = m(1, 3, 4, 5, 7, 9)$
0	1	0	1	0	1	0	0	1	0	0	5	$F' = m(1, 2, 3, 7)$
0	1	1	0	1	1	0	0	0	0	0	6	$G' = m(0, 1, 7)$
0	1	1	1	0	0	0	1	1	1	1	7	
1	0	0	0	0	0	0	0	0	0	0	8	
1	0	0	1	0	0	0	1	1	0	0	9	

[a]$m(10, 11, 12, 13, 14, 15)$ are "don't care" states. The 7490 will not allow these states.

FIGURE 4-26 Package information on SN7447 decoder driver.

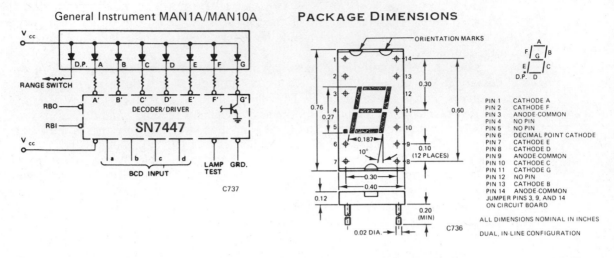

FIGURE 4-27 Luminosity versus forward current per segment.

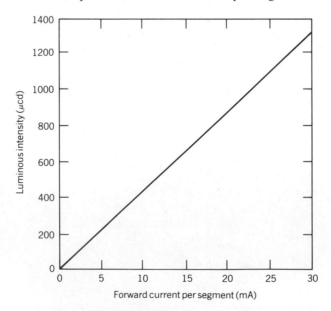

TABLE 4-3   Absolute Maximum Ratings for a Typical Display

Power dissipation at 25°C ambient	480 mW
Derate linearly from 25°C	6.4 mW/°C
Storage and operating temperature	−55°C to 100°C
Continuous forward current	
Total	120 mA
Per segment	30 mA
Reverse voltage per segment	10.0 V
Solder time at 260°C	5 sec

Table 4-3 gives the absolute maximum ratings for a typical display. This information is useful in determining the size of the current-limiting resistor for a particular value of *VCC*. The curves suggest limiting the current of an individual segment to about half the maximum (or 15 mA). This will serve to improve the life of the display. Seven-segment displays have found very common usage in the digital and microprocessor industries. Since displays are frequently used in multiples, it is useful to observe the connection of several displays. Only one 7447-type decoder chip is necessary in this type of application. All the seven-segment lines (A′–G′) are connected in common as shown in Fig. 4-28. The individual displays are strobed (pulsed) in sequence at a rate fast enough to elude detection by the human eye. The display will appear to be continuously illuminated if the displays are strobed (switched on) at a rate that exceeds the persistence of the human eye (approximately 15 pulses/sec). The data sent to the 7447 BCD inputs must of course be synchronized with the switching of the strobe lines. Since the strobe lines are the individual power lines for single displays, they must be able to supply sufficient current to illuminate the display. The timing diagram in Fig.

FIGURE 4-28 Using a single decoder to drive multiple digits.

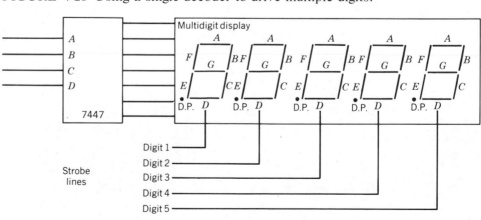

FIGURE 4-29 Strobed digit lines for a multidigit display.

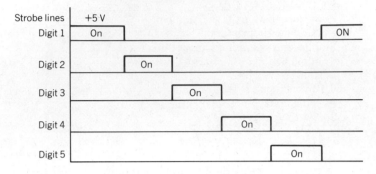

4-29 shows a set of sequential pulses that could drive the five-digit display illus-trated. As each digit is turned on, the appropriate seven-segment information is provided to that one digit. Since the other digits are not illuminated at this time, it does not matter that they get the seven-segment data too. This type of display is referred to as a *multiplexed* display. Each display shares the seven lines from a single decoder/driver circuit. This type of operation reduces the number of wires that are needed to operate a multidigit display.

Very often it is desirable to use a decade counter and to monitor its operation using a seven-segment LED type of display. It is so common that a 4-bit counter, a 4-bit latch, and a BCD-to-seven-segment decoder/driver are available in one DIP. Figure 4-30 shows the 74143 4-bit counter/latch, seven-segment LED/lamp driver. It is a very handy IC to have available.

FIGURE 4-30 74143 counter/latch/driver. (Courtesy of Texas Instruments Incorporated.)

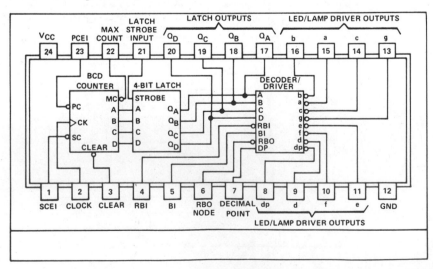

10 GATE ARRAY LOGIC

A gate array is a large (> 1000) number of uncommitted transistors fashioned into logic cells. These cells can be turned into flip-flops, shift registers, counters, and more. Instead of designing a printed circuit board using TTL logic or a microprocessor-based system, you can have your entire logic requirement implemented in a ''logic array.'' The advantage is speed not available in the microprocessor. In a microprocessor, each instruction requires time to execute; in a gate array, hard-wired logic is faster. It used to be very expensive to design a specific TTL-based system. The microprocessor made it more general and easier to change. However, certain applications require speed that the microprocessor cannot yet produce. Hard logic in an inexpensive form is now provided with the gate array.

The trick is that gate array logic does not become economically feasible until the 2000-gate level. Typical quantities range from 2000 to 200,000 arrays for a typical application. A logic array or gate array usually has an application requiring greater than 1000 gates. Gate arrays can have upward of 100 inputs and outputs. There can be 75 to 7500 gates per chip. Each chip performs a complete logic function. Gate arrays operate to 1 nsec/gate and 1 mW/gate. They are available in bipolar (TTL), CMOS, and ECL. Some of the literature suggests an ''economic opportunity envelope'' of 2000 to 200,000 units. In a gate array, the manufacturer of such devices connects the logic cells together in accordance with the purchaser's specifications. A single chip then replaces a standard PC board design using TTL- or microprocessor-based parts. Time and money are saved and a high-speed product is the result. It is a matter of designing the interconnects between standard ''cells.'' In this way the gate array represents an economic alternative to traditional design.

11 PROGRAMMABLE ARRAY LOGIC: THE PAL

The PAL (programmable array logic) is similar to the gate array but exists on a smaller scale.* In addition, the PAL concept allows the user (rather than a manufacturer) to perform the programming operation. The HAL is hard-array logic and is programmed at the factory for large-quantity use. This technology (PAL and HAL) is also referred to as fusible-link technology. Figure 4-31 shows a simple logic circuit using fusible links. By selectively blowing the fuses, different logic functions can be created. By opening fuses F_2–F_4 and F_5–F_7, the output logic function becomes

$$\text{output} = I_1 + \overline{I_2}$$

By opening fuses F_2, F_3, F_5, and F_8, the output logic function becomes

$$\text{output} = I_1 \cdot \overline{I_2} + \overline{I_1} \cdot I_2$$

Many functions could be designed using this simple array by opening the appropriate fuses. Figures 4-32 and 4-33 show the resulting logic. A more complex array

---

*PAL is a registered trademark of Monolithic Memories, Inc.

FIGURE 4-31 Simple fusible link circuit. (Used by permission of
Monolithic Memories Inc.)

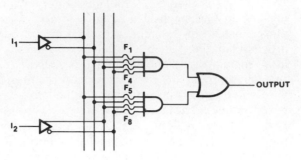

FIGURE 4-32 Simple logic function. (Used by permission of
Monolithic Memories Inc.)

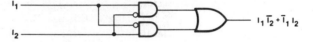

FIGURE 4-33 Implementation of logic function using PAL. (PAL is
a registered trademark of Monolithic Memories Inc. Used by
permission of Monolithic Memories Inc.)

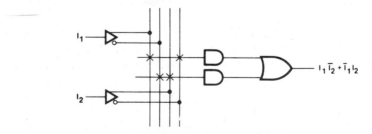

is shown in Fig. 4-34. It contains an FPLA (Field Programmable Logic Array).
These PAL-type arrays are available in many forms. A 12-input and 10-output
AND-OR-invert array appears in Fig. 4-35.

The PAL offers low-cost implementation of small logic functions as well as a
simple, high-speed method for developing MSI logic functions. A PAL programmer can burn the inexpensive PAL from a set of simple Boolean equations. The

FIGURE 4-34 Complex PAL-type arrays. (Used by permission of Monolithic Memories Inc.)

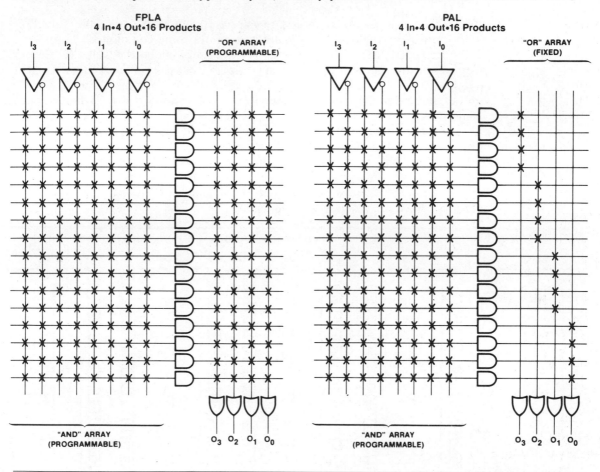

FIGURE 4-35 AND-OR-Invert array. (Used by permission of Monolithic Memories Inc.)

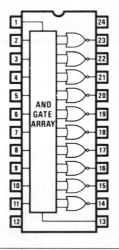

PAL offers logic gates, counters, registers, comparators, and other functions typically used by the designer. Figure 4-36 shows some common PAL products.

The PAL finds applications in memory decoders, video controllers, memory-mapped I/O, microprocessor interfacing, refresh clocks, scroll generators, UARTs, and security equipment. The PAL reduces hardware, increases speed, and decreases design and production time. Figure 4-37 shows a circuit implemented using a PAL.

Under development at this time is an electronically erasable programmable logic array (EEPLA or E^2PLA), which would allow the logic to be reconfigured from time to time. This means that the function of a circuit board could be altered without changing any of the hardware. The use of a logic array in the first place greatly simplified the design of a logic circuit, and now the prospect of altering circuitry already in place is very appealing.

FIGURE 4-36 Common PAL products. (Used by permission of Monolithic Memories Inc.)

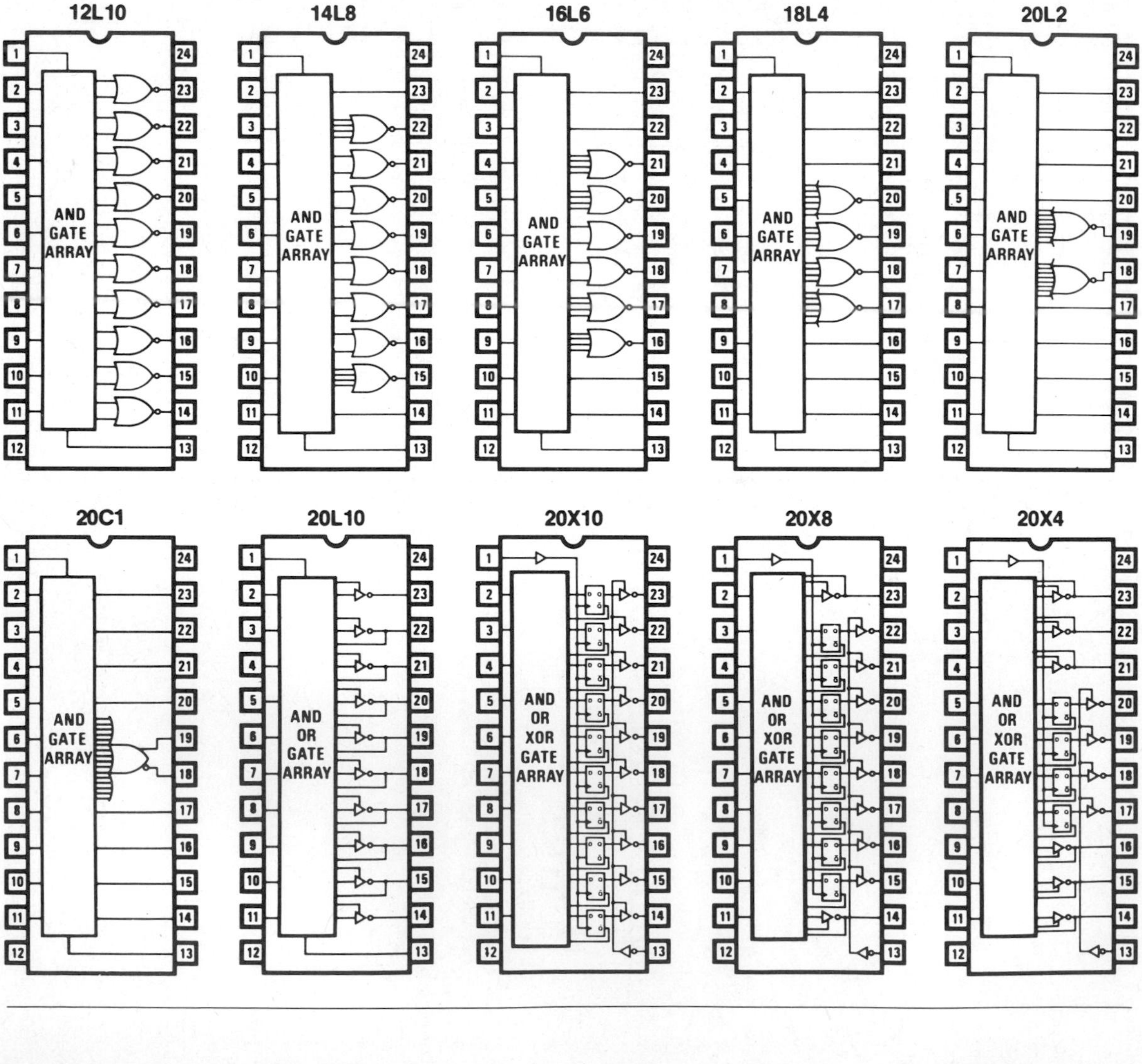

FIGURE 4-37 A typical PAL application. (Used by permission of PROLOG CORPORATION.)

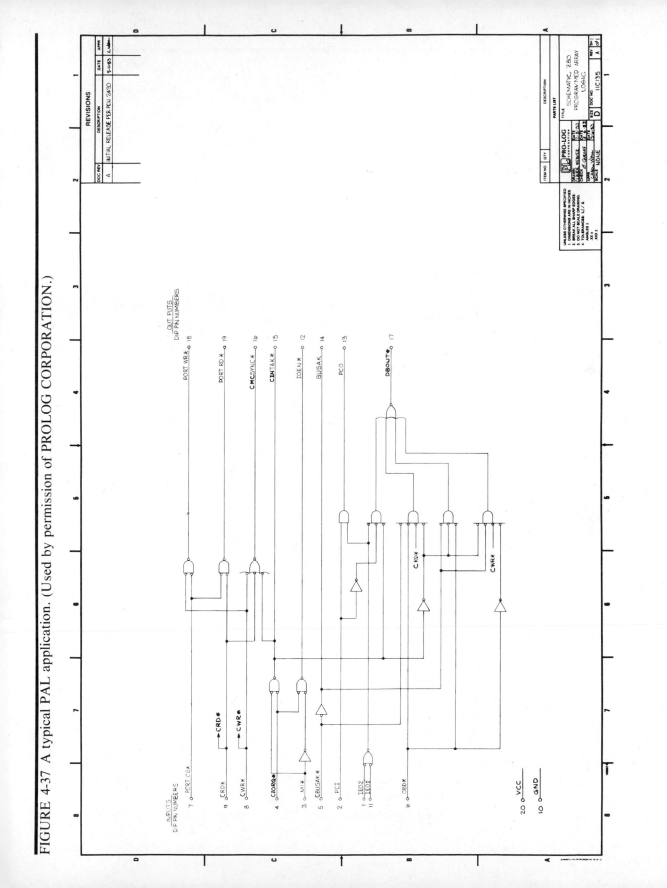

12 SUMMARY

Digital black box design is a very useful tool in the development of large-scale digital circuitry. While the numbers of inputs to a DBB can vary and become very large, simple methods are available to reduce the logic functions necessary to produce a valid output. These functions are Boolean reduction and Karnaugh mapping, and both serve to eliminate extra inputs or logic states. In both methods, similar terms are kept while different terms are eliminated.

13 GLOSSARY

Boolean reduction.   An algebraic method of reducing the number of logic gates necessary to perform a logic function.

Gate cost.   The number of gates required to perform a DBB function.

Package cost.   The number of ICs required to perform the DBB function. Usually less than the gate cost.

Product of the sums.   Boolean method utilizing all DBB inputs that produce a zero output.

Sum of the products.   Boolean method utilizing all DBB inputs that produce a one output.

PAL.   Programmable array logic.

PROBLEMS

4-1   Reduce each equation using Boolean reduction.
  (a) $f = A\,B + A\,\overline{B} + B\,C + \overline{B}\,\overline{C}$
  (b) $f = A\,B\,C + A\,\overline{B}\,C + \overline{A}\,\overline{B}\,\overline{C}$
  (c) $f = A\,\overline{B}\,\overline{C} + \overline{A}\,B\,\overline{C} + \overline{A}\,B\,C$

4-2 Show that each of the following can be reduced to fewer needed gates just by factoring.
  (a) $f = A\,\overline{B}\,\overline{C} + \overline{A}\,B\,\overline{C}$
  (b) $f = W\,\overline{X}\,\overline{Y} + W\,X + W\,Y$
  (c) $f = A\,\overline{B}\,\overline{D} + A\,\overline{B}\,\overline{C} + A\,\overline{B}\,D$

4-3   Design a three-input DBB using both sum of products and product of sums. Reduce the results using Boolean reduction.
  (a) $f = m(1, 3, 4, 6)$
  (b) $f = m(1, 3, 7)$
  (c) $f = m(2, 3, 5)$

4-4   Convert the results of Problem 4-3 to both all-NAND and all-NOR logic using DeMorgan's theorem.

4-5   Use a three-variable Karnaugh map to design the circuitry for $f = m(0, 1, 4, 6, 7)$. Be certain to use both S/P and P/S methods.

4-6 Use a four-variable Karnaugh map to implement each of the following functions.
(a) $f = m(0, 1, 4, 5, 6, 7, 10)$
(b) $f = m(2, 4, 6, 10, 12, 14)$
(c) $f = m(2, 3, 8, 9, 12, 13, 14)$

4-7 Write the simplest Boolean equation for the following Karnaugh maps.
(a) (Fig. 4-38)   (b) (Fig. 4-39)   (c) (Fig. 4-40)

FIGURE 4-38 Problem 4-7a.   FIGURE 4-39 Problem 4-7b.   FIGURE 4-40 Problem 4-7c.

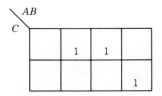 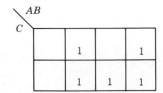 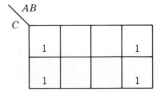

4-8 Repeat Problem 4-7 for the following maps:
(a) (Fig. 4-41)   (b) (Fig. 4-42)   (c) (Fig. 4-43)   (d) (Fig. 4-44)

FIGURE 4-41 Problem 4-8a.   FIGURE 4-42 Problem 4-8b.

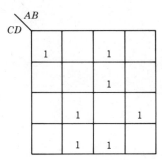 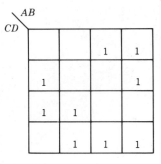

FIGURE 4-43 Problem 4-8c.   FIGURE 4-44 Problem 4-8d.

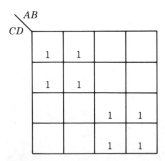 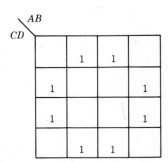

4-9 Draw the circuitry necessary to implement the following function in simplest form: $f = m(0, 1, 7, 15)$.

4-10 Convert the result of Problem 4-9 into all-NOR logic.

4-11 How can an inverter be made out of a NOR gate? How can an OR gate be made from a group of NAND gates? Prove with Boolean algebra.

4-12 Use Boolean algebra to prove that the following circuit behaves like an exclusive OR gate.

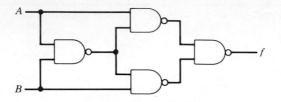

CHAPTER 5
FLIP-FLOPS, COUNTERS, AND SHIFT REGISTERS

1 INSTRUCTIONAL OBJECTIVES

When finished with this chapter, you should be able to:

1. Identify and describe the operation of the following 1-bit memory devices: $\overline{R}$-$\overline{S}$ flip-flop, type D flip-flop, J-K flip-flop.
2. Read and interpret the function table for standard 7400 series flip-flops.
3. Understand the function of different types of binary counters including divide by 2, divide by 3, divide by 4, divide by 5, divide by 10, divide by 16.
4. Draw a timing diagram for a binary counter.
5. Identify the basic shift register connection and describe its operation.

2 SELF-EVALUATION QUESTIONS

Keep the following questions in mind and try to answer them when you have completed the chapter.

1. How does a flip-flop store a zero or a one?
2. What do SET and RESET mean?
3. What do synchronous and asynchronous mean?
4. What is an illegal state in a flip-flop?
5. How are combinations of flip-flops able to form binary counters?
6. What is the difference between serial and parallel data?

3 INTRODUCTION

129

The purpose of this chapter is to acquaint you with flip-flops and circuits composed of flip-flops (such as counters and shift registers). Flip-flops are used in digital electronic circuits to store temporary results, to divide frequencies, and to rotate binary words, to name a few applications. While counters and shift registers are very useful, to understand them you must first be familiar with the operation of the flip-flop.

4 FLIP-FLOPS

4.1 The R-S Flip-Flop or d-c Latch

In Chapter 3 we saw how a switch was debounced by a simple circuit employing two-input NAND gates. Reproduced here in Figure 5-1, this circuit, called a *d-c Latch,* is more commonly known as an $\overline{R}$-$\overline{S}$ flip-flop. In an $\overline{R}$-$\overline{S}$ flip-flop, there are two-inputs, $\overline{R}$ and $\overline{S}$, and two outputs, Q and $\overline{Q}$. The bars over the $\overline{R}$ and $\overline{S}$ inputs indicate that the reset and set functions occur when either $\overline{R}$ or $\overline{S}$ is taken low. To reset the latch, place a zero on $\overline{R}$. To set the latch, place a zero on the $\overline{S}$ input. When not being used, $\overline{R}$ and $\overline{S}$ are normally left high.

The set and reset states of the flip-flop are defined as follows:

① When Q is high the flip-flop is "SET."

② When Q is low the flip-flop is "RESET."

The flip-flop can be used as a 1-bit memory device to remember one bit of a binary number. "SET" can be used to represent a stored one, "RESET" can be used to represent a stored zero. Since the flip-flop has only these two states, it is ideal for storing a binary bit.

To store a one in an $\overline{R}$-$\overline{S}$ flip-flop we simply ground the $\overline{S}$ input. The $\overline{S}$ input is then returned high, and the flip-flop is "SET" and a one considered to be stored. To store a zero in an $\overline{R}$-$\overline{S}$ flip-flop we ground the $\overline{R}$ input. The $\overline{R}$ input is then returned high, and the flip-flop is "RESET" and a zero stored.

FIGURE 5-1 $\overline{R}$-$\overline{S}$ latch used as a switch debouncer: pullup resistors should be used when a switch is in the input driver.

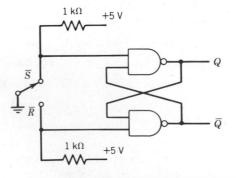

FIGURE 5-2 (*a*) Truth table for (*b*) a 74279 $\overline{S}$-$\overline{R}$ Latch. (Courtesy of Texas Instruments Incorporated.)

FUNCTION TABLE

INPUTS		OUTPUT
$\overline{S}^\dagger$	$\overline{R}$	Q
H	H	Q_0
L	H	H
H	L	L
L	L	H*

(*a*)

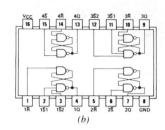

(*b*)

As we will see, all flip-flops have some kind of input and almost always have the Q and $\overline{Q}$ outputs. By Q and $\overline{Q}$ we mean that the outputs are complements of each other—when one output is a zero, the other is a one, and vice versa. Therefore, if the flip-flop is SET, Q is 1 and $\overline{Q}$ is 0.

It is very useful to have access to both outputs when designing digital circuits because other TTL gates often require both signals. The outputs of a flip-flop are *never* the same (either both one or both zero)! It is important to remember that this condition never occurs in flip-flop circuitry. If two 1s or two 0s are found, the flip-flop in use is operating incorrectly.

To gain an understanding of how the inputs ($\overline{S}$-$\overline{R}$) affect the outputs in our simple circuit, we will use a commercially available $\overline{S}$-$\overline{R}$ latch (Fig. 5-2). The 74279 is a standard TTL gate, and can be found in any TTL data book. The four circuits in the 74279 are identical to the one we have used to debounce our switch (except for two extra inputs: $1\overline{S}2$ and $3\overline{S}2$, which we will discuss later). Also, in this particular integrated circuit, we are provided with only one output, the Q output.

Let us take a look at the function table for the 74279 (Fig. 5-2*a*) and see if we can gain an understanding of the operation of the flip-flop. You will notice that when the $\overline{S}$ and $\overline{R}$ inputs are both high (H), the output is shown as Q_0. This means that the output Q is in the same state or level when both inputs are high as the state it was in when the inputs were different.

To prove this, let's trace through the operation of a single $\overline{S}$-$\overline{R}$ flip-flop, assuming the following initial conditions: The $\overline{S}$ and $\overline{R}$ inputs are at zero and one, respectively. From the truth table we know that the Q output should be high. Figure 5-3 shows this.

FIGURE 5-3 Operation of an $\overline{R}$-$\overline{S}$ latch.

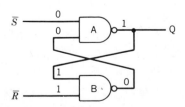

FIGURE 5-4 $\overline{R}$-$\overline{S}$ Latch with both inputs high.

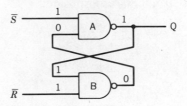

We know from Chapter 3 that the output of a NAND gate is a one whenever one or both inputs are low. In our example the $\overline{S}$ input is driving the input of NAND gate A low, producing a high at the output (which is also the Q output). This output also feeds one input of NAND gate B. Since the other input ($\overline{R}$) is also a one, the output of the gate becomes zero (this output would be $\overline{Q}$ if we were to use it). From this *steady state* condition, let's change the $\overline{S}$ input back to a logic "1" and see what happens (Fig. 5-4). Since we had a zero on input 2 of NAND gate A, changing input 1 ($\overline{S}$) from a zero to a one does not affect the output (Q). This is an example of the Q_0 principle. If we had used the $\overline{R}$ input instead, the Q output would have stayed low, just as the truth table predicts.

One last point before we move on to the next type of flip-flop. Notice that the bottom line of the truth table shows $\overline{S}$ and $\overline{R}$ both low, and the output as H* (Fig. 5-2a) (the asterisk means that the output is not guaranteed to stay high when both inputs are brought high again). This condition exists in most flip-flops and is a result of the d-c characteristics of the device. Most digital circuits are designed to ensure that they will never enter that condition. In other words, the circuit driving the flip-flop will never apply two low inputs to the $\overline{S}$-$\overline{R}$ latch.

4.2 The Type D Flip-Flop

Unlike the $\overline{S}$-$\overline{R}$ latch, the type D flip-flop has four inputs. These inputs are the data input (D), the clock input (CK), the preset input (PR) and the clear (CLR)

FIGURE 5-5 Clock waveform with rising edge causing a trigger of the flip-flop.

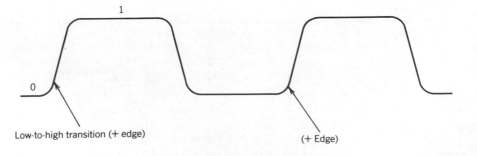

Low-to-high transition (+ edge)   (+ Edge)

FIGURE 5-6 Data transferred to Q when clock goes from low to high.

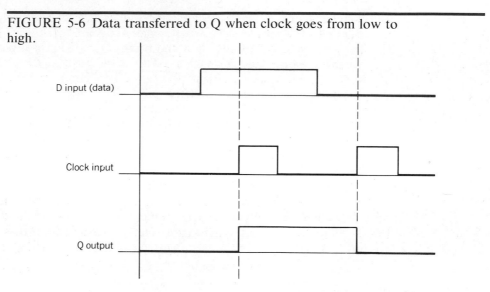

input. To get data into the flip-flop we use a combination of two inputs, the data input and the clock input. Unlike the $\overline{S}$-$\overline{R}$ flip-flop, data appears at the Q output only when we *clock* the flip-flop. To clock the flip-flop we apply a low-to-high transition at the clock input, which causes the flip-flop to *latch* onto the data present at the D input. Figure 5-5 shows the low to high transition in the form of a pulse. The D flip-flop actually clocks or triggers (reads data) on the rising or positive edge of the pulse (low to high transition). Figure 5-6 illustrates data transfer.

An operation that uses a clock line to control the flow of data into the flip-flop is called *synchronous*. That is, nothing happens until there is a change in the clock input. If we look at the function table for a 7474 type D flip-flop (Fig. 5-7), we will see this principle more clearly.

FIGURE 5-7 (*a*) Truth table for (*b*) a 7474 type D flip-flop. (Courtesy of Texas Instruments Incorporated.)

FUNCTION TABLE

INPUTS				OUTPUTS	
PRESET	CLEAR	CLOCK	D	Q	$\overline{Q}$
L	H	X	X	H	L
H	L	X	X	L	H
L	L	X	X	H*	H*
H	H	↑	H	H	L
H	H	↑	L	L	H
H	H	L	X	Q_0	$\overline{Q}_0$

(*a*)

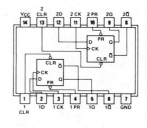

(*b*)

Two other inputs to the D flip-flop are preset and clear. The circles at the inputs indicate that the operations occur when the inputs are taken low. To clear (reset) the flip-flop, ground the CLR input and return high. To preset (set) the flip-flop, ground the PR line and return high.* The clock input (CK) is shown with the standard symbol for a clock input: $\dashv\!\!\!\triangleright\!\text{CK}$. The triangle indicates a clock input. This clock input responds to a positive edge. Other flip-flops respond to negative edges and have this symbol as an indicator: $\dashv\!\!\circ\!\!\triangleright\!\text{CK}$. The circle indicates a high-to-low transition causing a trigger, and the triangle again shows that it is a clock input.

Lines 4 and 5 of the function table show that the output Q will follow the data on input D when we receive a low-to-high (↑) transition on the clock line. Because this is a flip-flop, the data is also latched or "remembered" until we change it, so when the clock line goes low again, the outputs stay the same (see bottom line of the function table). By using synchronicity we can control the flow of data into the flip-flop. It is by this synchronous method of operation that we can design circuits in which flip-flops are used to divide digital signals and shift binary words.

EXAMPLE 5-1

The 7474 is used as a "divide-by-2" circuit in the digital circuit shown in Fig. 5 8. We will apply a square wave to the clock input and watch the Q and $\overline{Q}$ outputs. If we assume that the flip-flop is "*reset*" when we begin (or Q is low), the timing diagram for this simple circuit will look like Fig. 5-9. During the first low-to-high transition of the clock, a one gets clocked into the flip-flop (this is because D is tied to $\overline{Q}$). When this happens, Q goes to one and $\overline{Q}$ goes to zero. These outputs will remain this way until the next low-to-high transition.

When the next transition does occur, a zero gets clocked into the flip-flop. By doing so, the Q and $\overline{Q}$ outputs change back to zero and one again. We are now back where we started. If you take a close look at the timing diagram, you will notice that low-to-high transitions on the Q output occur once for every two low-to-high transitions on the clock input; thus we have a divide-by-2 circuit.

In addition to our two synchronous inputs, we also have two *asynchronous* inputs, PRESET and CLEAR (also called set and reset). When we say "asyn-

---

FIGURE 5-8 A 7474 Connected as a divide-by-2 circuit.

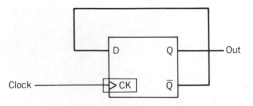

---

*"Preset" and "clear" operate asynchronously regardless of the state of the clock line.

FIGURE 5-9 Timing waveforms for 7474 divide-by-2 circuit. $\overline{Q}$ is high because we assumed that Q was low to begin.

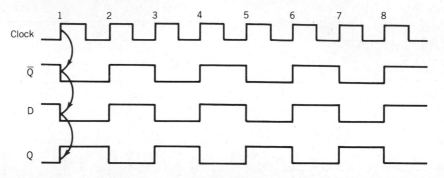

FIGURE 5-10 Frequency division circuits. (*a*) Divide by 3 and (*b*) divide by 5.

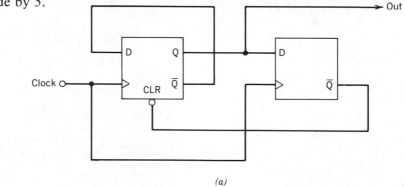

(a)

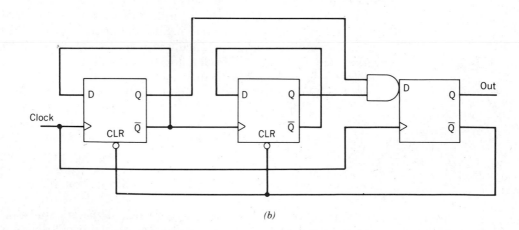

(b)

FIGURE 5-11 (*a*) and (*b*) Circuits for 7475 used as a data latch. (*c*)
Timing diagram. (Courtesy of Texas Instruments Incorporated.)

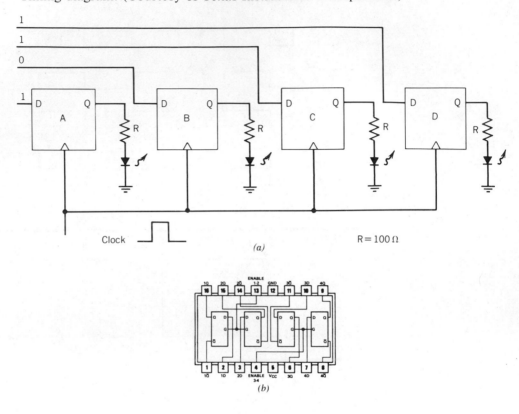

chronous," we mean that conditions in the flip-flop are affected at any time by
the PRESET and CLEAR inputs, regardless of what state the clock is in. A look
at the truth table will show this. During normal operation both the PRESET and
CLEAR lines are high (lines 4, 5, and 6 in Fig. 5-7*a*). This is the normal mode of
operation. However, to set the flip-flop (Q is high) at any time, merely ground
the PRESET line. In line with this, to clear the flip-flop at any time, pull the
CLEAR line low. Remember, the outputs of the flip-flop will change immediately
(Q = 1 for PRESET, Q = 0 for CLEAR) when the PRESET and CLEAR inputs
are changed, regardless of the clock or data inputs. (This is further illustrated in
the truth table by the letter X, meaning "don't care" what is happening on the
clock and data inputs—we simply set Q to a logic "1.") As in the $\overline{S}$-$\overline{R}$ latch, there
is also an illegal state in the 7474 (see line 3 of the truth table—both PRESET
and CLEAR low). Care must be taken to avoid this condition.

Before we finish with the 7474, look at Fig. 5-10, which presents two more
applications of the flip-flop used in frequency division. Study the circuits and
determine exactly how each one works. (*Note*: An oscilloscope would be very
useful!)

FIGURE 5-11 (continued)

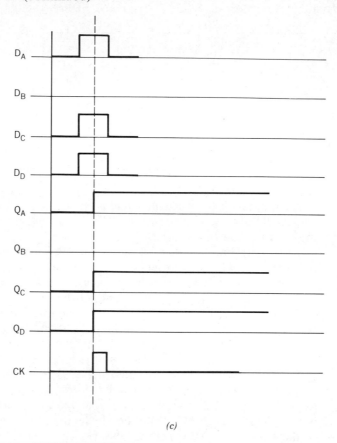

(c)

The D-type latch is very useful in storing binary numbers that may exist only briefly. Figure 5-11 shows a binary number being clocked or "gated" into a 4-bit data latch using D-type flip-flops. The outputs of the latch drive four indicator LEDs. Even when the binary number is no longer available, the "latch" remembers and continues to show the data. A 7475 quad latch is ideal for this purpose.

4.3 The J-K Flip-Flop

The last flip-flop we will look at, the type J-K flip-flop (Fig. 5-12), comes in a standard 16-pin DIP (SN7476) and has a few more functions than the type D flip-flop. Like the other two flip-flops we have studied, the J-K flip-flop has two outputs, Q and $\overline{Q}$. It also has PRESET and CLEAR inputs that operate just as they do in the type D flip-flop. The three exceptions in the J-K flip-flop are the J, K, and clock inputs.

The first thing to notice is the inverting circle on the clock input. We know from Chapter 3 that this means that we are inverting the input. By doing this, we will need a high-to-low (↓) transition on the clock input to operate the flip-flop.

FIGURE 5-12 (*a*) and (*b*) The 7476 J-K flip-flop truth tables and (*c*) package information. (Courtesy of Texas Instruments Incorporated.)

INPUTS					OUTPUTS	
PRESET	CLEAR	CLOCK	J	K	Q	Q̄
L	H	X	X	X	H	L
H	L	X	X	X	L	H
L	L	X	X	X	H*	H*
H	H	⊓	L	L	Q₀	Q̄₀
H	H	⊓	H	L	H	L
H	H	⊓	L	H	L	H
H	H	⊓	H	H	TOGGLE	

(*a*)

INPUTS					OUTPUTS	
PRESET	CLEAR	CLOCK	J	K	Q	Q̄
L	H	X	X	X	H	L
H	L	X	X	X	L	H
L	L	X	X	X	H*	H*
H	H	↓	L	L	Q₀	Q̄₀
H	H	↓	H	L	H	L
H	H	↓	L	H	L	H
H	H	↓	H	H	TOGGLE	
H	H	H	X	X	Q₀	Q̄₀

(*b*)

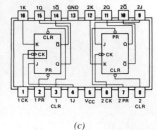

(*c*)

Second, let us consider the J and K inputs. They are slightly different from the $\overline{S}$-$\overline{R}$ inputs and the D input on the 7474. Lines 4 through 7 on the truth table show the operation of these inputs.

When both J and K are low, the flip-flop does not change state. The logic levels present at the Q and $\overline{Q}$ outputs *never* change, as long as both inputs are low. No amount of clocking the flip-flop will produce a change (assuming the PRESET and CLEAR inputs are not used and left high).

When J and K are different—that is, when they complement each other—data from J and K transfer straight through to Q and $\overline{Q}$ on the *falling edge* of a clock pulse. (It is this mode of operation that enables us to make shift resisters.)

Last, when both J and K are high, the outputs *toggle* on each successive clock pulse. This means that on the first pulse, Q and $\overline{Q}$ may be set to one and zero, respectively. On the next clock pulse, they toggle, or switch to zero and one. On the third pulse, they switch back to one and zero, and so on. This operation was accomplished in the 7474 D flip-flop by feeding the $\overline{Q}$ output back into the D input. In the J-K flip-flop, this isn't necessary. To divide by 2, we simply tie both J and K high, enabling the toggle mode.

5 OTHER FLIP-FLOPS

In the last section we looked at only three different flip-flops. Of course, there are many more flip-flops available. However, in this study we are simply learning how to use the most common TTL devices so that we will understand their role in the digital computers encountered in later chapters. For more information, study the following flip-flop devices:

7470 AND-gated J-K positive-edge triggered flip-flops with preset and clear

7472 AND-gated J-K master/slave flip-flop with preset and clear

7473 dual J-K flip-flop with clear

7475 4-bit bistable latch

7477 4-bit bistable latch

7478 dual J-K flip-flop with preset, common clear and common clock

6 ENABLES

Before we start our study of counters and shift registers, recall that the $\overline{S}$-$\overline{R}$ latch has two extra inputs, the $1\overline{S}2$ and $3\overline{S}2$ inputs. If we look at any of the flip-flop packages listed in Section 5, we might see other inputs (J1, J2, G, etc.). These enables are all important and serve to control the device from a second level. They provide us with a second input for controlling the flip-flop. This is often very useful when we want to control the J input to a flip-flop from two or more sources. The availability of standard TTL DIPs with extra inputs built in are advantageous when designing a digital circuit because they help to lower the package count and cost of the circuit.

7 COUNTERS

We have many different types of counters. We will first look at three of them in detail, and then talk about some of the other, more exotic counter chips. Three of the more common counters are the 7490 (divide by 10), the 7492 (divide by 12), and the 7493 (divide by 16). Counters are slightly more complex TTL packages composed of flip-flops and control gates (now you see why we covered flip-flops and gates first).

7.1 The BCD or Divide-by-10 Counter

The 7490 (Fig. 5-13) is a standard TTL 14-pin DIP that is specifically designed to be used as a divide-by-10, or BCD counter circuit. To get a count of 10 out of the device, we will need four binary outputs. On the 7490 (and the other ICs we will look at), these outputs are called A, B, C and D, where A is the LSB and D is the MSB of the counting chain.

The 7490 is really a two-part package. It consists of a divide-by-2 circuit and a divide-by-5 circuit. For normal divide-by-10 operation, we connect the output of the divide-by-2 stage to the input of the divide-by-5 stage. The A output is the output from the divide-by-2 stage and the B_{in} input is the clock for the divide-by-5 stage.

FIGURE 5-13 The 7490 BCD Counter. (Copyrighted material reprinted with permission of National Semiconductor Corporation.)

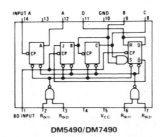

DM5490/DM7490

EXAMPLE 5-2

Figure 5-14 shows a 7490 set up as a divide-by-10 circuit. In this example we
notice how a 7490 is set up to divide by 10. The counting sequence of Fig. 5-14b
will result when the IC is connected as in Fig. 5-14a. By connecting the A output
to the B_{in} input, we connect the divide-by-2 output to the divide-by-5 input. If we
first divide by 2, then next divide by 5, overall we are dividing by 10! The R_0 and
R_9 inputs are grounded to enable a 0 to 9 count. Whenever a high pulse appears
at these inputs, the counter is reset. If the pulse appears at the R_0 inputs (both
must be "1" to enable a reset), the 7490 resets to zero. If it appears at the R_9
input, the 7490 resets to nine.

The 7490 is very versatile and can be made to divide by 3, 4, 6 or 7 just as
easily as it divides by 10. By controlling the reset pulse to the R_0 inputs, we can
control the count present at the outputs. Figure 5-15 shows the counter outputs
when a series of clock pulses is applied to A_{in}.

FIGURE 5-14 A 7490 set up as a divide-by-10 circuit. (a) Counter.
(b) Truth table. (Copyrighted material reprinted with permission of
National Semiconductor Corporation.)

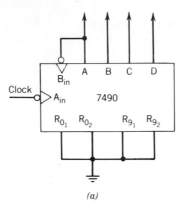

(a)

COUNT	OUTPUT			
	D	C	B	A
0	0	0	0	0
1	0	0	0	1
2	0	0	1	0
3	0	0	1	1
4	0	1	0	0
5	0	1	0	1
6	0	1	1	0
7	0	1	1	1
8	1	0	0	0
9	1	0	0	1

(b)

FIGURE 5-15 Timing diagram for the 7490 decade counter.

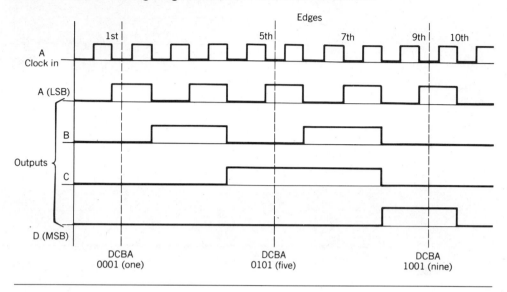

EXAMPLE 5-3

Derive circuits for divide by 4, divide by 6, and divide by 8 using the 7490. (*Note:* The R_9 inputs are not used, so they should be grounded.)

SOLUTION
To divide by 4 (Fig. 5-16) we simply feed the C output back to the R_0 inputs. Thus, every four clock pulses, the 7490 will reset to zero.

FIGURE 5-16 Divide-by-4 counter.

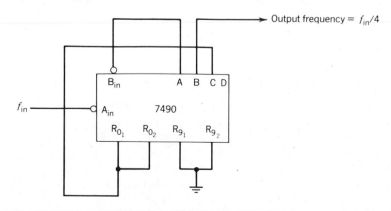

FIGURE 5-17 Divide-by-4 counter.

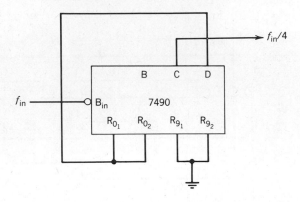

FIGURE 5-18 Divide-by-4 timing diagram.

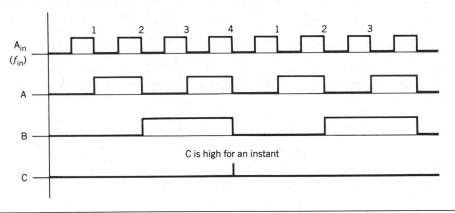

FIGURE 5-19 Divide-by-6 counter.

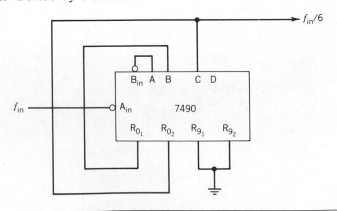

FIGURE 5-20 Divide-by-6 timing diagram.

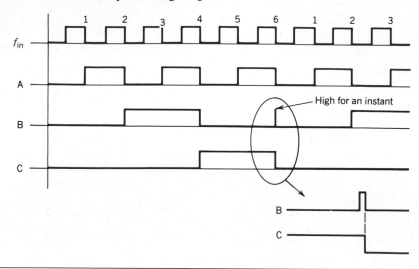

FIGURE 5-21 Divide-by-8 counter.

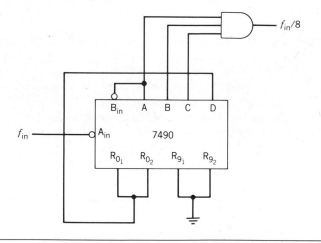

Note: If for some reason we want to use the divide-by-2 section for another purpose, and *still* have a divide-by-4 left over, the circuit and timing diagram shown in Figs. 5-17 and 5-18 will do the same.

To get a divide-by-6 circuit working (Figs. 5-19 and 5-20), we have to feed back two outputs, the B and the C outputs. Whenever both these are high, the 7490 will reset.

Last, to divide by 8, we simply feed the D input back to both reset inputs (Figs. 5-21 and 5-22).

FIGURE 5-22 Divide-by-8 timing diagram.

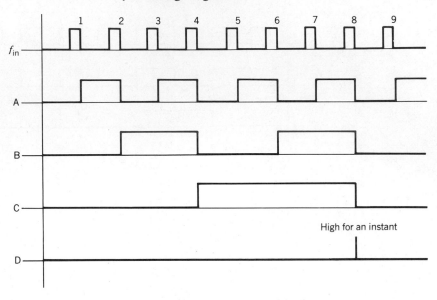

7.2 The 7492 Divide-by-12 Counter

While it was easy enough to use a 7490 to divide by 6 by feeding two outputs back into the reset inputs, we have an integrated circuit (Fig. 5-23) that does this automatically. The 7492 divide-by-12 counter comprises two stages, a divide by 2 and a divide by 6. As in the 7490, the two stages can be connected together or left operating as separate counters.

Unlike the 7490, the 7492 has only an R_0 input function. Even so, we can still use the 7492 as a divide by 12 or any other number up to 12 (see Fig. 5-24). Since the operation of the 7492 is so similar to the 7490 in principle, we will look at just one example of a division scheme.

FIGURE 5-23 A 7492 divide-by-12 integrated circuit. (Copyrighted material reprinted with permission of National Semiconductor Corporation.)

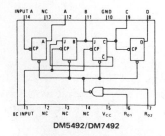

DM5492/DM7492

FIGURE 5-24 (*a*) Normal operation and (*b*) truth table for the 7492. (Copyrighted material reprinted with permission of National Semiconductor Corporation.)

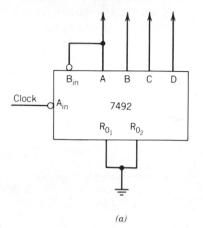

COUNT	OUTPUT			
	D	C	B	A
0	0	0	0	0
1	0	0	0	1
2	0	0	1	0
3	0	0	1	1
4	0	1	0	0
5	0	1	0	1
6	1	0	0	0
7	1	0	0	1
8	1	0	1	0
9	1	0	1	1
10	1	1	0	0
11	1	1	0	1

(*a*)   (*b*)

EXAMPLE 5-4

Connect a 7492 as a divide-by-11 counter.

SOLUTION

Because we have 3 bits to check for (11 D → 1011 B), we cannot simply connect them to the R_0 inputs, of which we have only two. Therefore, we combine two of them with an AND gate, as shown in Fig. 5-25. In this way, the 7492 will reset only when A, B, and D are all high. Because D goes high only once during the 11 clock pulses, we can use it as our divide-by-11 output.

FIGURE 5-25 A 7492 set up as a divide-by-11 counter.

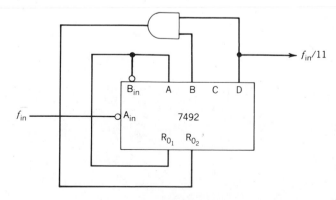

7.3 The 7493 4-Bit Binary Counter

Our last counter (Fig. 5-26) is a full binary counter. With its divide-by-16 operation, the 7493 gives a full hexadecimal count at its output from zero to F. With the 7493, counts (or divisions) of numbers up to 16 are available.

The 7493 is especially useful when binary counting schemes are desired, as in the addressing of memory locations. For example, if two 7493s are *cascaded* together, a count of 256 is achieved. This would take many more 7490s or 7492s with complex reset circuitry to achieve the same result. Each counter type, however, has its own usefulness. Figures 5-27 and 5-28 give additional information about the 7493. We will see in the next section how most of the reset circuitry for getting custom divisions (divide by 6, 7, 9, or 11) can be eliminated with other counters.

FIGURE 5-26 The 7493 divide-by-16 counter. (Copyrighted material reprinted with permission of National Semiconductor Corporation.)

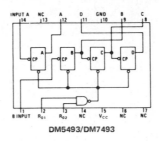

DM5493/DM7493

FIGURE 5-27 (*a*) Normal operation and (*b*) truth table for the 7493. (Copyrighted material reprinted with permission of National Semiconductor Corporation.)

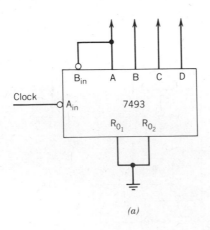

(a)

COUNT	OUTPUT			
	D	C	B	A
0	0	0	0	0
1	0	0	0	1
2	0	0	1	0
3	0	0	1	1
4	0	1	0	0
5	0	1	0	1
6	0	1	1	0
7	0	1	1	1
8	1	0	0	0
9	1	0	0	1
10	1	0	1	0
11	1	0	1	1
12	1	1	0	0
13	1	1	0	1
14	1	1	1	0
15	1	1	1	1

(b)

FIGURE 5-28 The 7493 timing diagram.

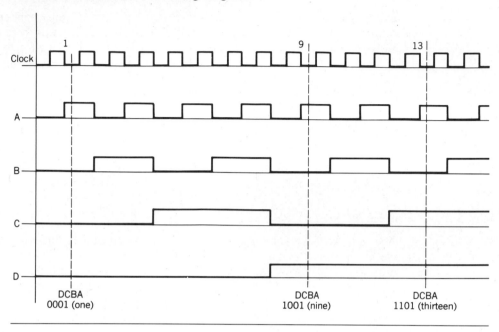

8 COMPLEX COUNTERS

A few of the more complex counters are:

74390 dual decade

74393 dual 4-bit counter

74160,162 decade counter with load and clear

74190,192 decade up/down with load, clear

74161,163 4-bit binary with load, clear

74191,193 4-bit binary up/down with load, clear

The last four groups of counters have special circuitry that aids in the development of custom dividing chains. For example, if we want to divide by 11, we first choose a 4-bit binary counter with load and clear. Next, we find the 2's complement of 11, which is 5. We place this binary number at the load inputs, A B C D of the counter chip, and connect a carry signal to the pin that controls the load function. In this way, the counter is "reset" to 5. Then after 11 counts, it "resets" to 5 again. This is because after 11 counts, it has gone from 0101 to 1111, producing a *ripple carry* out. When this happens, the 5 gets automatically loaded back into the counter. The output would look like this: . . . 5 6 7 8 9 A B C D E F 5 6 7 8. . . .

These complex counters can greatly simplify the design of a digital circuit. Complex reset networks can be eliminated with the proper choice of presettable counters.

9 SHIFT REGISTERS

Another important device used in digital circuits is the shift register. As we have seen in Chapter 2, shift registers can be used to multiply numbers (or at least to aid in the multiplication process). We have many other uses for them, however. We use them to control the transmission of information from computers to electronic displays and printers. They also were used in early video games and pinball machines for sequencing lights. All in all, the shift register is a very important device in digital electronics.

In this section we will look at three different types of shift registers (there are more) and become familiar with their operation.

10 SHIFT REGISTER OPERATION

10.1 Simple Shift Register

The shift register shown in Fig. 5-29 is made up of J-K flip-flops. To operate it, we simply place the data to be shifted on the *serial* input, and look at the four outputs as we clock the data into and through the circuit. Keep the truth table for the 7476 (Fig. 5-12*a*) in mind as we discuss this circuit. Assuming that all flip-flops are reset to begin with, the timing diagram (Fig. 5-30) shows what happens in our simple 4-bit shift register as we change the input and look at the output. The B, C, and D outputs are all shifted one bit at a time to the right of the A output. This is the primary operation of the shift-register. As we will find in our next shift register, the user has a choice of shifting these 4 bits right or left.

This basic shift register has a serial data input at which the bits are applied one at a time during the loading process. We will see later that data may be loaded into another type of register in parallel. In a parallel load shift register, the inputs to each flip-flop are all available at once. The register is loaded in one operation—all bits at once.

This basic shift register also has a serial output (the right-most flip-flop's Q output). To see all the stored data, it must be clocked through the right-most flip-

FIGURE 5-29 Simple 4-bit shift register.

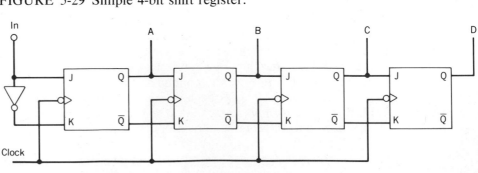

FIGURE 5-30 Output waveforms of a 4-bit shift register.

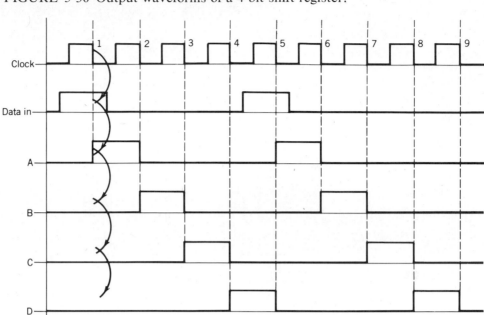

flop and observed one bit at a time. Many shift registers have all the Q outputs available, and this is a parallel output. Basically, "serial" means one bit at a time, whereas "parallel" means all bits at the same time.

Shift registers come in a variety of styles: serial in, serial out, parallel in, parallel out, and shift left/shift right. Parallel inputs use a *load* line to enter the data.

10.2 The 7495 4-Bit Right/Left Shift Register

In the 7495 we have the capability of shifting the 4 bits right or left in the shift register (see Fig. 5-31). In the operation of the shift register we place the 4-bit word on the A, B, C, D inputs. Then, by clocking the shift register either right or left (through the use of the mode control input), we obtain our data off of the A, B, C, D outputs. A shift register of this kind would be very useful if we were designing a TTL circuit to perform the shift operations for our BCD multiplication and division routines.

10.3 The 7491 8-Bit Shift Register

In microcomputers, we are frequently dealing with 8-bit words or bytes, so instead of using two 7495 shift registers, we look up a substitute (Remember, we always have lots of duplicates in the TTL data book!) that can handle 8 bits in one DIP.

FIGURE 5-31 A 4-bit shift right/left shift register. (Copyrighted material reprinted with permission of National Semiconductor Corporation.)

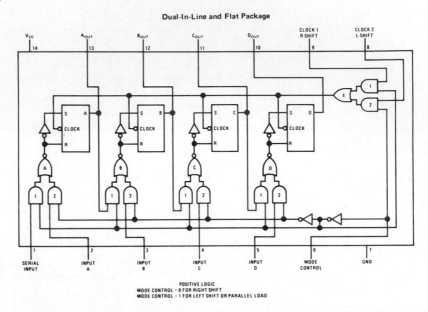

One integrated circuit that we might find is the 7491 8-bit shift register (Fig. 5-32).

In this integrated circuit, we place the data to be shifted on the A and B inputs, and clock the shift register. After eight clock pulses, we will begin to see the data

FIGURE 5-32 The 7491 8-bit shift register. (Copyrighted material reprinted with permission of National Semiconductor Corporation.)

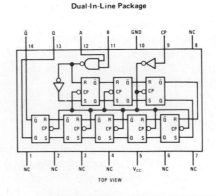

FIGURE 5-33 Timing diagram for a 7491 shift register. (Copyrighted material reprinted with permission of National Semiconductor Corporation.)

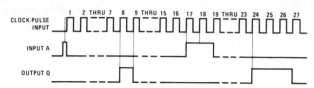

on the Q and $\overline{Q}$ outputs. For reasons that may not be apparent at this point, we can use this shift register to obtain a *delay*. It sometimes becomes necessary to delay a pulse from arriving at its destination for a short while (as in the case of the carry bit in some binary adders). To learn how to achieve such a delay, let's start by looking at the timing diagram for this shift register (Fig. 5-33).

We notice that the input A goes high for a short while before clock pulse 1. The output Q stays low for seven more clock pulses. In other words, it takes eight clock pulses to shift the A input to the Q output. Knowing this, we can make a delay circuit very easily.

EXAMPLE 5-5

Use a 7491 to make an 80-μsec delay circuit.

SOLUTION
To get an 80-μsec delay, we know that it will take eight clock pulses to get our signal out of the shift register (Fig. 5-34). So, if we want to wait 80 μsec for our signal, and it takes eight clock pulses to get our signal out of the shift register, each clock pulse should take 80/8 or 10 μsec!

FIGURE 5-34 The 7491 as an 80-μsec delay circuit.

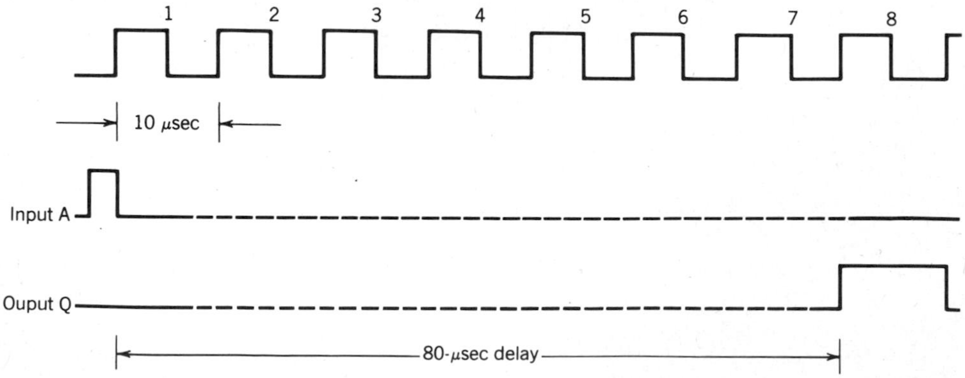

10.4 The 8-Bit Parallel Load Shift Register

The last shift register we want to look at is the 74166 8-bit parallel load shift register. The 74166 enjoys widespread use in video games for sending out video information to television monitors. A game field image is stored in binary in a microcomputer's memory. Since most microcomputers operate on 8 bits at a time, control circuitry is used to clock out video information in 8-bit segments. The 74166 (Fig. 5-35) is ideal for this application because all 8 bits can be presented at its inputs at once, and then clocked out individually.

Timing circuits take the 8 data bits from the computer's memory and place them at inputs A through H of the shift register. The shift/load input is then toggled to get the data into the shift register. Once this has been done, the data bits are clocked out one by one and mixed with other video signals to become white and black dots on the television screen.

FIGURE 5-35 (*a*) Logic and (*b*) connection diagrams for the 74166. (Copyrighted material reprinted with permission of National Semiconductor Corporation.)

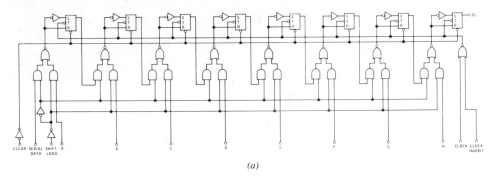

(*a*)

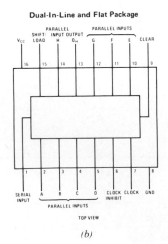

(*b*)

11 OTHER SHIFT REGISTERS

We have covered three different types of shift register, but many more exist. Most of these extend, however, into large bit counts (32-, 64-, dual 80-bit, etc.) and are covered in Chapter 7.

12 RACE CONDITIONS

A race condition exists in a digital circuit when the operation of the circuitry depends on the speed of the gates involved. Circuits of this type are *not* guaranteed to work because the speed of individual gates can vary due to production line variables. In the circuit shown in Fig. 5-36 we see how a poor design resulted in a race condition.

In this case, the MIX output goes high whenever the AL and BL outputs are different. Because of different propagation delay times of the flip-flops (see Fig. 5-37), it is possible for MIX to quickly pulse high as a result of different switching times at the Q outputs. To correct this possible race (error) condition, circuitry must be added or a complete redesign done. It is best to keep an eye out for circuits that might produce race conditions.

FIGURE 5-36 Circuit producing a race condition.

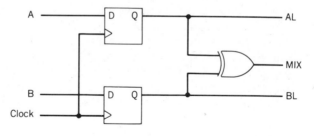

FIGURE 5-37 Error pulse produced by a race condition.

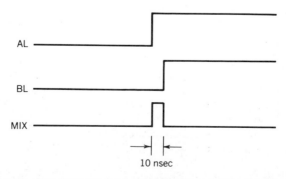

13 SUMMARY

In this chapter we looked at three different, but related digital devices: the flip-flop, the counter, and the shift register. We learned about the 74279 $\overline{S}$-$\overline{R}$ latch, the 7474 type D flip-flop, and the 7476 J-K flip-flop. We saw that most flip-flops come with PRESET and CLEAR inputs, in addition to their data inputs. As for counters, we looked at the operation of the 7490 decade counter, and the 7492 divide-by-12 and the 7493 divide-by-16 counters. We saw that all counters are composed of flip-flops connected in a dividing chain with some kind of reset logic. We also saw how to externally reset counters for obtaining custom counters. Last, we covered shift registers: the 7495 4-bit right/left shift register, the 7491 8-bit serial in shift register, and the 74166 8-bit parallel load shift register. All shift registers employ some method of clocking bits into or out of them for use by other digital circuits.

14 GLOSSARY

Asynchronous input.   This type of input causes immediate action by itself.

Bistable flip-flop.   Another name for the two-state flip-flop.

Clock.   A signal or input line that must be used to cause a flip-flop to toggle, a counter to count, or a shift register to shift.

Counter.   A connection of flip-flops that produces a binary count from a single input of pulses.

D-type flip-flop.   A flip-flop with a data input. Data is clocked into the flip-flop by the clock line (also has asynchronous preset and clear inputs).

D-C latch.   A flip-flop. A 1-bit memory device.

Don't care.   A state denoted by the symbol "X," meaning that the line so designated has no effect. It can be high or low with no effect on the circuit.

Edge triggering.   Most flip-flops toggle on a transition (edge) (zero to one or one to zero). When the pulse travels from one state to the other, however, toggling is initiated (triggered) by the edge.

Enable.   A line that is used to allow a signal to pass to a circuit.

Flip-flop.   A 1-bit memory device that can be either SET or RESET.

J-K Flip-flop.   A flip-flop with synchronous inputs J and K, and asynchronous inputs, clear and sometimes preset.

Load.   A line that is used to enter a binary number into a register or counter in parallel.

Parallel.   The presentation of all bits of a binary number on an equal number of wires at the same time.

RESET.   A reset flip-flop has Q = 0.

Serial.   Describing the presentation of one bit at a time on a single wire.

SET.   A set flip-flop has Q = 1.

Shift register.   A connection of flip-flops used to store long binary numbers and shift them to the left or right as necessary.

Synchronous.   Describing input that requires a clock pulse to cause activity. Such input cannot act on its own.

Timing diagram.   A drawing of the simultaneous waveforms that exist as a device (counter, shift register, etc.) proceeds through its various states.

Toggle.   When a flip-flop alternates from set to reset, it is said to toggle.

PROBLEMS

5-1   Explain the meaning of a low-to-high transition as applied to a clock pulse.

5-2   What is the meaning of a circle on the input to a flip-flop?

5-3   What is the difference between a synchronous input and an asynchronous input?

5-4   Describe in words the operation of the J and K inputs on a J-K flip-flop.

5-5   Design a divide-by-7 and a divide-by-13 counter using a 7493.

5-6   Sketch the timing diagram for the two counters designed in Problem 5-5.

5-7   Determine the effect on the binary count of a 4-bit (7493) counter when the ABCD outputs are inverted ($\overline{A}\ \overline{B}\ \overline{C}\ \overline{D}$).

5-8   Describe in your own words the operation of the simple 4-bit shift register.

5-9   How can an 8-bit shift register be used to produce a 240-msec delay?

5-10 Use two 7490 decade counters to make a divide-by-88 circuit.

5-11 Use two divide-by-10 counters (7490) to make a divide-by-48 circuit.

5-12 A certain circuit will divide by 88 when its control line DIV is high, and by 48 when DIV is low. Modify the circuits from Problems 5-10 and 5-11 to divide by 48 or 88 depending on DIV; use only two 7490s and some support TTL gates.

5-13 What is the maximum count (or division) that can be obtained with each of the following counters?
(a) A 7490 and a 7492
(b) A 7490 and a 7493
(c) A 7490 and a D flip-flop
(d) A 7492 and a 7493
(e) Three 7490s
(f) Four 7493s

5-14 Show how a 5-bit shift register can be made out of J-K flip-flops.

5-15 Show how an 11-bit shift register can be made out of an 8-bit shift register and some flip-flops.

5-16 If the clock line on a shift register is 10 MHz, what is the time between bits at the output?

CHAPTER 6
DATA CONTROL DEVICES

1 INSTRUCTIONAL OBJECTIVES

When finished with this chapter you should be able to:

1. Use selectors/multiplexers to switch large quantities of data into smaller, manageable groups.
2. Use decoders/demultiplexers to scan large quantities of data with smaller control words.
3. Configure transceivers and latches to switch bus-type data.

2 SELF-EVALUATION QUESTIONS

Keep the following questions in mind and try to answer them when you have completed the chapter.

1. When does an enable line become useful?
2. What is the difference between an open collector and tri-state output?
3. What is meant by (a) multiplexing, (b) demultiplexing?
4. What is a bidirectional bus driver?
5. What is an octal latch?

3 INTRODUCTION

All the various TTL gates and logic packages we have looked at thus far were used to manipulate data. Now, in our last chapter on TTL gates that are used extensively in the design process, we will study the data control TTL gate. Data control gates come in two types: selectors/multiplexers and decoders/demultiplexers. Selectors/multiplexers are very useful in parallel to serial conversion, and also in the switching of parallel data busses. Decoders/demultiplexers are very useful for converting coded data back into its original form (e.g., binary-to-decimal conversion). We will study four packages from each group and also consider some possible applications.

4 SELECTORS/MULTIPLEXERS

Having learned how you can use simple NAND or AND gates to *select* certain inputs in a circuit, you should recognize the schematic shown in Fig. 6-1. Although there appear to be only five gates in this circuit, there are actually eight because we need three inverters to make up the ⎯⎯⎯◠◡ function. We are able to

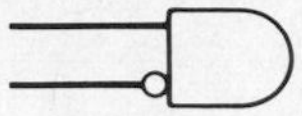

simplify the circuit by eliminating one inverter as indicated in Fig. 6-2. But we

FIGURE 6-1 Simple four-input selector.

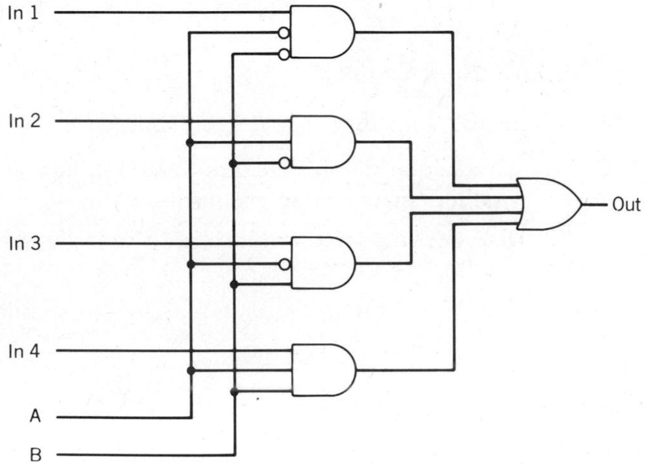

FIGURE 6-2 Simplified circuit of Fig. 6-1.

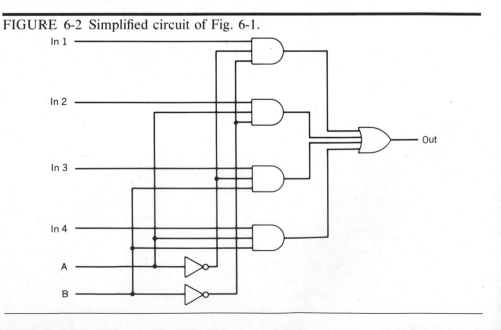

still have a lot of gates to deal with. If we had more than four inputs, say eight, this circuit would become a tangle of gates and wires. Fortunately we have available special integrated circuits designed specifically to perform this function. The following integrated circuits are used to convert large amounts of data into smaller, more manageable quantities. We actually *select* the particular portion of data we wish to look at. This becomes a very important and necessary task when one starts dealing with microcomputers.

4.1 The 74150 One-of-16 Data Selector

The 74150 is a 24-pin DIP that contains the logic needed to select one of 16 different inputs and transfer it to an output pin (see Fig. 6-3). The desired input

FIGURE 6-3 The 74150 one-of-16 data selector. (*a*) pinout, (*b*) truth table, and (*c*) block diagram. (Courtesy of Texas Instruments Incorporated.)

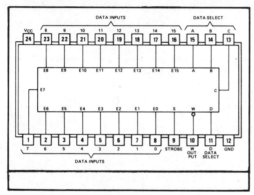

(*a*)

INPUTS					OUTPUT
SELECT				STROBE	
D	C	B	A	S	W
X	X	X	X	H	H
L	L	L	L	L	$\overline{E0}$
L	L	L	H	L	$\overline{E1}$
L	L	H	L	L	$\overline{E2}$
L	L	H	H	L	$\overline{E3}$
L	H	L	L	L	$\overline{E4}$
L	H	L	H	L	$\overline{E5}$
L	H	H	L	L	$\overline{E6}$
L	H	H	H	L	$\overline{E7}$
H	L	L	L	L	$\overline{E8}$
H	L	L	H	L	$\overline{E9}$
H	L	H	L	L	$\overline{E10}$
H	L	H	H	L	$\overline{E11}$
H	H	L	L	L	$\overline{E12}$
H	H	L	H	L	$\overline{E13}$
H	H	H	L	L	$\overline{E14}$
H	H	H	H	L	$\overline{E15}$

(*b*)

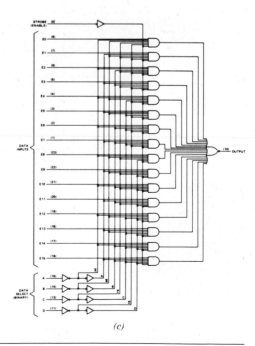

(*c*)

source (0–15) is selected by placing the appropriate 4-bit binary word on the data select (ABCD) inputs. Once this has been done, the complement ($\overline{D_{in}}$) of the selected data input will appear on the output pin W as soon as the strobe pin (S) goes low. Whenever the S pin is high, the W output remains high, regardless of the data present on the data source or select pins. The 74150 could be used to convert 16-bit parallel into serial data with the addition of a 4-bit binary counter on the data select inputs. It could also be used to select serial (TTL only) data from one of 16 sources. Or, it could be used to monitor the status of up to 16 switches, as in a home burglar alarm.

EXAMPLE 6-1

Figure 6-4 shows the 74150 used to select one of 16 serial data inputs. If the serial data signals are related to each other by integer multiples or other small increments, this could be a building block for an electronic music machine.

FIGURE 6-4 The 74150 1-of-16 selector.

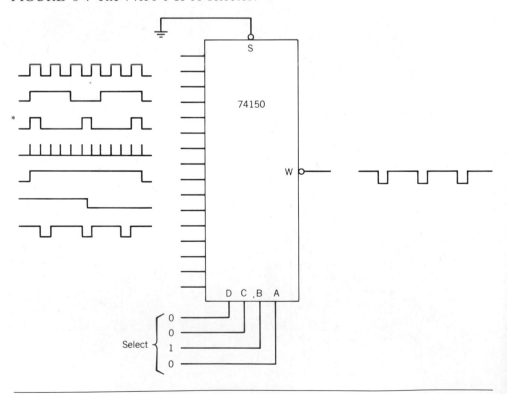

EXAMPLE 6-2

In Fig. 6-5, the 74150 is used to monitor the status of 16 doors and windows in a simple burglar alarm.

The 74150 does not find much use in microcomputer circuitry, but it is none-theless, a valuable integrated circuit. If tri-state outputs are desired, the 74351 dual eight-to-one (one-of-eight) selector can be used to perform the same operation and in addition provide OR-tying capability.

FIGURE 6-5 Simple 16-input burglar alarm.

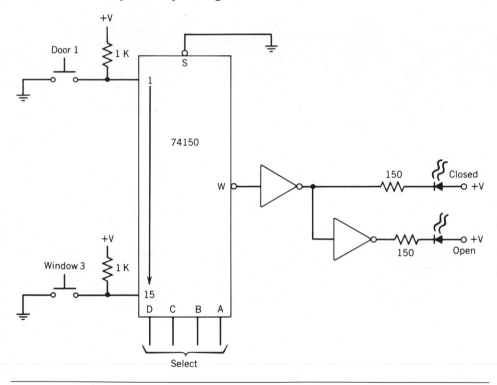

4.2 The 74151 One-of-Eight Selector

The 74151 (Fig. 6-6) is a 16-pin DIP that is used to select one of eight data inputs. The 74151 is particularly useful for converting 8-bit parallel data to serial data (ideal for microcomputer work) and finds application in some popular video games.

To select an input, the 3-bit binary word is placed on the data select (ABC) inputs. Then as soon as the strobe (S) line is pulled low, the data present on the

FIGURE 6-6 The 74151 one-of-eight selector. (*a*) pinout, (*b*) truth table, and (*c*) schematic diagram. (Courtesy of Texas Instruments Incorporated.)

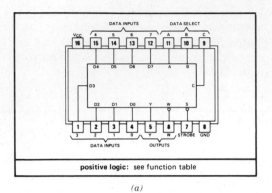

INPUTS				OUTPUTS	
SELECT			STROBE	Y	W
C	B	A	S		
X	X	X	H	L	H
L	L	L	L	D0	$\overline{D0}$
L	L	H	L	D1	$\overline{D1}$
L	H	L	L	D2	$\overline{D2}$
L	H	H	L	D3	$\overline{D3}$
H	L	L	L	D4	$\overline{D4}$
H	L	H	L	D5	$\overline{D5}$
H	H	L	L	D6	$\overline{D6}$
H	H	H	L	D7	$\overline{D7}$

(a)   (b)

(c)

selected input gets routed to the Y output. The complement of the same data appears on the W output. The 74151 could be used in a white noise generator (see Example 6-3). You should realize that a 74150 can be (and should be) used whenever two 74151s are needed to look at 16 inputs. It may be desirable, though, to use two at the same time. That is one advantage of having such a large variety of selectors to choose from.

EXAMPLE 6-3

In Fig. 6-7, 8-bit parallel data is sent to the 74151 (possibly from a computer's memory) and converted into serial data, which is amplified to produce white noise.

The W output is provided (probably because there was an extra pin) and saves an inverter if complement data is needed. If tri-state outputs are desired, the 74251 can be used to provide the same function with high-impedance (tri-state) output mode provided when the S pin is high.

FIGURE 6-7 Simple white noise source.

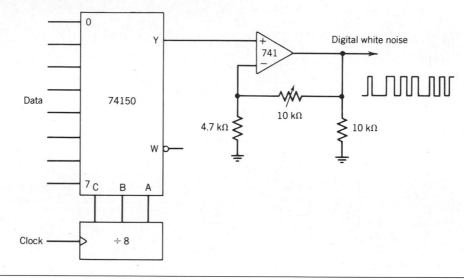

4.3 The 74153 Dual One-of-Four Selector

The 74153 (Fig. 6-8) is a 16-pin DIP that contains two four-to-one line selectors. Each selector has its own noninverting output and enable strobe. Both selectors share the same data select (AB) pins.

The data is selected by placing a 2-bit word on the data select inputs (AB) and pulling the appropriate strobe line low. Remember that the units operate separately but may still be strobed or enabled at the same time. The 74153 is commonly used in dynamic RAM circuitry to switch the appropriate address lines into the chip.

EXAMPLE 6-4

The 74153 is used to select one of four address busses for a dynamic RAM. We will soon study the dynamic RAM operation, but Fig. 6-9 shows schematically how a 74153 can be used to select different busses.

A tri-state version of the 74153 exists and is listed as the 74253. The operation is identical to the 74153 except that there is a high-impedance output when S is high.

4.4 The 74157 Quad Two-to-One Line Selector

Our last selector is the 74157 quad two-to-one data selector (Fig. 6-10). The 74157 is a 16-pin DIP that contains four data selectors, each capable of selecting one of two lines.

The data present on the A inputs appears on the Y outputs when S (select) is low and enable ($\overline{G}$) is low. The data on the B inputs appears when S is high and $\overline{G}$ is low. The idle state exists when $\overline{G}$ is high. In this case, all Y outputs are low

FIGURE 6-8 The 74153 dual one-of-four selector. (*a*) pinout, (*b*) truth table, and (*c*) schematic diagram. (Courtesy of Texas Instruments Incorporated.)

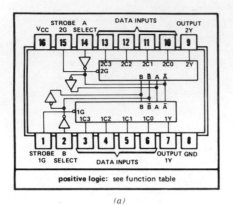

(a)

SELECT INPUTS		DATA INPUTS				STROBE	OUTPUT
B	A	C0	C1	C2	C3	G	Y
X	X	X	X	X	X	H	L
L	L	L	X	X	X	L	L
L	L	H	X	X	X	L	H
L	H	X	L	X	X	L	L
L	H	X	H	X	X	L	H
H	L	X	X	L	X	L	L
H	L	X	X	H	X	L	H
H	H	X	X	X	L	L	L
H	H	X	X	X	H	L	H

Select inputs A and B are common to both sections.
H = high level, L = low level, X = irrelevant

(b)

(c)

FIGURE 6-9 Selection of different busses by the 74153, which selects 24 bits of address into 6 bits at a time.

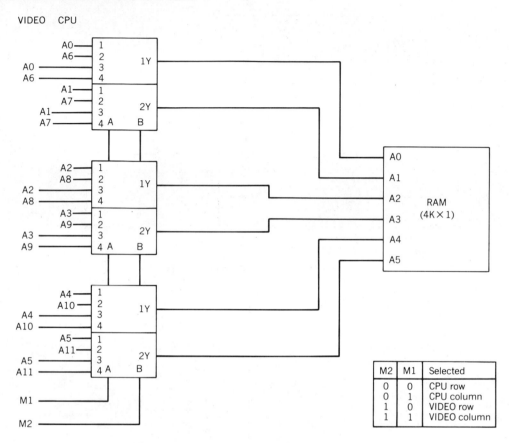

M2	M1	Selected
0	0	CPU row
0	1	CPU column
1	0	VIDEO row
1	1	VIDEO column

FIGURE 6-10 The 74157 quad two- to one-line selector. (*a*) pinout, (*b*) truth table, and (*c*) schematic diagram. (Courtesy of Texas Instruments Incorporated.)

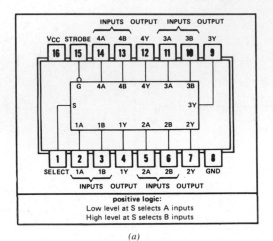

INPUTS				OUTPUT Y	
STROBE	SELECT	A	B	'157, 'L157, 'LS157, 'S157	'LS158 'S158
H	X	X	X	L	H
L	L	L	X	L	H
L	L	H	X	H	L
L	H	X	L	L	H
L	H	X	H	H	L

H = high level, L = low level, X = irrelevant

(*b*)

(*c*)

(high if the 74158 is used). If tri-stating is desired, the 74257 integrated circuit is also available.

The 74157 is very useful to applications in electronic circuits where two control circuits desire access to another circuit (although not at the same time) such as in an electronic video game. The central processing unit (CPU) may want to look at the screen memory (to update it) or the video circuitry may want to look at the screen memory (to send some video out to a monitor). The 74157 very easily accomplishes this task.

EXAMPLE 6-5

The 74157 can be used to switch between CPU and VIDEO address busses, as shown in Fig. 6-11.

To summarize, we use a selector whenever we want to narrow large amounts of data to more manageable sizes. We "multiplex" the data by selectively looking at the portions that are important to us. In the next section, we will see how we then "demultiplex" the data whenever we want to control a large number of data paths from a much smaller control word.

FIGURE 6-11 Circuitry for simple address bus switches.

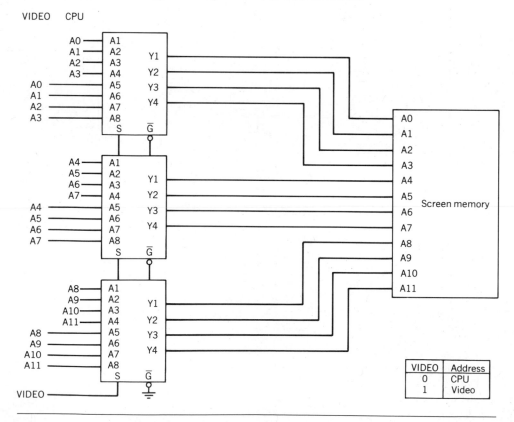

5 DECODERS/DEMULTIPLEXERS

This section dealing with decoders will become very important when we start our study of computer memory circuits and of computer applications in general. A decoder is used to obtain many outputs from a small source. For instance, with a decoder, we can use two inputs to control 4 outputs, three to control 8, and even four to control 16 different outputs. This serves to cut down on all the extra wiring that would be needed in an electronic circuit, and usually the number of gates, too!

5.1 The 74139 Dual Two-Line to Four-Line Decoder

The 74139 is a 16-pin DIP that enables two inputs to control four outputs. There are two separate two- to four-line decoders in the package, and each has its own enable line.

FIGURE 6-12 The 74139 dual two-line to four-line decoder. (*a*) pinout, (*b*) truth tables, and (*c*) schematic diagram. (Courtesy of Texas Instruments Incorporated.)

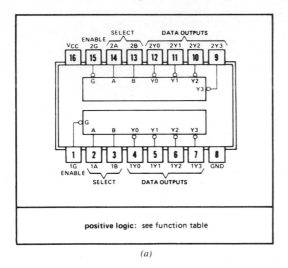

(*a*)

'LS139, 'S139
(EACH DECODER/DEMULTIPLEXER)

INPUTS		OUTPUTS			
ENABLE	SELECT				
G	B	A	Y0	Y1	Y2	Y3
H	X	X	H	H	H	H
L	L	L	L	H	H	H
L	L	H	H	L	H	H
L	H	L	H	H	L	H
L	H	H	H	H	H	L

H = high level, L = low level, X = irrelevant

(*b*)

'LS139, 'S139

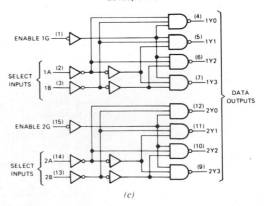

(*c*)

To select one of the four outputs (which are normally high), we simply place the 2-bit binary number corresponding to the output line we wish to decode (0, 1, 2, or 3) on the A, B select inputs, and pull the enable line (G) low. If G is high, the outputs remain high; otherwise, the selected output will go low. Please note that none of the decoders that we will study pass data (as the selectors did). They simply enable a certain output (although there are devices that *do* pass data to a selected output).

EXAMPLE 6-6

The 74139 is used in two ways in Fig. 6-13. One half is used to drive a relay through a Darlington pair, and the other half is used as an indicator. This could be part of a simple control circuit. With A and B low, relay 1 and LED 1 are turned on, and so on. Note that only one relay and LED may be on at any instant.

FIGURE 6-13 Simple control circuit using the 74139.

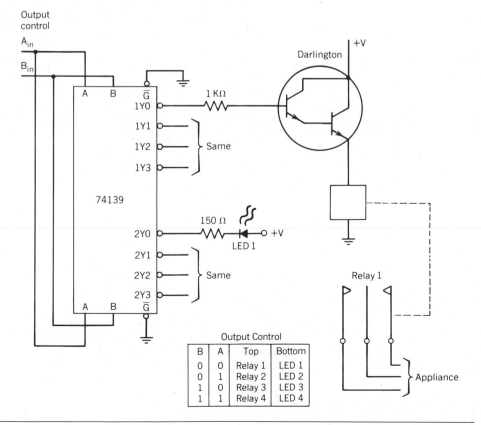

B	A	Top	Bottom
0	0	Relay 1	LED 1
0	1	Relay 2	LED 2
1	0	Relay 3	LED 3
1	1	Relay 4	LED 4

5.2 The 74138 Three-Line to Eight-Line Decoder

If four outputs aren't enough, we can turn to the 74138 (Fig. 6-14), a 16-pin DIP that is capable of decoding one of eight output lines from a 3-bit input word. (You should see that the outputs increase by the function 2^N, where N is the number of inputs.) As usual, the binary equivalent of the output we wish to decode is placed on the select inputs A, B, and C. Once the two enable inputs G2A and G2B have been pulled low, and the third enable input G1 has been pulled high, the selected output will go low.

FIGURE 6-14 The 74138 three-line to eight-line decoder. (*a*) pinout, (*b*) truth table, and (*c*) schematic diagram. (Courtesy of Texas Instruments Incorporated.)

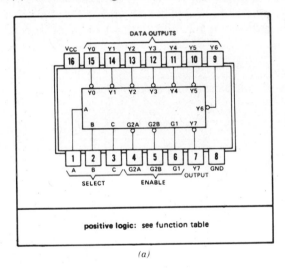

(a)

INPUTS					OUTPUTS							
ENABLE		SELECT										
G1	G2*	C	B	A	Y0	Y1	Y2	Y3	Y4	Y5	Y6	Y7
X	H	X	X	X	H	H	H	H	H	H	H	H
L	X	X	X	X	H	H	H	H	H	H	H	H
H	L	L	L	L	L	H	H	H	H	H	H	H
H	L	L	L	H	H	L	H	H	H	H	H	H
H	L	L	H	L	H	H	L	H	H	H	H	H
H	L	L	H	H	H	H	H	L	H	H	H	H
H	L	H	L	L	H	H	H	H	L	H	H	H
H	L	H	L	H	H	H	H	H	H	L	H	H
H	L	H	H	L	H	H	H	H	H	H	L	H
H	L	H	H	H	H	H	H	H	H	H	H	L

*G2 = G2A + G2B

H = high level, L = low level, X = irrelevant

(b)

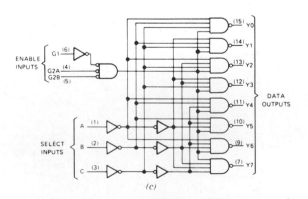

(c)

The enabling of the 74138 is complicated but also very useful. It is often desirable, and sometimes essential, to be able to enable a device from several locations. Figure 7-16 in the chapter on memories demonstrates this point well; however, we may still look at an example utilizing eight outputs.

EXAMPLE 6-7

The 74138 is used to create one of eight tones via a voltage-controlled oscillator (VCO). A voltage-controlled oscillator simply changes frequency as a function of input voltage (Fig. 6-15). Thus lower voltages, in this case, produce lower frequencies. In Fig. 6-16, we use the 74138 as an eight-position switch to send one of eight different voltages to the voltage-controlled oscillator.

Since only one of the outputs of the 74138 is low at any one time, seven resistors are pulled to ground, and the last one sources the $+5$ V. Thus, we have a voltage divider through an op amp with a gain set by the setting of the 100-kΩ potentiometer. The op amp output drives the VCO to produce a tone dependent on the op amp voltage. Changing the binary word on the ABC input changes the output selected, hence the voltage applied to the VCO.

5.3 The 7442 BCD to Decimal Decoder

The 7442 (Fig. 6-17) is a 16-pin DIP that converts a 4-bit input word into one of 10 selected outputs. If the input word is a valid binary representation of one of the output numbers (0–9), the selected output will go low. If the input word is higher than 9 (A–F), all outputs will remain high.

FIGURE 6-15 Frequency versus voltage in a voltage-controlled oscillator.

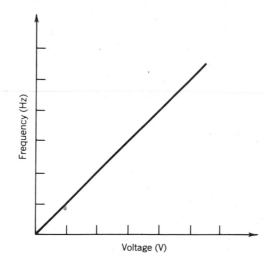

FIGURE 6-16 Simple tone generator.

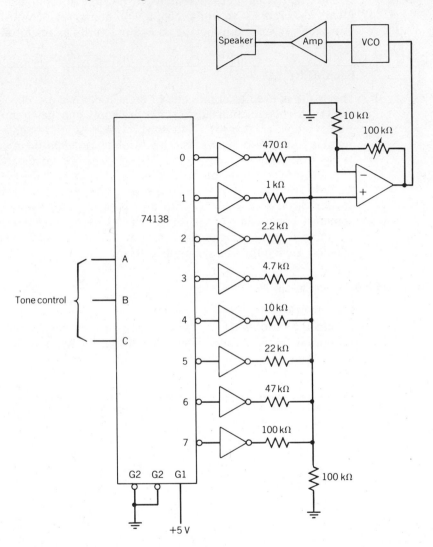

FIGURE 6-17 The 7442 BCD-to-decimal decoder. (*a*) pinout, (*b*) truth table, and (*c*) schematic diagram. (Courtesy of Texas Instruments Incorporated.)

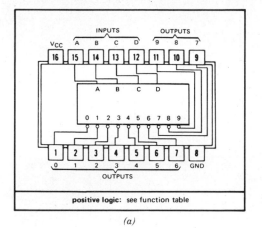

(a)

FUNCTION TABLE

NO.	'42A, 'L42, 'LS42 BCD INPUT				ALL TYPES DECIMAL OUTPUT									
	D	C	B	A	0	1	2	3	4	5	6	7	8	9
0	L	L	L	L	L	H	H	H	H	H	H	H	H	H
1	L	L	L	H	H	L	H	H	H	H	H	H	H	H
2	L	L	H	L	H	H	L	H	H	H	H	H	H	H
3	L	L	H	H	H	H	H	L	H	H	H	H	H	H
4	L	H	L	L	H	H	H	H	L	H	H	H	H	H
5	L	H	L	H	H	H	H	H	H	L	H	H	H	H
6	L	H	H	L	H	H	H	H	H	H	L	H	H	H
7	L	H	H	H	H	H	H	H	H	H	H	L	H	H
8	H	L	L	L	H	H	H	H	H	H	H	H	L	H
9	H	L	L	H	H	H	H	H	H	H	H	H	H	L
INVALID	H	L	H	L	H	H	H	H	H	H	H	H	H	H
	H	L	H	H	H	H	H	H	H	H	H	H	H	H
	H	H	L	L	H	H	H	H	H	H	H	H	H	H
	H	H	L	H	H	H	H	H	H	H	H	H	H	H
	H	H	H	L	H	H	H	H	H	H	H	H	H	H
	H	H	H	H	H	H	H	H	H	H	H	H	H	H

H = high level, L = low level

(b)

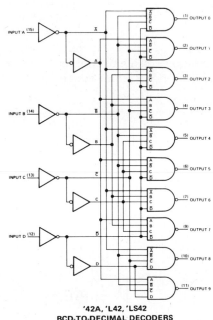

'42A, 'L42, 'LS42
BCD-TO-DECIMAL DECODERS

(c)

FIGURE 6-18 A BCD-to-decimal nixie display.

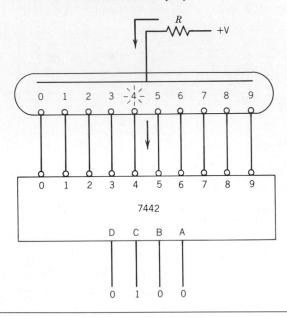

The 7442 was initially used as a decoder for a now almost obsolete display tube called the nixie tube (Fig. 6-18), which contained a filament, and 10 loops of wire, shaped to represent the digits 0 through 9. The selected output completed a current path in the nixie tube, causing the proper element (wire number) to glow. The nixie tube has been almost universally replaced by the seven-segment display and the even newer 5 × 7 dot matrix display.

FIGURE 6-19 The 74154 4-line-to 16-line decoder. (*a*) pinout, (*b*) truth table, and (*c*) schematic diagram. (Courtesy of Texas Instruments Incorporated.)

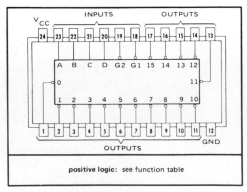

(*a*)

5.4 The 74154 Four-Line to Sixteen-Line Decoder

The 74154 is a 24-pin DIP that decodes a 4-bit word present on its input to one of 16 unique outputs. There are two enable pins (G1, G2) to help control the outputs. The desired output is pulled low by placing the correct 4-bit word for that output on the input pins (ABCD) and then grounding both the G pins. If either of the G pins is high, or if both are, all outputs will remain high.

FIGURE 6-19 (continued)

INPUTS						OUTPUTS																
G1	G2	D	C	B	A	0	1	2	3	4	5	6	7	8	9	10	11	12	13	14	15
L	L	L	L	L	L	L	H	H	H	H	H	H	H	H	H	H	H	H	H	H	H
L	L	L	L	L	H	H	L	H	H	H	H	H	H	H	H	H	H	H	H	H	H
L	L	L	L	H	L	H	H	L	H	H	H	H	H	H	H	H	H	H	H	H	H
L	L	L	L	H	H	H	H	H	L	H	H	H	H	H	H	H	H	H	H	H	H
L	L	L	H	L	L	H	H	H	H	L	H	H	H	H	H	H	H	H	H	H	H
L	L	L	H	L	H	H	H	H	H	H	L	H	H	H	H	H	H	H	H	H	H
L	L	L	H	H	L	H	H	H	H	H	H	L	H	H	H	H	H	H	H	H	H
L	L	L	H	H	H	H	H	H	H	H	H	H	L	H	H	H	H	H	H	H	H
L	L	H	L	L	L	H	H	H	H	H	H	H	H	L	H	H	H	H	H	H	H
L	L	H	L	L	H	H	H	H	H	H	H	H	H	H	L	H	H	H	H	H	H
L	L	H	L	H	L	H	H	H	H	H	H	H	H	H	H	L	H	H	H	H	H
L	L	H	L	H	H	H	H	H	H	H	H	H	H	H	H	H	L	H	H	H	H
L	L	H	H	L	L	H	H	H	H	H	H	H	H	H	H	H	H	L	H	H	H
L	L	H	H	L	H	H	H	H	H	H	H	H	H	H	H	H	H	H	L	H	H
L	L	H	H	H	L	H	H	H	H	H	H	H	H	H	H	H	H	H	H	L	H
L	L	H	H	H	H	H	H	H	H	H	H	H	H	H	H	H	H	H	H	H	L
L	H	X	X	X	X	H	H	H	H	H	H	H	H	H	H	H	H	H	H	H	H
H	L	X	X	X	X	H	H	H	H	H	H	H	H	H	H	H	H	H	H	H	H
H	H	X	X	X	X	H	H	H	H	H	H	H	H	H	H	H	H	H	H	H	H

H = high level, L = low level, X = irrelevant (b)

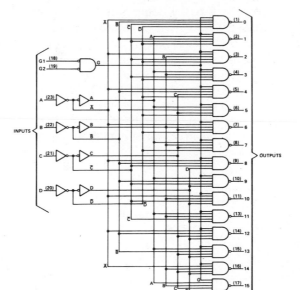

(c)

6 MICROCOMPUTER INTERFACE CIRCUITS

The great popularity of microcomputers has caused a need for a different type of integrated circuit. This is due to the parallel nature of the microcomputer and the need to move data from one point to another. Generally, in a microcomputer, there are eight wires for transmitting data. This is the data bus. The address bus, which is similar, requires at least 16 wires. Many different sections of the computer share these busses with one simple but important restriction, namely, that only one section may drive the bus at a time. This is why tri-state logic is so important.

7 TRANSCEIVERS AND LATCHES

Special circuits are needed by microcomputers to make the transfer of data convenient and fast.

1. A transceiver can pass data in one direction or another or in two directions.
2. A latch can accept and hold data for future use (it may also be called a flip-flop).

There are a number of common configurations available. Some packages contain separate tri-state buffers as in Fig. 6-20. Each buffer is tri-state and may have its output enabled using OE_1 or OE_2. Input lines are "data in 1" (DI_1) and "data in 2" (DI_2). Output lines are "data out 1" (DO_1) and "data out 2" (DO_2). Figures 6-21 and 6-22, respectively, show a set of buffers with one side of the buffers internally connected and a bidirectional transceiver. The enable lines in Fig. 6-22 are internally decoded to become a direction control line. As we shall see, a data bus may pass data in either of two directions.

Table 6-1 lists the common interface chips that are available (all are tri-state). Figure 6-23 shows a representative sample of a few of these logic gates. Remember, the idea is to switch groups of data on or off simultaneously.

For more information, consult an appropriate data manual. Let us examine the 74LS245 (Fig. 6-23c). There are eight A lines (A_0–A_7) and eight B lines (B_0–B_7). Data will flow from A (input) to B (output) if $S/\overline{R}$ is high (send mode). Data will flow from B (input) to A (output) is $S/\overline{R}$ is low (receiver mode). None of this takes place unless the chip enable line ($\overline{CE}$) is low. So $\overline{CE}$ turns the chip on, and $S/\overline{R}$ determines the direction of data flow. Hence, this is called a tri-state bidirectional bus transceiver.

FIGURE 6-20 Separate bus drivers.

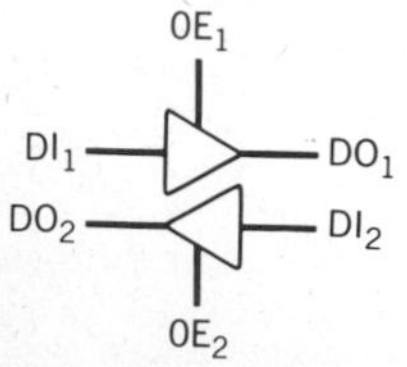

FIGURE 6-21 Internally connected buffers.

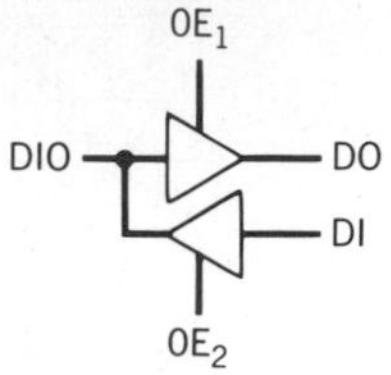

FIGURE 6-22 Bidirectional transceiver.

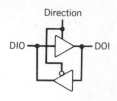

FIGURE 6-23 Bus transceivers. (*a*) and (*c*) octal; (*b*) hex. (© 1984
Signetics Corp.)

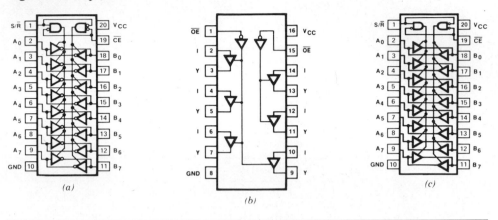

(*a*)

(*b*)

(*c*)

TABLE 6-1   Common Interface Chips

8T26A	S	Quad transceiver	'242	74LS	Octal transceiver
8T28	S	Quad transceiver	'243	74LS	Octal transceiver
8T95	S	Hex buffer	'244	74LS	Octal buffer
8T96	S	Hex buffer		74S	
8T97	S	Hex buffer	'245	74LS	Octal transceiver
8T98	S	Hex buffer	'273	74LS	Octal D flip-flop
8T125	LS	Octal transceiver	'363	74LS	Octal transparent latch
8T126	LS	Quad transceiver	'364	74LS	Octal D flip-flop
8T127	LS	Quad transceiver	'365	74	Hex buffer
8T128	LS	Quad transceiver		74LS	
8T129	LS	Quad transceiver	'366	74	Hex buffer
8TS805	S	Octal transparent latch		74LS	
8TS806	S	Octal D flip-flops	'367	74	Hex buffer
8TS807	S	Octal transparent latch		74LS	
8TS808	S	Octal D flip-flop	'368	74	Hex buffer
8TS809	S	Octal transparent latch		74LS	
'125	74	Quad buffer	'373	74LS	Octal transparent latch
	74LS			74S	
'126	74	Quad buffer	'374	74LS	Octal D flip-flop
	74LS			74S	
'240	74LS	Octal buffer	'377	74LS	Octal D flip-flop
	74S		'534	74S	Octal D flip-flop
'241	74LS	Octal buffer			
	74S				

FIGURE 6-24 Latching interface chips. (*a*) 273 octal D-type flip-flop and (*b*) 373 octal three-state latch. (© 1984 Signetics Corp.)

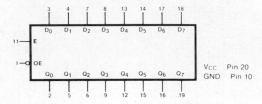

OPERATING MODE	INPUTS			OUTPUTS
	$\overline{MR}$	CP	D_n	Q_n
Reset (clear)	L	X	X	L
Load "1"	H	↑	h	H
Load "0"	H	↑	l	L

(a)

OPERATING MODES	INPUTS				OUTPUTS
	$\overline{OE}$	E	D_n	INTERNAL REGISTER	Q_0-Q_7
Enable & read register	L	H	L	L	L
	L	H	H	H	H
Latch & read register	L	L	l	L	L
	L	L	h	H	H
Latch register & disable outputs	H	L	l	L	(Z)
	H	L	h	H	(Z)

Figure 6-24 shows some of the common latching circuits, useful for holding either input data or output data. Two interface chips designed for use with the 8080A/85 microprocessors are the 8212 and the 8216/8226, shown in Fig. 6-25.

FIGURE 6-25 8080A/85 Interface elements. (*a*) 8212 and (*b*) 8216/8226.
(Reprinted by permission of Intel Corporation, copyright 1981.)

PIN CONFIGURATION

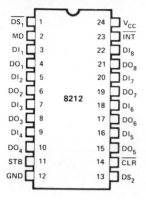

PIN NAMES

DI_1-DI_8	DATA IN
DO_1-DO_8	DATA OUT
$\overline{DS_1}$-DS_2	DEVICE SELECT
MD	MODE
STB	STROBE
$\overline{INT}$	INTERRUPT (ACTIVE LOW)
$\overline{CLR}$	CLEAR (ACTIVE LOW)

LOGIC DIAGRAM

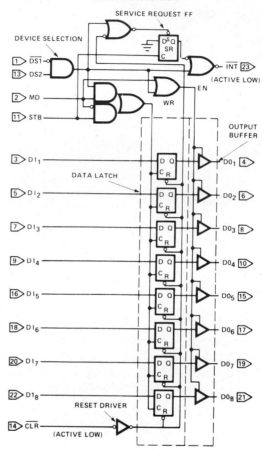

(*a*)

(*continued on next page*)

FIGURE 6-25b

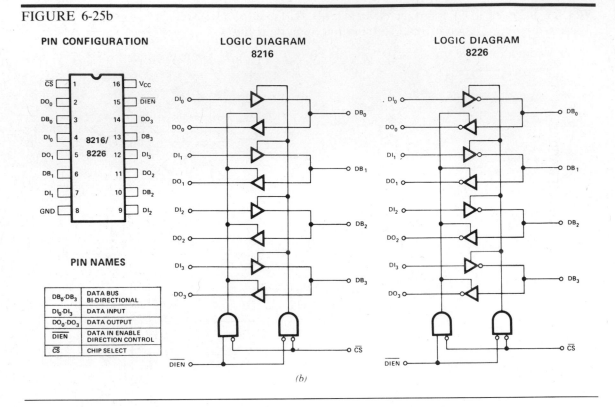

(b)

8 BRINGING IT ALL TOGETHER

There are many possible applications for the two types of data control devices we have just studied. Let's look at one that uses a handful of gates to drive, or multiplex, an 8-digit display. The display can be used for computer output, as in the status of the address and data busses, or as output for an electronic clock or calendar.

In Fig. 6-26 we see how the display is generated. A 500-Hz oscillator drives a divide-by-8 counter. For every count of 0 to 7, one seven-segment display is enabled by the 74138. In addition, each of the four 74151s looks at one of eight inputs. The inputs are separated so that each 74151 gets one bit of the input word (4 bits for a BCD word, etc.). The outputs of the 74151s are fed to a 7448 BCD/seven-segment decoder, which drives the display.

9 TIMING DIAGRAM FUNDAMENTALS

The simple timing diagrams of Chapter 5 were used to demonstrate how counter circuitry functions. They showed the states of the various outputs with time as the counters proceeded through their incremental counts. In the next chapter we

FIGURE 6-26 Circuitry to drive an 8-digit multiplexed display.

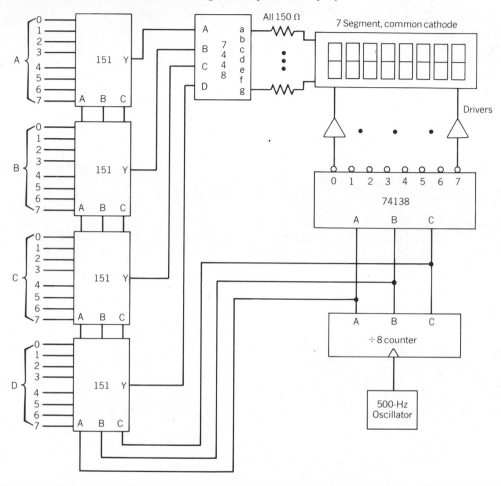

will be observing some more sophisticated timing diagrams as we study memory devices.

Now we present some of the definitions we will need when we examine these different memory circuits. Figure 6-27 shows the basic signals that can be present on a line and includes a definition of transitions that may be present. Figure 6-28 shows the convention used when a signal in one wire causes a signal in another.

The delay time between signals is represented in Fig. 6-29. Figure 6-30 represents signals on multiple wires or a parallel bus: all the signals on a parallel bus will change at nearly the same time, but they may be either high or low. Although not intended to show the specific signal on a wire, Fig. 6-30 does indicate that a group of wires has high and low levels on it. You may want to come back to Figs. 6-27 to 6-30 as you observe some of the manufacturers' drawings of the various memory circuits in the next chapter.

FIGURE 6-27 Basic levels on a single wire: transition in signal A causes transition in signal B.

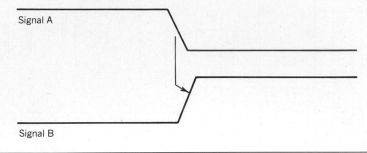

FIGURE 6-28 Signal definitions.

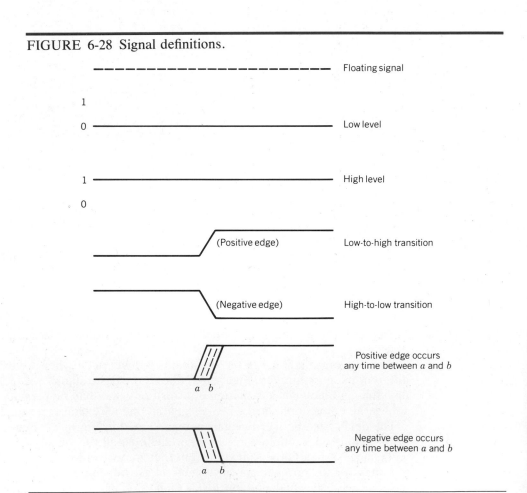

FIGURE 6-29 Time delay t_d between transitions in signals A and B is measured between the 50% points on each edge.

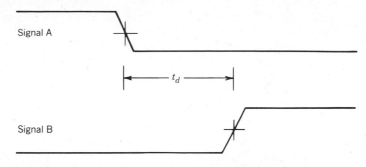

FIGURE 6-30 Signals on parallel paths. (*a*) Parallel bus: all signals change simultaneously. (*b*) Floating bus. (*c*) Logic states of bus signals are not important before signal change.

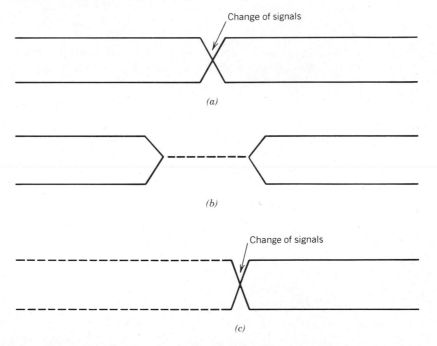

10 SUMMARY

In this chapter we studied the integrated circuit types that perform a very special function: selectors/multiplexers and decoders/demultiplexers. Through the use of selectors, we are able to handle large quantities of data by looking at smaller, controlled portions of it as we desire. Decoders allow us to control large amounts of outputs with a small number of inputs. As we saw, the number of inputs or outputs that we can control or look at is a function of the form 2^N, where N is the number of control bits. We have also studied the available tri-state micro-computer interface integrated circuits. These include bus drivers and bidirectional bus drivers called transceivers. These circuits are necessary because it frequently happens that many sections of a computer share a common data path or bus. Occasionally data must be latched for use by the bus as an outside source, and such integrated circuits are also common. Once we understand the operation of a microcomputer chip, the application of decoders and selectors, as well as all the logic gates we have seen so far, will become more apparent and also *very* interesting.

11 GLOSSARY

Data selector. See Selector.

Decoder/demultiplexer. A circuit that converts a binary (parallel) signal to a few lines.

Latch. Essentially a flip-flop or binary storage device.

Selector/multiplexer. A circuit that reduces many lines to a few; it selects one from many sources.

Transceiver. A circuit that can be switched to move data in one of two directions.

VCO. Voltage-controlled oscillator. The output frequency is proportional to input voltage. VCOs are used when frequency must be a function of voltage.

PROBLEMS

6-1 How many inputs can be multiplexed with a 5-bit control word? How many outputs can be multiplexed with an 8-bit control word?

6-2 Show that the circuit in Fig. 6-31 is a two-line to one-line multiplexer.

6-3 Which circuit is more practical, Fig. 6-32*a* or 6-32*b*? Why?

6-4 Which LEDs in Fig. 6-33 will light when:
(a) A is low, B is low
(b) A is high, B is low
(c) A is low, B is high
(d) A is high, B is high
    (Assume that all devices are properly enabled.)

6-5 Draw at least four cycles of the output waveforms $\emptyset A$ and $\emptyset B$ in Fig. 6-34. [Assume counter at 0 and both latches cleared $(Q = 0)$ to start.] *Hint:* Study the R-S flip-flop of Chapter 5 and first draw the output waveforms of the 7442 only.

FIGURE 6-31 Circuit for Problem 6-2.

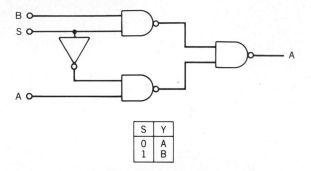

FIGURE 6-32 Circuits for Problem 6-3 (all LEDs). (*a*) One of 16 for 74138. (*b*) One of 16 for 74154.

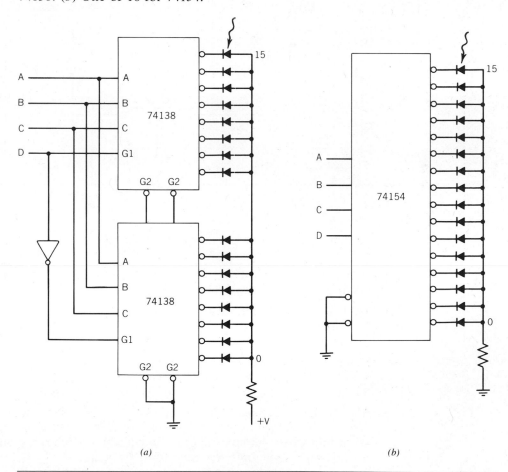

(*a*)   (*b*)

FIGURE 6-33 Circuit for Problem 6-4.

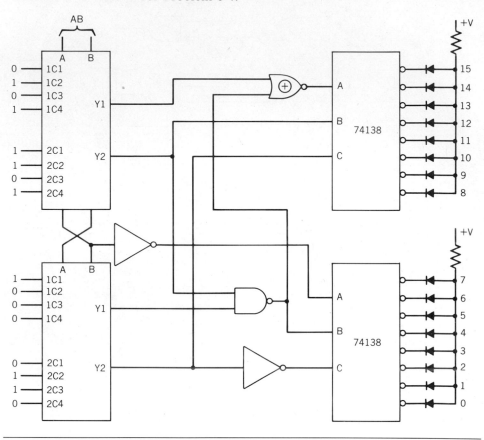

6-6 Design a circuit to pass a digital signal to one of four outputs. The outputs are selected by two binary inputs A and B.

6-7 Add an enable input to the circuit of Problem 6-6 so that the circuit decodes when $\overline{EN}$ is low and the outputs are all high when $\overline{EN}$ is high.

6-8 Find the smallest number of TTL chips that will select one of 256 outputs. The input select word is 8 bits long and the circuit must contain some kind of enable.

6-9 Design a circuit that will look at input bits 1 and 5 when X_1 is low and bits 3 and 7 when X_1 is high. Bits 1 and 3 are selected when X_2 is low and bits 5 and 7 are selected when X_2 is high. (*Hint:* Use an 8-to-1 device.)

6-10 A control word consists of 9 bits A0 through A8. Design a decoding circuit that will pull $\overline{IN}$ low if the control word equals 0BC hex and will pull $\overline{OUT}$ low when the control word equals 0BD hex. (*Hint:* Use an 8-input NAND gate, some inverters, and some OR gates.)

FIGURE 6-34 Circuit for Problem 6-5.

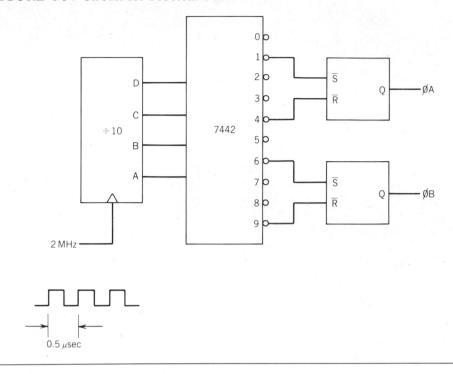

6-11 Modify the circuit of Problem 6-10 so that $\overline{\text{IN}}$ and $\overline{\text{OUT}}$ will go low only if the control word is correct AND an additional input $\overline{\text{STROBE}}$ is also low.

6-12 Design a circuit that will pull $\overline{\text{RD}}$ low when an 8-bit control word is any of the following: 00, 10, 20, 30, . . . , E0, F0. All numbers are in hex.

6-13 Modify the circuit of Problem 6-12 so that an additional output $\overline{\text{WR}}$ is pulled low when the control word is any of the following: 08, 18, 28, 38, . . . , E8, F8. Make sure that $\overline{\text{RD}}$ and $\overline{\text{WR}}$ are never low at the same time.

6-14 Pick any decoder/demultiplexer from the TTL data book and trace one signal from input to output.

6-15 Repeat Problem 6-14 for any selector/multiplexer.

CHAPTER 7
MEMORIES

1 INSTRUCTIONAL OBJECTIVES

When finished with this chapter, you should be able to:

1. Identify the various attributes of the different memory types.
2. Tell the difference between RAM and ROM.
3. Understand the uses of static versus dynamic memories.
4. Show how a byte-oriented memory is constructed from single-bit RAM.
5. Program or erase an EPROM.
6. Describe the operation of a decoder or selector in selecting a particular memory.
7. Understand magnetic memories.

2 SELF-EVALUATION QUESTIONS

Keep the following questions in mind and try to answer them when you have completed the chapter:

1. Why are shift register memories inherently slow compared to random access memories?
2. What are the advantages and disadvantages of ROM, PROM, EPROM, and EAPROM?
3. After an address has been sent to a RAM or ROM, why must there be a delay before reading data?
4. Which lines of a memory (RAM, ROM) are tri-stated and why?
5. What is the purpose of a character generator?
6. How are shift registers similar to tape or disk memories?

3 INTRODUCTION

We have already covered a few of the simpler memory devices, namely, the flip-flop and the shift register. Although very useful in digital circuits for storing temporary results, these devices are not practical for use in microcomputers. For

FIGURE 7-1 Breakdown of solid state memory families.

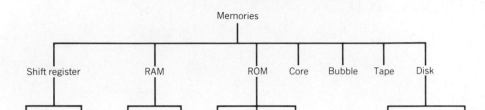

instance, in a digital circuit, we may use flip-flops to keep track of an 8-bit binary count, or some other word of information. In a microcomputer, where we frequently deal with thousands of bits of information, it becomes impractical and uneconomical to use single flip-flop packages for memory.

In this chapter we will see how a single integrated circuit is used to store these thousands of bits, and how they are used to implement a microcomputer memory.

Figure 7-1 shows a breakdown of the various types of memory. We will study each in detail and also learn about their function in the microcomputer. We will see how power requirements, speed, and density all become important when it is necessary to decide which type of memory is to be used.

4 SHIFT REGISTER MEMORIES

We begin our study of memories with the shift register. In Chapter 5 we saw how shift registers are used to delay signals and aid in the hardware multiplication process. Although not particularly useful as storage devices for microcomputers, shift registers do have some applications. Their relatively slow speed and low density are not suited for computers, but these properties are ideal for applications that call for repetitive data. Consider the operation of a television typewriter in which ASCII* data from an ordinary electronic keyboard is displayed on a television set. To maintain a constant image on the television screen, the data required to fill the screen must be sent to the television set 60 times a second. By using a shift register, we are able to *recirculate* the data to give a fresh image every 60th of a second. The term "recirculate" refers to the method of feeding the output of a shift register back into the input.

In Fig. 7-2, we see how a simple shift register memory is made, and how, through a control switch, we are able to recirculate the data in the shift register. With the switch in the Load position, data present on the Data In line is loaded or shifted into the shift register. Once we have completed loading the data into

*ASCII is a standard communications code that is covered in Chapter 8. For now, think of it as a 7-bit code.

FIGURE 7-2 Simple shift register memory.

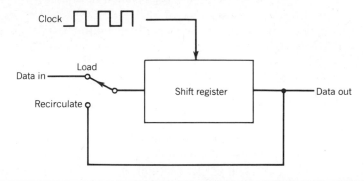

the shift register, we flip the switch into the Recirculate position. Now when we clock data out of the shift register, we load or recirculate the same data back into the shift register. By combining seven shift registers and the necessary recirculate logic, we can very easily make a memory able to store the ASCII data needed to make our TV typewriter function.

FIGURE 7-3 Seven 512-bit shift registers.

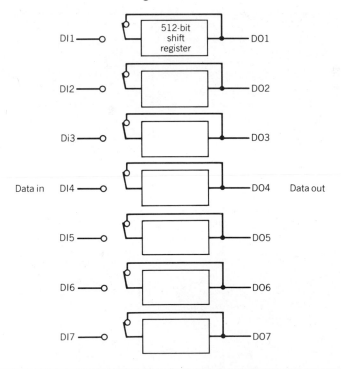

A TV typewriter may be capable of showing data on a television screen in a variety of formats. Let's assume that a 16-line display of 32 characters per line is desired. This means that $32 \times 16 = 512$ ASCII characters can be displayed. Since a television screen must be continuously updated to maintain a constant display (30 frames/sec, or 60 fields/sec), the ASCII data must be repeatedly available. Storage in shift registers is ideal because the data can be recirculated and is continuously available for display.

Seven 512-bit shift registers can be used to form the memory as shown in Fig. 7-3. The clock line of each register is connected to a common clock, to rotate all data at the same rate. This principle is used in many data terminals and shows an important application of a shift register memory.

TV typewriters are among the few devices that use shift registers. For another example, we can design a circuit that employs special recirculate logic to produce a random pattern of ones and zeros.

EXAMPLE 7-1

The 4-bit shift register of Fig. 7-4 is used to make a random pattern of ones and zeros. In this example, we are using a 4-bit serial in, parallel out shift register. We assume that the shift register is cleared to start (i.e., all outputs are low), and we will now see the effect of the exclusive NOR gate used to recirculate the output data.

FIGURE 7-4 Four-bit shift register for Example 7-1.

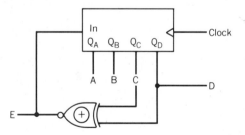

FIGURE 7-5 Initial state of shift register.

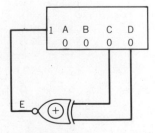

FIGURE 7-6 Shift register after first clock pulse.

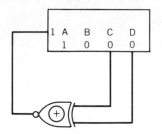

FIGURE 7-7 Shift register after two clock pulses.

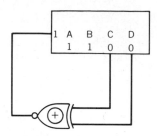

Initially (Fig. 7-5), with all the outputs at logic zero, the output E of the exclusive NOR gate will be a logic one. See Chapter 3 if you don't remember what a NOR gate is.

Since the output of the exclusive NOR gate is connected to the data input of the shift register, this one gets loaded into the register during the first clock pulse. When this happens, output A goes to a logic one, with B, C, and D unchanged (Fig. 7-6).

Since the C and D outputs did not change, the data input remains at logic one $(\overline{0 \oplus 0} = 1)$. This in turn gets loaded into the shift register during the second clock pulse (Fig. 7-7).

Now, after two clock pulses, the shift register outputs are 1 1 0 0. With one more clock pulse, they will be 1 1 1 0. Notice that outputs C and D are now different. The exclusive NOR of one and zero is zero! This zero gets clocked in on the fourth clock pulse. This sequence goes on and on and produces an output table like the one shown in Table 7-1.

This sequence repeats every 15 clock pulses. We know this to be true because

TABLE 7-1   Pseudo-random Sequence for a Four-Bit Shift Register

0 0 0 0	Initial state
1 0 0 0	1
1 1 0 0	2
1 1 1 0	3
0 1 1 1	4
1 0 1 1	5
1 1 0 1	6
0 1 1 0	7
0 0 1 1	8
1 0 0 1	9
0 1 0 0	10
1 0 1 0	11
0 1 0 1	12
0 0 1 0	13
0 0 0 1	14
0 0 0 0	15

the shift register (on the fifteenth pulse) goes back to 0 0 0 0. From here the sequence repeats again. By connecting the exclusive NOR inputs to different shift register outputs, different sequences can be made (they may not all be 15 words long, though).

Circuits like these (only containing more bits; hence, longer sequence lengths) are combined with summing amplifiers to make digital noise sources, music machines, and code scrambling devices.

5 STATIC VERSUS DYNAMIC

Before we continue with our discussion of memories, we take time out to define some important terms. When dealing with integrated memory devices, we fre-

FIGURE 7-8 (*a*) Static memory cell. (*b*) Dynamic memory cell.

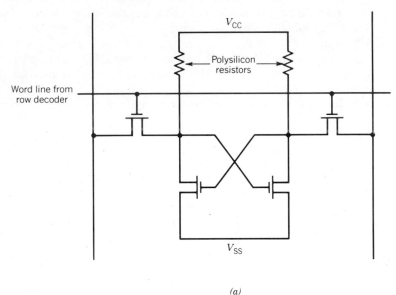

(a)

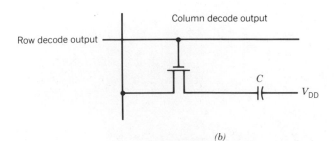

(b)

quently come across the terms *static* and *dynamic*. Static memory devices are composed of actual flip-flop packages, whereas dynamic devices contain capacitive elements for storage. The advantages of a dynamic device include its higher bit density, lower cost, and low power requirement. The static device has the advantage of increased speed. The speed of a memory is measured in terms of the time required to get data from a particular memory location. This is known as the *access time*. The access time for a static memory is as low as 20 nsec; the dynamic memory has an access time of 200–300 nsec. The designer is faced with making certain trade-offs when building a microcomputer system: high speed, but more power and space required; or lower speed, lower power, and less space. Larger systems seem to be moving toward dynamic memories favoring high density and lower power requirements.

Figure 7-8 shows a static memory cell (flip-flop) and a dynamic RAM cell (capacitor). The very small capacitors that make up the dynamic memory cell may be charged to represent a logic one and discharged to represent a logic zero. The capacitor cannot remain charged very long and must be recharged every 2 msec. The process is called a refresh operation, and dynamic RAM requires a separate refresh clock. The act of reading or writing can refresh the dynamic RAM, but separate refresh circuitry is usually used.

6 RAM AND ROM

Our study in microcomputer memories now takes us to the world of *random access memories* and *read only memories*. Unlike the shift register, which obliges us to clock the devices to get at the desired data, the random access and read only memories are organized as storage elements in sequential order. To get a piece of data out of a RAM or a ROM, we simply send the device a binary word that is actually an address of an internal storage area, and the device gives us our data.

RAMs and ROMs come in many sizes (number of bits or bytes that you can store) and are usually referred to as 256×4 bit, 1K by 1 bit (where $K = 1024$), or $1K \times 8$ bits. In the case of 256×4 bits, we have a memory consisting of 256 words of 4 bits each. To get at a particular word of 4 bits, we send the device one of 256 possible addresses, and the 4 bits stored at that location appear at the output.

In the $1K \times 8$ case, we can choose any of 1024 addresses to retrieve an 8-bit word (or byte). This kind of memory is usually referred to as a 1K byte RAM or ROM.

Now that we know a few basics, let's study the RAM in detail and see how we use it in a microcomputer memory.

7 RANDOM ACCESS MEMORY

For our study of the RAM, we will first look at an ideal RAM, then at a few of the most common RAMs used in microcomputer memory circuits. Notice the three RAM control functions shown in Fig. 7-9: Address, Read or Write, and Chip

FIGURE 7-9 An ideal 64-bit RAM.

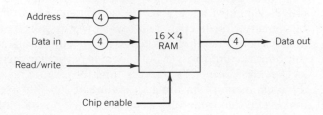

Enable. The chip enable function becomes *very* important when we combine many RAMs to make a large microcomputer memory. In most cases, this input to the RAM is called $\overline{CS}$, which means "not-chip select" or, essentially, that a logic zero must be present on this line for the chip to operate. (*Note:* $\overline{CS}$ is a digital signal and should not be confused with the usual power input pins to the device.) If this signal is not present, the outputs of the RAM usually tri-state (go to a high impedance) and are not seen any more, as far as other logic devices are concerned.

With the $\overline{CS}$ line low, the chip is prepared to accept address and data information and a read/write signal. In operation, the address is decoded inside the chip, and the particular storage element desired is selected. Once this has been done, the read/write line comes into play. Here we use only one line to carry out two functions. If we aren't *reading,* then we must be *writing.* In practical applications, this line is abbreviated $R/\overline{W}$, which implies that a logic one on this line will read RAM data and a logic zero will write RAM data. If we are reading, the data from the selected storage element or elements is sent to the output lines. If we are writing, the data present at the Data In lines is sent to the selected storage area, where it replaces the data stored there previously. Because the RAM is a static–volatile device, it will retain this data only as long as power is applied. Usually, upon power-up (the first application of power to a digital circuit), random patterns of ones and zeros appear in all RAMs, and data must be loaded into them before a useful memory is available to the user.

Thus far we have been looking at an "ideal RAM" where everything happens instantly. In other words, the desired data location was selected and either read or written into instantly after the chip was selected. Well, we know that *nothing* happens in zero time, and RAMs are no exception. We will now look at an actual memory chip and discover exactly what its time limitations are.

7.1 The 2102 1K × 1 Bit Static RAM

Figure 7-10 shows the connection diagram for a 2102 1K × 1 RAM and the internal block or logic diagram of the memory. The 10 address lines, A0–A9 are used to address or select any of 1024 ($2^{10} = 1024$) memory cells or locations. From the internal block diagram, we see that the 2102 RAM is organized as a memory cell array with 32 rows and 32 columns. The address lines are broken up into groups of five, with A0–A4 selecting one of 32 rows, and A5–A9 selecting one of 32 columns.

In addition to this, the input data control circuit, with the help of the column

FIGURE 7-10 (*a*) Pinout and (*b*) block diagram for the 2102.
(Copyrighted material reprinted with permission of National
Semiconductor Corporation.)

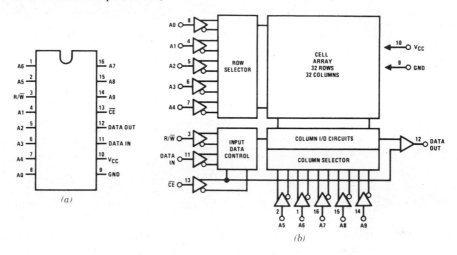

I/O circuits, controls the flow of data into and out of the RAM. Note that one
output of the $\overline{\text{CE}}$ buffer is used to control the data output. This control line tells
the data output buffer to either pass binary data or to tri-state.

Figure 7-11 shows the truth table for the 2102, which is essentially the same as
that of our ideal RAM. With the $\overline{\text{CE}}$ pin high, the output is tri-stated. With $\overline{\text{CE}}$
low, and R/$\overline{\text{W}}$ high, we do a "read" operation. With $\overline{\text{CE}}$ low and R/$\overline{\text{W}}$ low, too,
we perform a "write." Since we are already familiar with these operations, let's
study the timing constraints of the 2102.

Figure 7-12 shows the timing diagram for a read operation in the 2102 (R/$\overline{\text{W}}$ is
high). Two things must happen before we can get data out of the 2102. First, we
must set up a valid address, and *not* attempt to change it for the duration of the
read cycle. This length of time, called t_{RC}, is never shorter than 250 nsec in the
fastest 2102s made. (Many versions of the 2102 are on the market, all with varying
times of each cycle or setup time.) If we do change the address during this time
span, we cannot be sure that the data appearing on the output line is correct.

FIGURE 7-11 Truth table for a 2102 RAM. (Copyrighted material
reprinted with permission of National Semiconductor Corporation.)

$\overline{\text{CE}}$	R/W	D_{IN}	D_{OUT}	MODE
H	X	X	Hi-Z	Not selected
L	L	L	L	Write "0"
L	L	H	H	Write "1"
L	H	X	D_{OUT}	Read

FIGURE 7-12 Timing for a 2102 read operation. (Copyrighted material reprinted with permission of National Semiconductor Corporation.)

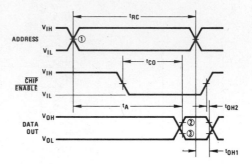

After our address has stabilized, we pull the $\overline{\text{CE}}$ line low and wait for the duration of the t_{CO} cycle, or the "chip enable to output" time, for our data to appear. This delay is a maximum of 100 nsec in fast 2102 memories. The access time T_A (the time to get data after presenting an address) is a maximum of 250 nsec. Human beings might have trouble understanding something that occurs in 250 *billionths* of a second, but computers have no trouble at all. Some computers today have to be slowed down to wait the 250 nsec for output data!

So much for taking data out of the 2102. Now let's see how we get data into the device. Figure 7-13 shows the necessary timing waveforms for a write cycle in the 2102. As for the read operation, we set up the address to be written into, and enable the chip. After waiting a minimum of 20 nsec (t_{AW}), we are allowed to pull the R/$\overline{\text{W}}$ line low, enabling a read operation. We must keep this line low for at least 100 nsec (t_{WP}) so that the internal circuitry has a chance to do its job. Notice from the timing waveforms that the data must be stable for a short time

FIGURE 7-13 Write cycle timing diagram for a 2102. (Copyrighted material reprinted with permission of National Semiconductor Corporation.)

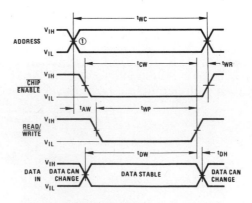

FIGURE 7-14 A 1K × 8 bit microcomputer memory.

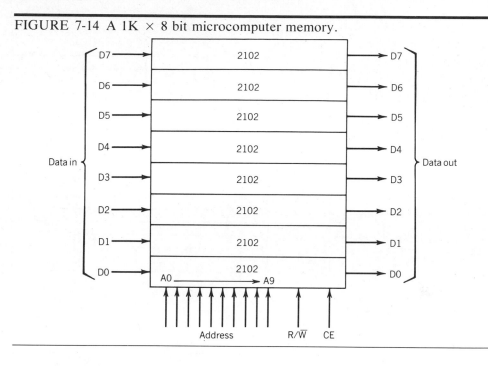

before and after both the onset and the completion of the write pulse. This ensures that the correct data gets stored at the desired location.

Now that we know how to read and write data using a 2102, let's see how this memory chip can be used to make a microcomputer memory. Remember that we need 8 bits of data at a time to run our microcomputer. We accomplish this by using eight 2102s in parallel. Each device gets the same address, R/$\overline{\text{W}}$, and $\overline{\text{CE}}$ lines. Only the data in and data out lines are separate, and they serve to form the eight input and output lines for our 1K × 8 (eight 2102s of 1K × 1 each yields a 1K × 8 bit memory, or a 1K-byte memory) bit microcomputer memory, as shown in Fig. 7-14.

The entire 1K × 8 bit block of memory circuitry has the same timing considerations as that of one 2102. We simply set up our address (and data if we are writing), enable the chip bank, and wait for our output data (or pulse the R/$\overline{\text{W}}$ line for a write operation).

We know that microprocessors can usually address 64K individual memory locations (16 address lines, $2^{16} = 65,536$). How then can we use 2102 RAMs to make 64K, 32K, or even 8K memory banks? The solution is simple and requires the use of the chip enable pin.

In Fig. 7-15 we see an 8K × 8 bit memory circuit that uses 2102 memory chips. There are eight groups of eight 2102s connected like the one in Fig. 7-14. In this circuit, all D0 input lines from the eight 1K-byte memory blocks are connected together, as are the D1–D7 lines. In addition, all the D0–D7 output lines are connected in parallel, too. Although this seems awkward, it actually works and is the way large computer memories are made from smaller units.

FIGURE 7-15 An 8K × 8 bit memory circuit using 2102s.

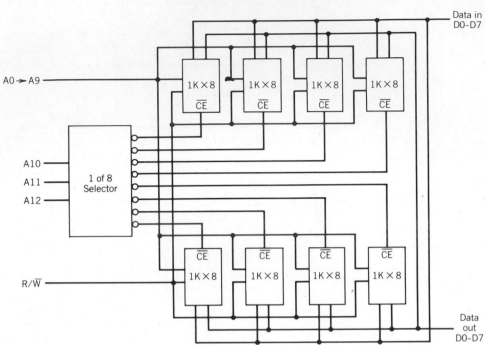

Remember that the output line of the 2102 is tri-stated if the $\overline{CE}$ line is high. Pulling the $\overline{CE}$ line low on any of the eight 1K blocks will enable that block and send out 8 bits of data on the data lines. Because the other seven blocks are tri-stated, they don't affect the 8 data bits on the output bus.

We decode one of the eight blocks by using the next three microprocessor address lines, A10, A11, and A12, and a one-of-eight demultiplexer chip. The one-of-eight decoder has three inputs and eight outputs (this works! $2^3 = 8$). The binary number present at the input tells the internal decoder circuitry to pull the selected output line low. This line then selects a group of eight 2102 memory

TABLE 7-2   Memory Map for the 8K RAM

Address	Function
0000-03FF	1st 1K of RAM
0400-07FF	2nd 1K of RAM
0800-0BFF	3rd 1K of RAM
0C00-0FFF	4th 1K of RAM
1000-13FF	5th 1K of RAM
1400-17FF	6th 1K of RAM
1800-1BFF	7th 1K of RAM
1C00-1FFF	8th 1K of RAM

FIGURE 7-16 Expanding 8K into 64K memory select.

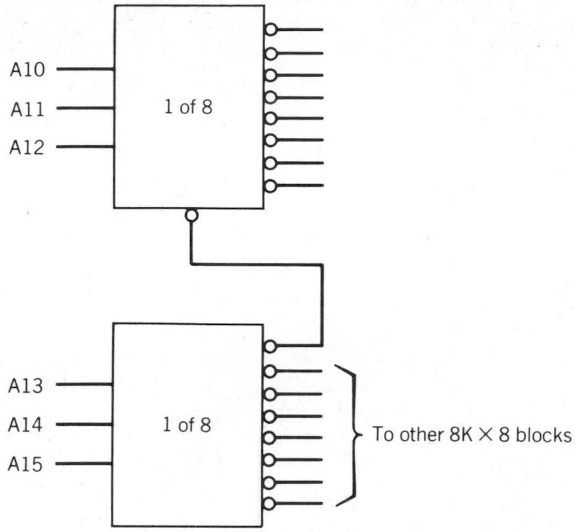

chips. Table 7-2 shows the memory map for the 8K RAM. A memory map is a table showing how the locations in a microcomputer memory are divided. They may be divided into areas for RAM, ROM, and I/O, and even left blank for future expansion. In Table 7-2, we have RAM from address 0000 through 1FFF (i.e., 8K locations). To build an even larger memory circuit for our microcomputer, we need modify our 8K × 8 memory circuit only slightly. Since we still have A13, A14, and A15 left, we can get eight more groups of 8K × 8 memory circuits to function. This puts us at 64K, which is the maximum our processor can handle (with 16 address lines).

In Fig. 7-16, we see how the basic 8K × 8 memory is modified to be selected as one of eight 8K-byte blocks of data. The 3 upper address bits A13–A15 enable one output of the one-of-eight decoder, which in turn enables the one-of-eight decoder on one of the 8K × 8 memory circuits.

When microcomputers were in their early stages, 1K × 8, 4K × 8, and 8K × 8 were the most common memory boards available, and most used the 2102 RAM. Today, 16K, 32K, and even 64K boards are the most common, as memory prices drop and the technology advances. Since it isn't practical or even easy to put 128 2102 memory chips together on a single board to make a 16K × 8 memory, we look for a chip with a higher bit density (and also a higher cost, unfortunately) to lower our package count. Our search continues with the 2114, the next most popular memory device.

7.2 The 2114 4K-Bit Static RAM

The 2114 is a 4096-bit RAM arranged as 1024 × 4 bits. The connection and block diagrams for the 2114 are shown in Fig. 7-17.

As with the 2102, we have 10 address lines (A0–A9), a $\overline{CE}$ line ($\overline{CS}$), and an

FIGURE 7-17 (*a*) Pinout and (*b*) block diagram for a 2114.
(Copyrighted material reprinted with permission of National
Semiconductor Corporation.)

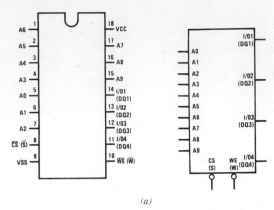

(*a*)

(*b*)

FIGURE 7-18 A 1K × 8 memory using the 2114.

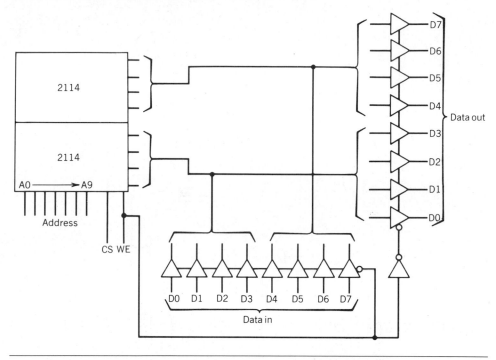

R/$\overline{\text{W}}$ ($\overline{\text{WE}}$). Unlike the 2102, we have four input/output lines. That's correct! The 2114 uses the same lines for input and for output. This is possible because we *never* read or write memory at the same time. We do one *or* the other. We need only add control circuitry to our memory logic to take care of this. Since the operation of the 2114 is very similar to that of the 2102 (except that it is slightly faster), we will learn right away how to use the 2114 as a microcomputer memory.

Figure 7-18 shows two 2114s connected to form a 1K × 8 memory. As with the 2102, we connect the A0–A9 address lines, the $\overline{\text{CS}}$ line, and the $\overline{\text{WE}}$ line to both chips. The outputs, however, serve two purposes. When $\overline{\text{WE}}$ is high, the outputs will contain *read* data. When $\overline{\text{WE}}$ is low, they must contain write data. To accomplish this, we use some commonly available buffers. The buffers are equipped with an enable line so that we may tri-state them when necessary, as we will see in our memory circuit. Since the buffer needs a zero to enable it, we see that the output buffers are disabled (or tri-stated) and the input buffers are enabled (simply pass information) when $\overline{\text{WE}}$ is low. Both buffers are never off or on at the same time due to the inverter on the enable line. When $\overline{\text{WE}}$ is high, the 2114 is set up for a read operation, and the output buffers are enabled (with the input buffers disabled). To make a larger memory, say, 16K by 8, we only have to set up some control circuitry for the $\overline{\text{CS}}$ line.

Figure 7-19 shows the circuitry for a 4K × 8 memory using 2114s. We utilize the next two address lines, A10 and A11, to select one of four 1K × 8 blocks. This may easily be expanded into a 16K memory with the addition of another one-of-four decoder on address lines A12 and A13.

FIGURE 7-19 A 4K × 8 memory using the 2114.

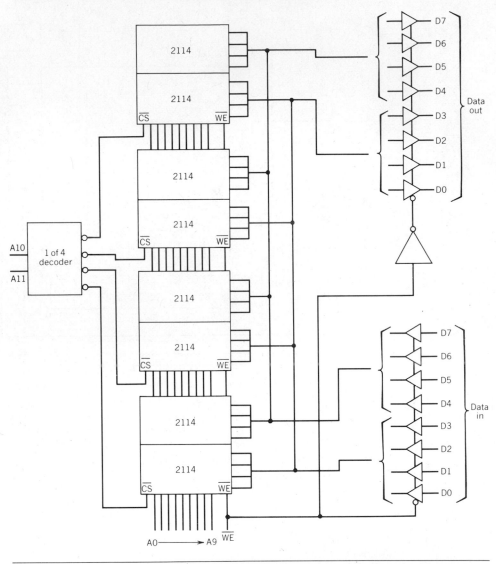

7.3 Newer RAMs

While this book was being written, RAMs were introduced that had much higher densities than existing random access devices. The 6116 (Fig. 7-20) is a 2K × 8 RAM with the same organization as the 2114 in that data comes in and goes out on the same lines. The 6116 offers a memory four times larger than the 2114 and 16 times larger than the 2102. Better yet, it is a CMOS chip, which means that it uses very low power. The advance in the semiconductor field constantly brings

FIGURE 7-20 Hitachi RAM. (Courtesy of Hitachi America, Ltd.)

HM6116FP-2, HM6116FP-3, HM6116FP-4

—Preliminary—

2048-word×8-bit High Speed Static CMOS RAM

■FEATURES

- High Density Small-Sized Package
- Projection Area Redueced to One-Thirds of Conventional DIP
- Thickness Reduced to a Half of Conventional DIP
- Single 5V Supply
- High Speed: Fast Access Time   120ns/150ns/200ns (max.)
- Low Power Standby   Standby: 100μW (typ.)
- Low Power Operation;   Operation: 180mW (typ.)
- Completely Static RAM: No clock nor Timing Strobe Required
- Directly TTL Compatible: All Input and Output
- Equal Access and Cycle Time

(FP-24)

■FUNCTIONAL BLOCK DIAGRAM

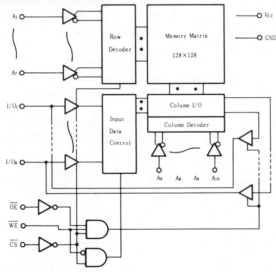

■PIN ARRANGEMENT

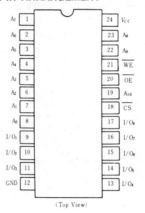

(Top View)

■ABSOLUTE MAXIMUM RATINGS

Item	Symbol	Rating	Unit
Voltage on Any Pin Relative to GND	V_T	−0.5* to +7.0	V
Operating Temperature	T_{opr}	0 to +70	°C
Storage Temperature	T_{stg}	−55 to +125	°C
Temperature Under Bias	T_{bias}	−10 to +85	°C
Power Dissipation	P_T	1.0	W

* V_{IN} min = −1.0V (Pulse Width ≦ 50ns)

■TRUTH TABLE

$\overline{CS}$	$\overline{OE}$	$\overline{WE}$	Mode	V_{cc} Current	I/O Pin	Ref. Cycle
H	×	×	Not Selected	I_{SB}, I_{SB1}	High Z	
L	L	H	Read	I_{cc}	Dout	Read Cycle(1)~(3)
L	H	L	Write	I_{cc}	Din	Write Cycle(1)
L	L	L	Write	I_{cc}	Din	Write Cycle(2)

newer, faster, and higher density memories to the market. At this time an 8K ×
8 RAM is under development and may now be available to you.

Thus ends our study of the random access memory. We now move on to a
different kind of memory, called the ROM, which is also useful in microcomputer
memories.

7.4 EPROMs, EAPROMs, PROMs, and ROMs

"EPROM" stands for an erasable programmable read only memory. In a programmable ROM (PROM), we put the information or data in just once. Every
time we read information from an EPROM after it has been programmed, we will
get the same data back. Even if power is turned off, and then back on again two
seconds, one day, even a year later, the same data remains in the EPROM. This
is possible by virtue of a process called *burning* in which the address and the data
to be stored are supplied to the EPROM and a high voltage is sent to a special
pin on the EPROM. This causes the internal circuitry to store the digital data as
an electrical charge, which is held permanently inside the memory. In most EPROMs,
this charge will remain indefinitely unless the EPROM is placed under an ultraviolet light for a short period of time (20 min). The ultraviolet light dissipates the
charge and essentially clears the ROM; hence the term "erasable".

The EAPROM (electrically alterable PROM) or EEPROM (electrically erasable
PROM, E^2PROM) is a new type of memory that can be erased by application of
a voltage instead of ultraviolet light. Either a specific location can be changed or
the entire memory cleared. The 2815 EEPROM is intended to be pin-for-pin
compatible with the 2716 EPROM (Fig. 7-21). With it comes the convenience of
a RAM-type memory, except that the data stored in the EEPROM is retained
when power is withdrawn from the device.

There are also ROMs that are programmable but not erasable. These ROMs,
simply called PROMs, may be burned only once. They are burned in the same

FIGURE 7-21 Block diagram and pinout for the 2815 EEPROM
(left) and the 2716 EPROM (right). (Reprinted by permission of Intel
Corporation, copyright 1983.)

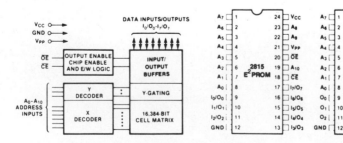

FIGURE 7-22 A character generator ROM.

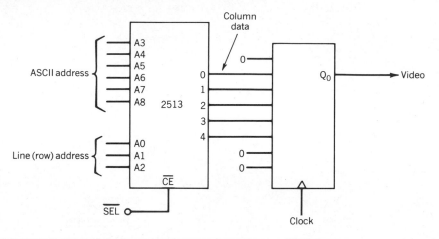

way as an EPROM except that instead of storing a voltage, the burn signal pops a small fuse at the desired storage location inside the PROM. Once popped, this fuse can never be fixed or replaced. So, if you make a mistake while burning a PROM, unless you can make a Christmas ornament out of it, you have to throw it out and start over with a new one. This is an advantage of an EPROM over a PROM, but EPROMs do cost more because they are erasable.

Least expensive is simply the ROM. ROMs are custom programmed at the factory by altering a mask in the design process. Like EPROMs and PROMs, the ROM retains its data with power off. Factory-programmed ROMs have a variety of functions and are very useful. They are used as character generators for TV typewriters (see Fig. 7-22), sine–cosine lookup tables for electronic trig function generators, code converters, and in many other applications. An EPROM or PROM may be used for specific applications that are desired by the computer hobbyist or industrial factory technician, as opposed to general uses.

We will now look at one of the most common ROMs available, and one of those best suited for microcomputers.

EXAMPLE 7-2

Figure 7-22 shows a 2513 character generator in a typical TV typewriter application. The 2513 contains all the letters of the alphabet, plus the numbers 0–9 and some special symbols, organized in a 5 × 7 dot matrix format. To display a character on a video display, each character is represented using a dot matrix. The 5 × 7 dot matrix of the 2513 offers the lowest resolution of available character generators, but it is simple to use. Other character generators offer better resolution of characters with 7 × 9, 9 × 11, and 11 × 13 dot matrices.

FIGURE 7-23 A character generator ROM. (*a*) A 5 × 7 dot matrix. (*b*) Outline for the 2513. (© 1984 Signetics Corp.)

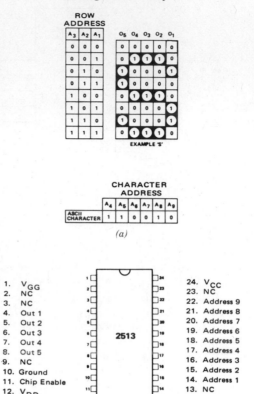

(a)

(b)

Figure 7-23 shows the 5 × 7 dot matrix (five columns and seven rows) and the 2513 outline drawing. The character generator contains the information to produce a character when an ASCII code is presented to the ROM. The inputs are the six address lines (ASCII data) and three row inputs. The circuitry that obtains information from the ROM must take it a row at a time. The 2513 character generator contains uppercase letters, and the entire contents of the ROM appear in Fig. 7-24. The ASCII code for a letter or symbol doubles as the letter (or symbols) address in the ROM (A3–A8). Output data is sent to an 8-bit shift register with the other three inputs set to zero to provide for blank spacing between characters.

EXAMPLE 7-3

In Fig. 7-25, a 74154 4-line to 16-line decoder is used to make a 16 × 8 bit ROM. The data are ordered in Table 7-3.

FIGURE 7-24 Contents of the 2513 ROM. (© 1984 Signetics Corp.)

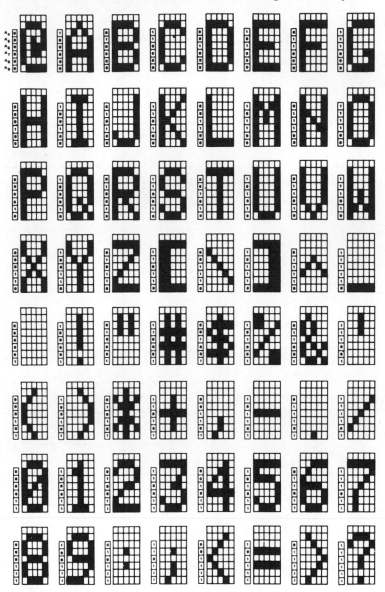

FIGURE 7-25 Diode matrix ROM for Example 7-3.

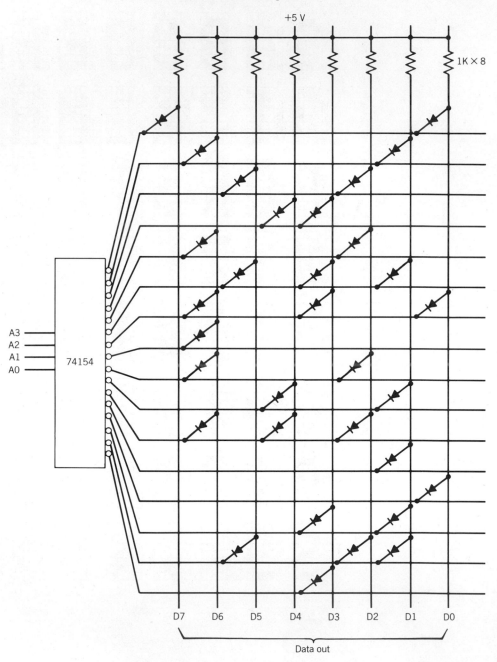

TABLE 7-3   Diode ROM Data

A3	A2	A1	A0	Data Out
0	0	0	0	7E
0	0	0	1	BD
0	0	1	0	DB
0	0	1	1	E7
0	1	0	0	BB
0	1	0	1	D5
0	1	1	0	B6
0	1	1	1	BF
1	0	0	0	BB
1	0	0	1	ED
1	0	1	0	AB
1	0	1	1	FD
1	1	0	0	FE
1	1	0	1	F5
1	1	1	0	D9
1	1	1	1	F7

When selected output goes low, it also pulls combinations of output lines low through the diodes connecting them. This type of memory is called a *diode matrix ROM* and is very useful when small ROMs are needed.

7.5 The 2708 1K × 8 Bit EPROM

Figure 7-26 shows the connection and internal block diagrams for a 2708 EPROM. Because it is a 1K × 8 ROM, it has 10 address lines (A0–A9) and 8 data output lines (O1–O8). The data output lines also double as the data input lines when the 2708 is being programmed or burned. The $\overline{CS}$ also has a double function, serving as the $\overline{WE}$ line during the burning process. The $\overline{CS}$ line serves the same purpose it does in RAMs: tri-stating the output lines when it is high, placing data on them when it is low. To read data from the ROM, we simply place the address for the desired memory location onto the address lines and pull the $\overline{CS}$ line low. The data appears on the data bus after a short delay (usually around 350 nsec) and is taken in by the microcomputer.

Figure 7-27 shows the switching time waveforms for reading the data from a location in a 2708 EPROM. Programming information into the EPROM is an involved operation that is an ideal task for a microcomputer. Each byte of data must be "burned" into each of the 1024 locations. This is done by tying $\overline{CS}$/WE high (write enabled), applying an address and byte of data to the EPROM, and applying a program pulse. The program pulse (26 V) must have an accumulated time of 100 msec. This may not be done with a program pulse width exceeding 1 msec. Hence the EPROM is programmed at least 100 times before the data is "burned," and this requires several minutes of computer time. Figure 7-28 shows the 2708 programming waveforms for this process.

The newer EPROMs require less external circuitry for level shifting and are easier to program. Let's look at the 2716.

FIGURE 7-26 (*a*) Pinout and (*b*) block diagram for the 2708
EPROM. (Reprinted by permission of Intel Corporation, copyright
1983.)

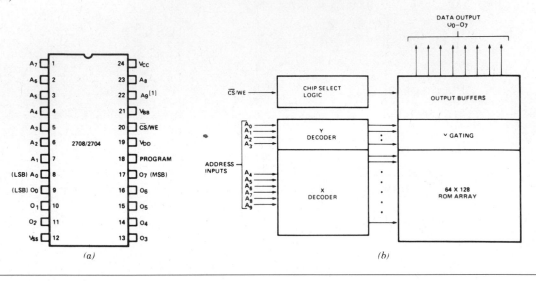

FIGURE 7-27 Timing waveforms for the 2708 read operation.
(Reprinted by permission of Intel Corporation, copyright 1983.)

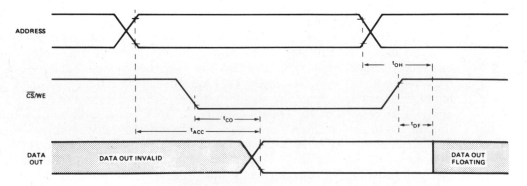

FIGURE 7-28 Burning a 2708. (Copyrighted material reprinted with permission of National Semiconductor Corporation.)

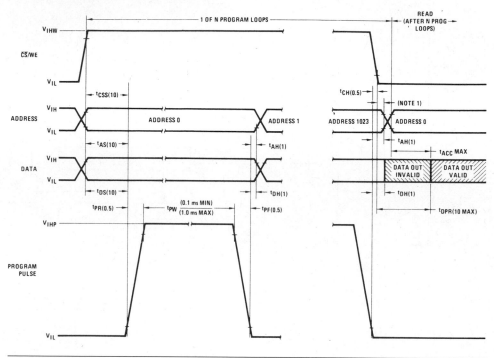

7.6 The 2716 16,384-Bit (2048 × 8) Ultraviolet Erasable PROM

The 2716 (Fig. 7-29) features twice the density of the 2708. The read operation is the same except that there is an additional address line to accommodate the increased memory locations. Programming is simpler in that each location is programmed completely before proceeding to the next. This requires 50 msec per location.

The 2716 EPROMs are full of logic ones when erased, and the programming process writes in zeros to the selected locations. An EPROM may be partially programmed by leaving "1"'s in unused locations. Additions may be made by reprogramming the original and also new data.

At this time, both 32K- and 64K-bit EPROMs are commercially available. Programming characteristics are similar to the 2708 and 2716.

FIGURE 7-29 (*a*) Pinout and (*b*) block diagram for the 2716
EPROM. (Copyrighted material reprinted with permission of
National Semiconductor Corporation.)

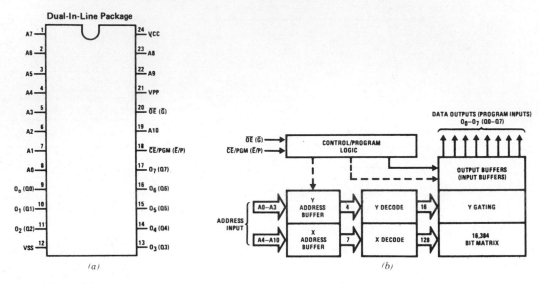

(*a*)   (*b*)

8 OTHER MEMORIES

In this section we will look at the other computer memories—some electronic,
some not. All have, or had, their uses in the computer field. Unlike semiconductor
memories, which have relatively low byte density, most of the devices discussed
here are very high density memories. We sometimes lose one thing to get another,
and in this case, to get higher densities we lose speed, money, or both.

8.1 Core Memories

Core memories are extremely small, donut-shaped pieces of ferromagnetic material. They are so small that hundreds of them would fit in a thimble (see Fig. 7-
30). They are usually arranged in a core plane similar to that of Fig. 7-30. Core
memories are used predominantly in large, general-purpose computers as high-
density, nonvolatile memory.

The ferrite core can be magnetized in either a clockwise or a counterclockwise
direction by a d-c current passing through a wire threaded through the center of
the core or "donut" (Fig. 7-31*a*, *b*). This is almost a mechanical type of memory
in that the state of a core is not affected if power is turned off for a period of
time. The principle by which these memories operate is based on a *BH* or magnetization curve. It takes a certain current to cause magnetization of a ferrite
material. If insufficient current is applied, the core cannot be affected. The min-

FIGURE 7-30 Closeup of an array of ferrite cores.

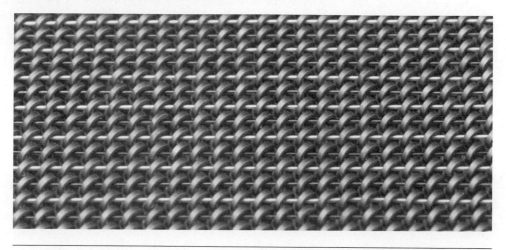

FIGURE 7-31 Magnetization of a ferrite core. (*a*) *BH* diagram. (*b*) Full current. (*c*) Half currents.

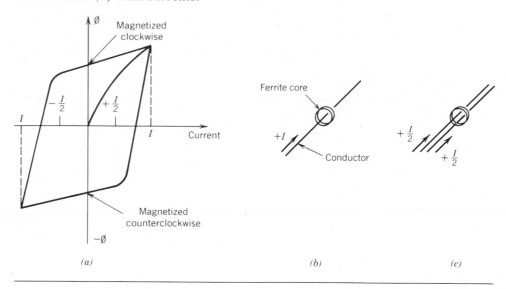

imum value of current needed to cause either magnetization or a change of magnetic state is called a full current. A current of insufficient amplitude to affect the core's state is called a half current.

The *BH* curve is actually a curve of flux or magnetization as a function of current. The core initially is demagnetized and the curve is just a point at the origin. As positive current is applied, a positive flux develops, pushing the point from the origin to the upper right-hand corner. When the current is turned off, the point returns to a point along the y or flux axis indicating magnetization in the clockwise direction. Further application of current ($+$) does not change or alter the present degree of magnetization. A negative current then can drive the point to a state of counterclockwise magnetization. A positive current will return the point to the top side of the curve. Once completed, the loop can be repeated over and over again. The point to note is that magnetization in either direction can be accomplished only by a "full" current. A half current is not sufficient to cause any change in magnetization.

The core represented in Fig. 7-31c shows a change of magnetization due to a full current on one wire. If two separate wires are used, we can obtain the effect of a full current by sending a half current down each wire. The magnetic fields add as though there were a full current on one wire! This is the key to addressing multiple cores.

Figure 7-32 shows a 16-core memory in a 4×4 arrangement or matrix. One X and one Y are needed to address a specific location. Thus, to address the first core in the upper left-hand corner, the X_1 and Y_1 address lines would be required.

Two wires, each carrying half the required current, have the same effect as a single wire carrying full current. The state of a core is determined by the voltage induced in a third wire called the sense line, when a change in magnetization occurs.

A core is addressed by sending a half current down the appropriate *pair* of address lines. This pulse of "full" current either will cause the addressed core to change state or will remain as it was. If a change of state occurs, a large change in magnetic flux results, inducing a pulse into the sense wire. If no change of state occurs, no change in magnetic flux results. There is no pulse produced in the wire.

FIGURE 7-32 A 4×4 arrangement of a 16-core memory.

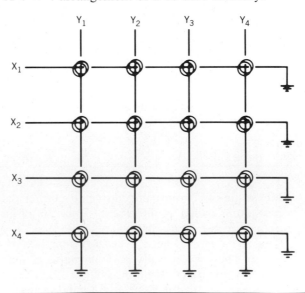

Let us assume that positive half currents are sent down the wire. These would magnetize the core in the clockwise direction. We interpret this direction to mean a binary "1." (A binary "0" would be represented by magnetization in the counterclockwise direction.) If the sense wire fails to produce a pulse, there has been no change in the direction of magnetization and we would conclude that a "1" was in the core. If the sense wire does produce a pulse, we conclude that the core was magnetized in the reverse direction representing a "0." This gives us a means of seeing what *was* in the core. Unfortunately, we have lost the information. If a zero was in the core, it was turned into a one during the read operation. We would have to rewrite the zero into the core. This type of memory is called a destructive readout (DRO) memory. The solid state memories studied were all of the nondestructive readout (NDRO) type.

This leads to the read/write operation. If in determining the contents of the memory, we destroy information, a write operation may be necessary. This then is called the "read/write" operation.

The output of the sense wire is extremely small. The area of the resulting pulse is integrated and measured to determine whether a "1" or a "0" was sensed. Special amplifiers (sense amplifiers) are used to convert to logic levels the small pulses on the sense wires.

This material on core memories is presented to contrast with the solid state versions. The core memories are complicated to work with because they require extensive circuitry to control the read/write operation along with the sense amplifiers. A 16-bit memory with eight address lines as shown in Fig. 7-32 would require only one sense wire threaded through each of the cores. Since only one core is addressed at a time, the sense wire can detect changes in all the cores. Such a 16-core matrix is called a core plane. Many planes would be used in a

FIGURE 7-33 Operation of the sense wire.

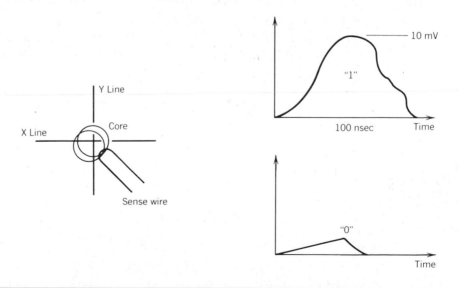

memory for a computer system to store binary words. For instance, 16 planes each with 16 cores would be stacked on one another to represent storage for 16 binary words of 16 bits each. The address lines from plane to plane would be "commoned" (i.e., all the X_1 lines tied together, etc.). One sense wire would be used for each plane (Fig. 7-33).

The operation of the sense wire depends on a change of magnetic flux in the ferrite core. When a half current is sent down each of the address lines (X and Y), a full current effect results. If the core changes state, a change in magnetic flux results that induces a small but measurable pulse in the sense wire. The area of this pulse—amplitude, in volts, × time—is determined by applying this to an electronic integrator. If the amplitude is above a minimum level, the conclusion is that a "1" existed. If the core does not change state, the pulse produced is much smaller, having minimal area. The integrator again measures the area of the pulse on the sense wire and concludes that it is too small to represent a change of state.

The integrator and a comparator circuit to produce either a logic one or zero level are contained in a "sense amp," which can be obtained commercially.

In reading the information out of a core, it is necessary to address the X and Y lines (see Fig. 7-34). This would normally write a one into the core. If the core was to have contained a logic zero, however, that piece of information would have been destroyed and must be written back into the core. If the core was to have contained a logic one, no problem. No need to write a zero in.

We will need circuitry to write a one or a zero into a core. Then, in reading the information, we must be prepared to correct lost data.

A circuit that could send a current either up or down a line in order to write either a one or a zero into a core might look like that in Fig. 7-35. The series resistor is selected to provide either a half or full current through the core, as required. If one wire is used, a full current is needed. If this represents only one address line, a half current will do. The other address line will apply the other half current.

The AND gates at either end of the wire threading the core will each have a zero out normally. To write a one, a current must travel the wire from left to right. A one from the left-hand AND gate and a zero from the right-hand AND

FIGURE 7-34 Sense amp circuit.

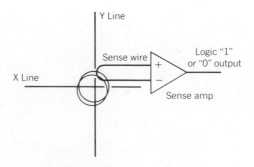

FIGURE 7-35 Circuit to write a "1" or a "0" into a core.

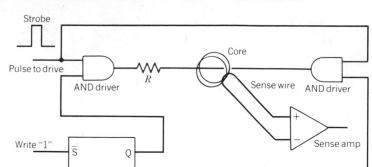

gate will accomplish this. The simple set or reset flip-flop turns one gate on at a time. The pulse to drive is applied after the "write one" or "write zero" signal has been established. The gates must be special drivers capable of holding a good one level or capable of accepting or "sinking" the current to ground (a good zero).

The process would be: select either write one or write zero and then pulse the strobe line. The strobe is a narrow pulse that indicates that the write operation is to take place.

Any number of cores could be on this single wire. A plane of cores could be made up by having this circuit drive the X_1 line with half current, and another circuit, threading cores for the Y_1 lines, could provide the other half current. Strobing both lines would write a one or a zero into the core.

Now let's return to the problem of "reading" the information in the core. In trying to read information, we will always write a one into a core. Hence, the beginning of the read operation consists of writing a one. If the sense amp produces a pulse, we know that we have changed the state of the core. It was a zero and must be put back. This calls for a write-a-zero operation. If the sense amp produces no pulse, we know that we have not changed the state of the core. That is, a one was stored, and no correction is needed.

Our procedure will be to write twice whenever a read operation is required. First we will write a one, then we will write a zero (assuming that we always try to correct the state of a core to a zero). If the correction is not required, we will *inhibit* the write-a-zero operation (Fig. 7-36).

Although the reading of the contents of a core is called a read/write operation, a better term would be "write/write operation." The read cycle:

1. Writes a one into the core.

2. Determines whether a one was already there.

3. If a zero was in the core, writes a zero in. If a one was in the core, inhibits the write-a-zero operation.

FIGURE 7-36 Inhibit operation.

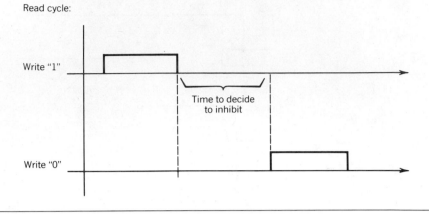

The inhibit operation is accomplished by threading yet another wire through the core (we're up to four wires now). This wire will have a half current passing through it for the inhibit. This negative current decreases the overall flux and prevents any writing operation. Some systems use the sense wire as the inhibit wire also.

The decision of whether to inhibit is made by the output of the sense amplifier. If the sense amp produces an output, the state of the core was changed and we must not inhibit the write-a-zero operation. If the sense amp produces no output, the state of the core was not changed and we must inhibit the write-a-zero operation and send a pulse down the inhibit wire to prevent the write-a-zero operation from having any effect.

FIGURE 7-37 The write-zero part of the read cycle.

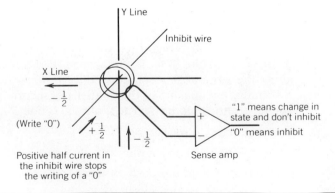

FIGURE 7-38 Circuit to inhibit core memory.

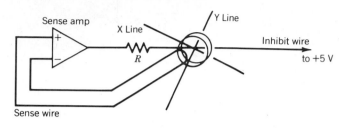

Figure 7-37 shows the write-zero part of the read cycle. Negative half currents are traveling down the X and Y lines. A positive half current up the inhibit wire would cancel the effect of the two negative half currents. This takes place only if the sense amp determines that it should. That is, a zero from the sense amp sends a negative half current down the wire; a one from the sense amp sends no current down the wire, and there is no inhibit operation. The core memory can be set up in any size required, but a fairly sizable circuit (Fig. 7-38) is needed to get it to function.

8.2 Magnetic Tape

The magnetic tape drive (Fig. 7-39) is essentially a very elaborate tape recorder. Binary data is stored on magnetic tape in the form of frequency bursts at very high densities—hundreds or even thousands of bits per inch. Because the unit is electronically controlled, the tape travels very fast across the R/W head and there are brakes on the disks that hold the tape reels. To keep the tape tight across the R/W head while the tape spins by at a high speed, the tape travels through an elaborate system of metal rollers and is also held in two vacuum chambers. This also keeps the tape from breaking when it reverses direction at high speeds. The tape has the advantage of being a high-density storage device, but is a very slow type of random access memory.

8.3 Bubble Memories

Bubble memories (Fig. 7-40) are one of the newest memory devices to hit the market. These low-power, high-density memories would be well suited for microcomputer use if they weren't so slow and cumbersome to use.

Bubble memories are set up as loops of magnetic bubbles rotated underneath a sensor. Control circuitry takes care of the rotating and addressing, but is somewhat slow. In one bubble memory system, the microprocessor can read or write only 18 bytes of data every 100 μsec. This is extremely slow compared to semiconductor memories. In addition, the bubble memory board circuitry needs advance warning of a computer power failure so that it can get all its bubble loops back in the right place; otherwise all data are lost. The bubble memory doesn't lose its data if powered off correctly, however, and it has the advantage of being useful as a type of reprogrammable ROM.

FIGURE 7-39 Magnetic tape drive.

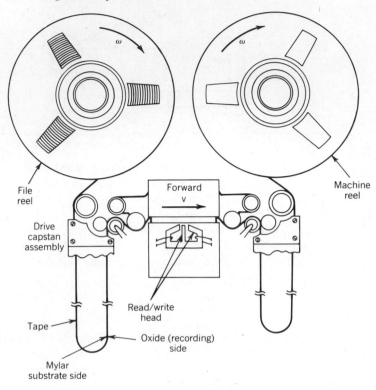

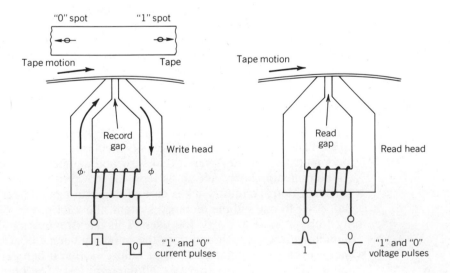

FIGURE 7-40 Bubble memory (GGG = Non-magnetic garnet substrate).

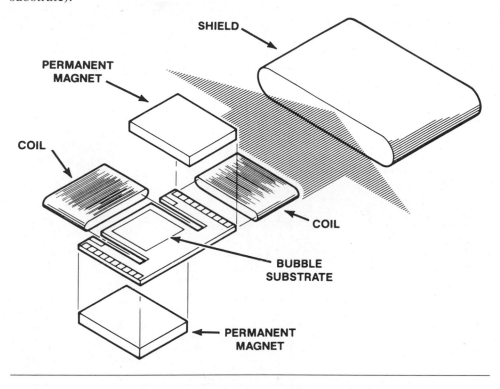

8.4 Disks

Disks, sometimes called hard disks, are magnetically coated devices similar to phonograph records. They sit on a spindle in the disk drive, usually stacked on each other separated by spacers, with a multiple read/write head bank able to move within them.

The disks spin so fast (thousands of revolutions per minute) that they create enough force to lift the read/write head and keep it suspended on a cushion of air created by the spinning action of the disk. Sometimes the R/W head hits the disk surface. This is known as a crash, and it is very bad for the disk. The usual result is a loss of data, sometimes a permanent loss.

While the disk is spinning, the R/W heads move in and out over the disks. Every minute increment or decrement of the R/W head selects a new track (a circular area on the disk where data is stored). On a hard disk, there are hundreds of tracks per side. Each track in turn contains many sectors (blocks of information), each containing many hundreds of bytes of information, stored in serial form. All told, it is not unusual for a hard disk to have a storage capacity of 10

FIGURE 7-41 A floppy disk drive.

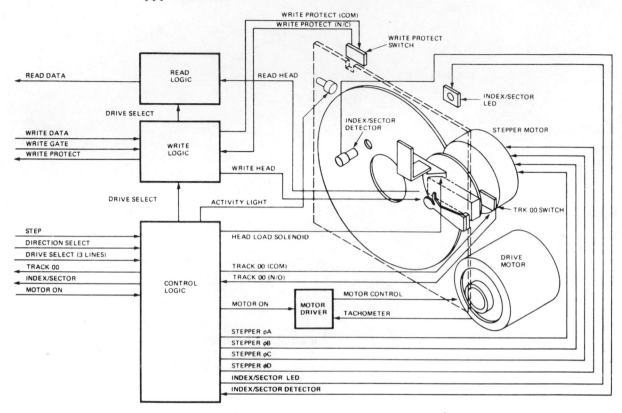

million to 200 million bytes! And to top all this off, the hard disk is fast! The only disadvantage is that even the cheapest hard-disk drives cost thousands of dollars, and the average computer hobbyist cannot afford one.

8.5 Floppy Disks

The last type of memory we will study is the floppy disk. The floppy diskette (Fig. 7-42) is made of a flexible material coated with a magnetic medium much like the hard disk. It is mounted inside the protective envelope with openings for the spindle and the read/write head. The floppy disk drive uses an opening for the sector hole to synchronize its operations.

Like the hard disk, the floppy diskette spins on its spindle, only at a much lower rate, usually 360 rpm. The floppy diskette is also broken up into tracks and sectors like the hard disk, although there aren't nearly as many. The two floppy diskettes popular today are the 5¼- and the 8-in. diskettes. (The diameter of the diskettes is given in inches.) Figure 7-41 shows a typical floppy disk drive.

The 5¼-in. diskette has usually 35 tracks with 10 sectors of 256 or 512 bytes per sector. A quick calculation shows that the total storage density of a 5¼-in.

FIGURE 7-42 Floppy diskette. There are 35 concentric tracks and
10 sector positions per track.

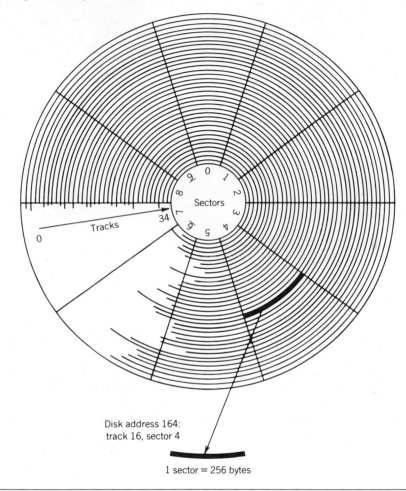

Tracks

0

34

Sectors

Disk address 164:
track 16, sector 4

1 sector = 256 bytes

diskette is around 90K bytes for the 256-byte sector and 180K bytes for the 512-byte sector. On the other hand, 8-in. diskettes have more tracks and sectors and can store around 360K bytes per diskette.

Data is stored in the form of frequency bursts on the magnetic surface. All the R/$\overline{\text{W}}$ circuitry is contained on the disk drive unit, and the computer has a controller board that controls the disk drive. Unlike the hard disk, which uses direct memory access (DMA: see Chapter 9) because it has such staggeringly high data transfer rates, the floppy disk is totally controlled by the computer and doesn't use DMA because of its slow speed. Most floppy disk drives can transfer 16K bytes of data to the computer in less than 3 sec, and the best thing about floppy disk drives is that almost everyone can afford them!

9 SUMMARY

Memory is a very essential part of every computer system. Fast memory is required to store both program and data information. Solid state RAM or ROM is used for this purpose. Core memory may also be used.

Slower types of memory, like tape, diskette, disk, or bubble, are capable of long-term program storage when power is removed from the computer. Generally, a library of programs is stored on slow memory. When a particular program is required, it is transferred to fast memory for use by the CPU.

Memory chips like the 2114 must be organized to form larger memories using decoders or selectors and data buffers. A program can then be stored and either executed or altered at the convenience of the programmer. Once a program has been judged complete and functional, it can be stored more permanently in a ROM.

10 GLOSSARY

Access time.   The time required by a memory to produce data after an address has been specified.

Bubble memory.   Magnetic bubbles in minor or major loops store binary data.

Burning (an EPROM).   Programming data into a ROM.

Character generator.   A special ROM that converts ASCII codes to dot matrix format.

Chip select or chip enable.   An input to a memory that connects the data lines to a data bus, hence activates the chip.

Core memory.   A magnetic memory shaped like a donut. Magnetization in the clockwise and counterclockwise directions represents a one or a zero, respectively.

Decoder.   A circuit to select one of several memory chips when an address is specified.

Disk.   A magnetic disk memory used in large computer systems.

Diskette.   Flexible disk magnetic memory in 5¼- or 8-in. diameter format.

Dot matrix.   Characters are represented on a display in a grid of dots of 5×7, 7×9, 9×11, or 11×13.

Dynamic memory.   A memory cell that is a charged or discharged capacitor that must be refreshed to maintain its charge.

EAPROM.   Electrically alterable programmable read only memory. Ultraviolet light is not needed to change data in a memory location.

EPROM.   Erasable programmable read only memory; programmed by user and erased by ultraviolet light.

Load/Recirculate.   The command whereby a shift register is first loaded with data and the data is then recirculated.

PROM.   Read only memory programmed by the user one time only.

Pseudo-random pulse generator.   A shift register circuit used to produce random bit sequences.

RAM.   Random access memory—read/write.

ROM. Read only memory; factory programmed.

Read/Write. Describing a memory that may be read from or written to.

Refresh. Dynamic memories must have the data stored in their small capacitive elements recharged or refreshed. The "Refresh" command triggers such recharging.

Selector. Same as decoder.

Shift register. A sequential memory that circulates stored data. The user must wait for data to come around the loop.

Static memory. A device in which each memory cell is a flip-flop.

Tape memory. Slow memory for long-term program storage.

TV teletype. A device that displays character information on an ordinary television screen.

PROBLEMS

7-1 A 32K memory is desired for use with an 8-bit microcomputer. If there are 32K memory locations:
  (a) How many locations are there actually?
  (b) How many *bits* are stored?

7-2 What is the purpose of a chip enable line?

7-3 (a) Does RAM lose data if power is turned off?
  (b) Does ROM?

7-4 Show how a single wire is used for read/write operations ($R/\overline{W}$).

7-5 How many bits are stored in (a) a 2102-type memory? (b) A 2114?

7-6 How many bytes may be stored on a 5¼-in. single-density, single-sided diskette?

7-7 How many address lines are required for a 4K × 8 memory (4096 × 8)?

7-8 How many address lines are required for a 2764-type memory? (64K bits = 8K locations.)

7-9 If a memory has 12 address lines, how many memory locations are there?

7-10 If a memory has 14 address lines, how many memory locations does it have?

7-11 If an 8K (8192) byte memory starts at address 2000 H, what is the address of its last location?

7-12 If a 16K (16,384) byte memory starts at address 5000 H, what is the address of its last location?

7-13 (a) How many address lines are needed for a 12K-byte memory?
  (b) If this 12K-byte memory starts at 0, what is the address of the last location?
  (c) If the memory starts at 5000 H, what is the address of the last location?

7-14 How many memory locations can be addressed by a microcomputer with (a) 16 address lines? (b) 20 address lines? (c) 24 address lines? (d) 28 address lines? (e) 32 address lines?

7-15 Can you think of applications for a computer that might require as many as 2^{24} memory locations?

7-16 Design a pseudo-random generator using a 5-bit shift register and an exclusive NOR gate. What output bits must be used to get a full 31-word sequence?

7-17 Repeat Problem 7-16 for an 8-bit shift register.

CHAPTER 8
DATA TRANSMISSION

1 INSTRUCTIONAL OBJECTIVES

When finished with this chapter you should be able to:

1. Describe the ASCII code and its use.
2. Convert 7-bit ASCII to 11-bit transmission code.
3. Explain the importance of "handshaking."
4. Show how modems and telephones are used in data transmission.
5. Describe the start, stop, and parity bits used in the 11-bit transmission code.

2 SELF-EVALUATION QUESTIONS

Keep the following questions in mind and try to answer them when you have completed the chapter:

1. Why is the "start" bit used in the 11-bit transmission code?
2. Why are CTS and RTS important?
3. What is the difference between a TTL and an RS232C transmitted code?
4. How are bits 6 and 7 used to separate the ASCII code into four groups of characters?
5. What is baud rate?
6. What are the advantages and disadvantages of serial and parallel transmission?

3 INTRODUCTION

To relay commands and data, it is essential to be able to communicate with a computer. Often this is done from a keyboard arranged in a manner similar to a typewriter. Each time a key is pressed, a binary code unique to that key is generated. The code is then transmitted in either parallel or serial form to the computer. Usually the code for the character is *echoed* back to the terminal by the computer so that it may be displayed on the screen of a computer terminal

FIGURE 8-1 Data terminal. (Courtesy of Applied Digital Data Systems Inc., a subsidiary of NCR Corporation.)

similar to that in Fig. 8-1. Two codes are in common use for the transmission of character data. These are the ASCII and EBCDIC codes, discussed in Sections 4 through 17, respectively.

4 THE AMERICAN STANDARD CODE FOR INFORMATION INTERCHANGE (ASCII)

The ASCII code (pronounced as-kee) is a 7-bit code for upper- and lowercase letters, digits, and special characters (!, $, %, etc.) and the control characters. Table 8-1 shows the 128-character ASCII code in binary format. The same code is given in Table 8-2 in a hexadecimal format.

TABLE 8-1   ASCII Code System and Character Set

Bits b7 b6 b5 →				0 0 0	0 0 1	0 1 0	0 1 1	1 0 0	1 0 1	1 1 0	1 1 1
b4	b3	b2	b1	COLUMN → ROW ↓	0	1	2	3	4	5	6	7
0	0	0	0	0	NUL	DLE	SP	0	@	P	`	p
0	0	0	1	1	SOH	DC1	!	1	A	Q	a	q
0	0	1	0	2	STX	DC2	"	2	B	R	b	r
0	0	1	1	3	ETX	DC3	#	3	C	S	c	s
0	1	0	0	4	EOT	DC4	$	4	D	T	d	t
0	1	0	1	5	ENQ	NAK	%	5	E	U	e	u
0	1	1	0	6	ACK	SYN	&	6	F	V	f	v
0	1	1	1	7	BEL	ETB	'	7	G	W	g	w
1	0	0	0	8	BS	CAN	(	8	H	X	h	x
1	0	0	1	9	HT	EM	)	9	I	Y	i	y
1	0	1	0	10	LF	SUB	*	:	J	Z	j	z
1	0	1	1	11	VT	ESC	+	;	K	[	k	{
1	1	0	0	12	FF	FS	,	<	L	\	l	|
1	1	0	1	13	CR	GS	–	=	M	]	m	}
1	1	1	0	14	SO	RS	.	>	N	∧	n	~
1	1	1	1	15	SI	US	/	?	O	_	o	DEL

▢ PRINTABLE CHARACTERS   ▨ PRINTER CONTROL CHARACTERS

▦ CODES GENERATED AND TRANSMITTED BY THE TERMINAL. BUT NO ACTION IS TAKEN.

USASCII CONTROL CHARACTERS
(From USA Standards Institute Publication X3.4–1968)

ACK	acknowledge	ETX	end of text
BEL	bell	FF	form feed
BS	backspace	FS	file separator
CAN	cancel	GS	group separator
CR	carriage return	HT	horizontal tabulation
DC1	device control 1	LF	line feed
DC2	device control 2	NAK	negative acknowledge
DC3	device control 3	NUL	null
DC4	device control 4 (stop)	RS	record separator
*DEL	delete	SI	shfit in
DLE	data link escape	SO	shift out
EM	end of medium	SOH	start of heading
ENQ	enquiry	STX	start of text
EOT	end of transmission	SUB	substitute
ESC	escape	SYN	synchronous idle
ETB	end of transmission block	US	unit separator
		VT	vertical tabulation

*not strictly a control character

TABLE 8-2   Hex-ASCII Table

00	NUL	21	!	42	B	63	c
01	SOH	22	"	43	C	64	d
02	STX	23	#	44	D	65	e
03	ETX	24	$	45	E	66	f
04	EOT	25	%	46	F	67	g
05	ENQ	26	&	47	G	68	h
06	ACK	27	'	48	H	69	i
07	BEL	28	(	49	I	6A	j
08	BS	29	)	4A	J	6B	k
09	HT	2A	*	4B	K	6C	l
0A	LF	2B	+	4C	L	6D	m
0B	VT	2C	,	4D	M	6E	n
0C	FF	2D	-	4E	N	6F	o
0D	CR	2E	.	4F	O	70	p
0E	SO	2F	/	50	P	71	q
0F	SI	30	0	51	Q	72	r
10	DLE	31	1	52	R	73	s
11	DC1 (X-ON)	32	2	53	S	74	t
12	DC2 (TAPE)	33	3	54	T	75	u
13	DC3 (X-OFF)	34	4	55	U	76	v
14	DC4	35	5	56	V	77	w
15	NAK	36	6	57	W	78	x
16	SYN	37	7	58	X	79	y
17	ETB	38	8	59	Y	7A	z
18	CAN	39	9	5A	Z	7B	{
19	EM	3A	:	5B	[	7C		
1A	SUB	3B	;	5C	\	7D	}
1B	ESC	3C	<	5D	]		(ALT MODE)
1C	FS	3D	=	5E	$\wedge$ ($\uparrow$)	7E	~
1D	GS	3E	>	5F	$-$ ($\leftarrow$)	7F	DEL
1E	RS	3F	?	60	`		(RUB OUT)
1F	US	40	@	61	a		
20	SP	41	A	62	b		

Let us look at the ASCII code for the capital letter A.

$$\text{ASCII ``A''} = \underset{\underset{\text{MSB}}{\uparrow}}{1}\ 0\ 0\ 0\ 0\ 0\ \underset{\underset{\text{LSB}}{\uparrow}}{1} = 41\ \text{H} = 65\ \text{D}$$

The code can be expressed in binary, hex, or decimal. The right-most bits represent a count with A = 41 H, B = 42 H, C = 43 H, and so on. Knowing that A is 41 H, the other capital letters can be determined without having to consult a reference. For instance, L is twelfth letter of the alphabet.

$$12\ \text{D} = \text{CH} = 1100\ \text{B}$$

Therefore,

$$L = 64\ \text{D} + 12\ \text{D} = 76\ \text{D} \quad \text{and} \quad L = 40\ \text{H} + \text{CH} = 4\text{CH}$$

and

$$L = 1000000 \ B + 1100 \ B = 1001100 \ B$$

The two most significant bits of the ASCII code (b_6, b_7) determine the type of characters being transmitted. You can verify this by consulting Table 8-3. The control characters are generated by the keyboard when the user simultaneously depresses a control key and a letter key. The control characters ($b_6 = 0$, $b_7 = 0$) and the capital letters ($b_6 = 0$, $b_7 = 1$) are very similar, the only difference being b_7.

TABLE 8-3 Effect of b_6, b_7

b_7	b_6	Character
0	0	Control characters
0	1	Digits, symbols
1	0	Capital letters
1	1	Lowercase letters

Depressing the control key forces b_6 and b_7 low. We can see from Table 8-1 that the letter A is in the same row as SOH (Start of Heading). The letter next to A is Q, which is in line with DC1 (Device Control 1). This juxtaposition points out that you can overlay the capital letter columns on the control columns. Table 8-4 lists some of the more important control codes. The symbol ↑ (up arrow) is used to indicate a control character. Most keyboards have a backspace key and a "new line" (carriage return) key. However, these functions may also be generated by depressing ↑ H and ↑ M, respectively.

TABLE 8-4 Important Control Codes

Function	Control Character	
BEL	↑ G	(Bell)
BS	↑ H	(Backspace)
LF	↑ J	(Line feed)
CR	↑ M	(Carriage return)
DC2	↑ R	(Device control 2)
DC4	↑ T	(Device control 4)
ETX	↑ C	(End of text)

5 THE PARITY BIT

Often in the transmission of data, errors occur. For example, startup of large motors or other electrical equipment generates noise that may interfere with the code being transmitted. The addition of an eighth bit to the ASCII code attempts to determine whether such an error has been produced. Figure 8-2 shows the addition of an unwanted "1" to the code for the letter "A." In this case, bit 4 is

FIGURE 8-2 Noise during data transmission of an "A."

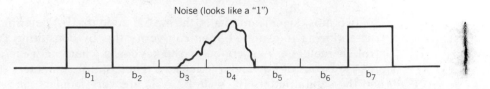

received as a one. The code 1 0 0 1 0 0 1 is received. An "A" is sent, but a letter "I" is received. The *parity* bit would catch this type of error. The idea is to use the eighth bit (parity) to indicate that the transmitted character will contain either an even number or an odd number of "1"s. If we assume that a computer was designed for even-parity checking, the parity bit is established as a zero or a one, as necessary to make the number of "1"s an even number. Since an "A" already has an even number of ones, the even-parity bit is zero. Table 8-5 lists a few characters and their parity bits.

TABLE 8-5   Parity Bits

Character		P_{even}	P_{odd}
A	1 0 0 0 0 0 1	0	1
B	1 0 0 0 0 1 0	0	1
G	1 0 0 0 1 1 1	0	1
L	1 0 0 1 1 0 0	1	0
W	1 0 1 0 1 1 1	1	0

If the transmitted parity bit is supposed to be even, and an odd number of ones is received (as in the "A" transmission example), it is assumed that an error has occurred and an error message will be issued. In noncritical applications, many computers do not bother to check the parity of received data. Most terminals, however, generate either an even or an odd parity bit and can also test incoming data for correctness.

To sum up this section, an eighth bit is added to the 7-bit ASCII code. This bit is called the parity bit, and the new code is sometimes referred to as 8-bit ASCII.

6 TRANSMISSION METHODS

The 8-bit ASCII data may be transmitted in two ways.

1. All 8 bits may be transmitted at once on eight separate lines. Such *parallel* transmission is used only for high-speed data transfer. Parallel transfer of data is very common *inside* computers. In this case the parallel paths are referred to as *busses*.

2. The 8 bits may be transmitted one at a time over a single wire. Serial transmission is used for lower data rates and has the advantage of taking only one

FIGURE 8-3 (*a*) Parallel ASCII transmission (all bits at once) and (*b*) serial ASCII transmission (one bit at a time).

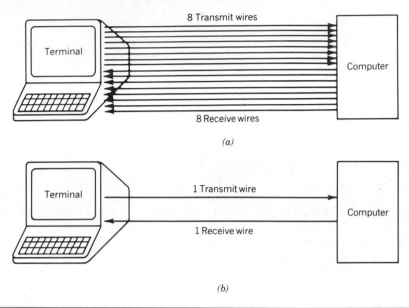

line. Keyboard data generally are transmitted serially to avoid having to run eight wires between a terminal and a computer. This would be a waste of wire for the low-speed data character information usually represented. Figure 8-3 illustrates serial and parallel transmission.

7 ELEVEN-BIT TRANSMISSION CODE

The 11-bit transmission code is used to transmit serial data. The line starts at a logical high level. If no character is being transmitted, the line is a one. The transmission begins with the line dropping low for a *start* bit. See Table 8-6 and Figure 8-4. The data bits follow the start bit beginning with the LSB (b_1), then the parity bits, and finally the *stop* bits.

FIGURE 8-4 Eleven-bit transmission waveform.

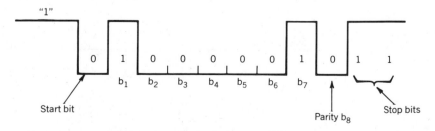

TABLE 8-6   Eleven-Bit Transmission Code

Bit	Code
1	Start bit (always = 0)
2	Data bit b_1 (LSB)
3	Data bit b_2
4	Data bit b_3
5	Data bit b_4
6	Data bit b_5
7	Data bit b_6
8	Data bit b_7 (MSB)
9	Parity bit b_8
10	Stop bit (always = 1)[a]
11	Stop bit (always = 1)

[a] May be 1, 1½, or 2 stop bits.

There are two ways to transmit serial data: synchronously and asynchronously (Fig. 8-5). The 11-bit transmission code is an asynchronous method, since the receiver uses the start bit to synchronize itself with the incoming data. Synchronous data transmission requires both a data line and a clock line. The separate clock line is used for synchronization and allows for higher data rates.

The 11-bit transmission code is sometimes modified in the following ways:

1. Stop bits: may be 1, 1½, or 2.

FIGURE 8-5 (*a*) Asynchronous and (*b*) synchronous data transmission.

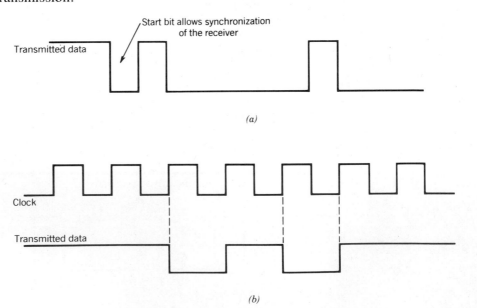

(a)

(b)

TABLE 8-7   Sample ASCII Characters

ASCII Letter	Even Parity	11-Bit Waveform
R = 1010010	1	
↑ G = BEL = 0000111	1	
5 = 0110101	0	

2. Parity: may be odd or even; may be mark (1) or space (0); or they may be eliminated.

3. Data bits: for non-ASCII data may be set from 5 to 8.

A common method of transmission is 8 data bits, no parity, and one stop bit. Let's look at the three characters in Table 8-7.

The next subject is the *rate* at which we transmit these bits. This is referred to as the baud rate, named for J. E. (Emile) Baudot, who did much work on an earlier code. Standard rates for transferring serial data are 110, 300, 600, 1200, 2400, 4800, 9600, and 19,200 baud. If data is transmitted at 300 baud (loosely 300 bits/sec), the time per bit is 1/300 sec or 3.333 msec. Since there are 11 bits used to define a transmitted character, the time to transmit a character is 11 $\times$ 1/300 = 11/300 or 36.6666 msec. If one character can be transmitted in 11/300 of a second, then at top speed a maximum of 300/11 or 27.2727 characters can be sent per second using the 11-bit transmission code. Table 8-8 lists these characteristics for the different baud rates.

The rate at which data is transmitted is frequently a matter of who is using the data. If a person is working at a data terminal, 1200 or 2400 baud generates data on the screen at a comfortable rate—not too fast, not too slow. High-speed data transfer (e.g., 4800 baud and above) is reserved for applications that do not involve display terminals read by human beings.

TABLE 8-8   Characteristics of Different Baud Rates

Baud	Time/Bit (msec)	Time/11 Bits (msec)	Maximum Number of Characters/Second
110	9.09	100.0	10
300	3.333	36.6666	27.2727
600	1.6666	18.333	54.5454
1200	0.8333	9.1666	109.09
2400	0.41666	4.5833	218.18
4800	0.20833	2.2916	436.36

8 RS232C: THE ELECTRONIC INDUSTRIES ASSOCIATION STANDARD

Standards for data transmission are set by the Electronic Industries Association (EIA). Standard RS232C establishes a means of transmitting ASCII data between a terminal and a computer. This includes the actual pin connections on a 25-pin connector common to most data terminals.

The first consideration is the problem of transmitting a TTL waveform down a long cable or line. The line looks like a capacitor, and a series of pulses (0 to +5 V) tends to charge the line. The result is a high level at the receive end (see Fig. 8-6).

A solution to this charging problem is to convert the one-sided waveform to a wave swinging plus and minus. The RS232C waveform converts the one-sided TTL signal (Fig. 8-7a) to a plus and minus swing and also inverts the waveform as shown in Fig. 8-7b. The voltage swing must be between ±3 and ±25 V. Generally, this voltage is selected based on what may already be available in the terminal. The normal state of an RS232C transmit line is a negative voltage when no transmission is in progress.

FIGURE 8-6 TTL waveform distortion on a line.

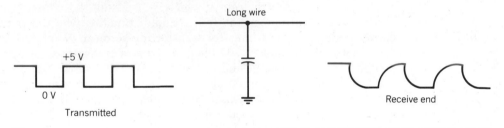

FIGURE 8-7 Serial data waveforms. (*a*) ASCII "A" in TTL and (*b*) RS232C "A."

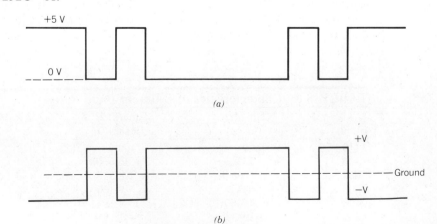

9 HANDSHAKING AND LINE PROTOCOL

Many applications involving a data terminal and a computer need only three wires: transmit, receive, and ground. In other applications, a computer may be too busy to receive data from a terminal at any random moment. In these cases it is

FIGURE 8-8 Connections for handshaking between the terminal and a computer.

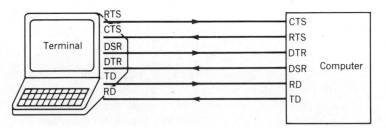

TABLE 8-9   EIA Standard DB25 Connector Pin Assignments

Pin Number	Circuit	Description
1	AA	Protective ground
2	BA	Transmitted data
3	BB	Received data
4	CA	Request to send
5	CB	Clear to send
6	CC	Data set ready
7	AB	Signal ground (common return)
8	CF	Received line signal detector
9	—	(Reserved for data set testing)
10	—	(Reserved for data set testing)
11		Unassigned
12	SCF	Secondary received line signal detector
13	SCB	Secondary clear to send
14	SBA	Secondary transmitted data
15	DB	Transmission signal element timing (DCE source)
16	SBB	Secondary received data
17	DD	Receiver signal element timing (DCE source)
18		Unassigned
19	SCA	Secondary request to send
20	CD	Data terminal ready
21	CG	Signal quality detector
22	CE	Ring indicator
23	CH/CI	Data signal rate selector (DTE/DCE source)
24	DA	Transmit signal element timing (DTE source)
25		Unassigned

necessary to establish a procedure or protocol for transmitting and receiving RS232C information. This is an electronic agreement or request to send information to the computer. If the computer is available and can receive information, then a Clear to Send signal is issued. Figure 8-8 demonstrates.

In the connections for "handshaking" between the terminal and a computer, the four signals involved are:

CTS	Clear to Send	DSR	Data Set Ready
RTS	Request to Send	DTR	Data Terminal Ready

Data Set Ready and Data Terminal Ready are used to determine essential conditions—whether the terminal or computer is turned on, whether a printer is out of paper, and so on.

The use of protocol requires four additional wires between a terminal and a computer. In noncritical applications it represents a high cost in additional cable and this "handshaking" may be dispensed with. Table 8-9 shows the standard connections for a DB25 connector. This standard is a part of the EIA RS232C document.

10 CABLE LENGTH

The length of an RS232C cable is limited by the baud rate. Longer lengths of line require a lower baud rate. This results from the effect of capacitance on the waveform at increased data rates. The EIA standard limits the capacitance of a line to 2500 pF (see Fig. 8-9 and Table 8-10).

TABLE 8-10   RS232C Parameters

Parameter	Value for RS232C
Line length (recommended maximum—may be exceeded with proper design)	50 ft
Input Z	3–7kΩ
	2500 pF
Maximum frequency	20K baud
Transition time (time in undefined area between "1" and "0") tr = 10–90%	4% of bit period or 1 msec
dV/dt (wave shaping)	30 V/msec
Mark (data "1")	-3 V
Space (data "0")	$+3$ V
Output Z	3–7kΩ
Open-circuit output voltage, V_0	$3\ V < |V_0| < 25$ V
$V_t = $ loaded V_0	$5V < |V_0| < 15$ V
	3–7kΩ load
Short-circuit current	500 mA
Power-off leakage	$> 300\Omega$
(V_0 applied to unpowered device)	$2\ V < |V_0| < 25$ V V_0 applied
Minimum receiver input for proper V_0	$> \pm\ 3$ V

FIGURE 8-9 Effect of capacitance on RS232C signal. (*a*) Desired
waveform. (*b*) Effect of capacitance.

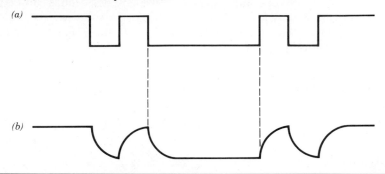

When a pulse that represents a bit is transmitted down a line, the edges of the
pulse begin to round off. The probability of correctly reading a bit deteriorates
as the line length increases. Table 8-11 lists recommended line lengths for given
data rates. These figures vary widely depending on the type of wire in use and
other considerations. In a particular application there may be no problem in
transmitting data at 9600 baud over lengths of 1000 ft. This is because receiver
circuitry always looks in the center of each bit position and does its best to
correctly distinguish between a zero and a one.

TABLE 8-11   Recommended Line Lengths for Given Data Rates

Baud Rate	Maximum Cable Length (ft)
300	4,000
1,200	1,000
2,400	500
4,800	250
9,600	125
19,200	65

11 SIMPLEX, HALF DUPLEX, AND FULL DUPLEX

In the serial transmission of data between two points, three kinds of (data) links
may be made. A link is the connection between the terminal and the computer.
A link that allows the transmission of data in only one direction is referred to as
a *simplex* data link. A two-way data link is called a *duplex* data link. In a *full
duplex* data link, data may be transmitted in both directions at the same time; for
this arrangement, a terminal may transmit to a computer and simultaneously
receive information from the computer. Most modern computer systems operate
in full duplex. A half duplex data link is also a two-way link, but data may not
be transmitted in both directions at the same time. One unit must be finished
transmitting before the other one can start.

12 CONVERTING BETWEEN TTL AND RS232C

The conversion from TTL to the RS232C signal can be done with a simple transistor or with an integrated circuit like the MC1488 designed for that purpose. Also, the MC1489 is designed to do the reverse conversion (RS232C to TTL). Figure 8-10 shows these circuits.

13 CURRENT LOOP TRANSMISSION

The RS232C standard involves transmitting voltage levels down a line. An older method still in use transmits a pulse of constant current instead. A 20- or 60-mA

FIGURE 8-10 TTL/RS232C Conversion circuitry. (*a*) Transistor TTL to RS232C. (*b*) 1488 Driver. (*c*) Transistor RS232C to TTL. (*d*) 1489 Receiver. (Copyrighted material reprinted with permission of National Semiconductor Corporation.)

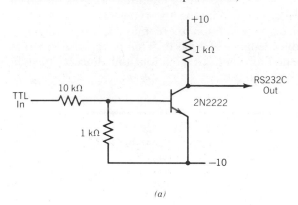

(a)

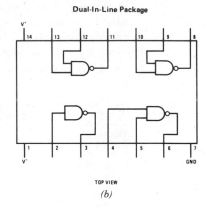

(b)

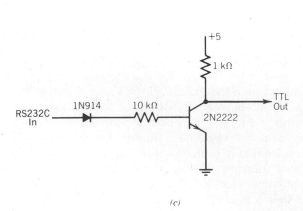

(c)

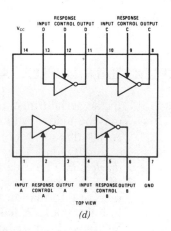

(d)

pulse is used to represent a binary "1" or mark. The 20-mA current is commonly used over the 60-mA current. A zero or space is represented using no current. This method was popular when many Teletype machines were used to receive the same information, as in news services like the Associated Press. The machines could be connected in series over a large area extending hundreds of miles. As long as the current pulse was constant, all machines received the data through a decoding relay. This was a half duplex data link. This method of transmission was quite slow (110 baud) because of the long cable lengths, but nonetheless very reliable. (A lot of news got delivered around the world!) Today this type of information is directed to individual radio and television stations directly by satellite, each station having its own receiver dish aimed at a stationary satellite. How things have changed!

14 DECODING ASCII CONTROL CHARACTERS

Certain control characters require decoding for many applications. For instance, frequently the BEL or Control G code is used to sound or alert an operator of computers. For decoding it, we need only design a simple black box. This can be done all at once or in two stages. Figure 8-11 shows both methods. The second uses the idea that control characters have both b_6 and b_7 at zero. Method 2 uses a separate gate to decode the control (CTL) function. This is useful if other control characters must be decoded—having two wires decoded saves on the number of input gates required.

Once the Control G (BEL) signal has been decoded, a sounding device can be activated. This signal can be used to drive a 555 timer connected as a single shot (pulse stretcher) driving a Sonalert. The Sonalert produces a piercing tone when a d-c level is applied. Figure 8-12 shows the circuitry.

The 555 timer is an integrated circuit capable of producing pulses to be used in oscillator applications; it also can serve as a one shot in this type of application

FIGURE 8-11 Control G decode. (*a*) Method 1 and (*b*) Method 2.

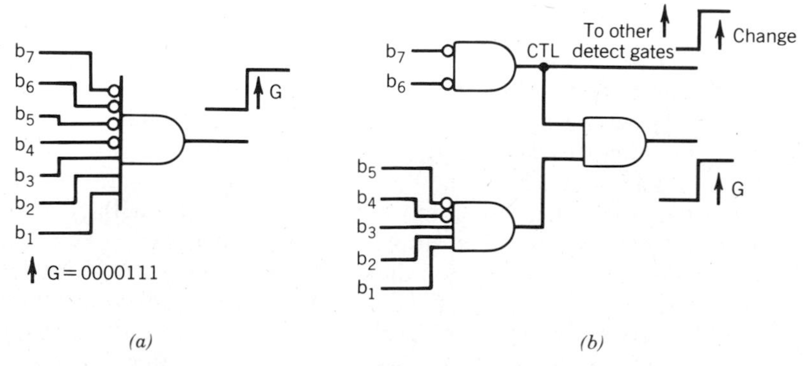

FIGURE 8-12 Control G (BEL) sounding circuit (SA = Sonalert).

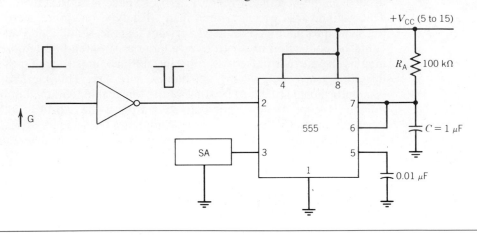

where a very narrow pulse is widened to allow a 100-msec tone at the Sonalert. Appendix C contains more information on the 555 timer.

15 GENERATING AN RS232C WAVEFORM USING A DIGITAL BLACK BOX

To further our understanding of the principles of data transmission, let's generate the ASCII code for the letter A at 300 baud. This signal can then be connected to a data terminal for display. Since we will drive the black box with a binary counter, the character will repeat over and over again.

EXAMPLE 8-1

Design a four-input DBB to produce the RS232C waveform for an "A."

SOLUTION

The code in ASCII for "A" is 1000001. Since there are an even number of ones, an even-parity system for error checking would make the parity bit a zero, thus keeping the number of ones an even number.

The standard method of data transmission begins with a start bit, followed by the seven data bits and the parity bit, ending with two stop bits. The start bit is always a zero. The stop bits are always logic "1"s. The normal state of the transmission line when data is not transmitted is a one level. Figure 8-13 shows the correct coding for the complete transmission of the letter "A."

We will need a digital black box to produce this pattern of bits at the correct rate to transmit an "A" to the computer or receiver unit. The design begins with a truth table and is simplified with a Karnaugh map (Fig. 8-14).

FIGURE 8-13 Coding for complete transmission of the letter "A" (SB = start bit, PB = parity bit).

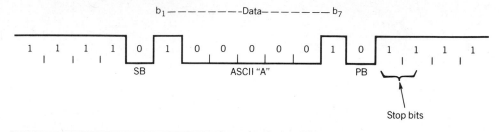

FIGURE 8-14 Design of digital black box to transmit an "A." (*a*) Truth table. (*b*) Karnaugh map.

A	B	C	D	f
0	0	0	0	1
0	0	0	1	0
0	0	1	0	1
0	0	1	1	0
0	1	0	0	0
0	1	0	1	0
0	1	1	0	0
0	1	1	1	0
1	0	0	0	1
1	0	0	1	0
1	0	1	0	1
1	0	1	1	1
1	1	0	0	1
1	1	0	1	1
1	1	1	0	1
1	1	1	1	1

(*a*)

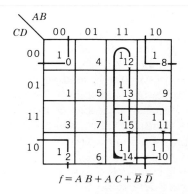

$$f = A\,B + A\,C + \bar{B}\,\bar{D}$$

(*b*)

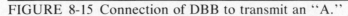

FIGURE 8-15 Connection of DBB to transmit an "A."

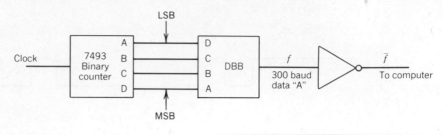

The DBB when constructed will follow the truth table and when driven by a binary counter will produce the correct waveform. The counter must be driven by exactly 300 Hz to produce 300 baud for the data terminal. In addition, the logic must be inverted to send $\bar{f}$ to the computer, since the transmission of this data is in negative logic (+5 V is a zero logical level). When properly "tuned," the terminal should show repetitive A's on the printout, because the counter keeps recycling. See Fig. 8-15.

16 LARGE-SCALE INTEGRATED RECEIVER/TRANSMITTER CHIPS

Digital black box design can be very helpful in experimenting with and using the RS232C standard for LSI receiver/transmitter chips. However, it is useful only in the transmission of a particular ASCII character. The use of a parallel load shift register greatly simplifies the conversion of an ASCII code in parallel to a series of binary bits for data transmission. The code is loaded in parallel and then clocked out at the desired baud rate.

In practice, and in microcomputer systems, this transmission and reception is accomplished in a specialized receiver/transmitter chip (Fig. 8-16). Several are available and in common use. Each contains a pair of separate shift registers. One is for transmitting an ASCII character (parallel to serial) and one is for receiving an ASCII character (serial to parallel).

In the example, the receiver is receiving an ASCII "A" in 11-bit transmission code. The transmitter is sending an ASCII "J." This illustrates the separateness of the receiver and transmitter sections. A character can be received at the same time as a different character is being transmitted. The receiver/transmitter (R/T) chip works with TTL levels and not the plus/minus swing of RS232C. That conversion must be done external to the chip using the converters (TTL to RS232C, etc.) previously mentioned. Both the transmitter and receiver sections have separate clock inputs (RxC = Receiver clock, TxC = Transmitter clock). These clock inputs determine the baud rate at which data is received or transmitted. Depending on the type of R/T used, the clock frequency is either 16 or 64 times

FIGURE 8-16 Receiver/transmitter chip.

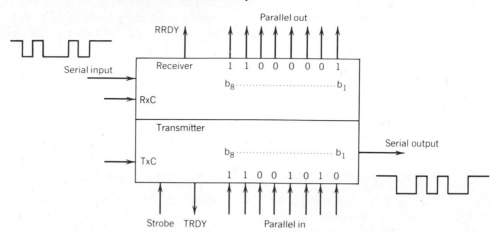

the desired baud rate. For example, if the desired baud rate is 300, the clock might be 16 × 300 or 4800 Hz.

Two status signals called "flags" are also available. These are the Receiver Ready Flag (RRDY) and the Transmitter Ready Flag (TRDY). When a character has been completely received, the RRDY flag is active (high in most applications). When the transmitter is available for use, the TRDY flag is active (high in most applications).

In addition, the transmitter has a strobe input, which is used to tell it that 8 bits of parallel data are ready to be entered and transmitted in serial.

The receiver/transmitter chip is of great value in data communication with terminals and computers. It automatically sends and receives serial data. Several common R/Ts are listed in Table 8-12.

All the chips are programmable. This means that certain aspects of the 11-bit transmission code may be changed. These involve the number of data bits, the number of stop bits, and the type of parity. The choices are: 5, 6, 7, or 8 data

TABLE 8-12   Large-Scale Integrated Circuit Receiver/Transmitters

Receiver/Transmitter (Manufacturer)	Name	Software Programmable
AY5-1013A (General Instruments)	UART: universal asynchronous receiver transmitter	No (hardware programmable)
8251 (Intel)	USART: universal synchronous–asynchronous receiver transmitter	Yes
6850 (Motorola)	ACIA: asynchronous communications interface adapter	Yes

bits; 1, 1½, or 2 stop bits; and odd, even, mark, space, or no parity. Typically, 8 data bits, 1 stop bit, and no parity are selected.

This allows a 10-bit transmission code after the addition of the standard start bit. At a given baud rate, character information transmits faster in 10 bits than 11 bits. At 300 baud:

10 bits: 300/10 = 30 characters/sec

11 bits: 300/11 = 27.27 characters/sec

The use of an R/T chip greatly reduces the circuitry needed for data transmission and simplifies design. The use of the R/T chip will be covered in more detail later in connection with microcomputers.

To return for a moment to the operation of the R/T, we make good use of the RRDY and TRDY "flags." To transmit a character:

1. Check the TRDY flag. If the transmitter is busy (not ready), *wait* until it is available.

2. When the transmitter is ready (available), place the data (ASCII and parity bits) on the parallel input and pulse the strobe line.

3. Data is transmitted automatically at a rate determined by TxC.

To receive a character:

1. Check the RRDY bit. It indicates when a character is received.

2. When RRDY is active, get the data from the parallel output.

You see, the R/T operates pretty much automatically!

17 THE EXTENDED BINARY CODED DECIMAL INTERCHANGE CODE (EBCDIC)

Although ASCII is an extremely popular code that is used widely in data communication and with microcomputers, another code exists and is used primarily in IBM (International Business Machines Corporation) equipment. This is an 8-bit code (in contrast with ASCII's 7 bits). The EBCDIC code uses 8 data bits and encodes data characters using zeros and ones, just as ASCII does. However, the codes are *not* the same. For instance, an ASCII space is 20 H while an EBCDIC space is 40 H. Some of the codes are compared in Table 8-13.

TABLE 8-13   Comparison of Some Codes

Character	ASCII	EBCDIC
Space	20 H	40 H
A	41 H	C1 H
B	42 H	C2 H
J	4A H	D1 H
S	53 H	E2 H
Carriage return	0D H	15 H

TABLE 8-14   EBCDIC Uppercase Letters

Character	Code	Character	Code	Character	Code	Character	Code
Space	40					0	F0
A	C1	J	D1			1	F1
B	C2	K	D2	S	E2	2	F2
C	C3	L	D3	T	E3	3	F3
D	C4	M	D4	U	E4	4	F4
E	C5	N	D5	V	E5	5	F5
F	C6	O	D6	W	E6	6	F6
G	C7	P	D7	X	E7	7	F7
H	C8	Q	D8	Y	E8	8	F8
I	C9	R	D9	Z	E9	9	F9

Clearly, the codes are quite different. Moreover, with 8 bits, 256 character codes are possible (as opposed to 128 with 7 bits). Table 8-14 shows the EBCDIC codes for uppercase letters.

18   COMMUNICATION OVER TELEPHONE LINES

Frequently data between a terminal and a computer must be communicated over standard telephone lines. In this case, the RS232C waveform is converted to pairs of tones. The device that performs this conversion is called a modem. The zero and one information is then transmitted using one frequency to represent a zero and another to represent a one.

The two pairs of frequencies in use—one pair to transmit and one pair to receive—are listed in Table 8-15. The use of four tones allows the simultaneous transmission and reception of data over a standard two-way (duplex) telephone line.

You will notice that two groups of four tones are shown, originate and answer. In the originate group, we transmit ones and zeros at 1270 and 1070 Hz, respectively. We receive ones and zeros at 2225 and 2025 Hz. This is the opposite of the answer group. This is because for the user, who *calls* the computer, we assign originate frequencies, and to the computer, which *responds,* we assign answer frequencies. Thus two modems (Fig. 8-17) are necessary. Most computer installations (at schools, banks, airports) employ such modulate/demodulate devices,

TABLE 8-15   Pairs of Frequencies

	Frequencies (Hz)	
Data	Originate	Answer
Transmit "0"	1070	2025
Transmit "1"	1270	2225
Receive "0"	2025	1070
Receive "1"	2225	1270

FIGURE 8-17 Originate and answer modems.

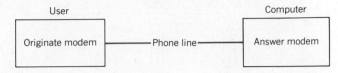

or modems. With this kind of connection, the user's transmit frequencies are the same as the computer's receive frequencies, and vice versa.

Sometimes the modem is wired directly to a telephone line and sometimes an acoustic coupler is used. The acoustic coupler uses the telephone's own mouthpiece and earpiece to couple the tones to the telephone line. Figure 8-18 shows an acoustic coupler, while Fig. 8-19 shows the data link in its entirety.

Since the frequency response of a telephone line is generally restricted to audio frequencies, this limits the rate at which binary data in tone form may be transmitted between two points via telephone. Typically data is transferred at 300 baud. More expensive modems can transmit at 1200 baud, but this type frequently costs more than a typical data terminal. The price for speed and accuracy is

FIGURE 8-18 (*a*) Acoustic modem. (Photo by John Young.)

always high! In any event, because the modem is a low-speed device, the transmission of high-speed data calls for such nonmodem media as direct wire, microwave, or laser communication.

FIGURE 8-18(b) Typical modem connections.

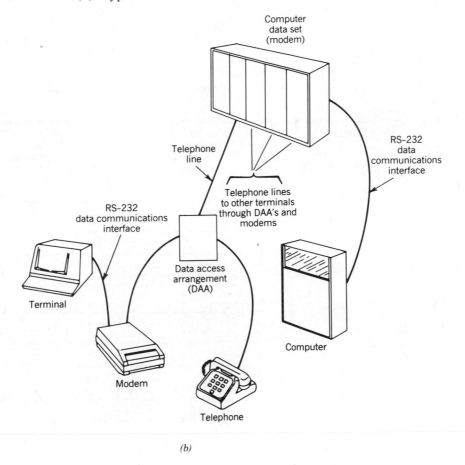

(b)

FIGURE 8-19 Duplex data link.

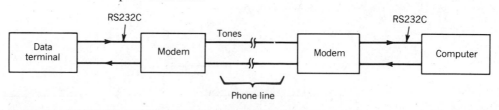

19 THE ELECTRONIC INDUSTRIES ASSOCIATION RS422 STANDARD

The RS232C standard specifies conditions for serial data transmission over relatively short distances using what is referred to as a single-sided or unbalanced line. Each signal has a single wire and the return is a ground conductor. Single-ended systems are prone to noise pickup, which becomes more severe with increasing distance. Telephone signals have been transmitted over a ''pair'' of wires for many years. Neither of the wires is ground; in fact, grounding of one of the wires introduces a large amount of a-c (60-Hz) hum into the signal.

A system that transmits signals over two wires is said to be ''balanced'' when there is no ground present. When one conductor is positive, the other is negative. Since neither wire has a ground reference, noise is induced equally in each wire

FIGURE 8-20 Connections for the RS422.

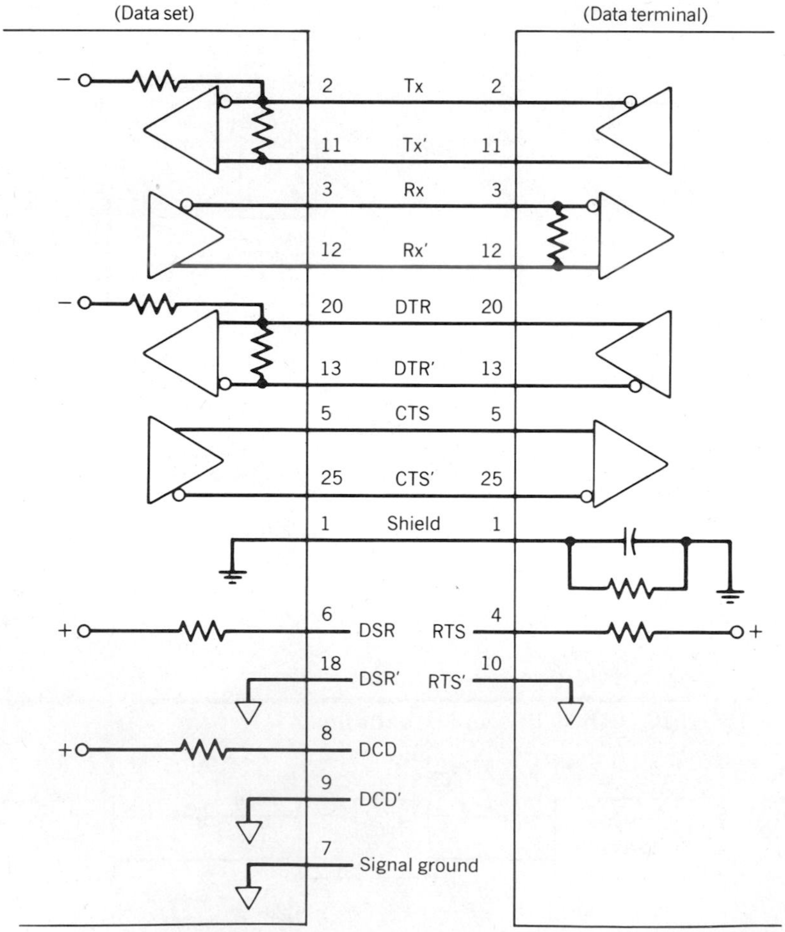

of the pair. The potential between the wires due to the noise then is zero. A balanced wire system can cover larger distances with great immunity to induced noise.

The EIA RS422 standard defines the balanced version of RS232C. A pair of wires is needed for each signal. The added wire in each case becomes the return line for the signal. These signals may include:

Transmitted Data	TxD
Transmitted Data Return	TxD'
Received Data	RxD
Received Data Return	RxD'
Request to Send	RTS
Request to Send Return	RTS'
Clear to Send	CTS
Clear to Send Return	CTS'
Data Set Ready	DSR
Data Set Ready Return	DSR'
Data Terminal Ready	DTR
Data Terminal Ready Return	DTR'
Signal Ground (one wire)	

One disadvantage of RS422 is the increased number of wires required. However, when distance and speed are essential with noise immunity a concern, RS422 defines an answer. Some connections for the RS422 are shown in Fig. 8-20.

20 SUMMARY

In this chapter we have learned about methods of transmitting data. The ASCII code was discussed extensively and a means of transmitting the code serially was described in detail. This method relies on the RS232C standard, which is widely used in the most modern data terminals. The transmission of data over telephone lines is another important factor in data communication.

21 GLOSSARY

ACIA. A type of receive/transmit integrated circuit (see Table 8-12).

ASCII. The (7-bit) American Standard Code for Information Interchange.

Answer. Name of the modem used by computer to communicate with user whose call it answered.

Asynchronous transmission. Serial transmission of data in which a start bit is used to synchronize the receiver clock.

Baud. Rate at which serial information is transmitted.

Bus. Parallel paths used for transmitting data at high speeds from any of several sources to any of several destinations.

Current loop. Method of serial transmission in which current pulses, not voltages, are sent on a wire.

DB25.   DB25P and DB25S are the standard plug and socket connectors used in serial data transmission facilities.

Data link.   The connection between a terminal and a computer. It can be direct wire, telephone, satellite, laser, or fiber optic.

Data terminal.   Electronic device used to communicate with a computer. Terminals may either print output on paper (like a typewriter) or display it on a television screen or custom display device.

EBCDIC.   The (8-bit) Extended Binary Coded Decimal Interchange Code.

EIA.   The EIA is the national trade association representing a full range of manufacturers in the electronics industry. (For more information, contact the Electronic Industries Association, Type Administration Office, 2001 Eye Street, N.W., Washington, DC 20006.)

Echo.   To transmit back to a display terminal.

Eight-bit ASCII.   ASCII code with the addition of a parity bit.

Eleven-bit transmission code.   Universal code used to transmit and receive serial data.

Even parity.   The number of ones transmitted in a particular word must be even.

Flags.   Signals or bits used during handshaking to determine the transmittal or reception of a character.

Full duplex.   Simultaneous transmission of data in two directions.

Half duplex.   Transmission of data in two directions, but only one direction at a time.

Handshaking.   Signals used by both the computer and the terminal to control the flow of information back and forth between them.

Mark.   Indicates a high level in serial transmission.

Modem.   Modulator/demodulator device used in the transmission of serial data over telephone lines.

Odd parity.   The number of ones transmitted in a particular word must be odd.

Originate.   Name of the modem used by the person calling a computer over telephone lines.

Parity.   Method used to help detect errors (during the transmission or reception of data).

RS232C.   A standard used in serial data communications.

Simplex.   Transmission of data in only one direction.

Space.   Indicates a low level in serial transmission.

Start bit.   Bit or signal used in serial transmission to indicate the beginning of new information.

Stop bit.   Bit or signal used to indicate the end of a serial transmission.

Synchronous transmission.   Serial transmission of data requiring a separate clock signal.

UART.   A type of receive/transmit integrated circuit (see Table 8-12).

USART.   A type of receive/transmit integrated circuit (see Table 8-12).

PROBLEMS

8-1   Write down the 8-bit ASCII code for the following characters using an even-parity bit:

(a) F		(f) R	
(b) J		(g) V	
(c) M		(h) ↑ G	
(d) n		(i) ↑ T	
(e) e		(j) <CR> (carriage return)	

8-2 Sketch the 11-bit transmission code (even parity) for each of the characters from Problem 8-1.

8-3 At 1800 baud, determine the maximum number of characters per second that can be transmitted using the 11-bit transmission code.

8-4 If only one stop bit is used, a character may be transmitted more quickly, since only 10 bits are required per character. Repeat Problem 8-3 using a 10-bit transmission code.

8-5 At 9600 and 19,200 baud, what are the times per bit required?

8-6 What is the maximum recommended length of cable for data transmitted at 9600 baud?

8-7 Design a parallel input circuit to decode the following control functions:
   (a) DC1   (c) Any control character (use NOR logic)
   (b) DC2   (d) ETX

8-8 Design a digital black box that will continuously generate an ASCII * (asterisk).

8-9 Can a single-parity bit be used to detect the change of more than one bit? Why or why not?

8-10 Why do UART's need clocks 16 or 64 times faster than the baud rate being used?

8-11 Fill in the following ASCII code table in the various bases.

ASCII	Binary	Hexadecimal	Decimal	P_{even}
L				
m				
!				
(ESC)				
LF				
H				
#				
6				
(EOL)				

8-12 Sketch the 10-bit RS232C waveform for each of the following ASCII characters (1 stop bit).
   X

   P

   7

8-13 Fill in the following table (10 bits = 1 character).

Baud Rate	Time/Bit	Time/Character	Maximum Number of Characters/sec
150			
400			
800			
900			
1000			

8-14 Draw a circuit to latch an LED in the on state when a control X character is received.

8-15 Design a digital black box to transmit a string of Ws.

8-16 Decode each of the following RS232C waveforms. (Use a ruler.)

(a)

(b)

(c)

8-17 Can a ROM (or a PROM) be used to convert ASCII to EBCDIC? If so, explain how.

8-18 Design a circuit with a minimum of TTL chips to decode the first eight control characters (ASCII 00-07). (*Hint:* Use a three-line to eight-line chip.)

8-19 Repeat Problem 8-17 but decode the first 16 control characters.

8-20 An exclusive OR gate can be used to compute parity. For two inputs A and B we know that the output will be low only when A equals B. In this fashion, the output represents "even" parity. If the inputs are 0,0 or 1,1 then the output is low. Can you design a circuit that uses these gates to compute the parity bit for a 7-bit input?

CHAPTER 9
ORGANIZATION OF COMPUTERS

1 INSTRUCTIONAL OBJECTIVES

When finished with this chapter you should be able to:

1. Identify:
   (a) The four main parts of a computer system.
   (b) The different busses in a computer system.
   (c) The basic instructions common to all microcomputers.
2. Explain:
   (a) Interrupts.
   (b) Direct memory access.
   (c) The use of mnemonics.
   (d) Addressing modes.

2 SELF-EVALUATION QUESTIONS

Keep the following questions in mind and try to answer them when you have completed the chapter:

1. What are the four basic parts of a computer and what are their functions?
2. What is a bidirectional bus?
3. What is the difference between assembly language and machine language?
4. What is an interrupt?
5. What are the basic instructions in a microcomputer?

3 INTRODUCTION

The use of computers has transformed the way in which we live. Although we all come in contact with these machines many times a day, we often fail to realize it when there is a computer at work performing some useful function for us. Computers are usually thought of as large, expensive, and complicated machines that either perform complex calculations or store vast quantities of information. We have come to regard these machines as being the heart of corporations, governments, or schools and universities. However, large, *general-purpose* computers are no longer the only kind. Now the word "computer" is also applied to smaller machines that are less complicated, less expensive, and more dedicated to a specific task. It is this type of computer, the "micro," that this book is really all about. Microcomputers tend to be dedicated to a specific application or purpose and are produced in far greater quantities than are the large, general-purpose machines. They touch our everyday lives in ways we seldom suspect, for microcomputers serve in *control*-type applications. They are designed into different types of equipment, making the equipment perform its function. The primary effect of the microcomputer has been to replace circuit design using standard TTL logic or CMOS logic with a few microcomputer LSI circuits. In this way, a single microprocessor can be programmed to control thousands of applications. Using the same components for many uses has eliminated the costly design of a different TTL circuit for every application.

The cost of a *dedicated* microcomputer has become so small that these devices are finding routine use in appliances and equipment of many types. (Remember, they are used to control things for us, and that is the interest of this book.) The first micro was used in the hand-held calculator. It was a 4-bit device intended to perform 4-bit (BCD) arithmetic. At the time (1971) this micro was not thought of as a computer. However, it started the revolution, and when the 8-bit micro (Intel 8008) was created in 1972, it shocked the industry. Table 9-1 lists 10 different microcomputers, their relative size, and their speed.

3.1 What is a Computer?

All computers have a common basis and as such have the same structure. There is a real difference between a *computer* and a *processor*: the computer is the whole machine; the processor is a small but very important part. Every computer contains four important elements:

TABLE 9-1   Different Microcomputers

Processor	Bits	Speed[a]	Maximum Memory (Bytes)
4004	4	500 kHz	8K
8008	8	500 kHz	16K
8080A	8	2 MHz	65,536
Z80	8	4 MHz	65,536
8085	8	4 MHz	65,536
6800	8	1 MHz	65,536
6502	8	1 MHz	65,536
8086	16	5 MHz	2^{20}
Z8001	16	4 MHz	2^{23}
68000	32	12.5 MHz	2^{24}

[a]Speed is subject to change by manufacturers.

FIGURE 9-1 Computer block diagram.

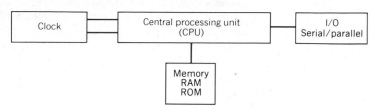

1. *The Processor or Central Processing Unit (CPU):* To add, subtract, multiply, divide, shift, AND, OR, XOR, interpret instructions, and make decisions. (In a microcomputer this is sometimes called the MPU.)

2. *The Clock:* To order functions in a desired sequence; the timing sections.

3. *The Memory Section:* To store the program or programs, and data.

4. *The Input and Output (I/O) Section:* To give the computer a way of communicating with the outside world; this means a data terminal or perhaps something to be controlled.

Figure 9-1 shows the four components of any computer, large or small. The four components are connected by wires or printed circuit paths that are kept as short as possible. Since electrical signals travel at a finite velocity, distance tends to slow things down. The velocity of a pulse is roughly two-thirds of the speed of light or $\frac{2}{3}$ of $(300 \times 10^6$ m/sec). This works out to 1.5 nsec/ft. In modern computers, distances down to a few thousandths of an inch remain our enemy! The designer must work carefully to reduce any unnecessary length in signal paths between the four elements of a computer. In a large computer, memory is placed as close to the processor as is physically possible. The I/O section, which may involve slow data terminals, can be placed farther away without noticeable effect.

3.2 The Clock

Each processor is designed to operate at some top speed. The clock frequency is chosen as close to maximum as possible. Most clock circuits use a quartz crystal to maintain the high frequency (as in digital wrist watches) and to produce

FIGURE 9-2 Typical waveform of a two-phase clock.

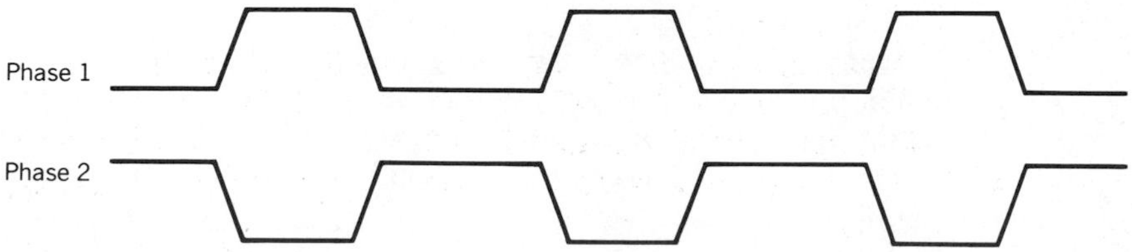

Phase 1

Phase 2

two clock signals. These are the Ø1 (phase one) and Ø2 (phase two) clocks. Often these are not available to look at, but they are there!

The purpose of these two signals is to make certain that the computer cannot read from and write to memory at the same time. The 2Ø clock (two-phase clock) is sometimes called a read/write clock. Figure 9-2 shows the typical waveform. Since the pulses are never up at the same time, this is called a two-phase (2Ø) nonoverlapping clock. In some computers the clock and crystal are a separate area. In others, the crystal connects directly to the CPU (or MPU) and the clock circuitry is on the same chip as the CPU.

3.3 The Central Processing Unit (CPU)

The CPU (or MPU) is the brain of the computer. It is here that instructions are interpreted and decisions are made. Arithmetic operations, logical operations, data moves, shifting, and compares are performed by the CPU. The CPU obtains instructions one at a time from a sequence of instructions stored in memory. The instruction is decoded to determine its meaning. If more information is required, the CPU gets it from memory. The CPU then performs the task required of it. The instructions of all computers may include the following categories:

1. Move data from one place to another.
2. Input or output.
3. Arithmetic operation.

4. Logical operation.
5. Jump to an address.
6. Compare data.

3.4 The Memory

There are many types of memory used by computers. All computers have some RAM and most have some ROM. Some computers have slower memories in the form of disk drives and even slower memories like magnetic tape. For the time being, we will concentrate on RAM and ROM, both solid state memories.

Two types of information are stored in the memory (RAM or ROM): the program itself and data required by the program. The program can consist of many main programs operating separately or main programs and subroutines that are used by the main programs. In reality, only one program can operate at a time, since there is only one CPU to perform the tasks each instruction requires (see Fig. 9-3). The data used by a program can take on many forms. It could be ASCII data or numerical data, possibly BCD information.

When a computer is first turned on, the RAM type of memory comes up in a random way. A program must be loaded in RAM before a computer can run it. Either it is entered manually or it is loaded from a slower type of memory like a disk storage drive.

Many programmers place programs in a ROM that does not lose information when power is turned off. When the computer is turned on, a program exists in ROM that can be immediately run, saving work for the operator of the computer.

The power of a computer is usually determined by two parameters related to memory:

1. The number of memory locations available.

FIGURE 9-3 Computer memory.

Address 0
Main program
Subroutine 1
Subroutine 2
Data
Address N

FIGURE 9-4 Address bus and data bus.

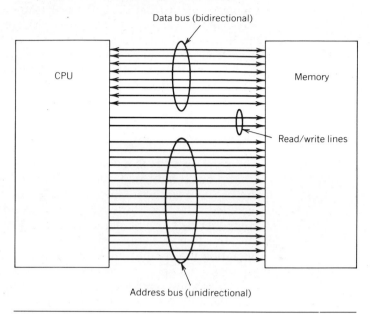

2. The number of bits a location can store.

As they say, the bigger the better! The data stored in the memory can be passed back and forth between the CPU and memory. The data can be read from memory by the CPU, or the CPU can send data to the memory. This is accomplished in the following way. The CPU must first decide which memory address is to be used. Then, if a write into memory is involved, the CPU sends the data to the desired location. If a read from memory is involved, the memory sends the data from the desired memory location to the CPU. The CPU and the memory communicate by two sets of signal paths. These are the *data bus* and the *address bus*. A "bus" is a group of wires having a common purpose. In the case of the 8080A microprocessor, eight data lines make up the data bus. The 8080A can operate a memory having 65,536 locations (8 bits per location) and there are 16 address lines in the address bus.

The address wires or lines run in one direction (unidirectional). The CPU sends an address down the bus to the memory. The data lines are bidirectional between the CPU and the memory, since data can be passed in either direction. Figure 9-4 shows the address bus and data bus between CPU and memory.

The memory must be able to decode the address information to determine which memory location is involved. Two additional lines, the "read" and "write" lines, are shown between the CPU and memory. These are used to determine which way data flows on the data bus (CPU to memory or memory to CPU). The bus structure allows high-speed *parallel* communication between the CPU and memory.

3.5 The Input/Output Section

Input or output from the processor to the outside world can be done in several ways. I/O is usually accomplished with serial or parallel type information. Data terminals usually are serial, and the I/O section may contain a USART to provide serial data. Switches, indicator lamps, and other devices to be controlled frequently require parallel information. This also is available in the I/O section. The components of the I/O section usually are also connected to the address and data busses, and two control signals are used. These are I/O Read and I/O Write.

3.6 The System Bus

Three busses exist in a computer system—the address bus, the data bus, and the control bus (Fig. 9-5). The *control bus* includes the memory read and write lines, the I/O read and write lines, and a reset line. Other control-type signals may be included. The three busses together are referred to as the system bus, and most components of a computer system communicate with one another over the system bus. In addition, many computers have a *status bus* that indicates the particular operation the computer is doing at a given moment. This information may be of interest to an operator or someone trying to debug a program.

As you know, many wires are needed in a computer to interconnect the parts. Lengths must be kept short to reduce the time required to communicate. In most computer systems, these signal paths or busses must be carefully designed so that they can operate at high frequencies. Another problem is the cross-talk between adjacent signal paths; that is, the induction by a signal on one wire of a similar

FIGURE 9-5 A computer and its bus structure.

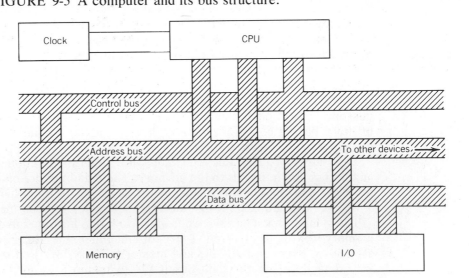

signal in nearby wires. Also, when many devices share a common bus, the circuit doing the driving must be capable of driving many other circuits. Hence, bus drivers may be needed to supply the necessary drive current. Since the CPU is prone to issuing address information, it may first drive a bus driver to produce the necessary current to drive large memory and I/O sections.

One final word: the address bus is generally unidirectional, the data bus is bidirectional, and the components of the control bus may be unidirectional or bidirectional.

4 THE OPERATION OF A COMPUTER

We have learned that the program (or sequence of instructions) is stored in the memory section of a computer and that the CPU reads the instructions one at a time from the memory.

We can now expand on these important points. Every CPU has a special-purpose register that is used to point to the current instruction address in memory. If our program is stored in memory, we need a way to point to each location that contains an instruction. This special-purpose register is called either the *program counter* or the *instruction pointer*. This register is a part of the CPU and drives the address bus.

A computer follows a very rigid sequence of operations. These are as follows:

1. The CPU places an address on the bus and reads an instruction from memory on the data bus. This is called "instruction fetch."

2. The CPU takes the instruction from the data bus (from memory) and *decodes* it. As this is done, the CPU finds out whether more information is needed to execute that particular instruction. If so, it gets the additional information from memory.

3. The CPU *executes* the instruction.

Thus the basic sequence is: FETCH an instruction, DECODE it to determine its meaning, and EXECUTE it.

4.1 Instruction Fetch

The CPU loads its program counter (PC or IP) with the address of the instruction in memory. The address appears on the address bus and memory picks it up. The address is decoded and the data from the desired memory location is placed on the data bus. The CPU picks up the data from the data bus. The program counter is incremented by one to point to the next memory location in preparation for the next item in the program.

4.2 Decode Cycle

The CPU decodes the instruction and determines whether more information is required to perform the instruction. If not, the CPU executes the instruction. If more information is required, it will be in the next memory location, which the program counter (PC or IP) is now pointing to. The CPU may have to get one, two, or more additional pieces of information from memory to complete the instruction. In any event, the information is sent to the CPU from memory on the data bus when an address is provided by the program counter.

4.3 Execute Cycle

The CPU performs the desired instruction, utilizing the additional data it obtained if this step was necessary. We shall see that different instructions require different numbers of memory locations. In an 8-bit microcomputer, there may be instructions that occupy only one memory location. Other types of instruction require two, three, or more locations. In any case, both the programmer and the CPU know how many locations an instruction requires.

4.4 Processor Registers

The processor (CPU) of any computer contains areas called registers for the storage of binary numbers. Registers vary both in size and in quantity: a large general purpose computer may have 16 registers, each 36 bits long. An 8-bit microcomputer may have two 8-bit registers, while another may have seven 8-bit registers. A 16-bit microcomputer may have eight 16-bit registers or more. Generally, the more registers you have available, and the larger the registers, the more powerful the computer. Again, the bigger the better!

5 COMPUTER INSTRUCTIONS

All computers have a basic set of instructions that cover similar functions. They may have different names or symbols and they may have different hexadecimal codes to represent them, but they perform similar operations. For instance, to put a piece of data into a register, we would *load* the register on one computer, but *move* data into the register on a different machine.

Load A, Data   (on computer A)

Move A, Data   (on computer B)

Both operations load register A with data. Computers use symbols or *mnemonics* to represent the operation:

LDA A, Data   or   MVI A, Data

LDA A stands for load accumulator A, while MVI A stands for move immediate data to A. Both result in the same operation.

Mnemonics are used by the programmer in writing a program. They are not used on the computer itself. The symbols are referred to as an *assembly language*, and each computer has its own assembly language (set of mnemonics).

The computer requires a code, usually in hexadecimal or binary, to represent the instruction.

$$MVI\ A\ =\ 3E\ H\quad(8080A)$$
$$LDA\ A\ =\ 86\ H\quad(6800)$$

These codes are called *machine language*. Hence, a programmer writes in assembly language (using symbols) and then must convert to machine language (using codes).

In the case of LDA A, Data, the machine code for LDA A is only part of the required information. The data is also required. When this instruction is stored in memory, the data must follow it. If Data = 55 H, then LDA A, 55 H becomes 8655 in machine language. These codes are also called operation codes or opcodes. Each computer has its own set of opcodes. The instruction to load accumulator A with 55 H requires two bytes to complete the instruction. We will find that instructions may require only one byte. Some will require two, three, or more to complete a task.

Programs called *assemblers* aid in writing machine language programs by converting mnemonics (assembly language) to opcodes (machine language). Assemblers are frequently written in a high-level language like Fortran. Programs called *disassemblers* perform the reverse operation; that is, they convert opcodes back into mnemonic form (e.g., 8655 is converted back to LDA A, #$55, where # represents immediate data and $ means hex data follows).

Suppose we intend to load some registers with data. The instructions might appear as follows:

$$Load\ A,\ 46\ H$$
$$Load\ B,\ 23\ H$$
$$Load\ C,\ 41\ H$$
$$Load\ D,\ 29\ H$$
$$Load\ E,\ 6A\ H$$

If the machine codes (opcodes) are:

$$Load\ A\ =\ 3E\ H$$
$$Load\ B\ =\ 06\ H$$
$$Load\ C\ =\ 0E\ H$$
$$Load\ D\ =\ 16\ H$$
$$Load\ E\ =\ 1E\ H$$

the program would appear in machine language as follows:

$$
\begin{array}{ll}
3E & 46 \\
06 & 23 \\
0E & 41 \\
16 & 29 \\
1E & 6A \\
\end{array}
$$

These instructions, each requiring 2 bytes of memory, would be placed sequentially in memory as follows:

Memory
3E
46
06
23
0E
41
16
29
1E
6A

Note that each memory location is 8 bits wide, and each location is numbered in hexadecimal. This would represent a segment of a larger program. The CPU would take each instruction and its data one at a time. The CPU would decode 3E to load register A with 46 H, and perform the operation. Then the CPU would move on to 06 and load register B with 23 H.

This is how a computer operates. The program is loaded into memory as a sequence of operation codes. The computer then follows the "instructions."

5.1 The MOVE Instruction

All computers must move data from one place to another. The move may be from one register (the source) to another (the destination), from a register to a memory location, or from a memory location to a register (see Fig. 9-6).

Data may also be moved to an input or output device. In any event, all computers must be able to move data around. Loading a register is considered to be a move of data to the register. This category then includes the load instruction.

5.2 Add or Subtract

It is common to instruct a computer to add data to a register or subtract data from a register:

ADD A, 26 H

SUB A, 46 H

ADD A, 26 H causes 26 H to be added to the A register. SUB A, 46 H causes 46 H to be subtracted from the A register. The result of these operations is left in the A register.

FIGURE 9-6 Types of MOVE. (*a*) A copy of the contents of A is moved to B. (*b*) A copy of the contents of A is moved to a memory location. (*c*) A copy of the contents of memory is moved to register C.

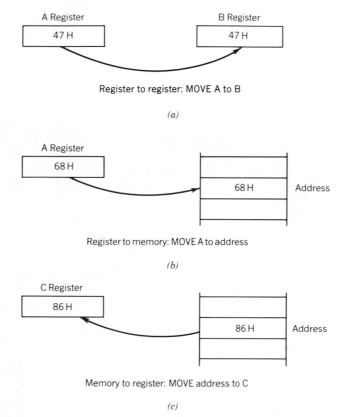

Register to register: MOVE A to B

(a)

Register to memory: MOVE A to address

(b)

Memory to register: MOVE address to C

(c)

Many computers have increment and decrement instructions. INC A means to add *one* to the contents of A. It is the same as ADD A, 1. DEC A, which subtracts one from A, is the same as SUB A, 1. If the increment and decrement instructions are not available, the ADD and SUB instructions can be used.

Keep in mind that each computer uses a different mnemonic for the operation involved. ADD A, INC A, INR A, and INX A may all mean increment.

5.3 Logical Operations

Logical AND, OR, invert, and exclusive OR are performed on most computers. These instructions may be between the contents of two registers or between a register and data or between a register and memory.

AND A, 0F H   AND A register with 0F H
XOR A, 06 H   Exclusive OR A register with 06 H

These logical operations are performed on binary numbers of the same size, one bit at a time.

EXAMPLE 9-1

AND 0F H with 46 H.

$$
\begin{array}{ll}
0\,0\,0\,0\,1\,1\,1\,1 & \text{0F H} \\
\underline{0\,1\,0\,0\,0\,1\,1\,0} & \underline{46\ H} \\
\text{Result:}\quad 0\,0\,0\,0\,0\,1\,1\,0 & \text{06 H}
\end{array}
$$

The second bits are both one (one and one = one). The third bits are both one (one and one = one). Only these bits come through.

EXAMPLE 9-2

OR 07 H with 52 H.

$$
\begin{array}{ll}
0\,0\,0\,0\,0\,1\,1\,1 & \text{07 H} \\
\underline{0\,1\,0\,1\,0\,0\,1\,0} & \underline{52\ H} \\
\text{Result:}\quad 0\,1\,0\,1\,0\,1\,1\,1 & \text{57 H}
\end{array}
$$

$\rightarrow 0 + 1 = 1$, etc.
$\rightarrow 0 + 0 = 0$
$\rightarrow 0 + 1 = 1$
$\rightarrow 0 + 0 = 0$

EXAMPLE 9-3

Exclusive OR 29 H with 0F H.

$$
\begin{array}{ll}
0\,0\,1\,0\,1\,0\,0\,1 & \text{29 H} \\
\underline{0\,0\,0\,0\,1\,1\,1\,1} & \underline{\text{0F H}} \\
\text{Result:}\quad 0\,0\,1\,0\,0\,1\,1\,0 & \text{26 H}
\end{array}
$$

$\rightarrow 0 \oplus 1 = 1$
$\rightarrow 0 \oplus 0 = 0$, etc.

Other common operations are shift or rotate to the left or shift or rotate to the right.

LSL = Logical Shift Left

RLC = Rotate Left

These instructions move the contents of a register *one* place to the left.

EXAMPLE 9-4

RLC A: Rotate register A left.

A Before | 0 0 1 1 1 0 1 0 |

A After | 0 1 1 1 0 1 0 0 |

Finally, the complement instruction is used to invert the bits of a number.

5.4 Jump or Branch Instructions

Jumping or branching to an address or branching to a part of the program a certain number of locations lower or higher is accomplished by:

JMP 2010 H   Jump to address 2010 H and execute the instructions beginning at that location.

or

BRANCH 28 H   Jump to an address 28 H bytes greater in memory (or add 28 H to current PC).

If we are executing the branch instruction at address 3000 H, then Branch 28 H may displace us or branch to address 3028 H, where execution of instructions continues. This type of jump instruction (i.e., by displacement rather than to an absolute address) is called relative addressing. Both types are common.

The other types of jump (and branch) are on condition, JZ or BRZ: jump or branch if a register contains zero as the result of some operation. We also have JNZ and BRNZ: jump or branch if a register does not contain zero as the result of some operation. Other conditions that may be met are related to the sign of a number (plus or minus), the carry bit (often an arithmetic operation), or the state of a parity bit. These conditions vary from computer to computer.

5.5 Compare Instructions

Programmers frequently are required to test data to determine whether a condition is true or false. When a key is depressed on a keyboard, you may need to know whether it is an ASCII carriage return (0D H). This is done using a compare instruction:

CMP A, 0D H

which tells the computer to compare the contents of the A register with the code 0D H. If they are the same, a zero flag may be set, allowing a conditional jump (JZ) to be executed. Many keyboard codes must be recognized by programs: control C (halt), line feed, and escape are a few examples. The compare instruction allows this to be done.

5.6 IN/OUT

The I/O section of the computer allows communication with the world outside the computer. Many devices may be connected to the computer's I/O section. Each device is connected to a port, and each port has a separate number for identification. If a data terminal is connected to port 47 H for example, a computer can do input or output from port 47 H.

IN A, 47 H   In from port 47 H to register A
OUT A, 47 H   Out to port 47 H from register

Port 47 H may actually be a USART needed to send serial data to a data terminal using RS232 signals.

Some computers do not use input/output ports and assign a memory location to another device (e.g., a data terminal). Thus a particular memory location may be associated with (wired to) the data terminal. MOVEing data to or from that memory location is equivalent to the in or out instruction. This type of I/O is called *memory-mapped I/O,* since memory locations (and a MOVE instruction) are used instead of IN/OUT-type instructions.

5.7 Summary of Instructions

Although other instructions exist to perform other operations, the basic instructions of a computer are:

1. MOVE instructions.
2. Arithmetic: add, subtract.
3. Logical: AND, OR, XOR, complement, rotate.
4. Jump or branch to an address.
5. Compare data.
6. Input/output.

Remember these instruction types as you study some real microcomputers in the chapters that follow.

5.8 Addressing Modes

We have seen that a microprocessor contains registers and that they are used to store and manipulate numbers. Locations in memory can be used to serve similar purposes. As assembly language programs are written, there are a number of ways of using the microprocessor's instructions to refer to data that may be stored in memory. Let us examine the different ways (modes) in which instructions can address memory. There are five basic *modes* in which most microprocessors refer to memory locations:

Implied   Register direct
Immediate   Register indirect
Absolute

1. *Implied:* In the implied mode, no information other than the instruction is required. Examples include incrementing a register and logical operations.

> INR A   Add one to A
> OR A, B   Logical OR of A and B

2. *Immediate:* The instruction is followed with the data.

> LDA A, 26   Load A with the data 26 D
> ADD B, 14 H   Add 14 H to register B

In some microprocessors, immediate data is preceded by a # symbol:

> LDA A, #26   The # symbol specifies the
>   immediate mode

3. *Absolute:* The data is found in a memory location specified as a part of the instruction.

> MOV A, (1000 H)   Load the A register with the data
>   at location 1000 H. 1000 H is not
>   going into A; the data *at* location
>   1000 H is going into A.

For certain MPUs, the parentheses may be used to show the absolute addressing mode.
(a) LDA A, #40 H may mean to put 40 H into the A register. This is the immediate addressing mode.
(b) LDA A, (40 H) may mean to put the data found at address 40 H into A. This is the Absolute addressing mode.

4. *Register direct:* In this mode, the register contains the data.

> LDA A, D6   Load B from the data in register D6.

The contents of one register is transferred to another.

5. *Register indirect:* In this more complex addressing mode, the data is not in the register. The contents of the register point to a memory location that contains the data.

> LDA C, A5   Load C from the data located at
>   the address specified in register
>   A5. A5 is used to point to the
>   data.

Different microprocessors use different modes or means of referring to memory addresses. These are the simpler ones common to most microprocessors. In Chapter 15 we will learn about more complex modes as they apply to specific microprocessors.

6 DIRECT MEMORY ACCESS

Before leaving this discussion of what a computer is all about, let us recall that each computer consists of four basic elements: clock, CPU, memory, and I/O.

Memory and I/O communicate with the CPU via the address, data, and control busses. Generally, the CPU issues address information. The address bus is unidirectional from the CPU.

In certain memory applications, however, it is sometimes useful for one type of memory to communicate directly with another type of memory. When a disk-type memory containing information (a program, perhaps) must load the program into memory (RAM), time would be wasted if each piece of data had to go through the CPU (i.e., CPU reads data from the DISK, CPU writes data to RAM).

Generally, the DISK sends data directly to RAM. (Refer to Chapter 7 for a description of DISKs.) In this instance of a large transfer of data between the DISK and RAM, the CPU disconnects itself from the address bus (tri-states itself) so that the DISK may directly control the address bus, hence the RAM. This operation is called direct memory access (DMA); it speeds things up by bypassing the CPU. Some computer systems that have disk memories perform the DMA operation to load or save programs. The DISK must have its own controller to perform this data transfer without the help of the CPU.

To be run, a program must be in the computer's main memory (RAM). Programs cannot be run on slow-memory media like disk or tape.

7 THE INTERRUPT

When a program is running on a computer, it is performing some task. It may be doing calculations or routinely controlling or operating a piece of equipment—that is, the computer is operating normally, which it usually does over long periods of time.

Infrequently, however, events occur that require a computer's immediate attention. Most computers are able to be *interrupted* from the program they are currently running. When a program or computer is interrupted, the computer is usually sent to a program stored in memory at a special location. This program, called an interrupt routine, handles the condition that causes the interrupt.

As an example, suppose a computer is routinely controlling the temperature and humidity in a large building. The computer also routinely monitors lighting in an attempt to save electrical energy (e.g., by turning off lights in unused areas). These are the main jobs of the computer and its program.

However, all the building doors and windows are wired for security purposes. In addition, heat and smoke detectors are placed around the building. If an emergency (fire, burglary) occurs, the computer is interrupted from its main program. At the time of the interrupt, the main program becomes unimportant. The interrupt causes the computer to run a special program that dials a telephone line to alert police and fire departments. The computer can also operate a tape-recorded message or, better yet, operate a commonly available speech synthesizer to exclaim, "Help!"

8 SUMMARY

In this chapter we studied the four basic parts of a computer: clock, CPU, memory, and I/O. We also studied the busses used to control the flow of information in a computer, namely, the data, address, control, and status bus.

We also learned about the basic instructions common to all computers: MOVE, logical arithmetic, input/output, jump or branch, and compare.

In addition, we studied DMA and interrupts, and were introduced to mnemonics and machine language.

9 GLOSSARY

Address bus. The parallel path in which the binary memory address from the CPU is sent to the memory section.

Assembler. A computer program that converts mnemonics (assembly language) into opcodes (machine language).

Assembly language. A group of mnemonics representing the instruction set of a computer.

Bidirectional bus. Any parallel group of wires that can carry signals both ways. A bidirectional data bus can *send* and *receive* information.

CPU. Central processing unit. The element of a computer that performs all mathematical and logical operations.

Control bus. The CPU bus used in controlling all sections of the computer.

DMA. Direct memory access. The process in which an external device takes over the computer's busses and performs operations on the computer's memory.

Data bus. The parallel path in which information from the CPU is sent to other parts of the computer.

Decode cycle. The CPU operation in which an instruction fetched from memory is decoded, to guide the CPU to its next cycle.

Dedicated microcomputer. A microcomputer programmed to perform only one task.

Disassembler. A computer program that converts machine language back into mnemonics.

Execute cycle. The cycle in which the CPU performs an operation on internal registers or an external memory location or I/O device.

General-purpose computer. A computer able to perform a variety of jobs or tasks.

Instruction fetch. The CPU operation used to fetch a word from memory to be interpreted as an instruction.

Instruction pointer. See Program counter.

Interrupt. An external signal (condition or set of conditions) causing the computer to immediately halt its current operation and perform a specific task.

Machine language. Actual binary or hexadecimal code representative of a computer instruction.

Memory-mapped I/O. An input/output exchange that uses a decoded memory address instead of a port.

Mnemonics. Abbreviated character representations of individual computer instructions.

Nonoverlapping Clock. An oscillator that produces two signals at the same frequency, but out of phase with each other, so that both signals are never positive at the same time.

Operation codes (opcodes). The sets of machine language codes that make up a computer's instruction set.

Port. An electronic device in a computer that allows information exchange between the computer and the outside world.

Processor. See CPU.

Program counter. The CPU register that contains the current instruction address in memory.

Status bus. CPU outputs showing the present state of operation in a computer.

PROBLEMS

9-1 List six applications of microprocessors found around you that were *not* listed in this chapter.

9-2 Draw the block diagram of a computer. Label each part and all busses.

9-3 A microprocessor chip may be 0.2 in. on a side. Calculate the time required for an electrical signal to travel the width of the chip, assuming a speed of $\frac{2}{3} c$ (two-thirds the speed of light).

9-4 If a memory unit is located 1.5 m from the CPU, calculate the delay in transmitting information from one to the other.

9-5 List the six basic instruction types of a computer.

9-6 Define assembly language and show its difference from machine language.

9-7 Describe DMA.

9-8 What is an interrupt?

9-9 List and define the basic addressing modes of a computer.

9-10 Describe the fetch, decode, and execute operation and tell how it is modified if further data is needed by the CPU.

9-11 Explain how the address bus is used. Explain how the data bus is used. What signals are traveling on these busses?

9-12 How is the CPU different from the memory section of a computer?

9-13 What is the purpose of a mnemonic?

9-14 How does port I/O differ from memory-mapped I/O?

9-15 What is the purpose of the timing section of a computer?

9-16 In human terms, when you receive a burn we might say that you are experiencing an interrupt. Why is this analogy valid?

9-17 What is the function of the program counter or the instruction pointer?

9-18 What is the difference between an assembler and a disassembler?

9-19 How do logical instructions differ from arithmetic instructions?

9-20 Find the logical AND of the following bytes:
   (a) AND 0E H with 8A H
   (b) AND 55 H with 40 H
   (c) AND 27 H with F0 H
   (d) AND CC H with 55 H

9-21 Find the logical OR of the following bytes:
   (a) OR 40 H with CC H
   (b) OR E3 H with 22 H
   (c) OR 49 H with 77 H
   (d) OR 55 H with AA H

9-22 Find the logical XOR of the following bytes:
   (a) XOR 22 H with 49 H
   (b) XOR FF H with 44 H
   (c) XOR DB H with 20 H
   (d) XOR AC H with 00 H

9-23 It is desired to determine the necessary masks to look at each bit in the data D6 H. The bits are numbered 1 to 8. (*Hint:* Bit 1 is a "0" in D6 H and bit 8 is a "1.") List the mask quantity to be used with the AND operation:

Bit 1 = ANI 01 (Here's the first answer.)

Bit 2 =

Bit 3 =

Bit 4 =

Bit 5 =

Bit 6 =

Bit 7 =

Bit 8 =

CHAPTER 10

A SOFTWARE LOOK AT THE 8080A/85 MICROPROCESSOR

1 INSTRUCTIONAL OBJECTIVES

When finished with this chapter you should be able to:

1. Draw a labeled diagram of the various registers available on the 8080A.
2. Identify the five basic types of instructions used by the 8080A.
3. Use the chapter as a reference to find a given instruction and determine its effect on the condition bits.
4. Understand the terms mnemonic, opcode, and assemble.
5. Identify 1-, 2-, and 3-byte instructions.
6. Assemble a simple 8080A program into machine language form.
7. Calculate the time required to perform 8080A instructions.
8. Use the following instructions: IN, OUT, MVI A, DCR A, JNZ XRA A, NOP, and HLT.

2 SELF-EVALUATION QUESTIONS

Keep the following questions in mind and try to answer them when you have completed the chapter:

1. What is the difference between a register and a register pair? How is each identified?
2. Which group of instructions is used to put data into registers? Into memory locations?
3. What are the condition bits (flags) used for?
4. Which instructions set flags, and which instructions use the flags (once set) to determine their own choice of action?
5. What is byte swapping?
6. How does the program fit into memory—in mnemonic form or in opcode form?
7. How can a "program loop" be used to produce delay?

3 INTRODUCTION

The first commercial microprocessor was developed by the Intel Corporation in 1971. It was a 4-bit microcomputer (the 4004) and had a few instructions that made it "programmable." This meant that it could be changed for use in many different applications, replacing to some extent traditional TTL logic design. The 4004, which was an outcropping of the first (and revolutionary) hand-held calculators, was followed by the 8008. The Intel 8008 was faster and introduced 8-bit opcodes. Its power revolutionized the industry. The next advance was the 8080A (Fig. 10-1), running at 2 MHz and featuring a wide variety of instructions. It has become an industry standard with its 244 mnemonics, contained in its instruction set. An improvement over the 8080A is the 8085, shown in Fig. 10-2.

It may surprise you to learn that these microprocessors are used in a multitude of electronic areas from fighter planes to kitchen blenders and are programmed for these applications by engineers and scientists and by high school and college students.

4 INTERNAL ORGANIZATION

The 8080A microprocessor is an 8-bit CPU fabricated with N-channel MOS technology. It can directly address 65,536 bytes of memory and has 11 internal data registers. These registers are called the accumulator (which is a *very* important

FIGURE 10-1 An 8080A microprocessor DIP. (Reprinted by permission of Intel Corporation, copyright 1981.)

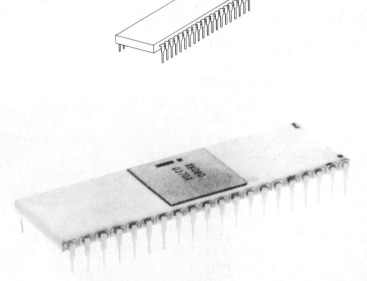

FIGURE 10-2 An 8085 8-bit microprocessor. (Reprinted by permission of Intel Corporation, copyright 1983.)

FIGURE 10-3 Internal register organization of the 8080A.

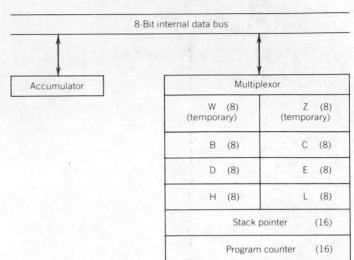

register); registers B, C, D, E, H, L, W, and Z; and the stack pointer and program counter registers (which both contain 16 bits of storage capacity). All the letter-designated registers are 8-bit registers, with BC, DE, and HL forming three 16-bit register pairs. [The first letter designates the register pair (RP), i.e., B = BC, H = HL, etc.] Registers W and Z, which are temporary registers used only by the internal CPU circuitry, are the only ones that may not be used by the programmer.

Figure 10-3 shows the basic register organization of the 8080A. (A more intense study of the 8080A is taken up in Chapter 13.) Note that all registers have access to the 8080A's internal 8-bit data bus. This data bus is also available to the outside world via eight pins on the 8080A package that contain the 8-bit bidirectional data bus. It is through this data bus that the 8080A gets its program information, in the form of 8-bit instructions from memory. We will not concern ourselves with *how* the 8080A gets its information from memory (this, too, is covered in detail in Chapter 13). Rather, we want to know how we can get the 8080A to perform a function for us via a carefully laid-out program.

5 THE INSTRUCTION SET

There are five groups of instructions in the 8080A instruction set: the data transfer, arithmetic, logical, branch, and stack-I/O-machine control groups. All in all, there are 93 basic instructions, and because of combinations of registers that can be specified for different instructions, we get 244 different instructions. It is combinations of these 244 instructions, coupled with other data, that control the operation of the 8080A. The tasks that the 8080A can perform through programming are limited only by the programmer's imagination.

We will look at a few programming examples in this chapter, with more extensive ones to come. But first, let's gain an understanding of the function in the CPU of the 8080A instructions.

5.1 Data Transfer Group

In the data transfer group, we have 13 instructions that control the flow of data between registers and memory. Each instruction operates on either a register or a memory address. As we will see, although there are only 13 basic instructions, the permutations number 70 or more.

1. *MOV R1, R2* (move register to register) R1 ← R2.   This command takes the data from register 2 and puts it in register 1. Both register 1 and register 2 can be any of the usable seven (A, B, C, D, E, H, and L), and R can stand for any of the seven registers.

EXAMPLE 10-1

MOV E, A

Before instruction:

After instruction:

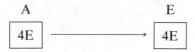

Note: All numbers in this chapter are expressed in hexadecimal; hence the H is not used where the context is unmistakable.

2. *MOV R, M* (move memory to register) R ← (HL).   This instruction takes the data in the memory location specified by the HL register pair and places it in register R.

EXAMPLE 10-2

MOV C, M

Before instruction:

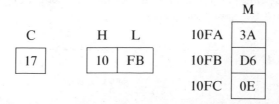

After instruction:

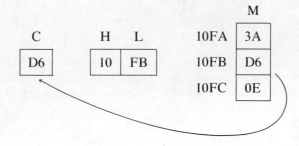

3. *MOV M, R* (move register to memory) (HL) ← R.   Data from register R is placed in the memory location specified by the HL register pair.

EXAMPLE 10-3

MOV M, A

Before instruction:

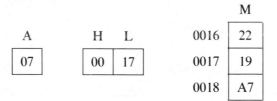

After instruction:

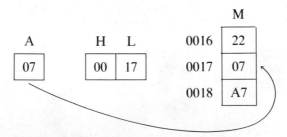

4. *MVI R, data 8* (move immediate data into register) R ← Data.   Place 8 bits of data into the specified register.

EXAMPLE 10-4

MVI D, 3FH

Before instruction:

' D

```
┌────┐
│ 99 │
└────┘
```

After instruction:

D

```
┌────┐
│ 3F │
└────┘
```

5. *MVI M, data 8* (move immediate data into memory) (HL) ← Data.   Place 8 bits of data into the memory location pointed to by HL.

EXAMPLE 10-5

MVI M, 0C0H

Before instruction:

			M
H L	3736	01
37 | 37	3737	19
	3738	3E

After instruction:

			M
H L	3736	01
37 | 37	3737	C0
	3738	3E

6. *LXI RP, data 16* (load register pair immediate) RP ← Data.   Place two bytes of data (16 bits) into the specified register pair.

EXAMPLE 10-6

LXI B, 3C07H

Before instruction:

```
  B    C
┌────┬────┐
│ 20 │ 8F │
└────┴────┘
```

After instruction:

B   C

| 3C | 07 |

7. *LDA addr* (load accumulator direct) A ← (ADDR).   Take the data byte stored in memory location addr and place it in the accumulator.

EXAMPLE 10-7

LDA 107BH

Before instruction:

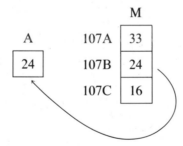

		M
A	107A	33
19	107B	24
	107C	16

After instruction:

		M
A	107A	33
24	107B	24
	107C	16

8. *STA addr* (store accumulator direct) (ADDR) ← A.   Store 8 bits of data in the memory location specified by addr.

EXAMPLE 10-8

STA 3033H

Before instruction:

		M
A	3032	01
FB	3033	9A
	3034	55

After instruction:

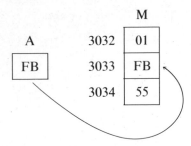

M

A   3032   01

FB   3033   FB

           3034   55

9. *LHLD* addr (load H and L direct) L←(ADDR), H←(ADDR + 1).   Take the data stored in memory at locations addr and addr + 1 and place it in the HL register pair.

EXAMPLE 10-9

LHLD 404FH

Before instruction:

H   L   M

21 | 09   404E 99

               404F 30

               4050 42

               4051 61

After instruction:

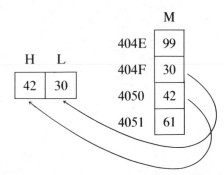

M

               404E 99

H   L   404F 30

42 | 30   4050 42

               4051 61

Note that the data in the first memory location gets put into the lower register in the pair, and the second byte gets put into the upper register. The reverse happens in the next instruction.

10. *SHLD addr* (store H and L direct) (addr) ← L, (addr + 1) ← H.   Place the data in the HL register pair into memory locations addr and addr + 1.

EXAMPLE 10-10

SHLD 0088H

Before instruction:

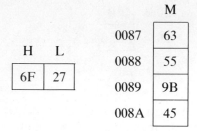

After instruction:

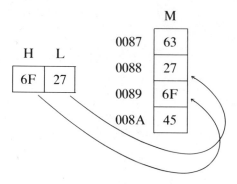

11. *LDAX RP* (load accumulator indirect) A ← (RP).   Place the data from the memory location specified by the value of the register pair into the accumulator.

EXAMPLE 10-11

LDAX D

Before instruction:

	A		D	E		M
	05		11	98	1197	23
					1198	45
					1199	67

After instruction:

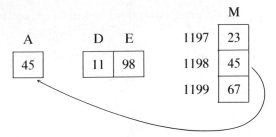

12. *STAX RP* (store accumulator indirect) (RP) ← A. Store the data in the accumulator in the memory location specified by the value stored in the register pair.

EXAMPLE 10-12

STAX B

Before instruction:

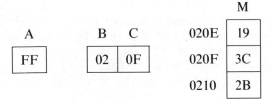

After instruction:

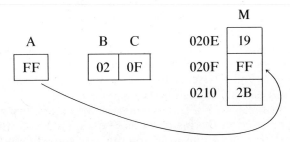

13. *XCHG* (exchange DE and HL) HL ⇆ DE. Exchange the contents of register pair DE with the contents of register pair HL and vice versa.

EXAMPLE 10-13

XCHG (This is a unique instruction; hence there are no permutations!)
  Before instruction:

After instruction:

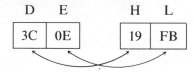

Example 10-13 concludes the data transfer group of 8080A instructions. When combined with other instructions that control the type and amount of data to be transferred, the data transfer instructions become invaluable tools to the programmer.

5.2 Arithmetic Group

Before we start our study of the arithmetic group, it is necessary to explain what a flag is. A flag is a signal or bit used to indicate that a desired operation has occurred. For instance, we would use a zero flag to indicate when the result of an operation or instruction (addition, subtraction, etc.) is zero. Flags or condition bits are very useful and necessary to the proper sequence of instructions in a program. The 8080A has five flags: the zero, sign, parity, carry, and auxiliary carry flags. The auxiliary carry flag, which is used inside the 8080A for some special instructions, is not directly accessible by the programmer. The first four flags can be tested directly, however.

The *zero* flag is set (meaning it has a logic "1" value) if the result of an instruction is a zero; otherwise it is cleared (meaning at logic "0").

The *sign* flag (which indicates the positiveness or negativeness of a result) is set when the MSB (bit 7) of a result is one; otherwise it is cleared.

The *parity* flag is set when the result of an operation has even parity. It is cleared when the result has odd parity.

The *carry* flag is set if the result of an operation was greater than 0FF H (in addition) or smaller than 00 H (in subtraction).

All these flags are affected when the arithmetic instructions are used, and some are even used in the instructions. You should note that no flags are affected in the data transfer group of instructions previously studied.

1. *ADD r* (add register or memory to accumulator).   The byte in the specified register is added to the contents of the accumulator. All condition bits are affected.

EXAMPLE 10-14

ADD B

Before instruction:

A		B		A	Z	S	P	C	
3E		28		0	1	0	0	0	Flags

After instruction:

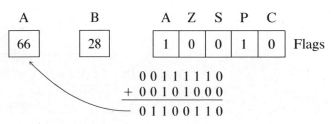

A		B		A	Z	S	P	C	
66		28		1	0	0	1	0	Flags

$$\begin{array}{r} 0\,0\,1\,1\,1\,1\,1\,0 \\ +\ 0\,0\,1\,0\,1\,0\,0\,0 \\ \hline 0\,1\,1\,0\,0\,1\,1\,0 \end{array}$$

The carry bit is not set, but the auxiliary carry is set, since there is a carry from bit 3 into bit 4.

EXAMPLE 10-15

ADD M

Before instruction:

A			M	A	Z	S	P	C	
3E		2900	40	1	0	1	0	1	Flags
		2901	20						
H	L	2902	3F						
29	00								

After instruction:

A	A	Z	S	P	C	
7E	0	0	0	1	0	Flags

$$\begin{array}{r} 0\,0\,1\,1\,1\,1\,1\,0 \\ +\ 0\,1\,0\,0\,0\,0\,0\,0 \\ \hline 0\,1\,1\,1\,1\,1\,1\,0 \end{array}$$

2. *ADC r*   (add register or memory to accumulator with carry). The byte in the specified register or memory location plus the content of the carry bit are added to the contents of the accumulator. All condition bits are affected.

EXAMPLE 10-16

ADC D

Before instruction:

A	D	A	Z	S	P	C	
6C	43	0	0	0	0	1	Flags

After instruction:

A	D	A	Z	S	P	C	
B0	43	1	0	1	0	0	Flags

$$\begin{array}{r} 0\ 1\ 1\ 0\ 1\ 1\ 0\ 0\ \text{A} \\ 0\ 1\ 0\ 0\ 0\ 0\ 1\ 1\ \text{D} \\ +\underline{\qquad\qquad 1}\ \text{Carry} \\ 1\ 0\ 1\ 1\ 0\ 0\ 0\ 0 \end{array}$$

Since the carry bit was previously set, it is added to the sum of A and D registers.

3. *ADI Data 8* (add immediate data to accumulator). The byte of data immediately following the instruction is added to the accumulator using 2's-complement arithmetic. All condition bits are affected.

EXAMPLE 10-17

ADI 66H

Before instruction:

A	A	Z	S	P	C	
26	0	1	1	0	1	Flags

After instruction:

A	A	Z	S	P	C	
8C	0	0	1	0	0	Flags

$$\begin{array}{r} 0\ 0\ 1\ 0\ 0\ 1\ 1\ 0 \\ +\underline{0\ 1\ 1\ 0\ 0\ 1\ 1\ 0} \\ 1\ 0\ 0\ 0\ 1\ 1\ 0\ 0 \end{array}$$

EXAMPLE 10-18

ADI −47H

Before instruction:

A

| 8C |

A Z S P C

| 0 | 0 | 1 | 0 | 0 | Flags

After instruction:

A

| 45 |

A Z S P C

| 1 | 0 | 0 | 0 | 1 | Flags

```
  1 0 0 0 1 1 0 0        8C H
+ 1 0 1 1 1 0 0 1      + B9 H   (−47 H)
  0 1 0 0 0 1 0 1        45 H
```

The carry bit is set as is the auxiliary carry bit. Two's-complement arithmetic was performed.

4. *ACI Data 8* (add immediate data to accumulator with carry). The byte of data immediately following the instruction is added to the contents of the accumulator plus the contents of the carry bit. All condition bits are affected.

EXAMPLE 10-19

ACI 26H

Before instruction:

A

| 96 |

A Z S P C

| 0 | 0 | 0 | 0 | 1 | Flags

After instruction:

A

| BD |

A Z S P C

| 0 | 0 | 1 | 1 | 0 | Flags

```
              1              1 H Carry
  1 0 0 1 0 1 1 0          96 H A
  0 0 1 0 0 1 1 0        + 26 H Data 8
  1 0 1 1 1 1 0 1          BD H
```

5. *SUB r.*   The byte in the specified register is subtracted from the accumulator using 2's-complement arithmetic. If there is no carry indicating that a borrow has occurred, the carry bit is set; otherwise it is reset. All condition bits are affected.

EXAMPLE 10-20

SUB C

Before instruction:

	A		C		A	Z	S	P	C	
	58		29		0	0	0	0	0	Flags

After instruction:

	A		C		A	Z	S	P	C	
	2F		29		0	0	0	0	0	Flags

```
   58 H     0 1 0 1 1 0 0 0  ⟶    0 1 0 1 1 0 0 0
 − 29 H   − 0 0 1 0 1 0 0 1         1 1 0 1 0 1 1 1 ← 2's Complement
                               1 0 0 1 0 1 1 1 1
                                  ↖ Carry
```

The carry was reset, so the carry flag is set indicating a "borrow."

EXAMPLE 10-21

SUB M

Before instruction:

	A			M		A	Z	S	P	C	
	4C		4689	21		0	0	0	0	0	Flags
	H	L	468A	6D							
	46	8A	468B	14							

After instruction:

	A		A	Z	S	P	C	
	DF		0	0	1	0	1	Flags

```
   4CH     0 1 0 0 1 1 0 0 ⟶    0 1 0 0 1 1 0 0
 − 6DH   − 0 1 1 0 1 1 0 1       + 1 0 0 1 0 0 1 1 ⟵ 2's Complement
                               0 1 1 0 1 1 1 1 1
                                  ↖ Carry
```

The carry was reset, so the carry flag is set indicating a "borrow."

6. *SBB r.* The carry bit is internally added to the contents of the specified byte. This value is then subtracted from the accumulator using 2's-complement arithmetic. All condition bits are affected.

EXAMPLE 10-22

SBB D

Before instruction:

A	D	A	Z	S	P	C	
06	03	0	1	0	1	1	Flags

After instruction:

A	D	A	Z	S	P	C	
02	03	1	0	0	0	0	Flags

03 H + Carry = 04 H
2's Complement of 04 H = 1 1 1 1 1 1 0 0

```
   06    0 0 0 0 0 1 1 0
 + FA    1 1 1 1 1 1 0 0
       1 0 0 0 0 0 0 1 0
          └── Carry
```

The carry is a "1" so the carry flag is reset, since this is a subtract operation.

7. *SUI Data 8.* The data byte immediately following the instruction is subtracted from the contents of the accumulator using 2's-complement arithmetic. All condition bits are affected.

EXAMPLE 10-23

SUI 4CH

Before instruction:

A	A	Z	S	P	C	
2F	0	1	1	0	1	Flags

After instruction:

A	A	Z	S	P	C	
E3	1	0	1	0	1	Flags

$$
\begin{array}{ccc}
\text{2FH} & 0\,0\,1\,0\,1\,1\,1\,1 & 0\,0\,1\,0\,1\,1\,1\,1 \\
-\text{ 4CH} & -\,0\,1\,0\,0\,1\,1\,0\,0 & +\,1\,0\,1\,1\,0\,1\,0\,0 \longleftarrow \text{2's Complement} \\
\hline
& & 0\,1\,1\,1\,0\,0\,0\,1\,1 \\
& & \qquad\text{Carry}
\end{array}
$$

The carry is "0"; therefore the carry flag is set indicating a "borrow."

8. *SBI Data 8.*   The carry bit is internally added to the byte immediately following the instruction. This value is then subtracted from the accumulator using 2's complement arithmetic. The carry bit will be set if there is no carry from bit 7 and reset if there is a carry. All condition bits are affected.

EXAMPLE 10-24

SBI 24H

Before instruction:

A		A	Z	S	P	C	
7F		0	0	1	0	1	Flags

After instruction:

A		A	Z	S	P	C	
5A		1	0	0	1	0	Flags

24 H + Carry = 25 H
2's complement of 25 H = 1 1 0 1 1 0 1 1

$$
\begin{array}{cc}
\text{7 F} & 0\,1\,1\,1\,1\,1\,1\,1 \\
+\text{ DB} & +\,1\,1\,0\,1\,1\,0\,1\,1 \\
\hline
& 1\,0\,1\,0\,1\,1\,0\,1\,0 \\
& \qquad\text{Carry}
\end{array}
$$

Since the carry is set, the carry flag is reset in this subtract operation.

9. *INR r.*   Increment the byte in the specified register or memory location by one. All condition bits but the carry are affected.

EXAMPLE 10-25

INR B

Before instruction:

B		A	Z	S	P	C	
2F		0	0	0	0	0	Flags

After instruction:

B		A	Z	S	P	C
30		1	0	0	1	0

10. *DCR r.* Decrement the contents of specified register or memory location by one. The carry bit is not affected, but all other condition bits are affected.

EXAMPLE 10-26

DCR L

Before instruction:

L		A	Z	S	P	C	
90		0	0	0	1	0	Flags

After instruction:

L		A	Z	S	P	C	
8F		0	0	1	0	0	Flags

EXAMPLE 10-27

DCR M

Before instruction:

H	L		M		A	Z	S	P	C	
28	C3		28C3	4B		0	0	0	0	0	Flags

After instruction:

H	L		M		A	Z	S	P	C	
28	C3		28C3	4A		1	0	0	0	0	Flags

11. *INX rp.* Increment the 16-bit quantity in the specified register pair by one. B = BC, D = DE, H = HL, SP. WARNING: *No* condition bits are affected!

EXAMPLE 10-28

INX H

Before instruction:

H   L

20	FF

After instruction:

H   L

21	00

12. *DCX rp*.   Decrement the 16-bit quantity in the specified register pair by one. *No* condition bits are affected.

EXAMPLE 10-29

DCX D

Before instruction:

D   E

00	00

After instruction:

D   E

FF	FF

13. *DAD rp* (double add instruction).   The 16-bit quantity in the specified register pair is added to the 16-bit quantity in the HL register pair. The result is placed in the HL pair! Only the carry bit is affected.

EXAMPLE 10-30

DAD B

Before instruction:

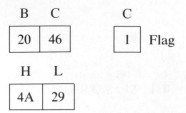

B   C   C

| 20 | 46 | | 1 | Flag
|----|----|

H   L

4A	29

After instruction:

B	C
20	46

C	
0	Flag

H	L
6A	6F

Note: The instruction DAD H will double the contents of the HL register pair.

14. *DAA* (decimal adjust accumulator). The 8 bits in the accumulator are adjusted to form a pair of BCD-encoded digits based on the contents of the accumulator and the state of the carry and auxiliary carry flags. This instruction is included for the implementation of BCD addition and subtraction.

1. If the least significant 4 bits of the A register are greater than 9 or if the auxiliary carry is set, 6 is added to the contents of the A register.

2. If the most significant 4 bits of the A register are greater than 9 or if the carry is set, 60 H is added to the contents of the A register.

All condition bits are affected by this operation.

EXAMPLE 10-31

DAA

Before instruction:

A
7B

A	Z	S	P	C	
0	0	0	0	0	Flags

After instruction:

A
81

A	Z	S	P	C	
1	0	1	1	0	Flags

For more information on this instruction, see the detailed example on BCD addition in Chapter 12, Advanced Programming.

5.3 Logical Group

The logical group of instructions (there are 19) perform Boolean operations (AND, OR, NOT) on data in registers or memory locations and also control the state of the condition flags.

1. *ANA r* (AND register).   The ANA r command ANDs the binary value of a register with the accumulator and saves the result in the accumulator. In doing so, the carry flag is cleared. The logical AND function of the two bits is one if and only if both of the bits equal one. All condition bits are affected except the auxiliary carry.*

EXAMPLE 10-32

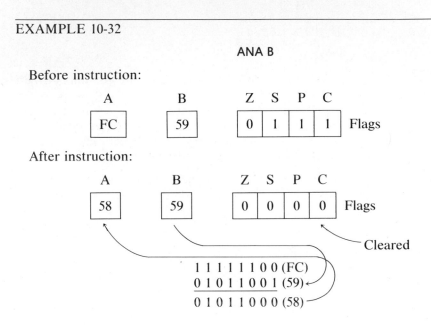

ANA B

2. *ANA M* (AND memory). The accumulator and the memory byte addressed by the HL register pair are ANDed together, and the result is placed in the accumulator. The carry flag is cleared. All condition bits are affected.

EXAMPLE 10-33

ANA M

Before instruction:

	A		H	L		M		Z	S	P	C	
	37		31	84	3183	26		0	0	0	0	Flags
					3184	59						
					3185	8F						

---

*The auxiliary carry is not affected by *any* of the logical instructions.

After instruction:

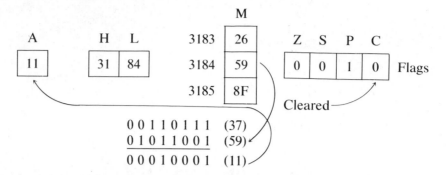

00110111 (37)
01011001 (59)
00010001 (11)

3. *ANI data 8* (AND immediate).   This AND's the accumulator with 8 bits of supplied data and places the result in the accumulator. The carry flag is cleared. All condition bits are affected.

EXAMPLE 10-34

ANI 7BH

Before instruction:

A   Z S P C
[B4]   [0][1][1][1] Flags

After instruction:

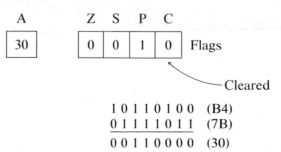

                          Cleared

10110100 (B4)
01111011 (7B)
00110000 (30)

4. *XRA r* (exclusive OR register).   The accumulator is exclusive ORed with register R and the result placed in the accumulator. The carry flag is cleared. All condition bits are affected.

EXAMPLE 10-35

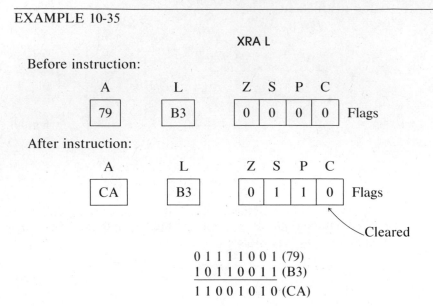

XRA L

Before instruction:

A	L	Z	S	P	C	
79	B3	0	0	0	0	Flags

After instruction:

A	L	Z	S	P	C	
CA	B3	0	1	1	0	Flags

Cleared

```
0 1 1 1 1 0 0 1 (79)
1 0 1 1 0 0 1 1 (B3)
1 1 0 0 1 0 1 0 (CA)
```

Note: In the special case of r = A, we have XRA A, which clears the accumulator. In this case, the zero flag is set because the result is zero. Proof of this is left as a problem for the reader.

5. *XRA M* (exclusive OR memory). The accumulator and the memory byte addressed by the HL register pair are exclusive ORed and the result placed in the accumulator. The carry flag is cleared. All condition bits are affected.

EXAMPLE 10-36

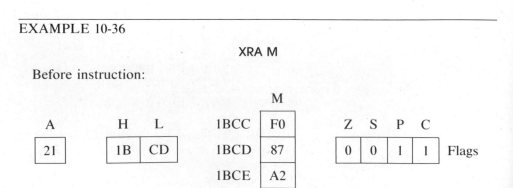

XRA M

Before instruction:

A	H	L		M	Z	S	P	C	
21	1B	CD	1BCC	F0	0	0	1	1	Flags
			1BCD	87					
			1BCE	A2					

After instruction:

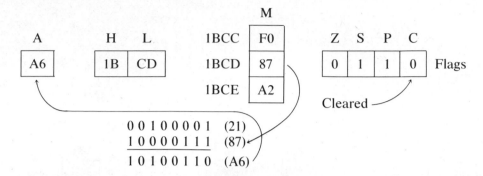

$$
\begin{array}{ll}
0\ 0\ 1\ 0\ 0\ 0\ 0\ 1 & (21) \\
1\ 0\ 0\ 0\ 1\ 1\ 1 & (87) \\
\hline
1\ 0\ 1\ 0\ 0\ 1\ 1\ 0 & (A6)
\end{array}
$$

6. *XRI data 8* (exclusive OR immediate).   The accumulator and the given 8 bits of data are exclusive ORed and the result placed in the accumulator. The carry flag is cleared. All condition bits are affected.

EXAMPLE 10-37

XRI 0C5H

Before instruction:

After instruction:

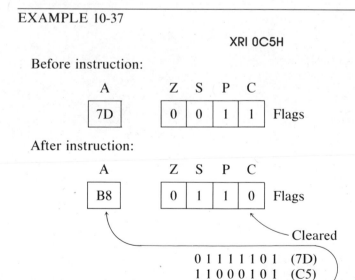

$$
\begin{array}{ll}
0\ 1\ 1\ 1\ 1\ 1\ 0\ 1 & (7D) \\
1\ 1\ 0\ 0\ 0\ 1\ 0\ 1 & (C5) \\
\hline
1\ 0\ 1\ 1\ 1\ 0\ 0\ 0 & (B8)
\end{array}
$$

7. *ORA r* (OR register).   The accumulator and register R are ORed and the result placed in the accumulator. The carry flag is cleared. All condition bits are affected.

EXAMPLE 10-38

ORA C

Before instruction:

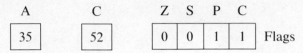

After instruction:

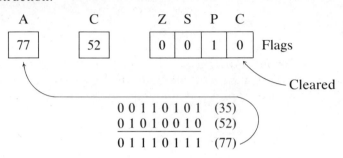

8. *ORA M* (OR memory).   The accumulator and the memory byte addressed by the HL register pair are ORed and the result is placed in the accumulator. The carry flag is cleared. All condition bits are affected.

EXAMPLE 10-39

ORA M

Before instruction:

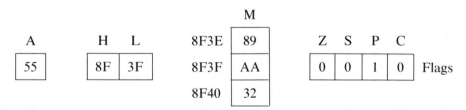

After instruction:

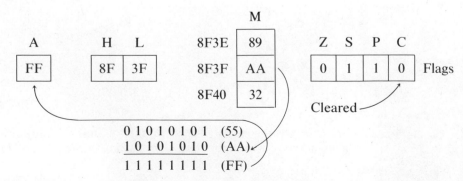

9. *ORI data 8* (OR immediate).   The accumulator and the supplied byte are ORed and the result placed in the accumulator. The carry flag is cleared. All condition bits are affected.

EXAMPLE 10-40

ORI 0BFH

Before instruction:

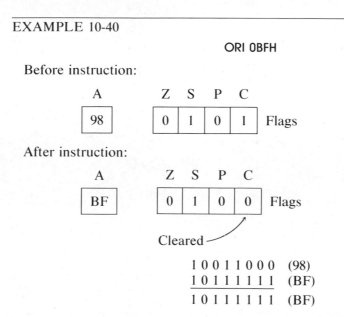

After instruction:

```
1 0 0 1 1 0 0 0   (98)
1 0 1 1 1 1 1 1   (BF)
1 0 1 1 1 1 1 1   (BF)
```

10. *CMP r* (compare register).   The contents of the accumulator are compared to the contents of register r. The comparison is performed by internally subtracting the contents of r from A, leaving both registers unchanged. The zero flag is set if A = r. The carry flag is set if A is less than r. The accumulator remains unchanged, although the condition flags change as a result of the subtraction performed in the compare. All condition bits are affected.

EXAMPLE 10-41

CMP H

Before instruction:

After instruction:

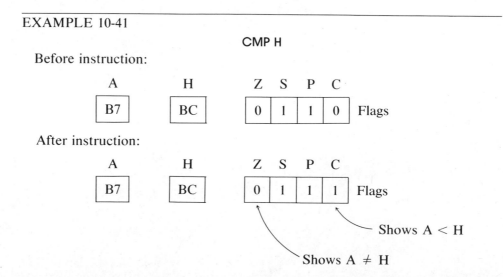

11. *CMP M* (compare memory).   The contents of the accumulator are compared to the memory byte specified by the address stored in the HL register pair. The zero flag is set if A = M (HL). The carry flag is set if A is less than M (HL). All condition bits are affected.

EXAMPLE 10-42

CMP M

Before instruction:

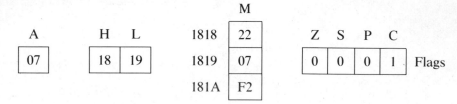

After instruction:

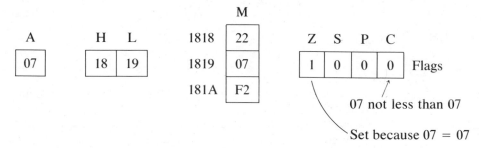

12. *CPI data 8* (compare immediate).   The contents of the accumulator are compared with the supplied byte. The zero flag is set if A = data 8. The carry flag is set if A is less than data 8. All condition bits are affected.

EXAMPLE 10-43

CPI 0DH

Before instruction:

	A		Z	S	P	C	
	30		0	0	1	0	Flags

After instruction:

	A		Z	S	P	C	
	30		0	0	1	0	Flags

The flags remain the same, since 30 H ≠ 0D H (zero flag) and 30 H > 0D H (carry flag).

13. *RLC* (rotate accumulator left).   The contents of the accumulator are all shifted one bit position to the left. Both the carry flag and the LSB are set to the value shifted out of the high-order position. Only the carry flag is affected.

EXAMPLE 10-44

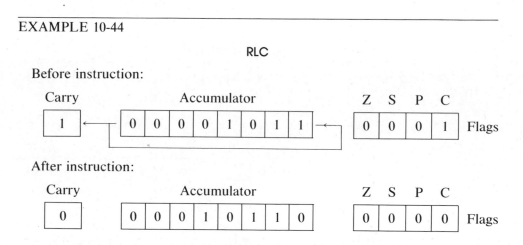

---

14. *RRC* (rotate accumulator right).   The contents of the accumulator are rotated one bit position to the right. Both the carry flag and the MSB are set to the value shifted out of the LSB. Only the carry flag is affected.

EXAMPLE 10-45

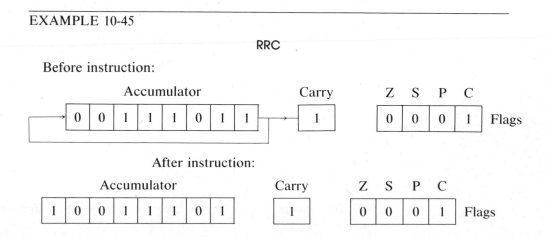

15. *RAL* (rotate left through carry).   The contents of the accumulator are rotated one bit position left through the carry flag. The LSB is set equal to the carry flag and the carry flag is set equal to the value shifted out of the MSB. Only the carry flag is affected.

EXAMPLE 10-46

RAL

Before instruction:

Carry		Accumulator								Z	S	P	C	
0	←	1	0	0	1	1	1	1	1	0	1	1	0	Flags

After instruction:

Carry	Accumulator								Z	S	P	C	
1	0	0	1	1	1	1	1	0	0	1	1	1	Flags

16. *RAR* (rotate right through carry).   The contents of the accumulator are rotated one bit position right through the carry flag. The MSB is set to the value of the carry flag. The carry flag is set to the value rotated out of the LSB.

EXAMPLE 10-47

RAR

Before instruction:

Accumulator								Carry	Z	S	P	C	
1	1	1	0	1	0	1	1	1	0	1	1	1	Flags

After instruction:

Accumulator								Carry	Z	S	P	C	
1	1	1	1	0	1	0	1	1	0	1	1	1	Flags

17. *CMA* (complement accumulator).   The contents of the accumulator are complemented or inverted. No flags are affected.

EXAMPLE 10-48

CMA

Before instruction:

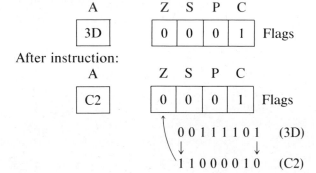

After instruction:

18. *CMC* (complement carry). The carry flag is inverted. No other flags are affected.

EXAMPLE 10-49

CMC

Before instruction:

Z	S	P	C	
1	0	1	0	Flags

After instruction:

Z	S	P	C	
1	0	1	1	Flags

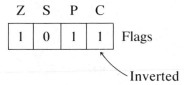

19. *STC* (set carry). The carry flag is set to a one. No other flags are affected.

EXAMPLE 10-50

STC

Before instruction:

Z	S	P	C	
1	1	0	0	Flags

After instruction:

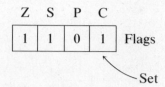

The logical group of instructions is very useful in manipulating and testing data for specific results. When combined with the next group, the branch group, the logical group provides the programmer with a large resource of instructions in which to write programs.

5.4 Branch Group

There are eight basic instruction types in the branch group. This does not sound like many, but we do have *conditional* jumps and subroutine calls. A conditional jump is a jump that does *not* occur unless a certain condition is met. (The real use of our flags!)

The conditional jumps in the 8080A are as follows:

JNZ	Jump if not zero (Z = 0)
JZ	Jump if zero (Z = 1)
JNC	Jump if no carry (C = 0)
JC	Jump if carry (C = 1)
JPO	Jump if parity odd (P = 0)
JPE	Jump if parity even (P = 1)
JP	Jump if positive (S = 0)
JM	Jump if minus (S = 1)

EXAMPLE 10-51

JNZ 3000H

Before instruction:

After instruction:

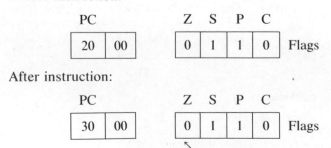

If the condition for the jump is not met, the program counter is merely incremented to the next instruction address.

1. *JMP* addr (unconditional jump).   Always jump to addr.

EXAMPLE 10-52

JMP 2171H

Before instruction:

PC

40	02

After instruction:

PC

21	71

In addition to conditional jumps, the 8080A has conditional calls. A call is used to execute a "program within a program," more commonly called a *subroutine*. To execute a call instruction, the 8080A first places the return address (the address to come back to after the completion of the subroutine) into a special memory area called the stack. It then jumps to the address specified in the call instruction.

The stack pointer is a special 8080A 16-bit register used to point to a specific area in memory where temporary values are stored. It can be set by using the LXI SP, data 16 instruction from the data transfer group. In a call instruction, the address of the instruction following the call is "pushed" or stored in the stack area.

EXAMPLE 10-53

119A CM 2113H
119D next instruction after call if minus

Before instruction:

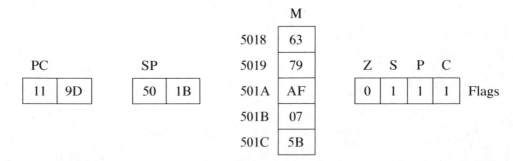

After instruction:

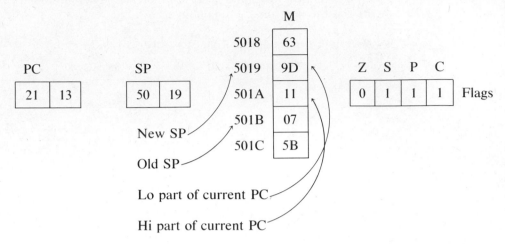

PC

| 21 | 13 |

SP

| 50 | 19 |

M

5018	63
5019	9D
501A	11
501B	07
501C	5B

New SP
Old SP
Lo part of current PC
Hi part of current PC

Z S P C

| 0 | 1 | 1 | 1 | Flags

In this example, the current PC (119D) H got pushed onto the stack area. The lo byte (9D) was stored in M (SP−2). The hi byte of the PC (11) was stored in M (SP−1); then the stack pointer was replaced by SP−2, and the call address was transferred to the PC register.

The eight conditional calls available to the programmer are as follows:

CNZ
CZ
CNC
CC
CPO
CPE
CP
CM

In addition, there is an unconditional call instruction, CALL addr, which does everything that the conditional calls do, regardless of the condition flags. The CALL instruction always calls a subroutine.

After the 8080A has executed a subroutine, it needs some way of knowing that it must go back, or "return" to the address right after the call instruction. The instruction that accomplishes this is called the RET instruction, and simply means "return unconditionally."

As usual, however, we have conditional instructions of this type; these are:

RNZ
RZ
RNC

RC

RPO

RPE

RP

RM

EXAMPLE 10-54

RET

Before instruction:

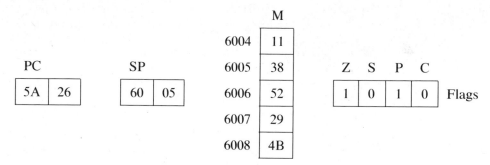

After instruction:

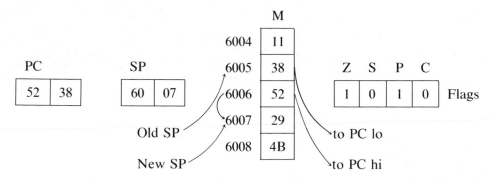

The unconditional (independent of condition flags) return tells the 8080A to take M (SP) and put it into the lo byte of the PC. Then it takes M (SP + 1) and puts it into the hi PC byte. The SP is then replaced by SP + 2.

2. *RST n* (restart).   The restart instruction is a type of unconditional subroutine call in which the 8080A specifies the subroutine address instead of the programmer. The eight RST instructions are listed in Table 10-1; each one has its own unique address.

TABLE 10-1 Restart Instructions and
Their Subroutine Addresses (in hex)

Instruction	Address
RST 0	0000
RST 1	0008
RST 2	0010
RST 3	0018
RST 4	0020
RST 5	0028
RST 6	0030
RST 7	0038

You may notice that there are only 8 bytes between the different subroutine addresses for the RST instruction. Well, 8 bytes is not a very large space for a subroutine. Thus a programmer usually uses the RST instruction for a subroutine that is called *very* often and puts an unconditional jump at the restart address to the real address of the subroutine. To clarify this, let's consider a programmer who has a subroutine that is called 50 times (from different points) in the program. This means that a total of 150 bytes (one for the instruction and two for the address times 50) have been used to accomplish all the calls. By replacing all the 3-byte call instructions with a 1-byte RST instruction and putting an unconditional JMP to the subroutine at the restart address, only 53 bytes need be used, thus saving 97 bytes of program memory. Often it becomes important to get a program to fit into a certain number of bytes (e.g., to be usable in a 2708 or 2716 PROM), and methods like these help to accomplish the task.

3. *PCHL* (jump indirect to HL).   The program counter is replaced by the 16-bit value in the HL register pair.

EXAMPLE 10-55

PCHL

Before instruction:

H   L   PC

| 11 | 08 | | 30 | 17 |

After instruction:

H   L   PC

| 11 | 08 | | 11 | 08 |

This concludes the branch group of instructions. By using conditional calls and jumps, we are able to control sequences of events in our program, hence to perform desired tasks for certain events.

5.5 Stack-I/O-Machine Control

There are 14 instructions that are used to alter the stack, send and receive data, and control some internal flags. However, two (RIM and SIM) can be used only on the 8085.

1. *PUSH RP* (push). Push the contents of the specified register pair onto the stack. The hi byte of the pair goes into M (SP − 1), the lo byte goes into M (SP − 2), and the stack pointer (SP) gets replaced by SP − 2.

EXAMPLE 10-56

PUSH H

Before instruction:

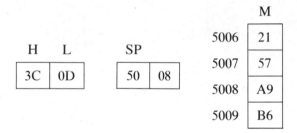

After instruction:

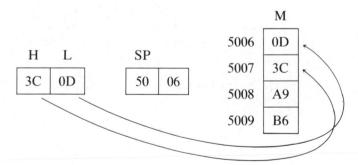

The PUSH instruction is used to temporarily save the HL register pair so that the HL registers may be used for something else.

2. *PUSH PSW* (push the program status word). The same as PUSH RP except that the PSW, program status word, is pushed onto the stack. The PSW is composed of the accumulator and a special byte that contains the condition flags.

The lo byte is the condition flag byte and the hi byte of the PSW is the accumulator (see Fig. 10-4). The lo byte contains 1, 0, and 0 in bit locations 1, 3, and 5, and condition flags C, P, AC, Z, and S in locations 0, 2, 4, 6, and 7. We will see in Chapter 11 how useful the PUSH commands are when we interrupt the 8080A.

FIGURE 10-4 PSW. (Program Status Word).

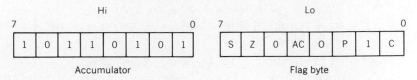

3. *POP RP* (pop).   The register pair specified is loaded with data from the stack. The lo byte gets loaded from M (SP). The hi byte gets loaded from M (SP + 1). Then the stack pointer is replaced by SP + 2.

EXAMPLE 10-57

POP B

Before instruction:

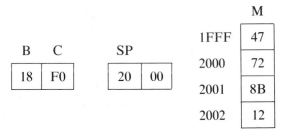

After instruction:

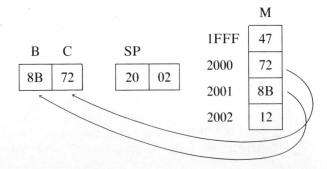

4. *POP PSW* (pop the program status word).   This is the same as a regular pop except the accumulator gets loaded with the hi byte popped off the stack and the condition flags get set to the lo byte popped.

5. *XTHL* (exchange top of stack with HL).   Register L changes places with M (SP) and register H changes places with M (SP + 1).

EXAMPLE 10-58

XTHL

Before instruction:

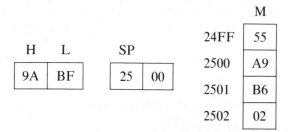

After instruction:

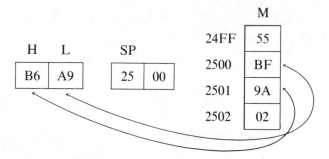

6. *SPHL* (move HL to SP).   The contents of the HL register pair are copied into the SP register, thus defining a stack area in memory starting at address HL.

EXAMPLE 10-59

SPHL

Before instruction:

After instruction:

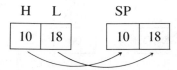

All the instructions we covered in the data transfer section showed various ways to transfer data from register to register and from register to memory, but we do not yet know how to get data from the outside world into the computer, or vice versa. We have two means available: one is the port method, and the other, called memory-mapped I/O, is covered in Chapter 11.

A port is an electronic device composed of TTL gates that is connected to the 8080A's data bus and address bus. The 8080A can output or input to any of 256 ports. It does this by sending a binary number 0–255 on the low part of the address bus. The port circuitry decodes this (ports are designed to work with only one of the 256 different numbers) and then waits for a signal to tell it to place data onto the data bus, or take data off the data bus.

It is through ports that computers communicate with terminals, modems, X–Y plotters, disk drives, and other hardware.

Ports constitute the computer's method of communicating with the outside world.

7. *IN port* (input).   Data from the specified port is placed into the accumulator.

EXAMPLE 10-60

IN 17H

Before instruction:

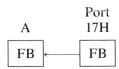

After instruction:

8. *OUT port* (output).   Data from the accumulator is sent to the desired port (via the external data bus.)

EXAMPLE 10-61

OUT 07H

Before instruction:

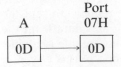

After instruction:

FIGURE 10-5 8085 RIM instruction.

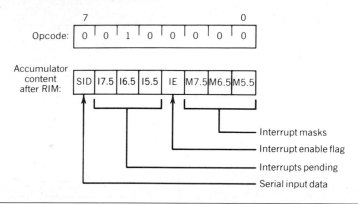

9. *EI* (enable interrupts).   Set the internal 8080A interrupt enable flip-flop. The flip-flop is enabled at the completion of the EI instruction.

10. *DI* (disable interrupts).   Clear the internal 8080A interrupt enable flip-flop. The flip-flop is cleared at the completion of the DI instruction.

11. *HLT* (halt).   Stop the processor. All flags and registers remain unchanged. The CPU remains halted until either an interrupt or a reset occurs.

12. *NOP* (no operation).   Do nothing. The flags and registers remain unchanged.

13. *RIM* (Read Interrupt Mask: 8085 only).   Upon execution of the RIM instruction (Fig. 10-5), information related to serial input and the interrupts is loaded into the A register. The following information is loaded:

- Current interrupt mask status for the 8085 restarts 5.5, 6.5, and 7.5. A "1" indicates that the interrupt is disabled.
- Interrupt enable flag status ("1" = interrupts enabled) except the TRAP interrupt.

    Hardware interrupts pending (5.5, 6.5, 7.5).
- Serial input data (into bit 7).

14. *SIM* (Set Interrupt Mask: 8085 only). The execution of the SIM instruction (Fig. 10-6) uses the contents of the A register to set up any of the following conditions:

- Disable any of the RST 5.5, 6.5, and 7.5 hardware interrupts. (A "1" in the bit position disables the interrupts.)
- Load the serial output data output latch (from bit 7).
- Reset the edge-triggered RST7.5 input latch.

FIGURE 10-6 8085 SIM instruction.

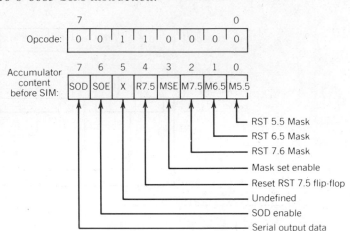

6 THE COMPLETE 8080A/85 INSTRUCTION SET

The full 244-entry instruction set for the 8080A is presented in Table 10-2. You should recognize most of the instructions from the examples that were given in Section 5. The symbols NOP, MOV A, B, JNZ, and POP H are all mnemonics—abbreviated names of the instruction.

In addition, you should notice the opcode column. An opcode is a hexadecimal number representing the number of the instruction. We see that the opcode for NOP is 00 H, for MVI H is 26 H, and for CALL is 0CD H. This is the way the instruction is actually stored in memory. When the 8080A reads one of these instruction bytes, it decodes it internally to determine what kind of instruction it is. Essentially, there are three types of instructions: 1-, 2-, and 3-byte instructions.

One-byte instructions include NOP, XRA A, and STC, and are represented in memory by the bytes 00 H, AF H, 37 H.

Two-byte instructions (e.g., MVI C, 09 H, IN 02 H, ORI 0FF H) are represented in memory as 0E 09, DB 02, F6 FF.

As the 8080A reads the opcodes, it identifies the number of bytes in each set of instructions. Hence, it also looks at the memory location right after the one holding the instructional opcode. It then executes these 1- or 2-byte instruction and also increments the program counter so that it does not try to execute the ''data'' as an instruction.

Finally, we have the 3-byte instructions: CALL 20F7 H, JMP 3000 H, LXI H, 2F2F H are examples. To execute an instruction, the 8080A looks at the next two bytes and skips over them to get to the instruction opcode that follows. The bytes are stored in memory as: CD F7 20, C3 00 30, 21 2F 2F. Notice that the order of the last two bytes in 3-byte instructions is reversed. We store an address or number in memory lo byte first, and then the hi byte. In this way, SHLD 1234 H becomes 22 34 12, with 34 H being the lo byte and 12 H being the hi byte. Many micro-

TABLE 10-2 8080 Instruction Set: 2-MHz Clock Rate

Instructions	Opcode	Cycle States	Execution Time (μsec)	Memory Accesses
MOV A,A	7F	5	2.5	1
MOV A,B	78			
MOV A,C	79			
MOV A,D	7A			
MOV A,E	7B			
MOV A,H	7C			
MOV A,L	7D			
MOV B,A	47			
MOV B,B	40			
MOV B,C	41			
MOV B,D	42			
MOV B,E	43			
MOV B,H	44			
MOV B,L	45			
MOV C,A	4F			
MOV C,B	48			
MOV C,C	49			
MOV C,D	4A			
MOV C,E	4B			
MOV C,H	4C			
MOV C,L	4D			
MOV D,A	57			
MOV D,B	50			
MOV D,C	51			
MOV D,D	52			
MOV D,E	53			
MOV D,H	54			
MOV D,L	55			
MOV E,A	5F			
MOV E,B	58			
MOV E,C	59			
MOV E,D	5A			
MOV E,E	5B			
MOV E,H	5C			
MOV E,L	5D			
MOV H,A	67			
MOV H,B	60			
MOV H,C	61			
MOV H,D	62			
MOV H,E	63	5	2.5	1

Instructions	Opcode	Cycle States	Execution Time (μsec)	Memory Accesses
MOV H,H	64	5	2.5	1
MOV H,L	65			
MOV L,A	6F			
MOV L,B	68			
MOV L,C	69			
MOV L,D	6A			
MOV L,E	6B			
MOV L,H	6C			
MOV L,L	6D	5	2.5	1
MOV M,A	77	7	3.5	2
MOV M,B	70			
MOV M,C	71			
MOV M,D	72			
MOV M,E	73			
MOV M,H	74			
MOV M,L	75			
MOV A,M	7E			
MOV B,M	46			
MOV C,M	4E			
MOV D,M	56			
MOV E,M	5E			
MOV H,M	66			
MOV L,M	6E			
MVI A,DDD	3E			
MVI B,DDD	06			
MVI C,DDD	0E			
MVI D,DDD	16			
MVI E,DDD	1E			
MVI H,DDD	26			
MVI L,DDD	2E	7	3.5	2
MVI M,DDD	36	10	5.0	3
INR A	3C	5	2.5	1
INR B	04			
INR C	0C			
INR D	14			
INR E	1C			
INR H	24			
INR L	2C	5	2.5	1
INR M	34	10	5.0	3
DCR A	3D	5	2.5	1

Instructions	Opcode	Cycle States	Execution Time (μsec)	Memory Accesses
DCR B	05	5	2.5	1
DCR C	0D			
DCR D	15			
DCR E	1D			
DCR H	25			
DCR L	2D	5	2.5	1
DCR M	35	10	5.0	3
ADD A	87	4	2.0	1
ADD B	80			
ADD C	81			
ADD D	82			
ADD E	83			
ADD H	84			
ADD L	85	4	2.0	1
ADD M	86	7	3.5	2
ADC A	8F	4	2.0	1
ADC B	88			
ADC C	89			
ADC D	8A			
ADC E	8B			
ADC H	8C			
ADC L	8D	4	2.0	1
ADC M	8E	7	3.5	2
SUB A	97	4	2.0	1
SUB B	90			
SUB C	91			
SUB D	92			
SUB E	93			
SUB H	94			
SUB L	95	4	2.0	1
SUB M	96	7	3.5	2
SBB A	9F	4	2.0	1
SBB B	98			
SBB C	99			
SBB D	9A			
SBB E	9B			
SBB H	9C			
SBB L	9D	4	2.0	1
SBB M	9E	7	3.5	2
CMP A	BF	4	2.0	1

(continued on next page)

TABLE 10-2 8080 Instruction Set: 2-MHz Clock Rate

Instructions	Opcode	Cycle States	Execution Time (μsec)	Memory Accesses
CMP B	B8	4	2.0	1
CMP C	B9			
CMP D	BA			
CMP E	BB			
CMP H	BC			
CMP L	BD	4	2.0	1
CMP M	BE	7	3.5	2
ADI DDD	C6	7	3.5	2
ACI DDD	CE	4	2.0	1
SUI DDD	D6			
SBI DDD	DE			
CPI DDD	FE	7	3.5	2
ANA A	A7	4	2.0	1
ANA B	A0			
ANA C	A1			
ANA D	A2			
ANA E	A3			
ANA H	A4	4	2.0	1
ANA L	A5	4	2.0	1
ANA M	A6	7	3.5	2
ANI DDD	E6	7	3.5	2
ORA A	B7	4	2.0	1
ORA B	B0			
ORA C	B1			
ORA D	B2			
ORA E	B3			
ORA H	B4	4	2.0	1
ORA L	B5			
ORA M	B6	7	3.5	2
ORI DDD	F6	7	3.5	2
XRA A	AF	4	2.0	1
XRA B	A8			
XRA C	A9			
XRA D	AA			
XRA E	AB			
XRA H	AC	4	2.0	1
XRA L	AD			
XRA M	AE	7	3.5	2
XRI DDD	EE	7	3.5	2
IN DDD	DB	10	5.0	2
OUT DDD	D3	10	5.0	2

Instructions	Opcode	Cycle States	Execution Time (μsec)	Memory Accesses
HLT	76	7	3.5	1
NOP	00	4	2.0	1
DI	F3			
EI	FB			
RLC	07	4	2.0	1
RAL	17			
RRC	0F			
RAR	1F	4	2.0	1
LXI B,ADDR	01	10	5.0	3
LXI D,ADDR	11			
LXI H,ADDR	21			
LXI SP,ADDR	31	10	5.0	3
STA ADDR	32	13	6.5	4
LDA ADDR	3A	13	6.5	4
CMA	2F	4	2.0	1
DAA	27			
STC	37			
CMC	3F	4	2.0	1
POP B	C1	10	5.0	3
POP D	D1			
POP H	E1			
POP PSW	F1	10	5.0	3
PUSH B	C5	11	5.5	3
PUSH D	D5			
PUSH H	E5			
PUSH PSW	F5	11	5.5	3
XCHG	EB	4	2.0	1
XTHL	E3	18	9.0	5
SPHL	F9	5	2.5	1
DAD B	09	10	5.0	1
DAD D	19			
DAD H	29			
DAD SP	39	10	5.0	1
STAX B	02	7	3.5	2
STAX D	12			
LDAX B	0A			
LDAX D	1A	7	3.5	2
INX B	03	5	2.5	1
INX D	13			
INX H	23			
INX SP	33	5	2.5	1

Instructions	Opcode	Cycle States	Execution Time (μsec)	Memory Accesses
DCX B	0B	5	2.5	1
DCX D	1B			
DCX H	2B			
DCX SP	3B	5	2.5	1
SHLD ADDR	22	16	8.0	5
LHLD ADDR	2A	16	8.0	5
JMP ADDR	C3	10	5.0	3
JC ADDR	DA			
JZ ADDR	CA			
JM ADDR	FA			
JPE ADDR	EA			
JNC ADDR	D2			
JNZ ADDR	C2			
JP ADDR	F2			
JPO ADDR	E2	10	5.0	3
PCHL	E9	5	2.5	1
CALL ADDR	CD	17	8.5	5
CC ADDR	DC	17/11	8.5/5.5	5/3
CZ ADDR	CC			
CM ADDR	FC			
CPE ADDR	EC			
CNC ADDR	D4			
CNZ ADDR	C4			
CP ADDR	F4			
CPO ADDR	E4	17/11	8.5/5.5	5/3
RET	C9	10	5.0	3
RC	D8	11/5	5.5/2.5	3/1
RZ	C8			
RM	F8			
RPE	E8			
RNC	D0			
RNZ	C0			
RP	F0			
RPO	E0	11/5	5.5/2.5	3/1
RST 0	C7	11	5.5	3
RST 1	CF			
RST 2	D7			
RST 3	DF			
RST 4	E7			
RST 5	EF			
RST 6	F7			
RST 7	FF	11	5.5	3

FIGURE 10-7 Sample 8080A application.

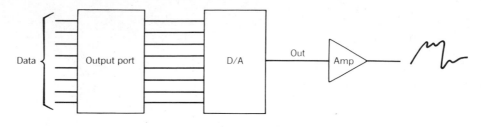

processors use this convention, (called Intel byte swapping), so become *very* familiar with it.

Before we look at some actual programming examples, let's consider the number of cycles per instruction, which is important in our instruction table. We see that NOP, DAD D, and JMP 3007 H take, respectively, 4, 10, and 10 clock cycles. This means that an NOP instruction takes four clock cycles to execute. Since the 8080A has a clock frequency of 2 MHz, the NOP instruction takes:

$$4 \left(\frac{1}{2 \text{ MHz}}\right) = 4 (0.5 \times 10^{-6} \text{ sec}) = 2 \text{ } \mu\text{sec}$$

to execute. By the same method, DAD D and JMP both take 5 μsec to execute. The exact time of an instruction must be known when timing is critical in a program. As an example, the 8080A might be programmed to send data out to a port, which in turn gets fed into a digital-to-analog converter to generate a waveform (Fig. 10-7). If the programmer ignored the fact that the instructions in the 8080A take a specific time to execute, the time between samples for the desired waveform would not be correct; hence the waveform would be distorted, off frequency, or otherwise wrong.

7 SIMPLE 8080A/85 PROGRAMMING EXAMPLES

Now that the instruction set for the 8080A has been introduced, we present a few simple programming examples using new instructions. You are encouraged to compare their use to the description of the instruction in Sections 5 and 6 of this chapter. These examples make use of input/output ports and the need to produce delay loops in most microcomputer programs.

7.1 Input/Output Ports

The 8080A CPU can communicate with outside devices through its IN and OUT instructions. The 8080A can support 256 input/output ports if the necessary decoding hardware has been provided by the designer. A port can be both an input *and* an output port, with the direction differentiating between the two ports. Let us assume that we are using a computer with port number FF (255D) connected to perform input from eight SPST switches and output to eight LEDs as shown in Fig. 10-8.

FIGURE 10-8 Circuitry for an 8-bit input/output port (addr = FF) for an 8085 CPU system.

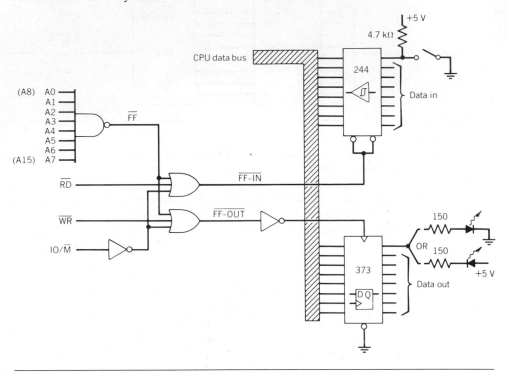

The switch port contains eight switches, which provides a byte of information based on the position of all the switches. An up position yields a "1" in a given bit position. The light port contains eight light-emitting diodes that indicate the contents of a byte of information sent to that port.

A simple program would be to read or input the setting of the switches to the accumulator and output that quantity from the accumulator to the light port. The light port would then show the state of the eight switches.

EXAMPLE 10-62

```
IN 0FFH      ;In from switch port
OUT 0FFH     ;Out to light port
HLT          ;Stop the computer
```

In this example, an input is done from port FF, the switch port. An output is done to port FF, the light port. Both ports are FF H (255 D), the difference being that one is an input and one is an output. It doesn't make much sense to input from a set of lights! Figure 10-9 shows the flow of data. Notice that the accumulator is involved in the operation. This brings up the very important point that the A

FIGURE 10-9 Flow of data.

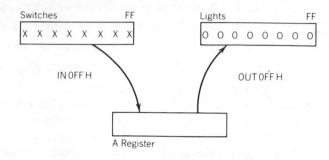

register in the 8080A is central to the operation of the computer. It is involved in *all* I/O operations using ports. It is also used in all logical and arithmetic operations.

Since this is a program that must run on the computer, we must have a logical stopping point, and the HLT or halt instruction assures that the computer will not run away executing random data in memory as instructions.

The operation codes (opcodes) for the instructions IN and OUT are DB H and D3 H, respectively.

$$IN\ Port\ Number\ =\ DB\ Port\ Number$$
$$OUT\ Port\ Number\ =\ D3\ Port\ Number$$

The opcode for HLT is 76 H.

The program in machine language (opcodes) would appear as a string of hexadecimal pairs:

$$DB\ FF\ D3\ FF\ 76$$

The program is loaded into the computer memory in order to run the program. Each byte is loaded into a sequential memory location beginning at a starting location or origin (ORG) specified by the programmer.

A program is written in mnemonic form using a well-planned syntax. This allows a computer to read the information in mnemonic form and convert the mnemonics to a string of hexadecimal opcodes. This process, whether done by a person or by a computer, is referred to as "assembling" a program. The programmer writes an 8080A program in terms of symbols and mnemonics and then the program is assembled or made ready for the 8080A in opcode form. Table 10-3 shows the correct format for writing programs in mnemonic form as used in this text.

TABLE 10-3 Assembler Syntax

LABEL	MNEMONIC	OPERAND	COMMENTS
	ORG	1000H	;Starting Address
START:	IN	0FFH	;In from Port FF
	OUT	0FFH	;Out to Port FF
	HLT		;Stop
	END		;End of Program

The four "fields" are used as follows:

> LABEL   A name or symbol used by jump, call instructions
>
> MNEMONIC   The instruction in symbolic form
>
> OPERAND   Information required by the instruction, if any
>
> COMMENTS   Additional information, preceded by a semicolon.

ORG and END are called *pseudo*-opcodes. They are not for use by the 8080A but by the assembler. The assembler must know the starting address or ORG to calculate any jump or call addresses. The assembler must also know when to stop assembling; hence the END instruction.

The program designed to transfer switch information to light information is stored in memory as shown in Fig. 10-10, beginning at ORG = 1000 H. The first instruction occupies two 8-bit memory locations (1000 H and 1001 H). This is a 2-byte instruction. The third instruction is a 1-byte instruction. The 8080A has 1-, 2-, and 3-byte instructions. The opcodes are placed in memory. There will be other random hex pairs in unused memory locations. The computer will start running our program at location 1000 H, proceeding from one sequential instruction to another until it is told to halt, which occurs at location 1004 in this program. The computer will never get to the random bytes at locations 1005, 1006, and so on.

This program will run only once. The current switch positions will be read and transferred to the lights. Then the computer will enter the halt mode. Any further changing of the switches will not be seen on the lights, since the only connection between the lights and the switches consists of the accumulator and our program. To reflect any change of switch setting on the lights, we must repeat the program.

FIGURE 10-10 Program to transfer switch information to light information.

Address	Data	
0FFF	C2	Old random data
1000	DB	} First instruction and data
1001	FF	
1002	D3	} Second instruction and data
1003	FF	
1004	76	Third instruction
1005	21	} Other data
1006	42	

This is done in a better version of this program by replacing the *halt* (HLT) instruction with a jump (JMP) to the start of the program at location 1000 H. This appears as follows:

```
ORG 1000H
IN 0FFH
OUT 0FFH
JMP 1000H;Run program again
END
```

1000	DB
1001	FF
1002	D3
1003	FF
1004	C3
1005	00
1006	10

DB FF D3 FF C3 00 10

Notice that the bytes following the jump instruction (C3) are reversed:

C3 00 10

Address 1000 H

This is standard for all 8080A 2-byte quantities (Intel byte swapping).

Between inputting the switch information and outputting to the lights, we find the data in the accumulator. This is a chance to alter the switch data.* For instance, we could invert or complement the data to produce an opposite effect on the lights. At the same time, we will change the program to jump to a label instead of an address:

```
        ORG 1000H
TOP:    IN 0FFH
        CMA   ;Complement accumulator
        OUT 0FFH
        JMP TOP
        END
```

The assembler realizes that the input instruction (IN) has the opcode DB stored at 1000 H. TOP, a label, is identified as the memory location containing the address of the IN instruction or 1000 H.

The opcode for CMA is 2F H. Thus the opcode string is as follows:

---

*This particular version looks at the switches almost 67,000 times every second!

1000	DB
1001	FF
1002	2F
1003	D3
1004	FF
1005	C3
1006	00
1007	10

DB FF 2F D3 FF C3 00 10

This new program inverts the data going from the switches to the lights and also runs continuously.* If a switch is changed, the effect is seen 12 μsec later (almost immediately). This is calculated by referring to the time each instruction takes to complete. The total program time is 17 μsec.

$$
\begin{array}{ll}
\text{IN} & 5\ \mu\text{sec} \\
\text{CMA} & 2\ \mu\text{sec} \\
\text{OUT} & 5\ \mu\text{sec} \\
\text{JMP} & \underline{5\ \mu\text{sec}} \\
& 17\ \mu\text{sec}
\end{array}
$$

$\left.\begin{array}{l} \\ \\ \end{array}\right\}$ 12 μsec

7.2 A Simple Eight-Bit Binary Counter

The accumulator can be used to produce a simple 8-bit binary count at the light port (FF). The accumulator may be zeroed in one of three ways:

$$
\begin{array}{ll}
\text{MVI A, 0} & 3.5\ \mu\text{sec} \\
\text{XRA A} & 2\ \mu\text{sec} \\
\text{SUB A} & 2\ \mu\text{sec}
\end{array}
$$

MVI A, 0 moves 0 into the A register. XRA A exclusive ORs the A register contents with itself, producing zero.

EXAMPLE 10-63

If A contains 7E H, then XRA A produces

$$
\begin{array}{c}
0\ 1\ 1\ 1\ 1\ 1\ 1\ 0 \\
\underline{0\ 1\ 1\ 1\ 1\ 1\ 1\ 0} \\
0\ 0\ 0\ 0\ 0\ 0\ 0\ 0
\end{array}
$$

---

*This one looks at the switches only 59,000 times a second.

Since:

$$0 \oplus 0 = 0$$
$$1 \oplus 1 = 0$$

Subtracting the contents of A from itself also zeros the A register (see Fig. 10-11).

```
        TITLE "Simple Binary Counter Program"
        ORG 1000H
        MVI A, 0    ;Zero the A register
TOP:    OUT 0FFH    ;Out to the light port
        INR A       ;Add one to A
        JMP TOP     ;Repeat the output sequence
        END
```

Figure 10-12 shows the result of using a computer system assembler to assemble the program. Note the addresses on the left, then the opcodes, label, and mnemonics. The printout is referred to as a list file. A corresponding hex file is produced and looks like this:

3E 00 D3 FF 3C C3 02 10

Notice that the jump instruction, C30210, causes a jump to address 1002 H (remember byte swapping!). This is the address of the label TOP. This program (shown in memory in Fig. 10-13) has one problem. It operates too fast for the eye to discern a binary count! The loop part of the program:

OUT	5 μsec
INR	2.5 μsec
JMP	5 μsec
	12.5 μsec

requires 12.5 μsec to function. The LSB of the light port changes every 12.5 μsec. The MSB changes every 128×12.5 μsec or every 1.6 msec. This corresponds to approximately 300 Hz, too fast to see! To make the counter easier to see, some delay must be introduced. The concept of delay is very important in writing

---

FIGURE 10-11 Register A contents being transferred to the light port.

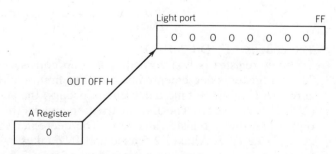

FIGURE 10-12 Simple binary counter program (DEFLT = not otherwise specified).

```
Bcc 8080A/8085 Assembler Ver. 13(40)        File:KOUNT                      Page   1
Title: ' SIMPLE BINARY COUNTER PROGRAM '



        LOC     OBJ     LINE     PROGRAM

                        1              TITLE' SIMPLE BINARY COUNTER PROGRAM '
                        2              ORG  1000H
                        3
        1000    3E00    4              MVI A , 0
        1002    D3FF    5      TOP:    OUT OFFH
        1004    3C      6              INR A
        1005    C30210  7              JMP TOP
                        8              END




Bcc 8080A/8085 Assembler Ver. 13(40)        File:KOUNT                      Page   2
Title: ' SIMPLE BINARY COUNTER PROGRAM '




        ----------------------------- SYMBOL TABLE -----------------------------


        ::  $    1008    PC   ::  B   0000  DEFLT ::  C   0001  DEFLT ::  D   0002  DEFLT ::
        ::  E    0003    DEFLT ::  H   0004  DEFLT ::  L   0005  DEFLT ::  M   0006  DEFLT ::
        ::  A    0007    DEFLT ::  PSW 0006  DEFLT ::  SP  0006  DEFLT ::  TOP 1002  ADR  ::




End of 8080A/8085 Assembly,     0 Error(s)
```

microcomputer programs and is covered in detail in the next chapter. The basic idea is to write a program that takes up CPU time. An example would be:

```
        ORG 1000H
        XRA A       ;Zero the A register
LOOP:   DCR A       ;Decrement the A register
        JNZ LOOP    ;Jump if not zero to loop
        END
```

The code is AF 3D C2 01 10.

The A register is first zeroed. The loop consists of DCR A and JNZ LOOP. The A register is decremented by one. The first time the register decrements from zero to FF H. Each time a decrement occurs, the accumulator contains one less (255, 254, 253, etc.). The JNZ instruction checks the zero flag, which will be set only when the A register hits zero. The decrementing loop continues until A is zero. Since DCR A takes 2.5 μsec and JNZ takes 5 μsec, each cycle of the loop takes 7.5 μsec. A complete cycle will take 256×7.5 μsec = 1920 μsec. A shorter

FIGURE 10-13 Simple counter in memory.

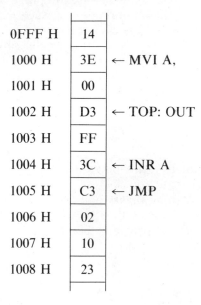

0FFF H	14	
1000 H	3E	← MVI A,
1001 H	00	
1002 H	D3	← TOP: OUT
1003 H	FF	
1004 H	3C	← INR A
1005 H	C3	← JMP
1006 H	02	
1007 H	10	
1008 H	23	

time can occur by prefilling the A register with a byte other than zero. For example, move 48 H into A and then enter the decrementing loop. This produces a delay of 48 H × 7.5 μsec or 72 × 7.5 μsec = 540 μsec.

The loop can be lengthened a bit by adding a few NOP instructions, which require 2 μsec each.

```
            ORG 1000H
    LOOP:   DCR A        2.5 µsec
            NOP          2.0 µsec
            NOP          2.0 µsec
            NOP          2.0 µsec
            JNZ LOOP     5.0 µsec
            END          13.5 µsec
```

The code is 3D 00 00 00 C2 00 10.

This loop requires 13.5 μsec for one pass. If A is zeroed to start with, 256 loops will produce a delay of 256 × 13.5 μsec = 3456 μsec. For longer delays, a better method is required. This can be accomplished with the use of more registers, as shown in the next chapter.

8 SUMMARY

This chapter has introduced you to a very powerful microprocessor, the 8080A. The organization of the processor into registers (A, B, C, D, E, H, L, PC, SP) and instructions for manipulating data in those registers were presented in detail. These instructions, when used in proper and clever order, form a program to be followed by the microprocessor. Such programs make the 8080A one of the most powerful controllers ever produced, far outranking those using traditional logic design (TTL). Simple microcomputer programs were introduced using input and output ports, and the concept of a delay loop was presented. The chapters that follow use the instructions presented here to great advantage. As we develop our programming skills, we will learn to use more instructions with each new program.

9 GLOSSARY

Accumulator.   An 8-bit register used by the 8080A for all arithmetic, logical, and in/out instructions.

Addr.   A 16-bit binary address.

AND. $0 \times 0 = 0, 0 \times 1 = 0, 1 \times 1 = 1$.

Assembler.   A program that converts mnemonics to opcodes, thus saving the programmer work.

Auxiliary carry bit.   A bit or flag indicating a carry from the fourth to fifth position of a byte.

Branch.   A relative jump. If the program counter is at address 4000 H and a branch 20 H is performed, the PC now contains 4020 H. The 8080A does *not* use this type of instruction.

Carry bit.   A bit indicating a carry resulting from a logical or arithmetic operation.

Comments.   The comment field in a program is used to indicate the purpose of each line. Good programmers use lots of comments.

Complement.   To change zeros to ones and ones to zeros. Example: the complement of 11011011 = 00100100. Also called the 1's complement.

Condition bits.   The 5 bits used in the 8080A to indicate the state of a register after an instruction has been performed. Also called flags.

Cycle time.   The minimum time unit in a microprocessor related to the clock (0.5 μsec in 8080A).

Data 8.   An 8-bit data byte.

Data 16.   Two 8-bit data bytes.

Disassembler.   A program that converts opcodes back into mnemonic form.

END.   A statement needed at the end of a program by an assembler.

Exclusive OR (XOR).   A logic function between two variables.

$$0 \oplus 0 = 0, \quad 0 \oplus 1 = 1, \quad 1 \oplus 1 = 0$$

Execution time.   The time required by the CPU to perform an instruction.

Flags.   See Condition bits.

Instruction set.   A list of mnemonics and opcodes used by the microprocessor.

Jump.   The processor goes to the specified address by loading that address into the program counter. The program continues execution from that address.

Label.   An abbreviation used by programmers to identify an address. If TOP = 1000 H, JMP TOP is preferred to JMP 1000 H.

Loop.   A series of repeating instructions.

Memory. A place for storage of hexadecimal pairs in binary form, representing a program or data.

Mnemonic. An abbreviation used to represent an instruction (MVI A, IN, HLT, etc.).

Opcode. A hexadecimal pair representing an instruction (3E, DB, 76, etc.).

Operand. Required information for an instruction. For example, in MVI A, 7 and OUT 0FF H: A, 7 and 0FF H are operands.

OR. $0 + 0 = 0,\ \ 0 + 1 = 1,\ \ 1 + 1 = 1.$

ORG. The origin or starting address of a program.

Parity flag. A conditional bit that is set to one if a byte contains an even number of ones; otherwise it is reset

Pop. To retrieve 16 bits of information from temporary storage in a place in memory called the stack. This function is controlled by the stack pointer.

Port. One of 256 8-bit input or output areas used by the CPU to communicate with the outside world (I/O section).

Program. A logical sequence of mnemonics or opcodes (instructions) followed by the computer.

Program counter. A 16-bit register used to keep track of the location in memory of the next instruction to be performed by the CPU.

Pseudo-opcode. Directions used by the assembler, but not used on the 8080A itself (e.g., ORG, END, TITLE, DB, DS, ASCII, EQU, SET).

Push. A way of pushing 16 bits into temporary storage in a place in memory called the stack. Controlled by the stack pointer.

r (register). One of the registers available on the 8080A (A, B, C, D, E, H, L, M). *Note:* M is not a register, but a memory location.

Register. An 8-bit storage location available to the CPU.

Return. An instruction used to return from a subroutine.

Rotate. To shift the bits of a register one position, either left or right, depending on the instruction.

rp (register pair). One of four possible register pairs holding 16-bit information. B = BC, D = DE, H = HL.

Sign flag. Set to "1" if the MSB of a byte in the accumulator has been set as the result of an operation.

Stack. An area of memory reserved by the programmer for use by the CPU for temporary storage of variables.

Stack pointer. A 16-bit register used to keep track of the location of the stack in memory.

Two's complement. Take the 1's complement and add one. Example: 2's complement of 10001101 is 01110010 + 1 = 01110011. The 8080A uses 2's complement math to perform subtractions.

Zero flag. A conditional bit indicating that a register has just been cleared (or that the result of the last operation was zero).

PROBLEMS

10-1 Sketch a block diagram of the internal structure of the 8080A registers.

10-2   Assemble the following 8080A program using the instruction set:

```
              ORG 2000H
              MVI A, 20H
              MVI B, 30H
      DEC:    DCR B
              JNZ DEC
              HLT
              END
```

10-3   Calculate the time required to perform the program in Problem 10-2.

10-4   Disassemble the following hexadecimal string of opcodes. (Put back in mnemonic form, ORG 2200 H.)

D3 20 3E 14 3C AF 21 00 30 C2 02 22 76

10-5   Determine the state of the five flags after this program sequence.

S	Z	P	C	A
0	0	0	0	0

```
MVI A, 55H
DCR A
```

10-6   Determine the state of the five flags after this program sequence.

S	Z	P	C	A
1	0	1	1	0

```
STC
LXI B, 0000H
DCX B
INR B
HLT
```

10-7   Determine the time delay produced by the following program loop:

```
        MVI A, 7FH   ;Load A register
DECR:   DCR A        ;Decrement A
        JNZ DECR     ;Jump if not zero and decrement again
        HLT
```

10-8   Repeat Problem 10-7 for this program loop:

```
              MVI A, 0A0H
      DECR:   DCR A
              NOP
              NOP
              NOP
              JNZ DECR
              HLT
```

10-9   Determine the closest hexadecimal value to put into the A register to produce a 2500-μsec delay in this delay loop.

```
              MVI A, _____
       DECR:  DCR A
              NOP
              NOP
              NOP
              NOP
              NOP
              NOP
              JNZ DECR
              HLT
```

10-10 Assemble each program in Problems 10-5 through 10-9, assuming an origin of 5000 H.

10-11 Calculate the difference between an ORG of 1000 H and 1000 D.

10-12 Show that XRA A clears the accumulator.

10-13 Show at least four ways of clearing the accumulator.

10-14 Show how the DAD H instruction can be used to multiply HL by 16.

10-15 Write a set of instructions to subtract the hex value 3BF7 from the HL register pair. Destroy NO registers.

10-16 Write a set of instructions to add 7FF7 to the DE register pair. (*Hint:* XCHG would be very useful.) Destroy NO registers.

10-17 Write a set of instructions to perform the following operations:
(a) DE = 8 * DE
(b) HL = BC + (3 * HL)
(c) A = B + E + (7 * D) (may use a loop)
(d) BC = BC/16 (ignore remainder)

CHAPTER 11

8080A/85 PROGRAMMING APPLICATIONS

1 INSTRUCTIONAL OBJECTIVES

When finished with the chapter you should be able to:

1. Write a delay routine (using a single register, double registers, or more) to delay the 8080A/85 for a specified amount of time.

2. Understand and be able to use the 8080A/85 stack operation instructions, including subroutine calls.

3. Be able to employ specific instructions to test, flip, clear, or set bits or bytes in the 8080A/85's registers or memory.

4. Write a routine to control the transfer of data to and from an I/O port employing the method of status bit checking.

5. Derive the control or mode word for a USART or PIA for a desired function.

6. Understand the methods involved in converting between ASCII data and hex, or BCD data, and vice versa.

7. Write a program from scratch in mnemonics (including labels, etc.) and assemble it, or assemble an existing program.

8. Write routines able to recognize single-letter commands and perform different tasks.

2 SELF-EVALUATION QUESTIONS

Keep the following questions in mind and try to answer them when you have completed the chapter:

1. What is "overhead"?

2. Explain the "push" and "pop" operations on the stack.

3. What is meant by "last in, first out"?

4. What is a subroutine?

335

5. What occurs in the first pass of an assembler?

6. What is a symbol table?

7. What are some of the advantages of serial data transmission over parallel? What are some disadvantages?

8. What is a keypressed signal used for?

9. What is scrolling?

3 INTRODUCTION

Chapter 10 introduced the 8080A instruction set, and we investigated a few programming examples using 8080A registers. This chapter concentrates on improving your use of the 8080A instructions by examining a number of microcomputer applications. (*Note:* The 8080A and the 8085 microprocessors have identical register structures and use the same instruction set. They are, however, different electrically.) Each succeeding idea or program introduced uses more of the instructions than the last. Table 11-1 shows another form of the instruction set for use in this chapter.

We begin with the delay loop and discuss different ways of producing delay. Topics covered in this chapter include:

1. Double-register delay.

2. Use of the stack.

3. Variable delay.

4. Caterpillar simulation.

5. Teletype/CRT I/O routines.

6. Traffic Signal controller.

7. Integrated circuit tester.

8. Disco light show.

9. Memory-mapped video display.

10. Simple adder routine.

4 DOUBLE-REGISTER DELAY

The single-register delay of Chapter 10 is limited in the amount of delay it can produce. The delay is caused by sending a short program through a loop many times.

$$
\text{Loop} \longrightarrow \text{DECR:} \quad
\begin{array}{ll}
\text{XRA A} & 2.0\ \mu\text{sec} \\
\text{DCR A} & 2.5\ \mu\text{sec} \\
\text{JNZ DECR} & 5.0\ \mu\text{sec}
\end{array}
\left.\right\} \ 7.5\ \mu\text{sec}
$$

This loop took $256 \times 7.5\ \mu\text{sec} = 1920\ \mu\text{sec}$. Longer delays are possible by using a pair of 8-bit registers arranged such that a nested loop (a loop within a loop) occurs. The idea is shown in Fig. 11-1. A pair of registers are selected for use, and each register is initially filled with ones (0FF H). We decrement register C until it is empty (255 decrements); then register B is decremented by one ($255 - 1 = 254$). Now register B is used to count the number of times register C is emptied. Register C is filled again, and another 255 decrements of C occur. Register B is again decremented ($254 - 1 = 253$). It will take 255×255 total decrements (of B and C) to have both registers end up at zero. The program looks like this:

TABLE 11-1 The 8080A/85 Instruction Set

DATA TRANSFER GROUP

Move (MOV) / Move (cont)

MOV			MOV			MOV		
A,A	7F		E,A	5F		H,A	67	
A,B	78		E,B	58		H,B	60	
A,C	79		E,C	59		H,C	61	
A,D	7A		E,D	5A		H,D	62	
A,E	7B		E,E	5B		H,E	63	
A,H	7C		E,H	5C		H,H	64	
A,L	7D		E,L	5D		H,L	65	
A,M	7E		E,M	5E		H,M	66	

MOV			MOV			MOV		
B,A	47		L,A	6F		C,A	4F	
B,B	40		L,B	68		C,B	48	
B,C	41		L,C	69		C,C	49	
B,D	42		L,D	6A		C,D	4A	
B,E	43		L,E	6B		C,E	4B	
B,H	44		L,H	6C		C,H	4C	
B,L	45		L,L	6D		C,L	4D	
B,M	46		L,M	6E		C,M	4E	

MOV			MOV		
M,A	77		D,A	57	
M,B	70		D,B	50	
M,C	71		D,C	51	
M,D	72		D,D	52	
M,E	73		D,E	53	
M,H	74		D,H	54	
M,L	75		D,L	55	
			D,M	56	

XCHG EB

Move Immediate (MVI)

MVI		
A, byte	3E	
B, byte	06	
C, byte	0E	
D, byte	16	
E, byte	1E	
H, byte	26	
L, byte	2E	
M, byte	36	

Load Immediate (LXI)

LXI		
B, dble	01	
D, dble	11	
H, dble	21	
SP, dble	31	

Load/Store

LDAX B	0A
LDAX D	1A
LHLD adr	2A
LDA adr	3A
STAX B	02
STAX D	12
SHLD adr	22
STA adr	32

ARITHMETIC AND LOGICAL GROUP

Add* (ADD) / ADC

ADD			ADC		
A	87		A	8F	
B	80		B	88	
C	81		C	89	
D	82		D	8A	
E	83		E	8B	
H	84		H	8C	
L	85		L	8D	
M	86		M	8E	

Subtract* (SUB) / SBB

SUB			SBB		
A	97		A	9F	
B	90		B	98	
C	91		C	99	
D	92		D	9A	
E	93		E	9B	
H	94		H	9C	
L	95		L	9D	
M	96		M	9E	

Double Add† (DAD)

DAD		
B	09	
D	19	
H	29	
SP	39	

Increment** (INR) / INX

INR			INX		
A	3C		B	03	
B	04		D	13	
C	0C		H	23	
D	14		SP	33	
E	1C				
H	24				
L	2C				
M	34				

Decrement** (DCR) / DCX

DCR			DCX		
A	3D		B	0B	
B	05		D	1B	
C	0D		H	2B	
D	15		SP	3B	
E	1D				
H	25				
L	2D				
M	35				

Specials

DAA*	27
CMA	2F
STC†	37
CMC†	3F

Rotate†

RLC	07
RRC	0F
RAL	17
RAR	1F

Logical* (ANA) / XRA

ANA			XRA		
A	A7		A	AF	
B	A0		B	A8	
C	A1		C	A9	
D	A2		D	AA	
E	A3		E	AB	
H	A4		H	AC	
L	A5		L	AD	
M	A6		M	AE	

ORA / CMP

ORA			CMP		
A	B7		A	BF	
B	B0		B	B8	
C	B1		C	B9	
D	B2		D	BA	
E	B3		E	BB	
H	B4		H	BC	
L	B5		L	BD	
M	B6		M	BE	

Arith & Logical Immediate

ADI byte	C6
ACI byte	CE
SUI byte	D6
SBI byte	DE
ANI byte	E6
XRI byte	EE
ORI byte	F6
CPI byte	FE

BRANCH CONTROL GROUP

Jump

JMP adr	C3
JNZ adr	C2
JZ adr	CA
JNC adr	D2
JC adr	DA
JPO adr	E2
JPE adr	EA
JP adr	F2
JM adr	FA
PCHL	E9

Call

CALL adr	CD
CNZ adr	C4
CZ adr	CC
CNC adr	D4
CC adr	DC
CPO adr	E4
CPE adr	EC
CP adr	F4
CM adr	FC

Return

RET	C9
RNZ	C0
RZ	C8
RNC	D0
RC	D8
RPO	E0
RPE	E8
RP	F0
RM	F8

Restart (RST)

RST		
0	C7	
1	CF	
2	D7	
3	DF	
4	E7	
5	EF	
6	F7	
7	FF	

I/O AND MACHINE CONTROL

Stack Ops

PUSH			POP		
B	C5		B	C1	
D	D5		D	D1	
H	E5		H	E1	
PSW	F5		PSW*	F1	
			XTHL	E3	
			SPHL	F9	

Input/Output

OUT byte	D3
IN byte	DB

Control

DI	F3
EI	FB
NOP	00
HLT	76

New Instructions (8085 Only)

RIM	20
SIM	30

ASSEMBLER REFERENCE

Operators

- NUL
- LOW, HIGH
- *, /, MOD, SHL, SHR
- +, –
- NOT
- AND
- OR, XOR

---

byte = constant, or logical/arithmetic expression that evaluates to an 8-bit data quantity. (Second byte of 2-byte instructions).

dble = constant, or logical/arithmetic expression that evaluates to a 16-bit data quantity. (Second and Third bytes of 3-byte instructions).

adr = 16-bit address (Second and Third bytes of 3-byte instructions).

* = all flags (C, Z, S, P, AC) affected.

** = all flags except CARRY affected; (exception: INX and DCX affect no flags).

† = only CARRY affected.

All mnemonics copyright ©Intel Corporation 1976.

FIGURE 11-1 Two-register delay program flow chart.

```
                        ┌───────────┐
                        │   Start   │
                        └───────────┘
                              │
                        ┌───────────┐
                        │Initialize │
                        │ counter 1 │
                        └───────────┘
                              │
                        ┌───────────┐◄──────────────┐
                        │Initialize │               │
                        │ counter 2 │               │
                        └───────────┘               │
                              │                      │
                        ┌───────────┐◄──────┐        │
                        │ Decrement │       │        │
                        │ counter 2 │       │        │
                        └───────────┘       │        │
                              │              │        │
                            ╱─────╲    No    │        │
                           ╱ Zero? ╲─────────┘        │
                           ╲       ╱                  │
                            ╲─────╱                   │
                              │ Yes                    │
                        ┌───────────┐                 │
                        │ Decrement │                 │
                        │ counter 1 │                 │
                        └───────────┘                 │
                              │                        │
                            ╱─────╲    No              │
                           ╱ Zero? ╲───────────────────┘
                           ╲       ╱
                            ╲─────╱
                              │ Yes
                        ┌───────────┐
                        │   Stop    │
                        └───────────┘
```

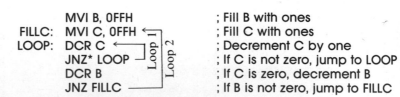

```
        MVI B, 0FFH          ; Fill B with ones
FILLC:  MVI C, 0FFH          ; Fill C with ones
LOOP:   DCR C                ; Decrement C by one
        JNZ* LOOP            ; If C is not zero, jump to LOOP
        DCR B                ; If C is zero, decrement B
        JNZ FILLC            ; If B is not zero, jump to FILLC
```

*The zero flag is set or reset by any register math operation. The
register most recently used is assumed to have affected the flag.

The effect of using this "nested loop" is that the registers multiply the delay
produced by a single register by 255 (255 × 255 = 65,025). The delay time is
calculated as follows:

```
        MVI B, 0FFH      3.5 μsec × 1           =            3.5
FILLC:  MVI C, 0FFH ←    3.5 μsec × 255         =          892.5
LOOP:   DCR C ←          2.5 μsec × 255 × 255   =      162,562.5
        JNZ LOOP ┘       5.0 μsec × 255 × 255   =      325,125.0
        DCR B            2.5 μsec × 255         =          637.5
        JNZ FILLC        5.0 μsec × 255         =        1,275.0
                                                     490,496.0 μsec
```

The delay comes to about a half second! Producing this type of delay is useful in learning programming and also important in applications requiring long time delays. In Section 12 we examine the operation of a traffic signal in which a half second of delay will *not* be long enough!

Keep in mind this one point: producing time delay in this manner completely ties up the microprocessor, which cannot do *two* things at once. We will discover a better way when we examine a new integrated circuit, the interval timer (Chapter 14.8). Meanwhile, the required time delay can be appreciably increased with the use of a few NOP instructions. This program makes use of a double-register load instruction, LXI B, to replace MVI B and MVI C. It is introduced for variation and as an alternate method of loading two registers. As each new *instruction* is used, return to Chapter 10 and reread the material about that instruction. Do it *now* for LXI B, data 16. Note that for a double-register instruction, an X is included in the mnemonic. Also, register pair (rp) BC is involved, but only the first letter (B) is used. The program and associated times look like this:

```
        LXI B, 0FFFFH    ; Fill register pair BC with ones
LOOP:   DCR C*           ; Decrement C
        NOP              ; 2 μsec delay
        NOP              ; 2 μsec delay
        NOP              ; 2 μsec delay
        JNZ LOOP**       ; If C not zero, jump to LOOP
        DCR B            ; If C is zero, take one from B
        JNZ LOOP***      ; If B not zero, jump to LOOP
        HLT              ; Stop the processor
```

A flowchart for this program (Fig. 11-2) should be helpful.
The timing for this new loop is as follows:

```
        LXI B, 0FFFFH    5.0 μsec × 1           =              5
LOOP:   DCR C ←          2.5 μsec × 256 × 255   =        163,200
        NOP              2.0 μsec × 256 × 255   =        130,560
        NOP              2.0 μsec × 256 × 255   =        130,560
        NOP              2.0 μsec × 256 × 255   =        130,560
        JNZ LOOP ┘       5.0 μsec × 256 × 255   =        326,400
        DCR B            2.5 μsec × 255         =          637.5
        JNZ LOOP         5.0 μsec × 255         =          1,275
        HLT              3.5 μsec × 1           =            3.5
                                                          883,201
```

*Register C starts with 0FFH the first time, zero thereafter.
**Register C will have set or reset the zero flag.
***Register B will have set or reset the zero flag.

FIGURE 11-2 Program flowchart.

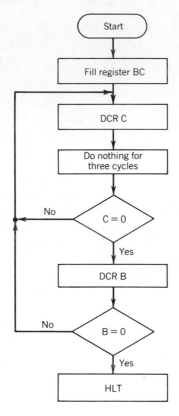

This loop has been extended to almost one second. The inner loop makes 256 trips, one more than the outer loop, because register C starts from 0, not FF, after the initial load.

For longer delays, a third register is required to avoid an endless string of NOPs and wasted program space in memory. Before looking at a three-register delay, let us improve the double-register delay using a double-register decrement instruction (DCX).

5 IMPROVED DOUBLE-REGISTER DELAY

The double-register decrement (DCX) decrements a register pair as a single 16-bit quantity. If the register pair contains 0FFFF H (65535) and a DCX is used, the 16-bit quantity becomes 0FFFE H (65534). This instruction does *not* affect any flags. It does *not* affect the zero flag. This means that we will not be able to use this flag immediately after a DCX to see whether the register pair is zero! We will explore a couple of methods of determining whether a register pair is at zero.

FIGURE 11-3 Checking to see whether DE = 0.

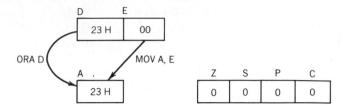

One way is to OR the two registers together. If even a single bit is set, the zero flag will show zero, meaning "not zero yet." If we decrement the DE register pair and then OR E with D (alternately D with E), the zero flag will be affected and a JNZ or JZ instruction can be used.

As the list of instructions indicates, there is no way to directly OR E with D. An indirect way (Fig. 11-3) is to move one of the registers to A and then OR the other register with A.

This is done as follows:

```
         LXI D, 43C1H   ; Delay quantity
LOOP:    DCX D          ; Decrement DE
         MOV A, E       ; Move E to A
         ORA D          ; ORA with D (D with E)
         JNZ LOOP       ; Jump if not zero to LOOP
```

The result of ORA D is placed in the A register, destroying the previous contents. The zero flag is set or reset, depending on the result of ORA D.

An alternate method is to ignore the contents of the least significant register (E) and check the contents of D versus the accumulator with a compare instruction (CMP D).

```
         XRA A          ;Zero the A register
         LXI D, 2943 H  ;Quantity for delay
DECRE:   DCX D          ;Decrement DE
         CMP D          ;Is D = 0 yet?
         JNZ DECRE      ;If not, continue
```

This program loop stops with the E register = 0FFH. Maximum delay may not be achieved, but this simpler loop offers an alternate delay method.

If you have not done it yet, look up the following new instructions in Chapter 10: LXI, DCX, CMP, ORA. Let us examine the delay produced by the first program presented using the DCX instruction.

```
         LXI H, 023CH   ;Quantity of delay
TRIP:    DCX H          ;Decrement HL
         MOV A, H       ;Ready to check H = L = 0
         ORA L          ;Is H = L = 0?
         JNZ TRIP       ;If not, continue decrementing
```

Note: 23CH = 572D.

```
         LXI H, Data 16   5.0 μsec × 1   =      5
TRIP:    DCX H            2.5 μsec × 572 =   1430
         MOV A, H         3.5 μsec × 572 =   2002
         ORA L            2.0 μsec × 572 =   1144
         JNZ TRIP         5.0 μsec × 572 =   2860
                                            7441 μsec
```

Since the LXI H, data 16 instruction does not figure in the loop and adds slightly to the delay time, it is called "overhead." It is necessary to the function of the loop, but because it is *outside* the loop, it does not add significantly to the delay.

Let us assemble this delay loop with an ORG of 4050 H.

```
         213C02    ;Note the byte-swapping
TRIP:    2B        ;address of TRIP = 4053 H
         7C
         B5
         C25340    ;The address is byte swapped too
```

Let us now work backward to write this delay loop to produce a delay of 5000 μsec.

```
         LXI D,   ?    ;Quantity unknown
DOWN:    DCX D                            2.5 μsec
         MOV A, E                         3.5 μsec
         ORA D                            2.0 μsec
         JNZ DOWN                         5.0 μsec
                                         13.0 μsec
```

The basic delay loop takes 13 μsec, and LXI D has an overhead of 5 μsec. Thus we need a delay of 5000 − 5 = 4995 μsec, and

$$\frac{4995 \ \mu\text{sec}}{13 \ \mu\text{sec}} = 384.2307692 \ \mu\text{sec}$$

We have a choice of 384 or 385 times through the loop.

$$384 \times 13 \ \mu\text{sec} = 4992 \ \mu\text{sec}$$
$$385 \times 13 \ \mu\text{sec} = 5005 \ \mu\text{sec}$$

Since 384 is closer (4992 μsec), we'll use that value in the DE register pair. Let's not forget to convert to hexadecimal! 384 D = 180 H. The instruction to load becomes LXI D, 180 H and the total delay is 4992 + 5 = 4997 μsec.

It's not 5000 μsec, but it's close! If you need 5000 μsec, add some overhead. Find an instruction that takes 3 μsec. (There isn't one, but you can come within a half microsecond!) Try adding INX D to the end of the program.

As a final version of this program, it may be desirable to have a variable delay set from a set of eight external switches connected to an input port. We choose port FF, used previously. Whenever the delay loop is used, the most significant part of the loop register is loaded from the setting of the port FF switches.

```
         MVI E, 0     ;Zero E register
         IN 0FFH      ;Get 8 bits from port FF switches
         MOV D, A     ;Transfer to D
```

```
DECRE:  DCX D       ;Decrement DE by one
        MOV A, E    ;Ready to test D = E = 0
        ORA D       ;D = E = 0?
        JNZ DECRE   ;If not, decrement again
```

Chapter 10 introduced a program to output a binary count to a set of lights. However, the lights moved too fast to permit observation of the binary counter. It would have been easier to see if the lights moved at a more reasonable speed. We conclude this section on producing delay by rewriting the following "fast" binary light program.

```
        ORG 1000H
        XRA A       ;Zero A register
OUTP:   OUT 0FFH    ;Out to the lights
        INR A       ;Add one to A register
        JMP OUTP    ;Output the new count
        END
```

The delay sequence (loop) is added after the output instruction. In addition, the delay loop uses the A register to test for zero, which means that it *cannot* also be used to keep track of the count appearing on the lights.

Let us use DE for the delay and B for the light counter.

```
        ORG 1000H
        MVI B, 0        ;Set B = 0
OUTP:   MOV A, B        ;Can output from A only
        OUT 0FFH        ;Count out to lights
DELAY:  LXI D, 0FFFFH   ;Maximum delay
DECR:   DCX D           ;One less in DE
        MOV A, D        ;D = E = 0?
        ORA E
        JNZ DECR        ;If not, decrement more
        INR B           ;Increment B
        JMP OUTP        ;Output it
        END
```

As an exercise, assemble the program and calculate the rate at which each light will flash at port FF.

6 USING THE STACK

The stack is an area of memory that is used for temporary storage. It can be located anywhere in memory ranging from 0 to FFFF H. Its actual location is selected by the programmer when the stack pointer is loaded or initialized (Fig. 11-4). It is important to place the stack in an unused area of memory where it will not interfere with the program to be run. As items are added to the stack it uses addresses lower than those last used; that is, the principal operation of the stack is to build toward zero. As items are added, the stack moves inevitably closer to 0. If it encroaches on program memory, the program will fail to function properly.

FIGURE 11-4 Setting up the stack area.

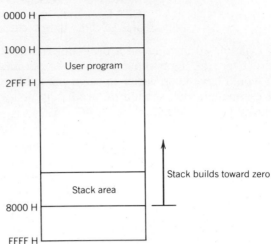

For example, the stack may be set to start at 8000 H. In correct operation address 8000 H will never be used. The first address used will be one less than 8000 H or 7FFF H. The stack memory locations are used two at a time by the following instructional types: PUSH, POP, CALLS, RETURNS, and a few others. If it is desired to save the contents of register pair DE so that D and E can be used for something else, PUSH D will place the contents of D at 7FFF H and the contents of E at 7FFE H. The registers are now free for other uses. Many items may be stored on the stack as long as they are retrieved from the stack in exact reverse order from which they were pushed onto the stack. To retrieve an item from the stack and put it back into the registers, a POP instruction is used: POP D puts the item at 7FFE H back in register E and the item (byte) at 7FFF H back in register D. The stack pointer now points at location 8000 H again. This memory structure is referred to as "last in, first out" (LIFO). In other words, the pair of items most recently added to the stack is the first pair to come off. The stack pointer is automatically advanced or decremented by the CPU when necessary. The programmer must remember to initialize the stack early in a program. However, enough space must be provided to prevent the stack from overrunning the program.

It is frequently necessary to save all the registers of the CPU so that they may be temporarily freed from use. This usually occurs with an "interrupt," as discussed in detail later. Four PUSH instructions put eight items on the stack.

```
PUSH PSW   ;Save A and flags
PUSH B     ;Save BC
PUSH D     ;Save DE
PUSH H     ;Save HL
```

Remember, the stack *must* be initialized before these instructions (LXI SP, 8000 H, or other appropriate address) can be used. These four PUSH instructions save the state of the machine. It is now completely available for other duties.

To return to the previous state, four POP instructions are required. They are done in exact reverse order:

```
POP H     ;Restore HL
POP D     ;Restore DE
POP B     ;Restore BC
POP PSW ;Restore A and flags
```

It is possible to swap register contents as follows:

```
PUSH H   ;Save HL
PUSH D   ;Save DE
POP H    ;DE now in HL
POP D    ;HL now in DE
```

Or the "XCHG" instruction may be used. Alternatively:

```
PUSH H
POP D
```

will duplicate HL in DE, destroying the contents of DE.

7 SUBROUTINES

A good programmer calls frequently on subroutines or subprograms to handle small jobs required by the main program. This is a very important use of the stack. A subroutine begins when a CALL instruction is encountered and ends when control returns to the main program.

```
MAIN:  Main program
          |
       CALL SUB1
          |
       End of main program
SUB1:  Subroutine
          |
       RET                        ;Return to main program
```

When the subroutine is called, the address of the next instruction in the main program is automatically pushed onto the stack. This return address will be needed when the subroutine is executing the RET instruction. *Subroutines* will not work reliably unless the programmer has initialized the stack. The 8080A/85 supports a large number of nested subroutines, limited only by available space in the stack memory.

Let us return to the binary counter program that sends a binary count to the port FF lights, rewriting the program to use a subroutine for the delay loop.

```
       ORG 1000H
       LXI SP, 2000H   ;Initialize stack pointer
       MVI B, 0        ;Start B = 0
OUTP:  MOV A, B
       OUT 0FFH        ;Out to light port
       CALL DELAY      ;Call delay subroutine
       INR B           ;Increment light counter
```

```
              JMP OUTP        ;End of main program
      DELAY:  LXI D, 0FFFFH   ;Start of subroutine
      DECR:   DCX D           ;Delay loop
              MOV A, D
              ORA E
              JNZ DECR
              RET             ;End of subroutine
              END
```

This program is really in two parts: a main program and a delay subroutine.

Let us first count from 1000 H and make a list of the addresses of the labels.

```
      OUTP:   MOV A, B → 1005H
      DELAY:  LXI D → 100FH
      DECR:   DCX D → 1012H
```

We begin counting in hexadecimal from the ORG = 1000H and write down the address that the labeled instructions will fit in, not forgetting to count in *hex* (8, 9, A, B, C, etc.). The assembler must do this for each program during the *first pass* through the program. This list of labels and associated addresses is called a *symbol table*. Be sure you can see where these addresses just came from! When the program is assembled, CALL DELAY becomes CALL 100F H, JMP OUTP becomes JMP 1005 H, and JNZ DECR becomes JNZ 1012 H.

Note that the instruction CALL DELAY is followed by INR B. When the subroutine is called, the address of INR B (100B H) is automatically pushed onto the stack for use by the subroutine when it must do a RETurn to the main program. Now study the assembly of the binary counter program (Table 11-2). Be sure to return to Chapter 10 and look up these new instructions and study them: PUSH, POP, CALL, RET.

Having covered most of our programming fundamentals, let's observe some programs that show the applications of microcomputers. Study each example in detail to improve your understanding and programming ability.

TABLE 11-2   Program Assembly

Address	Opcodes	Comments
1000H	310020	;Stack pointer
1003H	0600	;B = 0
1005H	78	;A ← B
1006H	D3FF	;Out to FF
1008H	CD0F10	;Call delay
100BH	04	;B = B + 1
100CH	C30510	;Jump OUTP
100FH	11FFFF	;Delay subroutine
1012H	1B	;Decrement
1013H	7A	;A ← D
1014H	B3	;ORA E
1015H	C21210	;JNZ DECR
1018H	C9	;Return

8 THE CATERPILLAR

The following programming example is to be completed by you as an exercise. It is another program that uses a set of eight lights, but whereas the examples in this chapter use port FF to operate the light port, the computer available to you may operate on a different port number.

This program (Fig. 11-5) turns on three adjacent lights at one end of the light port and, with a delay between moves, moves the three lights as a group to the other end of the port. The direction then changes and the lights move to the starting end. This process repeats over and over.

A new pair of instructions is needed to cause right and left shifting of data to occur. Both these instructions, RRC (0F) and RLC (07), affect the A register only. They are 8-bit rotate instructions that affect the carry bit but do not rotate through the carry bit. Two other shifting instructions, RAR (1F) and RAL (17), rotate through the carry and constitute a 9-bit rotate. These also affect the carry bit. Use of the instruction causes a 1-bit position shift left or right—just what we need for the caterpillar. You will use the 8-bit rotate instruction. Review all instructions in Chapter 10.

There are several ways to tell whether you have reached the end of the register.

1. You can shift five times with a delay between shift and output. You will count the shifts by setting up a counter register:

```
MVI B, 5
Shift  ←
Delay
Output
DCR B
JNZ Shift
```

FIGURE 11-5 Caterpillar program output.

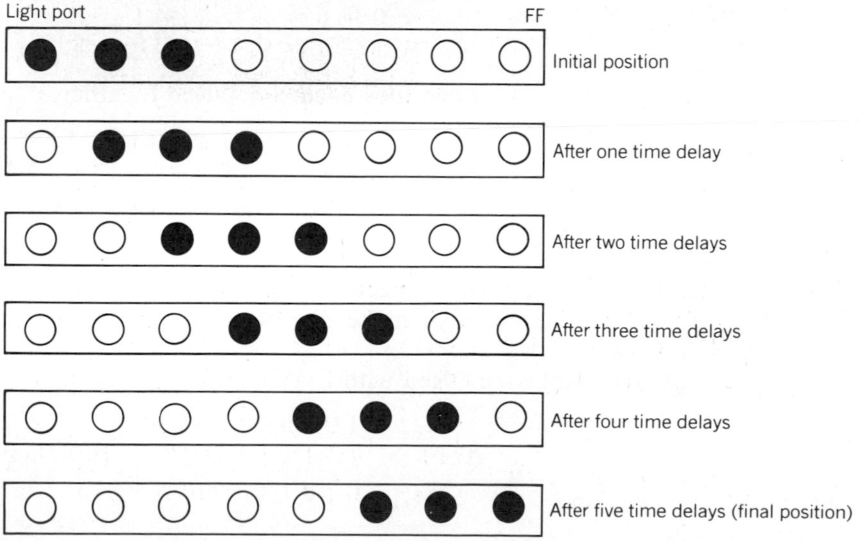

2. You can use a compare instruction (CPI) to see whether the contents of the accumulator are equal to the pattern at either end of the register (1F or E0). Assume here that a zero means the LED is on, and a one means the LED is off.

3. You can use the 9-bit rotate instructions and rely on the carry bits to tell when you are past either end.

You now have enough programming experience to write your first complete program. Needless to say, you will get the *best* experience if you have an 8080A/85-based microcomputer with a light port on which to run and debug the program. It is impossible to learn to program a microcomputer without both practice and the experience of trying to get a program to work correctly. Remember, what *you* program is what *you* get. The computer will *always* do exactly what you tell it (program into it). If you get a result you did not expect, then your expectations were a little off. Profit by the experience, and try again.

Now write your first program. Here are some further restrictions:

1. The program must use a variable delay subroutine. (Input delay from the port FF switches, so the caterpillar rate can change.)

2. There must be only *one* time delay between shifts, including a change in direction at each end. If the program pauses at an end for double time, you have more debugging to do.

Good luck, and have fun! You are about to become an 8080A/85 programmer!

9 BIT FLIPPING: A SOFTWARE FLIP-FLOP

Before proceeding to another programming example, let us investigate the exclusive OR ($\oplus$) function. This is very useful in flipping bits from one state to another. To review, the exclusive OR causes the following:

$$0 \oplus 0 = 0 \qquad 0 \oplus 1 = 1$$
$$1 \oplus 1 = 0 \qquad 1 \oplus 0 = 1$$

Consider two 8-bit quantities exclusive ORed together:

$$
\begin{array}{rl}
\text{F6} & 1\ 1\ 1\ 1\ 0\ 1\ 1\ 0 \\
\oplus\ \underline{\text{A2}} & \underline{1\ 0\ 1\ 0\ 0\ 0\ 1\ 0} \\
54 & 0\ 1\ 0\ 1\ 0\ 1\ 0\ 0 \\
& \qquad\uparrow\quad\uparrow\quad\uparrow
\end{array}
$$

Bit positions 3, 5, and 7 become one. These positions are the bits at which the two bytes *differed*. We make use of the exclusive OR function to produce a "software" flip-flop.

A byte exclusive ORed with FF will flip-flop the bits:

$$
\begin{array}{rl}
\text{A7} & 1\ 0\ 1\ 0\ 0\ 1\ 1\ 1 \;\leftarrow \\
\oplus\ \underline{\text{FF}} & \underline{1\ 1\ 1\ 1\ 1\ 1\ 1\ 1} \\
58 & 0\ 1\ 0\ 1\ 1\ 0\ 0\ 0 \;\leftarrow
\end{array}
\quad\Big] \;\text{Bits flipped}
$$

```
       58      0 1 0 1 1 0 0 0  ←
   ⊕  FF      1 1 1 1 1 1 1 1        Bits flipped back
       A7      1 0 1 0 0 1 1 1  ←
```

To build a *one*-bit flip-flop:

```
       01      0 0 0 0 0 0 0 1  ←
   ⊕  01      0 0 0 0 0 0 0 1        Bit 1 flipped or toggled
       00      0 0 0 0 0 0 0 0  ←

       00      0 0 0 0 0 0 0 0  ←
   ⊕  01      0 0 0 0 0 0 0 1        Bit 1 flipped or toggled back
       01      0 0 0 0 0 0 0 1  ←
```

This can be done as follows:

```
            MVI A, 1    ;Set the flip-flop
            XRI 01      ;Toggle the flip-flop
```

Another XRI 01 will toggle the bit back. Any of 8 bits can be toggled by using the correct quantity after the XRI.

EXAMPLE 11-1

Toggle the fourth bit of A7 H.

```
            A7 H        1 0 1 0 0 1 1 1
         ⊕ 08 H      ⊕ 0 0 0 0 1 0 0 0
            AF H        1 0 1 0 1 1 1 1
```

Now, toggle it back.

```
            AF H        1 0 1 0 1 1 1 1
         ⊕ 08 H        0 0 0 0 1 0 0 0
            A7 H        1 0 1 0 0 1 1 1
```

This allows the programmer the option of setting or resetting individual bits as needed. It also allows a software flip-flop to be set up to keep track of certain events in a program. Only the desired bit changes.

In the caterpillar program requested in Section 8, you should have ended up with two identical sequences in the main program, the only difference being the rotate instructions, RRC = 0F and RLC = 07.

This program can be shortened by modifying the code for the instruction *as the program runs*. Half of your original main program can be eliminated. There is a small difference between 0F and 07, and that is bit 4. An XRI 08H will toggle that bit from a "1" to a "0," thus modifying the program.

Suppose your RLC instruction is at address 1015 H.

```
            1015H RLC
            OUT 0FFH
            CALL DELAY
              ⋮
```

You can modify the sequence by adding the following steps to the appropriate part of the program:

```
LDA 1015H    ;Load A from 1015
XRI 08H      ;Flip the bit
STA 1015H    ;Store A back at 1015
```

This will cause the program to change the instruction at 1015 H from an RLC to an RRC and back again. You can considerably shorten your program. Bear in mind as you work that programmers and engineers consider good programs to be short programs, and also that a program that "self-alters" itself in this way cannot be put into ROM. It will execute properly only in RAM.*

10 SERIAL INPUT/OUTPUT TECHNIQUES

10.1 Teletype and CRT Input/Output Routines

One of the major requirements of a computer is that it be able to communicate with a user. This is usually done by way of teletype, printer, or video terminal. Since these devices are frequently serial data devices, a special receiver/transmitter area is required to perform the necessary conversion from the com-

FIGURE 11-6 Receiver/transmitter chip.

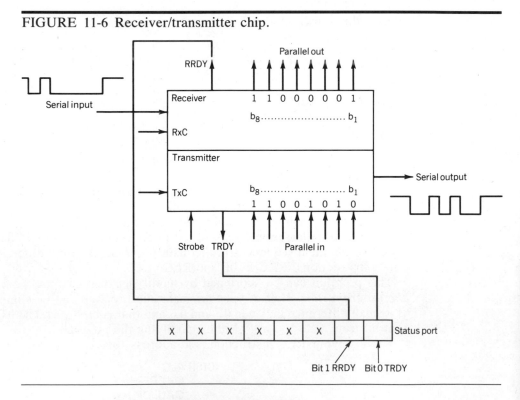

*The authors are not recommending self-modifying software.

puter's parallel data (data bus) to the serial data received/transmitted by the terminal (EIA data). We have already discussed the concept of the universal asynchronous receiver transmitter (UART). This area of the computer is organized as a pair of ports, one for the data and one for a few status bits (refer back to Chapter 8, Section 16). Figure 11-6 shows the basic receiver/transmitter circuitry in block diagram form.

Both the parallel inputs and parallel outputs are connected to a data port (41 H). The receiver ready (RRDY) and transmitter ready bits (TRDY) are connected to the lower two bits of a status port (40 H). The port numbers must be within the range of 0–255 (0–0FF H) and are determined when the R/T chip is wired to function with a computer. Port 40 H (status) and port 41 H are arbitrary numbers chosen for use in the programming examples to follow.

10.2 The 8251A USART

Figure 11-7 is the vendor's data sheet on an 8251A USART. This very versatile chip used with 8080A/85 microprocessor systems can handle both synchronous and asynchronous data, contains independent receiver and transmitter sections, and is software programmable. Baud rates are selectable to 64K. The operation of the 8251A is detailed in the manufacturer's specifications. These specifications explain the two initial bytes of data, the mode instruction and the command instruction, that must be sent to the 8251A to "set up" the desired parameters. These include baud rate, number of data bits, and type of parity.

Before using the USART (after a reset), the mode and command instructions are sent to the USART control port (status port when receiving), which in our example is port 40 H. Figure 11-8 shows the structure of these two bytes needed to "program" the USART before use. The asynchronous EIA RS232C standard is used by most terminals; hence we will not be using synchronous mode.

To establish the mode instruction, let's assume the following conditions: 8 data bits, 16 × baud rate factor, no parity (disabled), and a single stop bit. The mode instruction then is 4EH.

0	1	0	0	1	1	1	0

The correct command instruction is 27 H.

0	0	1	0	0	1	1	1

You may check this against the definition of the command instruction.

Any program or computer using a USART (8251A) for serial communication would have to output the following sequence before actually doing such data transfer. It must be done after a hardware reset.

```
MVI A, 4EH    ;Mode instruction
OUT 40H       ;Out to control port
MVI A, 27H    ;Command instruction
OUT 40H       ;Out to control port
```

FIGURE 11-7 USART data sheet for the 8251A. (Reprinted by permission of Intel Corporation, copyright 1982.)

8251A
PROGRAMMABLE COMMUNICATION INTERFACE

- Synchronous and Asynchronous Operation
- Synchronous 5–8 Bit Characters; Internal or External Character Synchronization; Automatic Sync Insertion
- Asynchronous 5–8 Bit Characters; Clock Rate—1, 16 or 64 Times Baud Rate; Break Character Generation; 1, 1½, or 2 Stop Bits; False Start Bit Detection; Automatic Break Detect and Handling
- Synchronous Baud Rate—DC to 64K Baud

- Asynchronous Baud Rate—DC to 19.2K Baud
- Full-Duplex, Double-Buffered Transmitter and Receiver
- Error Detection—Parity, Overrun and Framing
- Compatible with an Extended Range of Intel Microprocessors
- 28-Pin DIP Package
- All Inputs and Outputs are TTL Compatible
- Single +5V Supply
- Single TTL Clock

The Intel® 8251A is the enhanced version of the industry standard, Intel 8251 Universal Synchronous/Asynchronous Receiver/Transmitter (USART), designed for data communications with Intel's microprocessor families such as MCS-68, 80, 85, and iAPX-86, 88. The 8251A is used as a peripheral device and is programmed by the CPU to operate using virtually any serial data transmission technique presently in use (including IBM "bi-sync"). The USART accepts data characters from the CPU in parallel format and then converts them into a continuous serial data stream for transmission. Simultaneously, it can receive serial data streams and convert them into parallel data characters for the CPU. The USART will signal the CPU whenever it can accept a new character for transmission or whenever it has received a character for the CPU. The CPU can read the complete status of the USART at any time. These include data transmission errors and control signals such as SYNDET, TxEMPTY. The chip is fabricated using N-channel silicon gate technology.

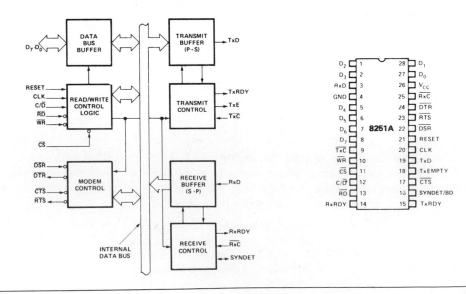

FIGURE 11-8 (*a*) Mode instruction format, asynchronous mode, and (*b*) command instruction format for the 8251A. (Reprinted by permission of Intel Corporation, copyright 1982.)

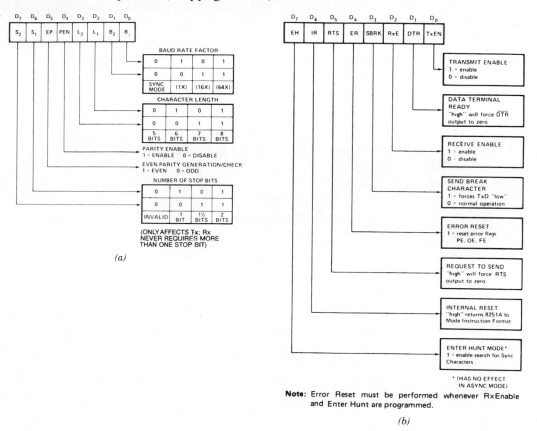

Once the preliminaries have been completed, the USART is ready for use. If the computer is using a UART (AY5-1013 or equivalent), this software setup is not needed. The UART is hand wired to set up parameters like baud rate and parity, making "instructions" unnecessary. However, the UART may then be harder to modify because of the wiring changes involved.

10.3 Using the Receiver/Transmitter Status Port

As noted before, two ports are involved: port 40 H (status) and port 41 H (data). Only two bits of the status port are of interest. Bit "0" will indicate whether the transmitter is ready for use and bit "1" will indicate whether the receiver is ready for use (i.e., has received a character).

Before attempting to transmit an ASCII character or perhaps receive an ASCII character, it is necessary to check the appropriate status bit by observing the status port. The following convention applies:

1. If TRDY is high, the transmitter is available for use (i.e., the transmitter buffer is empty); otherwise the transmitter is in the process of sending a character and unable to accept another character.

2. If RRDY is high, a character has been received and is available to the CPU at the USART data port.

To check the respective status bits, we must mask off the bits that are not of interest. For example, to check the TRDY bit:

```
STAT:  IN 40H    ;In from USART status
       ANI 01    ;Mask off TRDY bit
       JZ STAT   ;If zero, wait
```

This sequence brings in a quantity from the status port and looks only at bit "0" (the LSB). If low, the transmitter is busy, and we must wait until available (TRDY = 1).

Suppose 59 H is received from port 40 H. After the mask operation, we are left with 01, which is nonzero and indicates that the transmitter is *not* busy.

$$
\begin{array}{ll}
0\ 1\ 0\ 1\ 1\ 0\ 0\ 1 & \text{Status} \\
0\ 0\ 0\ 0\ 0\ 0\ 0\ 1 & \text{Mask} \\
\hline
0\ 0\ 0\ 0\ 0\ 0\ 0\ 1 & \text{Result}
\end{array}
$$

$$\uparrow$$
TRDY = 1 (not busy)

A similar procedure is used to check the receiver ready bit:

```
STAT2:  IN 40H    ;In from USART status
        ANI 02    ;Mask off RRDY bit
        JZ STAT2  ;If zero, no character has been received, so wait
```

This sequence examines the RRDY bit. If low, no character has been received and we continue to examine the bit until it goes high, indicating a received character.

10.4 Serial Input/Output Routines

Sending and receiving information is one of the most important tasks of a computer. If a user could not send commands to a computer, the machine would not know what operation to perform. Similarly, the computer must be able to output information to a display. All serial communication is done a character at a time. If a line of information is to be transmitted, it is sent one character followed by another. This results in the frequent use of a common programming sequence. Most computers then utilize an input and an output subroutine for handling serial data.

Let us first write a simple program to echo data from a terminal to a microcomputer. Figure 11-9 shows the setup. Essentially this program will cause the terminal to act as an electric typewriter. We will effectively connect the terminal keyboard to the terminal display. The computer will simply return or "echo" back to the display any characters it receives. In future programs, however, the

FIGURE 11-9 The electric typewriter.

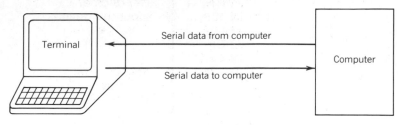

computer will not necessarily echo characters to the display—instead, they may be interpreted as instructions causing another operation to occur. We are concerned only with character data at this time, and will ignore the parity bit.

```
          TITLE "ELECTRIC TYPEWRITER"
          ORG 1000H
CIN:   IN 40H        ;Get USART status
       ANI 02        ;Mask RRDY bit
       JZ CIN        ;If zero, wait
       IN 41H        ;Otherwise, get character data
       MOV B, A      ;Save in B register
COUT:  IN 40H        ;Get USART status
       ANI 01H       ;Mask TRDY bit
       JZ COUT       ;If TRDY low, wait
       MOV A, B      ;Retrieve received character
       OUT 41H       ;Output to terminal
       JMP CIN       ;Go for another character
       END
```

The first part of the program (CIN) is the character input routine. When RRDY goes high we leave the CIN loop and get the received character from the data port (IN 41H). When the data byte is read from the data port, the USART automatically clears or resets the RRDY bit. The character is then temporarily saved in B register, since A will be needed in the COUT loop. In the COUT loop, we wait for TRDY to be high, then we get the received character back from B register and output it to the terminal (display). The program then repeats. In this case, keep in mind that the rate at which characters are printed is equal to the rate at which they are typed. Testing RRDY slows down the process of sending and receiving data. You should now assemble the preceding program.

We can easily and conveniently rewrite this program in terms of subroutines (a common practice).

```
          TITLE "IMPROVED ELECTRIC TYPEWRITER"
          ORG 1000H
          LXI SP, 2000H    ;Always establish stack for subroutines
NEXT:  CALL CIN          ;Get a character from keyboard
       MOV B, A          ;Save it in B
```

```
              CALL COUT        ;Output character
              JMP NEXT         ;Get another
       CIN:   IN 40H           ;Input subroutine
              ANI 02
              JZ CIN
              IN 41H
              ANI 7FH          ;Set parity bit = zero
              RET              ;End CIN
       COUT:  IN 40H           ;Output subroutine
              ANI 01
              JZ COUT
              MOV A, B
              OUT 41H
              RET              ;End COUT
              END
```

In this program, subroutines are used; hence the stack pointer is established. Both CIN and COUT are restructured as subroutines. Note that when CIN is called, the received character will be in the A register. Whenever CIN is called, we can expect to find the result in A. What happens after that is up to the programmer. The program also makes bit 7 (MSB), the parity bit, equal to zero. This feature is useful later in attempting to recognize command characters.

The output subroutine, COUT, expects to find the character to be output in the B register, since the A register is needed when testing the TRDY bit.

Frequently, the input and output routines are already available on a computer. For instance, in a disk-drive-based system, the I/O routines may be located as follows:

```
              CIN:   SET 2010H
              COUT:  SET 200DH
```

In this case, the user does not have to write I/O routines because they are already installed and available for use, saving on programming. In such a system, our program would look like this:

```
              TITLE "BETTER TYPEWRITER"
              ORG 1000H
              LXI SP, 2000H
       NEXT:  CALL CIN
              MOV B, A
              CALL COUT
              JMP NEXT
       CIN:   SET 2010H
       COUT:  SET 200DH
              END
```

The "SET" command is a pseudo-mnemonic used by assemblers. It merely states the location of the two subroutines in memory. The user *assumes* that they are present.

Let us assemble both the preceding programs for practice! Be sure to verify the hex code for yourself.

IMPROVED ELECTRIC TYPEWRITER

1000	31 00 20 CD 0D 10 47 CD 19 10 C3 03 10 DB 40 E6
1010	02 CA 0D 10 DB 41 E6 7F C9 DB 40 E6 01 CA 19 10
1020	78 D3 41 C9

Symbol Table

NEXT: 1003H
CIN: 100DH
COUT: 1019H

BETTER TYPEWRITER

1000	31 00 20 CD 10 20 47 CD 0D 20 C3 03 10

Symbol Table

NEXT: 1003H
CIN: 2010H
COUT: 200DH

10.5 Recognizing Commands from the Keyboard

Many control functions are used in the serial I/O section of a computer. The tab function, backspace, carriage return, and bel are just a few (see, e.g., Table 11-3). In the case of the hex code for such functions, the parity bit is always listed as zero. This can be verified by consulting the list of ASCII codes in Chapter 8. This is the main reason for forcing the MSB to zero in the input routine (ANI 7FH). This allows for a match when looking for a particular code.

Let us suppose that we have our BETTER TYPEWRITER program loaded at 1000H and our CATERPILLAR program loaded at 3000H. We will modify the former so that if an ESCape code is received, the computer jumps to address 3000H and runs the caterpillar.

TITLE "TYPEWRITER/CATERPILLAR"
ORG 1000H
LXI SP, 2000H

TABLE 11-3   Control Functions

Function	Control Code	Hex Code
Bel	↑ G	07
Backspace	↑ H	08
Carriage return	↑ M	0D
Tab	↑ I	09
Line feed	↑ J	0A
Escape	—	1B
DC 1 (start)	↑ Q	11
DC 3 (stop)	↑ S	13
ETX (half)	↑ C	03

```
NEXT:   CALL CIN
        MOV B, A
        CPI 1BH        ;Was character on escape?
        JZ 3000H       ;If so, jump to 3000H
        CALL COUT      ;Otherwise, continue
        JMP NEXT
        END
```

Let us pause and rewrite the same program with expanded assembler syntax, including more symbols.

```
        TITLE "TYPEWRITER/CATERPILLAR"
        ORG 1000H       ;Program starts at 1000H
        LXI SP, STACK   ;Need SP for subroutines
NEXT:   CALL CIN        ;Get a character from keyboard
        MOV B, A        ;Save it in B
        CPI ESC         ;Is it an escape character?
        JZ CATER        ;Done if it is
        CALL COUT       ;Otherwise, output to CRT
        JMP NEXT        ;Do another character
STACK:  SET 2000H       ;Here is the stack
CIN:    SET 2010H       ;Input subroutine address
COUT:   SET 200DH       ;Output subroutine address
ESC:    SET 1BH         ;Code for an escape
CATER:  SET 3000H       ;Location of caterpillar
        END             ;Pseudo-op for for assembler
```

Notice the use of the SET function to further use symbols (ESC, CATER) in the program. This kind of programming is the mark of a good programmer. The frequent use of descriptive comments is also important.

10.6 A More Advanced Typewriter

Our first printout shows a program that will accept data from the keyboard (a sentence, paragraph, short story, or letter to Mom). When the ESC key is pressed, the entire message will be played back. The program will not echo characters typed until the ESC key is pressed. We will use the HL register pair to point to a location in memory where the data will be stored (we'll start them at 4000 H). The commands MOV A, M and MOV M, A will be very useful here, so look them up and review their operation.

In this program, the C register is used to keep track of the character count. If 20 characters are input, we would want only 20 characters output. Now try to construct this program. We provide a symbol table to give you a head start in assembling the program.

Symbol Table

STACK:	2000H
START:	1003H
NEXT:	1008H
PLAY:	1016H
MORE:	1019H
CIN:	1024H
COUT:	1030H

PRINTOUT 11-1

```
                TITLE 'KEYBOARD MEMORY'
                ORG 1000H
                LXI SP , STACK          ;SET UP STACK AREA
        START:  LXI H, 4000H            ;SET HL TO 4000H FOR STORAGE
                MVI C, 0                ;CHARACTER COUNTER
        NEXT:   CALL CIN                ;GET A CHARACTER
                CPI 1BH                 ;IS IT AN ESCAPE?
                JZ PLAY                 ;IF SO, PLAY BACK MESSAGE
                MOV M, A                ;IF NOT, SAVE IN MEMORY
                INX H                   ;READY FOR NEXT MEMORY LOCATION
                INR C                   ;ADD ONE TO CHARACTER COUNTER
                JMP NEXT                ;GET ANOTHER CHARACTER
        PLAY:   LXI H, 4000H            ;RESET MEMORY POINTER
        MORE:   MOV B, M                ;CHARACTER TO B
                CALL COUT               ;OUTPUT IT
                DCR C                   ;ONE LESS TO DO
                INX H                   ;POINT AT NEXT CHARACTER
                JNZ MORE                ;IF NOT ZERO, DO ANOTHER
                JMP START               ;REPEAT PROGRAM
        CIN:    IN 40H                  ;INPUT ROUTINE
                ANI 02
                JZ CIN
                IN 41H
                ANI 7FH
                RET
        COUT:   IN 40H                  ;OUTPUT ROUTINE
                ANI 01
                JZ COUT
                MOV A, B
                OUT 41H
                RET
        STACK:  SET 2000H               ;SET THE STACK
                END
```

10.7 A Message Center

The program MESSAGE CENTER will reproduce one of six "canned" messages on request from the keyboard. It makes use of a data table at the end of the program, which contains ASCII code representing the messages. The digits 1 through 6 are used to select a message.

You should be sufficiently familiar with the CIN and COUT routines to be able to add them to the program in your own way. Each message ends with an exclamation point because this is the end mark on each message which tells us when to stop. The code for "!" is 21H; hence the CPI 21H instruction in the program. Study the program to be sure you understand its operation. Note that the data section of the program uses a DB or define-byte pseudo-opcode. It requests the assembler to *look up* the ASCII code for each character and place it in sequential memory locations.

For example, DAT1: DB 'THIS IS MESSAGE ONE!' would be encoded as follows:

```
54 48 49 53 20 49 53 20 4D 45 53 53 41 47 45 20
48 4E 45 21
```

and message two (DAT2) would follow immediately. Perhaps you should assemble the program to further point up the importance of an assembler (program) to help

PRINTOUT 11-2

```
                  TITLE 'MESSAGE CENTER'
                  ORG 1000H
                  LXI SP , 2000H              ;SET UP STACK
        START:    CALL CIN                    ;GET A CHARACTER
                  CPI 31H                     ;IS IT AN ASCII 1?
                  JZ ONE                      ;IF SO, JUMP TO ONE
                  CPI 32H
                  JZ TWO
                  CPI 33H
                  JZ THREE
                  CPI 34H
                  JZ FOUR
                  CPI 35H
                  JZ FIVE
                  CPI 36H
                  JZ SIX
                  JMP START                   ;DONE - TRY AGAIN
        ONE:      LXI H, DAT1                 ;LET HL POINT AT THE DATA
                  CALL OUTPT                  ;CALL SPECIAL OUTPUT ROUTINE
                  JMP START                   ;START OVER
        TWO:      LXI H, DAT2                 ;SECOND DATA
                  CALL OUTPT
                  JMP START
        THREE:    LXI H, DAT3                 ;THIRD DATA
                  CALL OUTPT
                  JMP START
        FOUR:     LXI H, DAT4                 ;FOURTH DATA
                  CALL OUTPT
                  JMP START
        FIVE:     LXI H, DAT5                 ;FIFTH DATA
                  CALL OUTPT
                  JMP START
        SIX:      LXI H, DAT6                 ;SIXTH DATA
                  CALL OUTPT
                  JMP START
        OUTPT:    MOV B, M                    ;READY TO OUTPUT
                  MOV A, M                    ;READY FOR END MARK
                  CPI 21H                     ;EXCLAMATION POINT?
                  JZ START                    ;IF SO, DONE
                  CALL COUT                   ;OUTPUT IT
                  INX H                       ;NEXT CHARACTER
                  JMP OUTPT
        DAT1:     DB 'THIS IS MESSAGE ONE!'
        DAT2:     DB 'THE COMPUTER IS BROKEN!'
        DAT3:     DB 'THE SECRETARY IS OUT!'
        DAT4:     DB 'TURN ME ON!'
        DAT5:     DB 'KWATZ!'
        DAT6:     DB 'I LOVE MY COMPUTER!'
        CIN:      SET 2010H
        COUT:     SET 200DH
                  END
```

you. The DB 'THIS IS MESSAGE ONE!' line causes the assembler to look up the ASCII codes for each character.

MESSAGE CENTER is *not* a very good programming example because many steps are repeated over and over. In other words, the program does not make very efficient use of available memory. The next example also involves the "message center" function but uses an important technique called lookup tables.

10.8 A Fancy Message Center: Lookup Tables

Better programming entails the use of lookup tables. Printout 11-3 shows a program that will behave in the same manner as its predecessor. There are two tables involved: the first is the command table, CMDS.

CMDS: DB '123456'

or, in ASCII,

31H, 32H, 33H, 34H, 35H, 36H

The second is a table of addresses, TABLE.

TABLE: DW DAT1
DW DAT2
DW DAT3
DW DAT4
DW DAT5
DW DAT6

The whole idea of the program is to accept a command from the keyboard (digits 1–6) and compare it to valid commands in the command table. Notice that there are six valid commands (NCMDS) and also six addresses in the address table.

There are a number of interesting features in this program:

CRLF: Outputs a carriage return and then a line feed to the terminal

STACK: Establishes an area at the end of the program as the stack area, thereby eliminating the need for the programmer to locate the stack. Use DS 30H to leave room for the stack.

STACK: DS 0

The DS pseudo-mnemonic defines storage of 30 hex bytes in memory. This stack area is larger than required, but the additional size will be useful in more advanced programs. Notice that the storage area is ahead of the stack address. Remember that the stack works toward lower memory locations.

MATCH: LDAX D

After a matchup between the incoming command (one to six), the DE register pair contains the address of the message address, not the message itself! LDAX D (Fig. 11-10) loads the A register with the data in the address specified by the DE register pair.

This type of program is used to match common commands like backspace (↑ H = 08 H), tab (↑ I = 09 H), halt (↑ C = 03 H), DC1 (↑ Q = 11 H), and DC3 (↑ S = 13 H).

Modifying the preceding program to recognize these five control characters is easy. We change the command table to contain the new codes and use the address table to point at the address of a subroutine to accomplish the desired purpose. If the TABLE contains a subroutine address, modify the program so that HL

PRINTOUT 11-3

```
                    TITLE 'ADVANCED MESSAGE CENTER'
                    ORG 1000H          ;SET ORIGIN
                    LXI SP,STACK       ;ESTABLISH THE STACK AREA
          START:    MVI B, NCMDS       ;NUMBER OF VALID COMMANDS
                    LXI H, CMDS        ;ADDRESS OF COMMAND TABLE
                    LXI D, TABLE       ;START OF ADDRESS TABLE
          TEST:     CALL CIN           ;GET A COMMAND CHARACTER
          CPR:      CMP M              ;IS IT IN THE TABLE?
                    JZ MATCH           ;IF SO, LETS GO PRINT
                    INX H              ;IF NOT, NEXT COMMAND
                    INX D              ;BUMP ADDRESS TABLE TOO
                    INX D
                    DCR B              ;ONE LESS COMMAND TO TEST FOR
                    JNZ CPR            ;LAST ONE?
                    JMP START          ;OTHERWISE START OVER
          MATCH:    LDAX D             ;GET ADDRESS OF  MESSAGE
                    MOV L, A           ;INTO HL REGISTER
                    INX D              ;PAIR
                    LDAX D
                    MOV H, A
          OUTPT:    MOV B, M           ;NOW LET'S OUTPUT CHARACTERS
                    MOV A, M
                    CPI 21H            ;TEST FOR END MARK
                    JZ CRLF
                    CALL COUT
                    INX H
                    JMP OUTPT
          CRLF:     MVI B, 0DH         ;OUTPUT A CRLF TO BE NEAT
                    CALL COUT
                    MVI B, 0AH
                    CALL COUT
                    JMP START

          NCMDS:    SET 6
          CMDS:     DB '123456'
          TABLE:    DW DAT1
                    DW DAT2
                    DW DAT3
                    DW DAT4
                    DW DAT5
                    DW DAT6
          DAT1:     DB 'THIS IS MESSAGE ONE!'
          DAT2:     DB 'THE COMPUTER IS BROKEN!'
          DAT3:     DB 'TURN ME ON!'
          DAT4:     DB 'KWATZ!'
          DAT5:     DB 'I LOVE MY COMPUTER!'
          DAT6:     DB 'TIME FOR LUNCH!'
          CIN:      SET 2010H
          COUT:     SET 200DH
                    DS 30H             ;LEAVE ROOM FOR STACK
          STACK:    DS 0
                    END
```

```
1000    31 DC 10 06 06 21 3C 10 11 42 10 CD 10 20 BE CA    1....!<..B.. ..
1010    1C 10 23 13 13 05 C2 0E 10 C3 03 10 1A 6F 13 1A    ..#.........o..
1020    67 46 7E FE 21 CA 2F 10 CD 0D 20 23 C3 21 10 06    gF".!./... #.!..
1030    0D CD 0D 20 06 0A CD 0D 20 C3 03 10 31 32 33 34    ... .... ...1234
1040    35 36 4E 10 62 10 79 10 84 10 8A 10 9D 10 54 48    56N.b.y.......TH
1050    49 53 20 49 53 20 4D 45 53 53 41 47 45 20 4F 4E    IS IS MESSAGE ON
1060    45 21 54 48 45 20 43 4F 4D 50 55 54 45 52 20 49    E!THE COMPUTER I
1070    53 20 42 52 4F 4B 45 4E 21 54 55 52 4E 20 4D 45    S BROKEN!TURN ME
1080    20 4F 4E 21 4B 57 41 54 5A 21 49 20 4C 4F 56 45     ON!KWATZ!I LOVE
1090    20 4D 59 20 43 4F 4D 50 55 54 45 52 21 54 49 4D     MY COMPUTER!TIM
10A0    45 20 46 4F 52 20 4C 55 4E 43 48 21 00 00 00 00    E FOR LUNCH!....
```

FIGURE 11-10 Use of LDAX D.

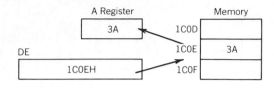

contains that. Then do a PCHL, and you're off! A few details remain, but these are left for you to attend to.

```
        NCMDS:  SET 5
         CMDS:  DB 8, 9, 3, 11H, 13H
        TABLE:  DW BSPC
                DW TAB
                DW CONTC
                DW STRT
                DW STOP
```

The subroutines BSPC, TAB, CONTC, STRT, and STOP are a part of the program. STRT and STOP are shown.

```
STOP:   IN 40H       ; Check status
        ANI 02       ; Check RRDY bit
        JZ TOP       ; Nothing, keep going
        IN 41H       ; Get a character
        ANI 7FH
        CPI 13H      ; Is it ↑S?
        JZ WAIT      ; Wait for ↑Q
        JMP TOP      ; No, keep going
WAIT:   IN 40H       ; Check status
        ANI 02       ; Check RRDY bit
        JZ WAIT      ; No wait
        IN 41H       ; Yes, get character
        ANI 7FH
        CPI 11H      ; Only start with ↑Q
        JZ TOP       ; Keep going
        JMP WAIT     ; Keep waiting
```

In this routine, ↑S stops a printout and ↑Q restarts it. TOP is somewhere in your program that is perhaps printing data or performing a task. The typical use of ↑S and ↑Q is to stop a terminal to give you time to do some reading.

This concludes our discussion of serial I/O, USARTs, and command tables. Be sure you understand the principles presented before proceeding to ''Advanced Programming,'' since these techniques are applied constantly in Chapter 12. Obviously, the preceding program segments dealing with ↑H, ↑I, and ↑C are incomplete. They are intended as food for thought for those who may be inclined to enjoy some heavy programming.

11 PARALLEL DATA HANDLING

11.1 Parallel Input/Output

The preceding section dealt with serial data transmission using a USART to and from a terminal. This section introduces you to parallel I/O using a peripheral interface adapter (PIA). The PIA provides parallel data paths between the outside world and the microprocessor. These parallel inputs and outputs can be used to connect a microcomputer to a variety of interesting applications. Among those we will be discussing are the following:

1. An integrated circuit tester,
2. A traffic signal controller,
3. A disco light show,
4. A security system.

Figure 11-11 shows the 8255A peripheral interface adapter. This very versatile LSI circuit is intended for parallel I/O operations. It produces 24 wires organized into three data ports. It also requires the use of a fourth port for control signals. For purposes of illustration, we will assign port numbers as follows:

FIGURE 11-11 Programmable peripheral interface data sheet for the 8255A. (Reprinted by permission of Intel Corporation, copyright 1981.)

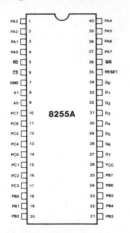

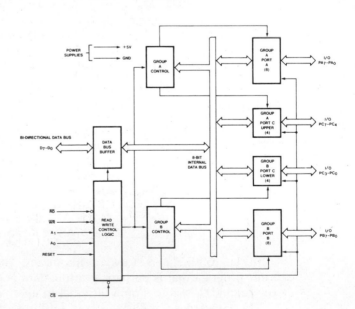

PIN NAMES

D_7, D_0	DATA BUS (BI-DIRECTIONAL)
RESET	RESET INPUT
$\overline{CS}$	CHIP SELECT
$\overline{RD}$	READ INPUT
$\overline{WR}$	WRITE INPUT
A0, A1	PORT ADDRESS
PA7-PA0	PORT A (BIT)
PB7-PB0	PORT B (BIT)
PC7-PC0	PORT C (BIT)
V_{CC}	+5 VOLTS
GND	0 VOLTS

Port 80 H   Port A   (8 lines)
Port 81 H   Port B   (8 lines)
Port 82 H   Port C   (8 lines)
Port 83 H   Control

The 8255A may be used in three different modes of operation:

Mode 0   Latched I/O
Mode 1   Strobed I/O
Mode 2   Bidirectional bus mode

We will be using mode 0 exclusively in all our programming examples. In mode 0 we find that the three ports (eight lines each) are organized into four groups. These groups *may be* declared as either input or output before using the PIA. With four groups, any one of which can be declared either input or output, there are $2^4 = 16$ combinations. Table 11-4 shows the matching control word for each of the 16 combinations. This control word *must be* sent to the PIA control port before using the PIA. After that, the operation is simple, as we shall see. If all ports are to be used for output (24 output lines), then the control word is 80 H and should be sent to the control port (83 H). This is done in the following way:

TABLE 11-4   8255A PIA Control Words (CW)

CW	Port A	Port C Upper	Port C Lower	Port B
80 H	Output	Output	Output	Output
82 H	Output	Output	Output	Input
81 H	Output	Output	Input	Output
83 H	Output	Output	Input	Input
88 H	Output	Input	Output	Output
8A H	Output	Input	Output	Input
89 H	Output	Input	Input	Output
8B H	Output	Input	Input	Input
90 H	Input	Output	Output	Output
92 H	Input	Output	Output	Input
91 H	Input	Output	Input	Output
93 H	Input	Output	Input	Input
98 H	Input	Input	Output	Output
9A H	Input	Input	Output	Input
99 H	Input	Input	Input	Output
9B H	Input	Input	Input	Input

Control word:

D_7	D_6	D_5	D_4	D_3	D_2	D_1	D_0
1	0	0	—	—	0	—	—

```
MVI A, 80H    ; Control word
OUT 083H      ; Out to control port
```

The PIA becomes an extremely valuable and versatile means of accomplishing I/O with a computer. Every microcomputer intended for control applications that does *not* use a PIA has some other form of I/O latch. In all our programs, we use port numbers to identify a parallel port, and it will be our job to do any initialization (control words) necessary.

11.2 An Integrated Circuit Tester

Our first IC tester program will be the testing of a 7400 quad two-input NAND gate. The inputs of the 7400 will be wired to a computer's parallel outputs. The outputs of the 7400 will be wired to the computer's parallel inputs. Figure 11-12 shows the connections. The 7400 gate inputs are wired to port 80 H (A_7–A_0). The 7400 gate outputs are wired to port 81 H (B_3–B_0).

A pair of indicator LEDs are wired to port 82 H (C_7, C_6). The LED connected to C_7 will be the pass LED, and the LED connected to C_6 will be the fail LED. After a test, one of the LEDs will be turned on. This is accomplished by sending a "1" (to turn the LED on) to a PIA output. A pass-or-fail sequence would appear like this:

```
PASS:   MVI A, 80H    ; Pass pattern
        OUT 82H       ; Out to port C
        HLT

FAIL:   MVI A, 40H    ; Fail pattern
        OUT 82H       ; Out to port C
        HLT
```

Either sequence leaves an LED on. But this is the end of the program.

Let's work on the beginning, starting with a review of the correct truth table for a two-input NAND gate (Fig. 11-13). We are going to test all four NAND gates at once. We will first output eight zeros and expect four ones back.

Table 11-5 shows the four tests to be performed by the program. The program will output 0 to port A (80 H) and do an input from port B (81 H). The incoming data should be 0F H. If so, we will proceed to the second test. If not, we will turn on the fail LED and end the test. In the second test, we will output 55 H to port A and expect to input 0F H from port B. The remaining tests are performed, and if successful, the pass LED is lit.

TABLE 11-5   7400 Gate Behavior

Test No.	Output to 7400[a]	Expected Response from 7400
1	0 0 0 0 0 0 0 0 or 0 H	1 1 1 1 or F H
2	0 1 0 1 0 1 0 1 or 55 H	1 1 1 1 or F H
3	1 0 1 0 1 0 1 0 or AA H	1 1 1 1 or F H
4	1 1 1 1 1 1 1 1 or FF H	0 0 0 0 or 0 H

[a]Gates 1, 2, 3, and 4 are represented by the first, second, third, and fourth pairs of bits outputted from any test.

FIGURE 11-12 Testing a 7400 quad two-input NAND gate.

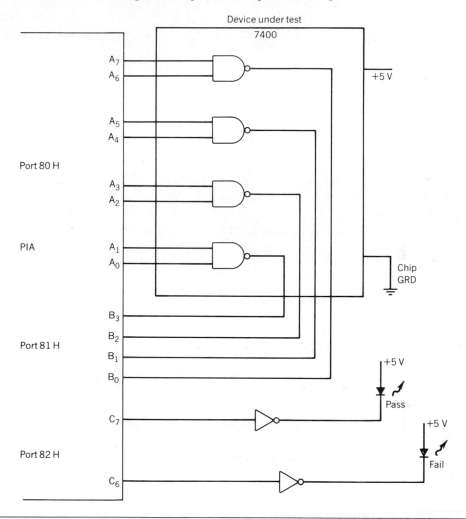

FIGURE 11-13 (a) NAND gate and (b) truth table.

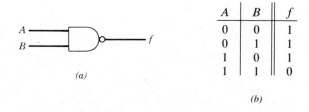

A	B	f
0	0	1
0	1	1
1	0	1
1	1	0

(a)

(b)

FIGURE 11-14 7400 IC test data.

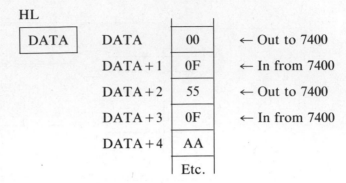

The program will use a data table that contains both output and expected input data, structured as follows:

DATA: DB 0, 0FH, 55H, 0FH, 0AAH, 0FH, 0FFH, 0

By comparing this list of data to Table 11-5, you will see its source. We will point to this data located in memory locations beginning at DATA by using the HL register pair and the M (memory) designation (see Fig. 11-14). After MOV A, M the A register should have zero in it. After INX H, HL will be pointing at DATA + 1 or 0F.

Let's take a look at the complete program (Printout 11-4). Keep in mind that this test is being performed very quickly. The speed is closely related to the time

PRINTOUT 11-4

```
        TITLE '7400 IC TESTER'
        ORG 1000H
START:  LXI H, DATA                 ;ADDRESS OF TEST DATA
        MVI B, 4                    ;NUMBER OF TESTS
MORE:   MOV A, M                    ;GET TEST DATA
        OUT 80H                     ;TO PORT A (7400 INPUTS)
        IN 81H                      ;GET TEST RESULTS
        ANI 0FH                     ;WANT TO BE SURE B7 - B4 ARE ZERO
        INX H                       ;POINT TO EXPECTED RESPONSE
        CMP M                       ;COMPARE THE TWO
        JZ FAIL                     ;FAIL IF NOT EQUAL
        INX H                       ;BUMP HL
        DCR B                       ;ONE LESS TEST
        JNZ MORE                    ;DONE IF B = 0
PASS:   MVI A, 80H                  ;PASS PATTERN
        OUT 82H                     ;OUT TO LED'S
        JMP START                   ;KEEP TESTING
FAIL:   MVI A, 40H                  ;FAIL PATTERN
        OUT 82H                     ;OUT TO LED'S
        JMP START                   ;KEEP TESTING
DATA:   DB 00 ,  0FH, 55H, 0FH, 0AAH, 0FH, 0FFH, 00
        END
```

FIGURE 11-15 IC tester Setup. (Photo by John Young.)

to perform a few instructions. Therefore, this is a dynamic test* and is preferable to a static test. Static testing would not catch a slow 7400. Study the program to understand its operation. Note that when inputting data (IN 81H), we mask the upper unused bits (B_7–B_4) to zero to assure a matchup to our data. An alternative approach is to ground those unused inputs, but a program step saves some wire! Figure 11-15 shows a setup for testing ICs. The patching arrangement allows for easily altering the setup for other ICs. The tester contains a power supply (7805 5-V regulator) and the inverters for driving the pass/fail LEDs. The output port drives the IC under test directly.

Here is the hex file for the simple IC tester program:

```
1000    21 24 10 06 04 7E D3 80 DB 81 E6 0F 23 BE CA 1D
1010    10 23 05 C2 05 10 3E 80 D3 82 C3 00 10 3E 40 D3
1020    82 C3 00 10 00 0F 55 0F AA 0F FF 00
```

Notice that the program never stops, but keeps testing over and over. This makes it easy to test a quantity of ICs. Just remove the unacceptable IC (fail light goes

---

*We do not test the gate at its full speed, however.

on) and put in another. You should assemble the program and verify the hex file.

Since the 7402 quad two-input NOR gate has the same type of setup as the 7400 (i.e., two inputs, one output), it should be easy to modify the DATA in this program to test a NOR-type IC with some slight wiring modifications to the tester. This is left as an exercise for you.

11.3 Print a Message, Too

The program could also print a simple pass-or-fail message on a display terminal using the serial I/O capabilities of the computer. This is done by either adding to or changing the PASS and FAIL routines. You may want to use both LEDs and the display terminal.

```
PASS:   LXI H, PMSG          ; HL has message address
MO:     MOV B, M             ; First character to B
        MOV A, M             ; Also to A
        CPI 21H              ; Is it end mark?
        JZ STOP              ; If so, stop
        CALL COUT            ; Otherwise, output it
        INX H                ; Next HL
        JMP MO               ; Keep going
PMSG:   DB 'THIS 7400 PASSES!'
STOP:   HLT                  ; Don't wait to have a million messages
```

It is left again to you to rewrite the FAIL sequence.

11.4 Generating Pulses

Many ICs, like counters (7490, 7493), shift registers (7491, 7494), and flip-flops or latches (7473, 7475, 7476), require a clock pulse for proper operation.

Figure 11-16 shows the two pulses needed for certain types of IC testing. It is a simple matter of sending out a one, then a zero, then a one to produce a down-going pulse.

```
MVI A, 01     ; Put a one in A
OUT 80H       ; Output a one
XRI 01H       ; Toggle the bits
OUT 80H       ; Output complement
XRI 01H       ; Toggle the bits
OUT 80H       ; Output a one
```

FIGURE 11-16 Generating pulses. (*a*) Down-going and (*b*) up-going.

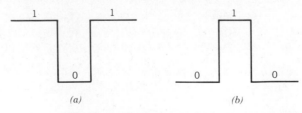

FIGURE 11-17 Test setups for the 7473.

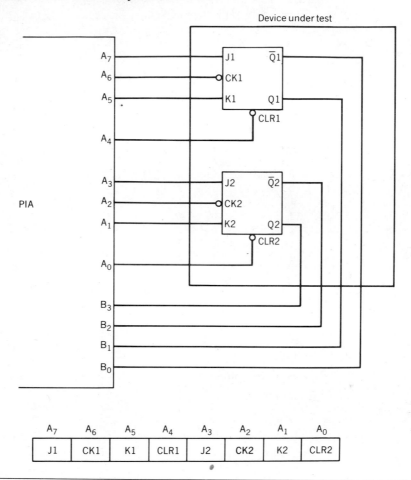

A_7	A_6	A_5	A_4	A_3	A_2	A_1	A_0
J1	CK1	K1	CLR1	J2	CK2	K2	CLR2

This sequence outputs the pulse! Let's test a 7473 dual J-K flip-flop, using the setup diagrammed in Fig. 11-17. Both clear (CLR) and clock (CK) lines respond to a down-going pulse. The requirements of the test are as follows:

1. Clear both FFs: $Q = 0$, $\overline{Q} = 1$.

2. Toggle test: $J = 1$, $K = 1$, pulse CK. $Q = 1$, $\overline{Q} = 0$.

3. Repeat: $J = 1$, $K = 1$, pulse CK. $Q = 0$, $\overline{Q} = 1$.

4. Test J-K function: $J = 1$, $K = 0$, pulse CK. $Q = 1$, $\overline{Q} = 0$ then $J = 0$, $K = 1$, pulse CK. $Q = 0$, $\overline{Q} = 1$.

Study the test requirements as formulated in Table 11-6 and write the program
    Continuous testing of ICs may be done under a number of different conditions. An IC would enter a test loop and the following conditions may be changed:

TABLE 11-6   Test Requirements for 7473 Flip-Flop

Test No.			Test Data (FF Inputs)								Test Data (Results)			
	J1	CK1	K1	CLR1	J2	CK1	K2	CLR2	Hex Data	Q2	$\overline{Q2}$	Q1	$\overline{Q1}$	Hex Data[a]
1, Clear	1	1	1	1	1	1	1	1	FF	X	X	X	X	X
	1	1	1	0	1	1	1	0	EE	0	1	0	1	05
	1	1	1	1	1	1	1	1	FF	0	1	0	1	05
2, Toggle	1	0	1	1	1	0	1	1	BB	1	0	1	0	0A
	1	1	1	1	1	1	1	1	FF	1	0	1	0	0A
	1	0	1	1	1	0	1	1	BB	0	1	0	1	05
	1	1	1	1	1	1	1	1	FF	0	1	0	1	05
3, Function	1	1	0	1	1	1	0	1	DD	0	1	0	1	05
	1	0	0	1	1	0	0	1	99	1	0	1	0	0A
	1	1	0	1	1	1	0	1	DD	1	0	1	0	0A
	1	1	1	1	1	1	1	1	FF	1	0	1	0	0A
	0	1	1	1	0	1	1	1	77	1	0	1	0	0A
	0	0	1	1	0	0	1	1	33	0	1	0	1	05
	0	1	1	1	0	1	1	1	77	0	1	0	1	05
	1	1	1	1	1	1	1	1	FF	0	1	0	1	05

X = don't care.

1. Temperature: 7400 series: 0–70°C, 5400 series: −55°C–125°C.

2. Vibration

3. Power supply voltage: 5 V +10%, 5 V −10%.

4. Altitude/humidity.

5. Long-term testing.

6. Testing at various radiation levels.

7. Combinations of these tests.

This type of testing is extremely common in industrial quality assurance applications. After all, bad parts make equally troublesome equipment. Most component insertion equipment routinely tests new ICs by these procedures, and an understanding of both the hardware and the software is essential.

12 A TRAFFIC SIGNAL CONTROLLER

More and more busy intersections are coming under microprocessor control. The simple sequencing of a traffic signal involves the use of timing loops and parallel output ports. The use of a relay, preferably a solid state relay, converts the 5-V logic levels to 120-V a-c control. It is essential that the low-voltage microcomputer be completely isolated from the 120 V a-c line, and this need contributes to the substantial cost of solid state relays. A real intersection is further complicated by

FIGURE 11-18 A typical intersection.

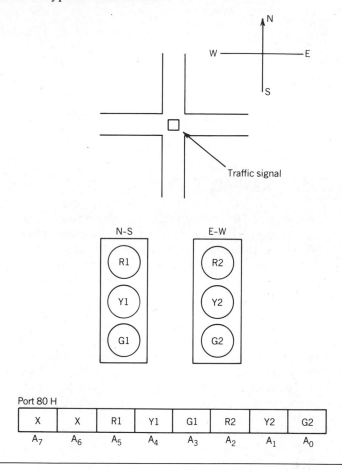

Traffic signal

the use of various inputs to the microcomputer. Ground loops (detectors under the pavement), ultrasonic detectors, emergency commands (via radio or hard wire), or walk buttons all affect the operation of signals in an intersection. We will look at a simple 12-light system (four sets of three: R, Y, G). Opposite lights will be tied together, thus allowing the control of six lights in the program.

Figure 11-18 shows how the port (80H) is connected to the lights. You may assume that the solid state relays are hidden in the traffic controller and that a "1" will turn a light on. As Table 11-7 indicates, bits A_7 and A_6 are not needed. When you write the controller program (you knew this was coming), allow 15 sec between each red-to-green transition. The 1/0 under the yellow lights means that to keep the program interesting, the yellow lights are to *flash* at a rate of 2 Hz for five flashes. You will also need two subroutines, one for a long delay and one for a short delay.

TABLE 11-7   The Sequence

| A5 | A4 | A3 | A2 | A1 | A0 |
R1	Y1	G1	R2	Y2	G2
1	0	0	0	0	1
1	0	0	0	1/0	0
0	0	1	1	0	0
0	1/0	0	1	0	0
1	0	0	0	0	1
		Repeat			

13 SCANNING A KEYPAD

A common but limited means of entering data into a microprocessor-based system is by a keypad. Figure 11-19 shows a hexadecimal keyboard containing 16 switches arranged in four rows and four columns. (This system is also used in larger typewriter-type keyboards.) The idea is to avoid the expense of a system that requires 16 wires or input bits, one per switch. The row lines are connected to four output bits on port A and the column lines are connected to four input bits on port B. (Note: If you use an 8255A PIA, you can split port C into an output group of four and an input group of four.) Also note that the column lines are pulled to +5 V by the 10-kΩ pullup resistors. If a button in the keypad is pressed, a row line and a column line are shorted together. For example, if the "B" key is pressed, the A_2 and B_3 lines will be shorted together.

To implement the reading of the keypad, we first place a zero on line A_0, and ones on A_1, A_2, and A_3. If any of lines B_0, B_1, B_2, or B_3 are low, we conclude that button 1, 2, 3, or 4 is depressed.

Next we check row 2. Line A_1 is pulled low (by the micro), leaving A_0, A_2, and A_3 high. Again, read lines B_0, B_1, B_2, and B_3 to see whether any are low. The process is repeated until all four rows have been "scanned." Notice that this system requires only 8 wires to scan 16 keys. Table 11-8 lists the requirements of the program. Note that the output data is 01, 02, 04, and 08 for the four tests. Printout 11-5 shows the keypad scanning program.

The program will output a character to the teletype or printer using COUT depending on which key is depressed. In the program keypad scanner, register B will keep track of which key has been depressed by incrementing as the program scans from zero through F keys. Register B starts at zero, if the "0" key was not depressed; then B is incremented and the "1" key is checked. If the "1" key has not been pressed, B is incremented again. If *no* key is pressed, B increments to F.

The DONE routine converts zero to F hex to the correct ASCII code; 0–9 is converted to 30H – 39H; A–F is converted to 41H – 46H. The delay routine is added to prevent crowding the screen with an excessive number of characters. If a key is held down, its character will repeat as the delay routine times out.

Most keypads and keyboards generate a keypressed signal (KP) when any key is pressed. This allows a computer to enter a key scan routine only when needed,

FIGURE 11-19 A hex keypad with keypressed line.

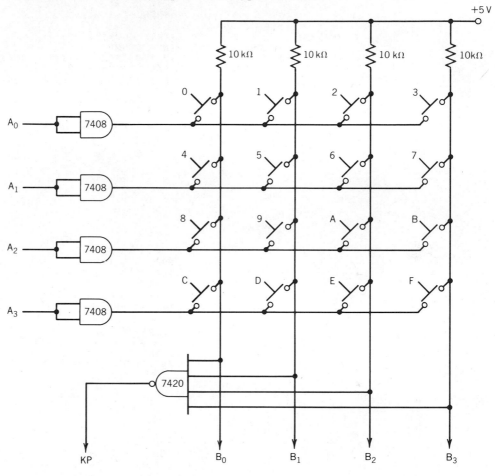

freeing itself for other things. The KP signal is usually tied to an interrupt line entering the processor. This is called a hardware interrupt. On 8080A/85-based systems, this hardware interrupt (INT = 1) line is pulled high and the processor stops its present program and goes to an interrupt service routine. More on this in the next section.

TABLE 11-8   Data for Keypad Scanning

Test No.,	Out to Port A				Input Data
	A3	A2	A1	A0	
1	1	1	1	0	Check B_0, B_1, B_2, B_3 for each test.
2	1	1	0	1	
3	1	0	1	1	
4	0	1	1	1	

PRINTOUT 11-5

```
                TITLE 'KEYPAD SCANNER'
                ORG   1000H
RSTRT:  MVI B, 0             ;DATA COUNTER
        MVI C, 4             ;ROW COUNTER
        MVI A, 0FEH          ;OUTPUT DATA = 11111110
        STA TEMP             ;SAVE FOR NEXT ROW
NEXT:   OUT 80H              ;OUT TO KEYPAD (A3-A0)
        IN 81H               ;GET COLUMN DATA (B3-B0)
        CMA                  ;COMPLEMENT DATA IE:1110 = 0001
        ANI 0FH              ;ONLY INTERESTED IN LSP
        CPI 01               ;IS IT B0?
        JZ DONE
        INR B
        CPI  02
        JZ DONE
        INR B
        CPI 04
        JZ DONE
        INR B
        CPI 08
        JZ DONE
        INR B
        LDA TEMP             ;GET ROW DATA
        RLC                  ;NEED DATA X 2
        STA TEMP             ;SAVE IT
        DCR C                ;ONE LESS ROW
        JZ RSTRT             ;NO KEYPRESSED - START AGAIN
        JMP NEXT
DONE:   MOV A, B             ;B CONTAINS HEX DATA
        CPI 09
        JNC LTR
        ADI 30H              ;ADD ASCII BIAS FOR DIGIT
        MOV B, A
        CALL COUT
        JMP DELAY
LTR:    ADI 37H              ;ADD ASCII BIAS
        MOV B, A
        CALL COUT
DELAY:  LXI H, 8000H         ;WANT TO EITHER PRINT
DEC:    DCX H                ;ONE CHARACTER OR REPEAT
        MOV A, H             ;IF KEY IS HELD DOWN,
        ORA L
        JNZ DEC
        JMP RSTRT            ;START OVER
TEMP:   DS 1
COUT:   SET 200DH
        END
```

Now let's produce a keypressed signal (Fig. 11-20). The four inputs to the NAND gate are tied to the B_0–B_3 lines along with the pullup resistors on the column lines. Normally, any line is high. As a key is pressed, one of the B_0–B_3 lines goes low, generating a "1" level from the output of the NAND gate. This can be the KP signal that activates the interrupt line on the processor. To use this method, all A lines must be held low when not scanning, so that a key hit on *any* line will work.

FIGURE 11-20 Keypressed signal.

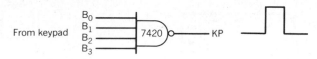

14 THE HARDWARE INTERRUPT

Both the 8080A and the 8085 have eight levels of interrupts. These are the restarts, and they are listed in mnemonic form as RST 0–RST 7. The 8085 has four additional interrupts: RST 5.5, 6.5, 7.5, and TRAP. Some of these are hard-wired interrupts and some are software interrupts:

RST 0–RST 7	Software 8080A/85
INT	Hardware 8080A/85
TRAP, RST 5.5, RST 6.5, RST 7.5	Hardware (8085 only)

The INT line on the processor generates an interrupt. What does all this mean? When any kind of interrupt is generated, the processor finishes its current instruction, acknowledges the interrupt, and looks to the data bus for an instruction telling it how to handle the interrupt. In 8080 systems, the CPU circuitry can be designed to jam a RST 7 onto the data bus after an interrupt has been received. Although there are many ways to interrupt a computer, at this time we discuss only the common hardware interrupt, INT.

Figure 11-21 shows both the 8080A and the 8085 MPUs. Find the INT line on each processor. This interrupt line, when taken high, can cause the processor to do an RST 7 and program execution continues at address 0038 H. (See Table 11-9 for a list of RST addresses.) It is common (appropriate) to have a jump installed (JMP) at that address to continue program execution at a memory location having more memory available for use. This JMP is usually to some type of interrupt-handling routine (e.g., Fig. 11-22).

We will see in Chapter 13 that any kind of instruction can be jammed onto the data bus after an interrupt request and that the 8080A/85 does not always execute an RST 7.

The beginning of most interrupt routines usually:

1. Disables the interrupt line so that the processor will not be interrupted again (DI instruction). This is redundant, however, because the 8080A disables its interrupt line after receiving an interrupt. No other interrupts may take place until the enable interrupt (EI) instruction has been executed.

2. Saves the current state of the machine by preserving the contents of all registers, including the flag register, as seen in Printout 11-6.

FIGURE 11-21 The 8080A and 8085 MPUs. (Reprinted by permission of Intel Corporation, copyright 1980.)

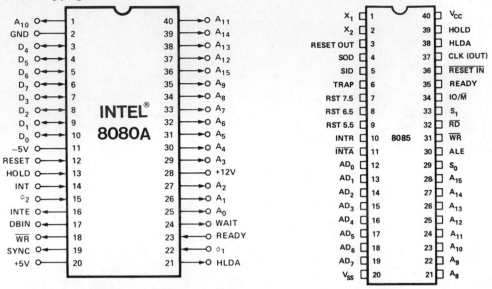

TABLE 11-9   Restart Addresses

8080A/85		8085 Only	
RST 0	00 H		
RST 1	08 H		
RST 2	10 H		
RST 3	18 H		
RST 4	20 H	TRAP	24 H
RST 5	28 H	RST 5.5	2CH
RST 6	30 H	RST 6.5	34 H
RST 7	38 H	RST 7.5	3CH

FIGURE 11-22 Interrupt-handling routine.

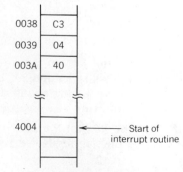

PRINTOUT 11-6

```
                TITLE     'INTERRUPT HANDLING ROUTINES'
                ORG 2000H
        INTR:   DI                      ;DISABLE INTERRUPTS
                PUSH  PSW                ;SAVE A + F
                PUSH B
                PUSH D
                PUSH H
                                        ;ROOM FOR WHATEVER
                                        ;YOU HAVE TO DO
                                        ;
                                        ;
                                        ;
                                        ;
                POP H                   ;RESTORE REGISTERS
                POP D
                POP B
                POP PSW
                EI                      ;ENABLE INTERRUPTS
                RET                     ;RETURN TO PROGRAM THAT
                                        ;WAS INTERRUPTED
                                        ;USING THE SAVE ADDRESS
                                        ;GENERATED BY THE
                                        ;ORIGINAL INTERRUPT.
                END
```

Once these steps are completed, you read the keypad, store the information, or whatever. After the "job" has been done, you restore the state of the registers, enable the interrupt line again, and RET. The RET gets the return address from the stack.

Interrupts are usually used for routines, sequences, or purposes that do not happen very often in terms of high-speed computers.

15 THE SOFTWARE INTERRUPTS

The main use of the software interrupt is for frequently used subroutines, which represent a means of saving space in memory. For example, the serial output routine (COUT) may be used many times in a program. That is, it may be called in many different places in a program.

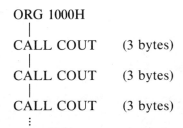

```
ORG 1000H
    |
CALL COUT     (3 bytes)
    |
CALL COUT     (3 bytes)
    |
CALL COUT     (3 bytes)
    ⋮
```

Each time it appears, three bytes of memory is used. However, if JMP COUT (C30D20) were installed at address 0018H, an RST 3 would accomplish the same purpose as all those CALL COUTs.

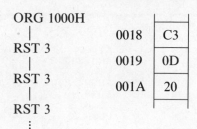

```
ORG 1000H
   |
RST 3
   |
RST 3
   |
RST 3
   ⋮
```

0018	C3
0019	0D
001A	20

The difference here is that an RST 3 (0DFH) requires only *one* byte of memory each time the output routine is needed.

Any program using this technique must be sure to install the JMP instruction at 0018. Here is a way to do it:

```
MVI A, 0C3H    ; JMP instruction
STA 18H
LXI H, 200DH
SHLD 19H
EI
```

A fancier way uses lots of labels:

```
        ORG 1000H
           |
        MVI A, 0C3H       ; JMP instruction
        STA REST3         ; Restart address
        LXI H, INTR       ; Address of interrupt routine
        SHLD REST3 + 1    ; Restart
        EI
           |
REST3:  SET 18H
 INTR:  EI
        PUSH PSW
        PUSH B
           ⋮
```

The interrupt is an essential method for getting the processor's attention when an infrequent, but important, job must be done.

The 8085 has only two additional instructions not found on the 8080A, namely, RIM (read interrupt mask) and SIM (set interrupt mask). These instructions either have to do with interrupt operations or are used to operate the serial input or serial output lines on the 8085 CPU (SID and SOD). By using the SID and SOD lines, implementation of serial data transfer can take place and you are referred to the software UART program in the next chapter (Printout 12-1) for an example of their use.

Let us concern ourselves with the use of the RIM and SIM instructions with respect to the hardware interrupts (RST5.5, 6.5, and 7.5). The 8085 has five hardware interrupt lines as shown in Fig 11-23.

The TRAP hardware interrupt on the 8085 has the highest priority and is unaffected by any mask or interrupt enable. Upon the TRAP interrupt, the CPU

FIGURE 11-23 Hardware interrupt for the 8085.

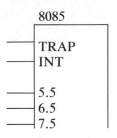

vectors to location 24 H. This occurs with the rising edge of the TRAP input. The priority of interrupts is shown in Table 11-10. Restarts 7.5, 6.5, and 5.5 have the next priorities, in that order. Their operation is affected by the RIM and SIM instructions. The RIM instruction allows reading of the current status of the interrupt masks (Fig. 11-24). The lower three bits (0, 1, 2) allow the programmer to read the state of the interrupt mask. If high, an interrupt is disabled. The RST 5.5, 6.5, and 7.5 instructions may be selectively disabled at the programmer's discretion, using the SIM instruction.

TABLE 11-10 Interrupt Priority, Restart Address, and Sensitivity[a]

Name	Priority	Address Branched To (1) When Interrupt Occurs	Type Trigger
TRAP	1	24H	Rising edge AND high level until sampled
RST 7.5	2	3CH	Rising edge (latched)
RST 6.5	3	34H	High level until sampled
RST 5.5	4	2CH	High level until sampled
INTR	5	See Note (2)	High level until sampled

NOTES:

(1) The processor pushes the PC on the stack before branching to the indicated address.
(2) The address branched to depends on the instruction provided to the cpu when the interrupt is acknowledged.

[a]Reprinted by permission of Intel Corporation, copyright © 1979.

FIGURE 11-24 Using the RIM instruction.

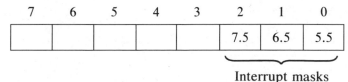

16 MEMORY MAPPING

16.1 A Memory-Mapped Video Display

Another aspect of microcomputer applications is the screening of a range of memory locations as a display area; this is called a memory-mapped video display. A certain amount of circuitry is required along with a character generator ROM. A typical display might have 16 rows of 64 characters each appearing on the screen, for a total of 1024 characters. The display circuitry is designed to continuously refresh the display 60 times per second in the same manner used by ordinary raster-scan television. However, programming is required to place the characters in a particular position on the screen. Let us assume that the 1024 character positions on the screen occupy memory space from address CC00 H to CFFF H. This range of memory is high enough to cause the least interference with large programs in lower address space. (In other words, it is out of the way.) The addresses CC00 H to CFFF H occupy 400 H or 1024 D memory locations, one byte for each character. With this type of display (Fig. 11-25), loading a memory location with an ASCII character is the equivalent of placing a character on the screen; hence the term memory-mapped video (also screen memory).

Each line of display corresponds to a range of memory locations, as indicated in Table 11-11. Placing a character in a memory location within the specified range puts a character on the screen in the chosen location.

16.2 Print a Simple Message

Whenever power is applied to a computer, the memory, including the memory used for the memory-mapped display, comes up in a random pattern. It is first necessary to *clear* the display by writing the ASCII code for a blank or space

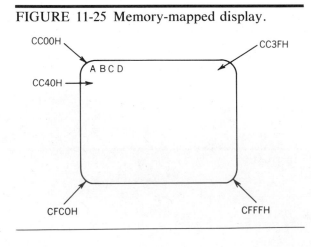

FIGURE 11-25 Memory-mapped display.

TABLE 11-11   Range of Memory Locations

Line	Hex Address
1	CC00 – CC3F
2	CC40 – CC7F
3	CC80 – CCBF
4	CCC0 – CCFF
5	CD00 – CD3F
6	CD40 – CD7F
7	CD80 – CDBF
8	CDC0 – CDFF
9	CE00 – CE3F
10	CE40 – CE7F
11	CE80 – CEBF
12	CEC0 – CEFF
13	CF00 – CF3F
14	CF40 – CF7F
15	CF80 – CFBF
16	CFC0 – CFFF

(20 H) into all 1024 D memory locations. This is easily accomplished with the following subroutine:

```
CLEAR:  LXI H, 0CC00H   ; Screen address
        LXI D, 400H     ; Need to do it 1024D times
   CLR: MVI M, 20H      ; Space code to M
        INX H           ; Ready for next address
        DCX D           ; Decrement DE
        MOV A, D        ; Test for zero
        ORA E
        JNZ CLR         ; If not, continue
        RET
```

In this subroutine, HL contains the address of screen memory. DE will down count 400 H = 1024 D locations. Since HL contains the address of memory (M), we move a space (20 H) to the memory location specified by HL for 1024 D times. All those blanks represent a blank screen.

Printout 11-7 shows how a simple message can be placed on video (display) line 7. In this program, the XCHG instruction is used to exchange the contents of HL registers with DE registers. This allows the M reference to refer to two locations in memory (data or display).

The message can be made to blink or flash by alternately clearing the screen and writing the message out to screen memory with appropriate delays between.

PRINTOUT 11-7

```
        TITLE 'MEMORY MAPPED DISPLAY DEMO PROGRAM'
        ORG 1000H
        LXI SP, STACK   ;SET UP STACK AREA
        CALL CLEAR      ;CLEAR THE SCREEN
        LXI H, MESG     ;HL HAS MESSAGE ADDRESS
        LXI D, 0CD80H   ;ADDRESS OF LINE 7
MORE:   MOV A, M        ;GET 1ST DATA CHARACTER
        XCHG            ;SWAP HL WITH DE
        MOV M, A        ;SEND TO SCREEN!
        XCHG            ;PUT HL AND DE BACK
        CPI 21H         ;WAS IT ENDMARK?
        JZ HALT         ;STOP IF ENDMARK
        INX H           ;NEXT DATA LOCATION
        INX D           ;NEXT SCREEN LOCATION
        JMP MORE        .
CLEAR:  LXI H, 0CC00H   ;SCREEN ADDRESS
        LXI D, 400H     ;NEED TO DO IT 1024D TIMES
CLR:    MVI M, 20H      ;SPACE CODE TO M
        INX H           ;READY FOR NEXT ADDRESS
        DCX D           ;DECREMENT DE
        MOV A, D        ;TEST FOR ZERO
        ORA E
        JNZ CLR         ;IF NOT, CONTINUE
        RET
HALT:   HLT             ;HALT/STOP!
MESG:   DB 'FRIENDSHIPS SHOULD LIVE LONG AND PROSPER!'
        DS 30           ;ROOM FOR STACK
STACK:  DS 1
        END
```

16.3 Scrolling the Display

Typically, the operation of a memory-mapped display is an inexpensive way of simulating a video terminal. Its operation is far from the sole function of the computer. It is, in fact, a small part of the computer's operation. It becomes a part of the utilities' routines, perhaps a part of COUT. A terminal will normally move its display from the bottom toward the top. The bottom line is cleared. New data is written on the bottom line. After a carriage return, all the data is moved up a line, discarding the top line and causing the entire display to move up, like a piece of paper being rolled through the platen of a typewriter. This is called a scrolled display. Since there are 64 D (40 H) characters on a line, all characters are moved back (up) this far beginning at the starting address of line two (CC40 H). Printout 11-8 is a sample scrolling routine.

16.4 A Scrolling Display Routine

A full scrolling display routine should be able to perform a number of operations on a memory-mapped video display:

1. Clear the display.
2. Scroll the display when a CR is received.
3. Place characters on the bottom line in sequence.
4. Respond to a backspace ($\uparrow$ H = 08 H).
5. Respond to a tab ($\uparrow$ I = 09).

PRINTOUT 11-8

```
            TITLE "SCROLLING ROUTINE"
            ORG 1000H
    SCROL:  LXI H, OCC40H      ;START LINE 2
            LXI D, OCCOOH      ;START LINE 1
            LXI B, 3COH        ;WILL MOVE 1024 - 64 CHARACTERS
    MOVE:   MOV A, M           ;MOVING ROUTINE
            XCHG
            MOV M, A
            XCHG
            INX H
            INX D
            DCX B
            MOV A, B
            ORA C
            JNZ MOVE
            MVI B, 40H         ;CLEAR LAST LINE
            LXI H, OCFCOH      ;START OF LINE 16
    CLRIT:  MVI M, 20H         ;ASCII BLANK CODE
            INX H
            DCR B
            JNZ CLRIT
            XRA A
            STA TEMP           ;SET CHAR COUNTER TO 0
            RET
    TEMP:   DS 1               ;FOR THIS SUBROUTINE
            END                ;ONLY
```

PRINTOUT 11-9

```
                TITLE 'FULL VIDEO ROUTINE'
                ORG 1400H
        VIDEO:  PUSH PSW          ;SAVE ALL REGISTERS
                PUSH B
                PUSH D
                PUSH H
                MOV A, B          ;READY TO COMPARE
                CPI ODH           ;CARRIAGE RETURN?
                CZ SCROL
                CPI 08            ;BACK SPACE?
                CZ BSPC
                CPI 09            ;TAB?
                CZ TAB
                ANI 60H           ;B6 = B7 = 0 MEANS CONTROL CHAR
                JZ REST           ;WON'T PRINT OTHER CONTROL CHARS
                LDA TEMP          ;GET CONTROL CHARACTERS
                INR A
                STA TEMP
                CPI 40H           ;ONLY ALLOW 64 CHARACTERS PER LINE
                CZ SCROL
                LXI H, OCFCOH     ;ADDRESS OF BOTTOM LINE
                ADD L             ;ADD L TO A
                MOV L, A          ;SAVE IT IN L
                MOV M, B
        REST:   POP H             ;RESTORE ALL REGISTERS
                POP D
                POP B
                POP PSW
                RET
        BSPC:   PUSH PSW          ;SAVE A FOR COMPARES ON RETURN
                LDA TEMP
                ORA A             ;WANT TO SEE IF A IS ZERO
                RZ
                LXI H , OCFCOH
                ADD L
                MOV L, A
                MVI M, 20H
                LDA TEMP
                DCR A
                STA TEMP
                POP PSW
                RET
        TAB:    LDA TEMP
                INR A
        CHECK:  ANI 07            ;WANT TO TAB OVER 8
                RZ
                LDA TEMP          ;GET COUNT BACK
                INR A
                STA TEMP
                JMP CHECK
        SCROL:  SET 1000H         ;TEMPORARY RELATED TO SCROLL ORG
        TEMP:   DS 1
                END               ;FOR SUBROUTINE ONLY
```

We assume that any character to be output will be in the B register and that all control characters other than backspace, tab, and carriage return (CR) will be ignored. A carriage return always is interpreted as a CRLF.

The scrolling video display routine of Printout 11-9 would be a part of a larger operating system for a computer. In fact, it would be a part of the output routine, COUT, as shown below.

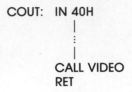

```
COUT:  IN 40H
        |
        ⋮
        |
       CALL VIDEO
       RET
```

17   DATA HANDLING

17.1 Data Conversion Techniques and Data Entry

Data that comes into a computer or, conversely, data being transmitted from a computer may take on various forms. It is usually necessary to convert from one form to another. For example, a display terminal transmits and receives ASCII data, which may have to be transformed from ASCII to BCD or hex for use in arithmetic-type routines.

17.2 ASCII to Hexadecimal

ASCII characters 0–9 (30 H to 39 H) and A–F (41 H to 46 H) may have to be converted to 4-bit hexadecimal (0000→1111B or 0 → F H).

Let us assume that a character is received from the keyboard (ASCII) using CIN. We can convert this to hexadecimal in the following subroutine:

```
CONVH:  CALL CIN    ;Get a character
        SUI 30H     ;Subtract ASCII bias
        CPI 10      ;Test for digit or letter
        RM          ;Return if minus—it's a digit and all converted
        SUI 7       ;Subtract 7 for letter
        RET         ;Done
```

Let's examine the operation of the subroutine using some sample ASCII data. Suppose a "7" is typed. The ASCII code 37 H is received. We must convert to 0111 B or 7 H.

37 H	(Call CIN)
− 30 H	(SUI 30H)
7 H	Result
− 10	Internally subtract 10
− 3	(RM) with 7 H in A register

Suppose a B is received. The ASCII code 42 H is received. We must convert to 1011 B or B H.

$$
\begin{array}{rl}
42\ \text{H} & \text{(Call CIN)} \\
-\ 30\ \text{H} & \text{(SUI 30 H)} \\
\hline
\end{array}
$$

$$
\begin{array}{rl}
\lceil\ 12\ \text{H} & \\
\mid\ -\ 10 & \text{(Internal subtract)} \\
\mid\ \overline{\ 2} & \text{Not minus, so continue} \\
\llcorner\!\!\rightarrow 12\ \text{H} & \\
-\ 7 & \\
\hline
11\ \text{D} & \text{or B H, which is correct!}
\end{array}
$$

Try some data of your own.

17.3 Hexadecimal to ASCII

After calculations have been performed or data manipulated, it may be necessary to print out the resulting data. Since internal data probably is in hexadecimal, it must be converted to ASCII. Assume that the data is in the A register.

```
CONVA:  CPI 10        ;Is it a digit?
        JM DIGIT      ;DIGIT if minus, otherwise letter
        ADI 37H       ;Add 37H for bias
        JMP OUT
DIGIT:  ADI 30H       ;Add 30H for bias
  OUT:  MOV B, A
        CALL COUT     ;Let's output it
        RET
```

Let's try some data, outputting first a 7 H, then a C H.

$$
\begin{array}{rl}
\lceil\ 7\ \text{H} & \\
\mid\ -\ 10\ \text{D} & \text{(Internal subtract)} \\
\mid\ \overline{\ -\ 3} & \text{(Perform jump on minus to DIGIT)} \\
\llcorner\!\!\rightarrow 7 & \\
+\ 30\ \text{H} & \text{ASCII bias for numbers 0–9} \\
\hline
37\ \text{H} & \text{Correct ASCII code for 7}
\end{array}
$$

Now let's output a C H.

$$
\begin{array}{rl}
\lceil\ \text{C H} & \\
\mid\ -\ 10\ \text{D} & \\
\mid\ \overline{\ 2} & \text{(Positive result)} \\
\llcorner\!\!\rightarrow \text{C H} & \\
+\ 37\ \text{H} & \text{ASCII bias for letters A–F} \\
\hline
43\ \text{H} & \text{Correct for ASCII C}
\end{array}
$$

Both types of data conversion are required frequently, so be sure you understand them.

17.4 Testing for Valid Digits (Hexadecimal)

On receiving data from a keyboard, it is often desirable to limit input to valid hexadecimal characters (0–F). If other characters are received, the data should be rejected, perhaps with an error message and a chance to reenter the data. A routine to test for valid data would be sure the ASCII code received was between 30 H and 39 H or between 41 H and 46 H. Assume that the ASCII data is in the A register.

```
VALDG:   CPI 30H    ;Is it less than "0"?
         JM ERR     ;Error if minus
         CPI 39H    ;Is it 9 or less?
         RM         ;OK if minus
         RZ         ;OK if zero ( = 9)
         CPI 41H    ;Is it less than "A"?
         JM ERR     ;Error if minus
         CPI 47H    ;Is it greater than "F"?
         JP ERR
         RET        ;Otherwise OK—within range 0–9 or A–F
ERR:     LXI H, ERMSG
ERRC:    MOV A, M
         MOV B, M
         CPI 21H    ;End mark?
         RZ
         INX H
         CALL COUT
         JMP ERRC
ERMSG:   DB 'ENTER 0-9 OR A-F ONLY!'
         END
```

17.5 Entering a Four-Character Hexadecimal Address

It is frequently necessary to enter a four-character hexadecimal value via the keyboard. Study Printout 11-10, which includes a routine to check for valid hexadecimal characters and does the conversion from ASCII to hexadecimal. The HL register ends up with the last four characters entered. It is assumed that some type of error routine (ERR) is available, similar to that just discussed. Care is necessary because the ERR message also uses the HL register pair. DAD H causes the contents of HL to be added to itself, thereby doubling HL. Remember that a doubling of a number in a register ($\times$ 2) is equivalent to a shift left. Since we are working with 4-bit numbers, four shift lefts (4 DAD H's) will move a hex character into the next of four positions.

18 BASE CONVERSIONS

18.1 Converting Decimal to Binary

Often, particularly if you are an accountant, you will have to accept decimal input from a keyboard. The method is similar to the one we employed to accept hex input characters; that is, we take a character, check to see if it is valid, then

PRINTOUT 11-10

```
INPUT:    LXI H, 0
NXT:      CALL CIN          ;INPUT ROUTINE FROM KEYBOARD
          CPI 0DH           ;CARRIAGE RETURN MEANS DONE
          RZ
          MOV B, A
          CALL COUT         ;OUTPUT ROUTINE TO ECHO
          CPI '0'           ;ERROR IF LESS THAN ZERO
          JC ERR
          CPI '9'+1         ;ERROR IF GREATER THAN 9
          JNC CH2
          JMP FIG           ;MUST BE A DIGIT
CH2:      CPI 'A'           ;ERROR IF LESS THAN A
          JC ERR
          CPI 'F'+1         ;ERROR IF GREATER THAN F
          JNC ERR
          SUI 07            ;REMOVE BIAS
FIG:      SUI 30H           ;REMOVE ZERO BIAS
          DAD H             ;ROTATE CURRENT HL 4 PLACES
          DAD H
          DAD H
          DAD H
          ADD L             ;ADD CURRENT A TO L
          MOV L, A          ;TRANSFER RESULT TO L
          JMP NXT           ;CONTINUE TILL CARRIAGE RETURN
```

perform an operation on our final value register. The procedure we now employ is slightly more complicated because, instead of multiplying by 16 (which was the result of four rotations), we have to multiply by 10. To do so, let's consider an example for a dedicated multiply-by-10 routine.

EXAMPLE 11-2

Multiply 7 by 10 by using only additions.

SOLUTION 1
The easiest approach is to add 7 to itself 10 times.

$$7 + 7 + 7 + 7 + 7 + 7 + 7 + 7 + 7 + 7 = 70$$

This obviously is not the best solution (it takes too long). We would like a simpler (and faster) method to do the problem.

SOLUTION 2
A far better and faster method of multiplication by 10, which also lends itself well to 8080A/85 register arithmetic, is as follows:

$$
\begin{array}{cccc}
7 & \nearrow 14 & \nearrow 28 & \nearrow 35 \\
+\,7 & +14 & +\,7 & +35 \\
(\times 2)\ 14 & (\times 4)\ 28 & (\times 5)\ 35 & (\times 10)\ 70
\end{array}
$$

Notice that we took the original number and added it to itself, thus multiplying it by 2. We added this result to itself, giving us four times what we started with.

PRINTOUT 11-11

```
        INPUT:    LXI H, 0
        NXT:      CALL CIN        ;INPUT ROUTINE FROM KEYBOARD
                  CPI 0DH         ;CARRIAGE RETURN MEANS DONE
                  RZ
                  MOV B, A
                  CALL COUT       ;OUTPUT ROUTINE TO ECHO
                  CPI '0'         ;ERROR IF LESS THAN ZERO
                  JZ ERR
                  CPI '9'+1       ;ERROR IF GREATER THAN 9
                  JC FIG
                  JMP ERR
        FIG:      SUI 30H         ;REMOVE ZERO BIAS
                  MOV D, H        ;PUT A COPY OF HL INTO DE
                  MOV E, L
                  DAD H           ;MULTIPLY HL BY 10
                  DAD H
                  DAD D
                  DAD H
                  ADD L           ;ADD CURRENT A TO L
                  MOV L, A        ;TRANSFER RESULT TO L
                  JMP NXT         ;CONTINUE TILL CARRIAGE RETURN
```

We then added our original number to this result and doubled the third result, which gave the desired answer. Now, to apply this to register arithmetic:

```
LXI D, 7 ⎫
LXI H, 7 ⎬  ;Both registers get the initial value
DAD H       ;Multiply H by 2
DAD H       ;Multiply H by 2 again
DAD D       ;Add the original number
DAD H       ;Multiply H by 2 again
```

The final result is in the HL register pair. We simply add error checking to provide a means of catching large numbers.

The preceding routine works only for numbers between 0 and 65,535 before giving an overflow.

The input routine of Printout 11-11 will accept numerical data from the user until a nondigit character is entered. The number must be in the range 0–65,535 or else the carry bit will be set [by inserting a JC OVFL (overflow) instruction after each DAD instruction, the carry bit will designate the overflow (i.e., # > 65535)] and the program will jump to an error-handling routine. Nonetheless, the input routine is very useful for smaller number entry.

18.2 Converting Binary to Decimal

We have a method of converting decimal to binary, but we need to be able to reverse the process. A routine to take a binary value and convert it into decimal is often a useful one (e.g., a routine for displaying the line numbers in a BASIC program). It would waste too many bytes to store the line numbers as ASCII characters, so a 2-byte binary number is used to store the line number, and a

PRINTOUT 11-12

```
NMBR:   LXI  D, 0D8F0H    ;LOAD  DE  WITH  -10000
        CALL SUBT
        LXI  D, 0FC18H    ;LOAD  DE  WITH  -1000
        CALL SUBT
        LXI  D, 0FF9CH    ;LOAD  DE  WITH  -100
        CALL SUBT
        LXI  D, 0FFF6H    ;LOAD  DE  WITH  -10
        CALL SUBT
        MOV  A, L         ;WHAT  IS  LEFT  MUST  BE
        ANI  0FH          ;LESS  THAN  10.
        ADI  30H          ;ADD  ASCII  BIAS
        MOV  B, A
        JMP  COUT         ;DISPLAY  LAST  DIGIT  AND  RETURN
SUBT:   MVI  A, 30H       ;INIT  A  TO  "ZERO"
        PUSH H            ;SAVE  CURRENT  HL
        DAD  D            ;SUBTRACT  DE  FROM  HL
        JNC  SKIP         ;QUIT  IF  WE  WENT  NEGATIVE
        POP  B
        INR  A            ;COUNT  NUM  OF  TIMES  WE  CAN  SUBTRACT
        JMP  SUBT+2
SKIP:   POP  H            ;GET  THE  LAST  GOOD  HL  BACK
        MOV  B, A         ;DISPLAY  THE  DIGIT  AND  RETURN
        JMP  COUT
```

routine, such as the one shown in Printout 11-12, is used to output the correct ASCII decimal representation.

Consider for a moment the number 587. No amount of bit flipping or rotating will give us the correct data for displaying this number in ASCII digits. The easiest method for recovering this number is to count multiples of powers of 10, and display them. In this example, we have 5 hundreds, 8 tens, and 7 ones. Since we work with 16-bit binary numbers, the largest number that we have to convert is 65,535. Our method then is to count how many 10,000s there are, how many 1000s, and how many 100s, 10s, and 1s. The method that accomplishes this task follows.

We subtract a known amount from the given number, and count the number of times we can do this until the result goes negative. We then display the count in ASCII decimal and proceed with the next known amount until we are done. The four amounts that are used in the multiple subtractions are given in Table 11-12. We obtain the hexadecimal equivalent by finding the 2's complement of the desired number. For instance, 1000 decimal is 3E8 H. To obtain the negative of 1000, we find the 2's complement by first complementing the number, and then by adding 1 to it:

TABLE 11-12   Multiple Subtractions

Decimal	Hex Equivalent
−10,000	0D8F0
−1,000	0FC18
−100	0FF9C
−10	0FFF6

$$\begin{array}{ll} 0\,0\,0\,0\,0\,0\,1\,1\,1\,1\,1\,0\,1\,0\,0\,0 & \text{03E8 H} \end{array}$$

$$\text{Complement} \rightarrow \quad \begin{array}{l} 1\,1\,1\,1\,1\,1\,0\,0\,0\,0\,0\,1\,0\,1\,1\,1 \\ \underline{+ \hspace{5.5cm} 1} \\ 1\,1\,1\,1\,1\,1\,0\,0\,0\,0\,0\,1\,1\,0\,0\,0 \quad\quad \text{FC18 H} \end{array}$$

Now, to subtract 1000 from a given quantity, we simply add FC18 H to it.

The routine SUBT adds the number in DE to the number in HL until the number in HL goes negative. As it does so, it increments the accumulator, which was previously set to an ASCII zero. If the routine does this five times, the accumulator goes from 30 H to 35 H, which is an ASCII five. The routine NMBR simply controls the value to be sent to SUBT in the DE register. First it sends − 10,000, then − 1000, then − 100, and finally − 10. When all this is done, anything in the HL register is obviously less than 10, so we simply get this value into the accumulator and add the ASCII bias onto it. As an exercise, write a routine to precheck the value in the HL register pair, and determine the correct entry point into NMBR so that there are no leading zeros displayed. Remember, the routine is good for numbers between zero and 65,535.

19 BINARY CODED DECIMAL ARITHMETIC

19.1 Adding Numbers Using BCD Arithmetic

There are several methods of representing numbers in computers. Large computers represent decimal numbers in pure binary, but must employ an algorithm to offset the rounding errors that are due to the imprecision of representing decimal numbers in binary. Microprocessor systems typically represent digits using 4-bit (BCD) codes. The use of BCD eliminates any errors of this type. Microprocessor chips also make available special instructions to easily use these techniques. The purpose of such decimal adjust accumulator (DAA) instructions is to correct the result of an add or subtract operation to a BCD result. For example, the sum of 8 D + 9 D should be 17 D. However, the microprocessor performs hexadecimal addition, not decimal addition.

$$\begin{array}{r} 08\text{ H} \\ +\,09\text{ H} \\ \hline 11\text{ H} \\ +\ \ 6 \\ \hline 17\text{ H} \end{array}$$

The sum of 8 H + 9 H = 11 H. The 6 is added by the DAA instruction to give an apparent result of 17D and 17H. In this way, decimal arithmetic may be implemented. (Reread the information on the DAA instruction in Chapter 10.)

19.2 Implementing an Addition Program

Figure 11-26 shows a routine for adding a pair of 6-digit, BCD-represented numbers, stored in two separate ranges of memory locations. The first number is stored in a range beginning at 0100 called NUM1, the second is stored starting at

FIGURE 11-26 Addition of BCD numbers.

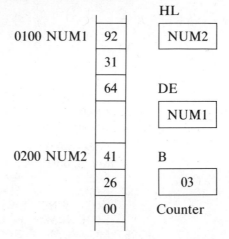

0200, called NUM2. The HL register pair is used to point at NUM2, and the DE register pair to point at NUM1. The numbers to be added are:

$$\begin{array}{ll} 643,192 & \text{NUM1} \\ +002,641 & \text{NUM2} \\ \hline 645,833 & \text{Result} \end{array}$$

The numbers are stored and added a pair at a time (i.e., 92 + 41, then 31 + 26, then 64 + 00). There are two rules:

1. Add any carries to the next pair.

2. Do a DAA if necessary.

Let us examine this problem in terms of adding hex numbers as the computer must do.

Step 1

$$\begin{array}{l} \quad 92 \text{ H} \\ +41 \text{ H} \\ \hline \text{D3 H} \quad \leftarrow \text{Should be 133} \\ +60 \text{ H} \quad \leftarrow \text{Use DAA} \\ \hline 133 \\ \uparrow\!\!\!\rule[0.4ex]{2cm}{0.4pt}\text{ Carry bit} \end{array}$$

Step 2

$$\begin{array}{l} \quad\;\; 1 \quad \leftarrow \text{Previous carry} \\ \quad 31 \\ +26 \\ \hline 058 \quad \leftarrow \text{OK!} \\ \uparrow\!\!\!\rule[0.4ex]{2cm}{0.4pt}\text{ No carry} \end{array}$$

Step 3

$$
\begin{array}{r}
0 \\
64 \\
+\,00 \\
\hline
64 \quad \leftarrow \text{OK!}
\end{array}
$$

The answer is 645,833.

The B register is used to count the number of *pairs* of numbers to be added, and the result of addition is restored in the range of memory addresses labeled NUM1. When done, we will look for the results in NUM1.

```
              ORG 1000H
              MVI B, 03H      ;Number of pairs to add
              STC             ;Sets carry bit
              CMC             ;Clears carry bit
              LXI D, NUM1     ;DE points to NUM1
              LXI H, NUM2     ;HL points to NUM2
       MORE:  MOV A, M        ;Get A pair
              XCHG            ;Swap DE with HL
              ADC M           ;Get hex sum of pair
              DAA             ;Convert to BCD
              MOV M, A        ;Save pair in memory
              XCHG            ;DE and HL back
              INX H           ;Point to next pair
              INX D           ;Point to next pair
              DCR B           ;Done yet?
              JNZ MORE
              HLT
```

Note: To be sure the carry bit starts reset, it must first be set (STC) and then complemented (CMC). (Or use XRA A!)

19.3 Entering the Numbers into Memory

Data from the keyboard is entered in ASCII and must be converted to BCD and packed in pairs. As the ASCII characters are entered, they must be within a valid decimal range (0–9 only). This requires a simpler valid digit routine:

```
      VALDG:  CPI 30H      ;Is it less than 0?
              JM ERROR
              CPI 3AH      ;Is it more than 9?
              JP ERROR
              RET          ;Digit OK—return
```

The procedure for converting and packing the ASCII pairs is fairly simple and requires four rotate instructions:

```
      GTCHR:  CALL CIN      ;Gets a character from TTY
              CALL VALDG    ;Checks to see if it's between 0 and 9
              ANI 0FH       ;Masks off MS nibble
              RLC           ;Moves BCD number to MS byte
              RLC           ;Moves BCD number to MS byte
              RLC           ;Moves BCD number to MS byte
```

```
            RLC                 ;Moves BCD number to MS byte
            MOV D, A            ;Temporary storage
            CALL CIN            ;Get another character
            CALL VALDG
            ANI 0F              ;Mask off MS nibble
            ORA D               ;Combine the two BCD numbers
```

This incomplete routine leaves a pair of ASCII characters in packed BCD form in the A register. It is left to you to store the A register contents in the appropriate memory location and prepare for the data to be placed in the next location.

The contents of register B consists of the number of pairs that can be used in memory. This can be determined during data entry, which is done for each number (NUM1 or NUM2) when a carriage return (0D H) is received.

19.4 Prompting for Data Entry

Any program that requires data entry from a keyboard should also be required to ask for that data. How else will the user know what data to enter? This requirement also reflects consideration of the feelings of computer users, which in turn is related to a field of study called human factors engineering. You already know how to enter this type of routine, but here is another sample:

```
            LXI H, MSG0         ;Address of message
    SEND:   MOV A, M            ;Get character
            CPI '*'             ;Last character yet?
            RZ
            MOV B, A            ;Move to B register
            CALL COUT           ;Out to TTY
            INX H               ;Point to next character
            JMP SEND
    MSG0:   DB 'ENTER FIRST NUMBER'
            DB 0DH, 0AH, '*'    ;0D = CR, 0A = LF
    MSG1:   DB 'ENTER SECOND NUMBER'
            DB 0DH, 0AH, '*'
```

To send the first message:

```
            LXI H, MSG0
            CALL SEND
```

To send the second message:

```
            LXI H, MSG1
            CALL SEND
```

This program is purposely incomplete. You cannot learn to program without some original brain power.

20 A DISCO LIGHT SHOW

One trendy application of microprocessors is a disco light show, consisting of 96 lamps (120 V a-c; 25 W) and 48 solid state relays. Not everybody can construct one of these, but it affords an excellent example of power control. Six output

FIGURE 11-27 Disco light show.

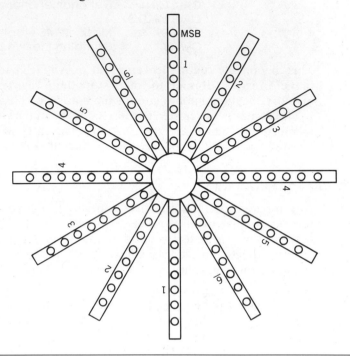

ports are required (48 lines). This can be accomplished by using a pair of PIAs. In the example shown in Fig. 11-27, the port numbers correspond to those of a National Semiconductor 80/10 Single Board Computer. This is not essential if six output ports are available to you. Initially you can consider writing a program based on the caterpillar software and output to all light strips at once. Experience has demonstrated that hundreds of different routines are possible when a group of avid programmers takes over. Strips across from one another are tied together; that is, the inside bulbs (LSBs) are common on strip one, and so on.

The light show uses an NSC board level computer (BLC). A standard monitor is in control at powerup. You may write your routines and run them in the 1K of on-board RAM from 3C00H to 3FFFH and load them using the on-board monitor. The two PIAs (six ports of output) are located at ports as follows:

PORT:	1	2	3	4	5	6
	0E4 H	0E5 H	0E6 H	0E8 H	0E9 H	0EA H

Control ports: 0E7 H and 0EBH

Before outputting patterns to the six ports, be sure to send the mode/control words to the two PIAs:

```
MVI A, 80H    ;All ports output
OUT 0E7H      ;To PIA (L)
OUT 0EBH      ;To PIA (R)
```

A sequence to send a pattern to all six ports is as follows:

```
MVI A, 80H   ;Mode set
OUT 0E7H
OUT 0EBH
MVI A, 55H   ;Alternate pattern
OUT 0E4H
OUT 0E5H
OUT 0E6H
OUT 0E8H
OUT 0E9H
OUT 0EAH
HLT
```

21 USING ONE-SHOTS

It is often necessary in a digital circuit (particularly those that contain microprocessors) to stretch out the length of a pulse to make it useful. For example, it is desired to keep the contacts of a certain relay closed for 100 msec. The digital logic driving the relay can provide a pulse of only 10 msec to the relay driver, which is only one-tenth of what we need. What can we do to eliminate the discrepancy illustrated in Fig 11-28?

The answer lies in the one-shot multivibrator, or monostable multivibrator as it is sometimes called. The purpose of a one-shot is to provide an output pulse

FIGURE 11-28 Required relay timing.

```
100 msec  ┌───────────────────┐
    ─────┘                     └─────   What we need

 10 msec  ┌┐
    ──────┘└────────────────────────    What we get
```

FIGURE 11-29 Simple one-shot operation.

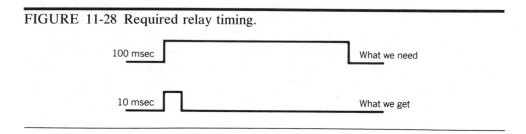

FIGURE 11-30 (*a*) Pinout and (*b*) truth table for the 74LS121 single-shot IC. (Courtesy of Texas Instruments Incorporated.)

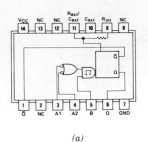

| FUNCTION TABLE | | | | |
| INPUTS | | | OUTPUTS | |
A1	A2	B	Q	Q̄
L	X	H	L	H
X	L	H	L	H
X	X	L	L	H
H	H	X	L	H
H	↓	H	⎍	⎍
↓	H	H	⎍	⎍
↓	↓	H	⎍	⎍
L	X	↑	⎍	⎍
X	L	↑	⎍	⎍

(*a*)   (*b*)

(when "triggered") of a desired width. A one-shot typically consists of a flip-flop and a resistance–capacitance network. In Figure 11-29 we see that the one-shot begins in a cleared state (Q = 0). When the signal T goes high, the flip-flop toggles and the Q output goes high. At the same time, capacitor *C* begins to charge through *R*. When the voltage Vx reaches an internal trip level, the flip-flop resets (Q = 0 again) and the circuit remains in this state until the next transition of T.

In an actual one-shot such as the 74LS121, the only external components that need to be added are *R* and *C*. Figure 11-30 gives the pinout and truth table for the 74LS121. Inputs A1, A2 and B are used to provide different triggering schemes. With input A2 grounded, a rising edge on the B input will trigger the '121. The pulse width of the output is defined by the following equation:

$$t_w = RC \ln 2$$

where *R* is measured in ohms, *C* in farads, and t_w in seconds.

In our previous problem we needed a pulse of 100 msec to drive a relay. If we choose *R* equal to 6.8 kΩ and *C* equal to 22 μF, we get:

$$t_w = (6800) \, (22 * 10^{-6}) \ln 2$$
$$= 104 \text{ msec}$$

which is very close to our requirement. When standard values of *R* and *C* do not mix to give us the pulse width we desire, a variable resistor is usually substituted

FIGURE 11-31 Functional relay driver.

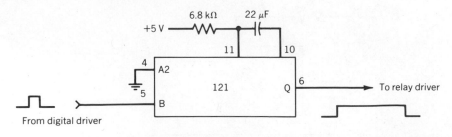

FIGURE 11-32 Temperature measurement with a one-shot.

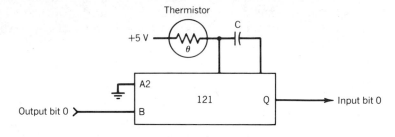

for R and the circuit is fine tuned. In our relay example, the pulse width is almost exactly what we want, so the final circuit contains a one-shot connected as shown in Fig. 11-31.

One-shots also can be applied in temperature measurement, as in Fig. 11-32, where temperature is measured by substituting a thermistor for R. A thermistor is a variable resistor whose resistance changes with temperature; thus, as the temperature changes, so will the pulse width of the '121's output pulse. We can set up a microprocessor to toggle a bit on an output port to trigger the one-shot, and we can use a bit on an input port to sample the one-shot's output. If we increment a counter every time we sample the output of the one-shot for as long as it remains high, we will get a count whose magnitude is a function of temperature. An example program to perform this task is as follows.

```
;Trigger is bit 0, output port 10H
;Input is bit 0, input port 10H
;We must sample the input port and continue to increment a counter
;until the input goes low
     START: LXI  B, 0      ;Clear counter
            MVI A, 01H     ;Trigger the one-shot
            OUT 10H
            XRA A
            OUT 10H
     LOOP:  INX  B         ;Increment counter
            IN   10H       ;Look at the '121's output
            RAR            ;Bit 0 into carry flag
            JC   LOOP      ;If carry set, then Q is high
              :            else BC represents temperature count
```

22 SUMMARY

In this chapter, we took advantage of the large variety of instructions in the instruction set of the 8080A/85 and combined them into programs designed to demonstrate their uses. We first introduced the concept of initializing registers and using them in counter loops to produce a desired, specifically timed delay. Next came the CALL and RET instructions when the stack and subroutine principles were presented. The idea of bit-flipping, and also the process of testing bits or bytes, was covered, for this is an important part of any large, task-oriented program. Specific examples of programs that support such devices as TTYs, memory-mapped video displays, USARTs, PIAs, and IC testers were given, and an understanding of them was greatly encouraged. The chapter served to introduce you to the power of machine-level programming, and also to show that the applications are endless.

23 GLOSSARY

Assembler.   A program that takes data in mnemonic form and converts, or "assembles," it into opcode, or machine code form, for direct loading into memory.

Command instruction.   Similar to the status port operation; one or two bytes sent to a receiver/transmitter chip to set the number of data bits, number of stop bits, baud rate factor, and parity information.

Data port.   A specific location that may be examined by a microcomputer and used to send or receive data to or from an external device.

DB.   The define-byte pseudo-op that reserves a place in memory for the data that follows it.

DS.   The define-storage pseudo-op that is used to reserve a desired number of memory locations during assembly.

DW.   The define-word pseudo-op that reserves 2 bytes in memory for the data that follows it.

Echo.   A method employed by programmers to send back to the user any character that the computer receives, thus showing the user that the proper character was received.

END.   The pseudo-op that tells the assembler that it has reached the end of the program undergoing assembly.

EQU.   An equivalent SET pseudo-op, usually reserved for 1-byte equates. SET usually uses 2 bytes for equates.

First pass.   In an assembler, a first pass is used to obtain the addresses for labels, and assign values to variables.

Interrupt.   An external signal that instructs the CPU to stop whatever it is doing and perform a requested task immediately. In an 8080A/85 interrupt, the program counter is pushed onto the stack before the interrupt service routine is executed.

Last in, first out.   Describing the stack procedure in which the last item "pushed" onto the stack is the first item to be "popped" off the stack.

Memory-mapped video.   A technique whereby a certain block of memory in a microcomputer is used as a display area for a monitor. Data placed in this memory is scanned, or mapped, by special circuitry, and placed in the appropriate place on a TV monitor.

Mode instruction.   In a USART, a byte sent to the mode port tells the USART what baud rate to use, the character length with parity bit, and the number of stop bits to use in serial transmission and receiving.

ORG.   An assembler pseudo-op that identifies for the assembler the address at which to begin assembling the program (i.e., the address of the first byte in memory for the program).

Overhead.   Instructions outside a repeated loop that contribute to the time to perform an operation.

Pop operation. The retrieval of stored data or addresses from the stack.

Pseudo-op. An instruction recognized by an assembler that directs the assembler to perform a specific function.

Push operation. A process of placing the contents of a register, or the return address in a subroutine call, into the stack area.

Scrolling. A method of moving all lines on a display up or down as required.

Second pass. In an assembler, the second pass places the values from the symbol table in the first pass whenever a label or variable is seen.

SET. A pseudo-op used in the first pass of an assembler to assign a value to the label specified by the SET function.

Solid state relay. An electronic relay that converts TTL logic levels to a-c voltage levels, with complete isolation.

Stack. A specified portion of a microcomputer's memory used for storing temporary values such as return addresses, counters, etc.

Status port. Actually a data port that gives the "status" or operating conditions of a specific data port (busy, transmitting, empty, clear, receiving, etc.).

Subroutine. A program within a program. Subroutines are generally used whenever a sequence of operations must be performed with great frequency.

Symbol table. Formed by the first pass of an assembler, the symbol table contains all labels and variables found in a mnemonic program listing, and their equivalent binary values for use in the program.

USART. A programmable communications interface (e.g., the 8251A) that can be configured, or programmed by software, and thus easily changed to different modes of operation.

PROBLEMS

11-1 Write a delay routine to give multiple delays of 4 msec. The binary value on port FF is multiplied by 4 to give the desired delay.

11-2 Write a delay routine to give multiple delays of 1 msec. The BCD value on port FF switch input is used as the multiplier.

11-3 At what rate (in tests per second) is the 7400 IC tested in the program of Printout 11-4?

11-4 Write a program to test:
    (a) A 7490 decade counter   (d) A 7491 shift register
    (b) A 7404 hex inverter   (e) A 7474 dual D flip-flop
    (c) A 7493 binary counter
Include message printouts for these tests.

11-5 Write a routine to echo characters from a keyboard only if they are capital letters or numbers.

11-6 Write a program to generate a square wave at 100 Hz, 1 KHz, and 5 kHz, for a 2-MHz CPU clock. Output is to bit 0, port 80H.

11-7 Modify the program in Problem 11-6 to accept numbers from a keyboard and use the number entered to select the duty cycle of the square wave (i.e., 1–10%, 2–20%, 7–70%).

FIGURE 11-33 Circuit for Problem 11-10.

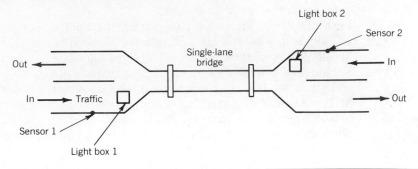

11-8   Write a program to *slide* a message from right to left across the top line of the memory-mapped video display used in this chapter.

11-9   Modify the program in Problem 11-8 to change the speed of sliding as the number on input port FF changes.

11-10 Write a traffic routine for a single-lane bridge (Fig. 11-33).
       Only one lane of traffic may cross the bridge at one time. The following additional rules apply:
       (a) If sensor 1 is on, indicating oncoming traffic, light box 2 must be red and light box 1 must switch from red to green.
       (b) If sensor 2 is on, light box 1 must be red and light box 2 must go from red to green.
       (c) Any lane has access to the bridge for 15 sec, once it gets a green light.
       (d) If both sensors are on, the one that has been on the longest gets priority.

11-11 Write a routine to output the BCD value (in ASCII) of a number in the BC register to a terminal. (Example: BC = 1234 H, so 31, 32, 33, and 34 hex must be sent.)

11-12 Write a routine to output the BCD value to eight places of four numbers stored in memory addressed by the DE register pair. Each byte contains 2 BCD digits.

11-13 Write a program to add two 6-digit BCD numbers, and display the result. Each set of numbers is stored in memory as three packed BCD pairs.

11-14 Write a program to normalize numbers into scientific notation.

$$\text{Examples:} \quad \begin{aligned} 6031.76 &\rightarrow 6.03176E03 \\ 0.0031415 &\rightarrow 3.1415E\text{-}03 \\ 1171.35 &\rightarrow 1.17135E03 \\ 0.00001 &\rightarrow 1.0000E\text{-}05 \end{aligned}$$

11-15 Show the code necessary to decode a control character received from a CIN routine and echo the appropriate control symbol back (via COUT). For example, if 07 is received, the following characters must be sent: ↑ and G.

11-16 Show that the following code converts 4-bit hex into ASCII (0–9 become 30–39 and A–F become 41–46).

```
                    ADI 90H
                    DAA
                    ACI 40H
                    DAA
```

11-17 What are the numbers in DE, HL, and BC after the following code has been executed? What is the value of the stack pointer?

```
                    LXI B, 0100H
                    LXI D, 7EF7H
                    LXI H, 3000H
                    LXI SP, 1000H
                    PUSH B
                    PUSH H
                    PUSH D
                    POP H
                    INX SP
                    INX SP
                    POP D
```

11-18 What is wrong with the following subroutine? (i.e.: Will it RET correctly.)

```
            SUB:    MVI B, 11H
                    ADD B
                    RLC
                    PUSH PSW
                    MVI A, 0FFH
                    RET
```

11-19 Write a short routine to add (using 16-bit addition, i.e., DAD) 16 numbers in memory, starting at location NUMS. Place the result in the HL register pair.

11-20 Modify the routine of Problem 11-19 to find the average of the 16 numbers by dividing the result by 16. Ignore the remainder.

11-21 Write a routine to find the absolute value of a number (stored in location NUM1) and place it in location NUM2.

11-22 Write a routine to test for lowercase characters a–z received from CIN and convert them to uppercase characters A–Z for echo to COUT.

11-23 Write a routine to do the following:
(a) Read a byte from port 66H.
(b) Jump on the following conditions: (1) to ACTA if bit 0 is set, (2) to ACTB if bit 2 is cleared, (3) to ACTC if bits 7, 6, and 4 are set. Assume that these conditions will not occur at the same time.

11-24 The 8080A is set up to vector to address 0038 H when it receives an interrupt. Where does it go to next if the following code is at 0038 H?

```
        0038    LHLD JUMP
                LXI D, 1234 H
                DAD D
                DAD H
                PCHL
        JUMP:   DW 3033 H
```

11-25 Show the activity of the stack pointer when the following code is executed. (The stack starts at 1000 H.)

```
BEGIN:  RST 7            0038  PUSH H
        JMP BEGIN              RST 2
                              POP H
                              RET
                        0010  ADD A
                              RLC
                              RET
```

11-26 Write a routine to convert 8-bit hex into 2-character ASCII (i.e., 7F H becomes 37 H and 46 H sent to COUT).

11-27 Write a routine to multiply a number in BC by 50. (*Hint:* Accumulate in HL with DADs.)

11-28 A BCD number (7,500,003) exists in memory as follows:

00	*07*	*50*	*00*	*03*
1	2	3	4	5

where byte 1 is at location TEST and byte 5 is at location TEST + 4. Write a routine to check these locations and jump to ERROR if the number is greater than 711,028,357.

11-29 Write a routine to:
(a) Multiply by 2 if greater than 100.
(b) Divide by 2 if less than 50.
(c) Subtract 23 from if between 50 and 100.
The number is in location DIGIT.

11-30 What is in the B register after the following code has executed?

```
        MVI D, 5
        MVI B, 0
        MVI A, 0B9H
LOOP:   RAR
        XRA B
        MOV B, A
        DCR D
        JNZ LOOP
```

11-31 Write a section of code to jump to a specific address depending on the value in the accumulator.

A	address
0	2000
1	3000
2	4800
3	9000
4	FC00

If the accumulator is greater than 4, jump to ERROR. All addresses are in hex.

CHAPTER 12
ADVANCED PROGRAMMING

1 INSTRUCTIONAL OBJECTIVES

When finished with this chapter you should be able to:

1. Use an 8085 to transmit and receive the 11-bit transmission code using the SID and SOD lines.

2. Understand the following advanced programming applications:
   A software-controlled security system using polling.
   A software/interrupt-driven real-time clock.
   A utility program to produce a hex/ASCII dump of a range of memory locations.
   A disassembler for 8080A, 8085, and 6502 microprocessors.
   A telephone dialer.
   A math package that does floating-point arithmetic $(+, -, \times, \div)$ using BCD.
   A bar code reader for the Universal Product Code.
   A 160-lamp multiplexed display.

3. Explain D/A and A/D conversion, especially as applied to a digital voltmeter and a speed control.

4. Transmit data of various forms between computers.

5. Save bytes by shortening programs.

2 SELF-EVALUATION QUESTIONS

Keep the following questions in mind and try to answer them when you have completed the chapter:

1. When an 8085 CPU is receiving data using a software UART, what happens if an interrupt occurs?

2. What values does hex dump have in aiding in the debugging of microcomputer software? What is the value of the ASCII data?

3. What is the function of a disassembler? How can it be used to find program errors?

4. What different types of math can a calculator perform? How does the use of BCD, as opposed to working in pure binary, benefit a computer user?

5. How could a laser be used to replace the optical scanner of the
   bar code reader? How does the software convert bars of variable
   width to ones and zeros?

6. What is the importance of conversions (D/A and A/D) in the
   world today?

7. How can each lamp be individually changed in a 160-lamp display
   that is driven from only 16 wires?

8. What different data formats might be transmitted from one
   computer to another? How can data be converted from one
   format to another?

3 INTRODUCTION

The first 11 chapters of this text have led you from the basics of digital electronics
through the use of the 8080A/85 microprocessors in typical applications of a
microcomputer to simple pragmatic situations. This chapter, which looks at advanced programming examples, is intended to serve as a guide to the in-depth
study of a programming language and at the same time to provide insight to the
tricks of the trade. Careful study of these examples will make you a better programmer in other situations. In every case, you are thoroughly encouraged to
execute each program on your own system, adapting to your own situation as
necessary.

In Chapter 11 we gave numerous examples of 8080A/85 programming that have
led to useful applications. All the functions that we study in this chapter, however,
have direct and immediate application, and are good examples of how the 8080A/85
can be used to solve specific business or academic problems. Of particular interest
are the WAND, MATH, and CLOCK programs, and you are encouraged to study
them in detail.

4 ADVANCED TOPICS

All the programs that follow contain some aspect of advanced programming. We
have briefly summarized them in Table 12-1. The programs cover a broad range
of applications and should serve to give you a good background in programming
skills.

5 THE SOFTWARE UART

In the preceding chapter we learned how to use a hardware receiver/transmitter
chip to handle serial I/O. The UART (or USART) did all the work with respect
to transmitting or receiving the 11-bit transmission code. Although a hardware
USART is not a very expensive item, it is not always necessary on microcomputers
that perform mundane control functions with little or no need to communicate
with a terminal; or perhaps the desire is to keep the cost of a computer to a

TABLE 12-1   Advanced Programming Examples

Program	Advanced Topic
SUART	Critical timing, data manipulation
CLOCK	Interrupts
HEXDUMP	Data manipulation
DISEMBLE	Mnemonic retrieval
PHONE	Pulse generation/conversion
MATH	Math routines
WAND	Light pen interfacing
A DIGITAL VOLTMETER	A-to-D and D-to-A conversions
A MULTIPLEX DISPLAY	Time-sharing data lines
DOWNLINE LOADING	Communicating between computers

minimum. In any event, a pair of simple subroutines can be written to produce the desired 11-bit transmission code. One bit of a parallel I/O port can be toggled at the correct rate for transmission purposes, and one bit can be examined to accomplish data reception. The software UART, however, will completely tie up the processor while sending or receiving data, and this should be considered when using the following ideas. Figure 12-1 shows the 11-bit transmission code.

Transmitting serial data requires a start bit (always zero) followed by the data bits (LSB first), then parity, and at least one stop bit. This pattern can be easily generated by a subroutine and an appropriate delay loop. The delay times needed are equal to the bit time (related to baud rate) and the half bit time.

Receiving serial data, which is also relatively simple, requires a set of delay times and begins when the line goes low (start bit).

5.1 Using the 8085 SID and SOD Lines

The 8085 was designed with serial I/O in mind. This particular CPU has a serial in-data (SID) line and a serial out-data (SOD) line. These lines are connected to the MSB of the accumulator when the RIM (20 H) and SIM (30 H) instructions are used. By manipulating bit 7 of the A register and using the SIM instruction, bit 7 is connected to the SOD line. [Bit 6 is the serial output enable (SOE) line and must also be high to transmit data on bit 7 (SOD).] By using the RIM instruction, data from SID is placed in bit 7 of the A register. Figure 12-2 shows an A register for the 8085.

FIGURE 12-1 Eleven-bit transmission code.

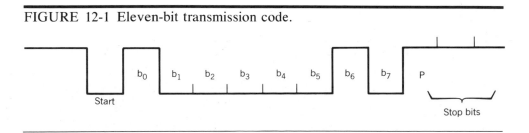

FIGURE 12-2 The 8085 Accumulator.

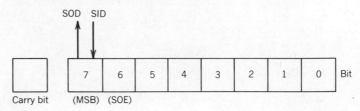

It is essential to disable the interrupt lines (DI) when using a software UART and enable (EI) when done. Otherwise, the serial I/O routines could be interrupted in midstream, losing data.

5.2 A Character Input Routine

Since the actual signals on an EIA RS232C serial I/O line can swing over an allowable range of ±3 to ±25 V, it is useful to have a level converter like the MC1489P available to convert to TTL levels. Figure 12-3 shows wiring to the SID line.

It will be the job of the program to do a RIM instruction and look for a bit equal to zero. The state of bit 7 is easily read by rotating the contents of A into the carry bit using an RLC. Printout 12-1 shows the routine (CI) for receiving 8 data bits.

The delay subroutine is a familiar one and assumes that DE contains the desired delay quantity. This is based on the desired baud rate as discussed in Chapter 8.

5.3 A Character Output Routine

Transmitting data is even easier. To send a bit from bit 7 to the SOD line, bit 6 must be high; then the SIM instruction is executed.

To send a 1: MVI A, 0C0 H; 11000000 SIM

To send a 0: MVI A, 040 H; 01000000 SIM

Printout 12-1 also shows the output routine (CO). The output line (SOD) is connected to an EIA converter (MC1488) to drive the actual line to the terminal. Figure 12-4 shows a TTL-to-EIA connection.

FIGURE 12-3 Wiring to the SID line.

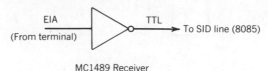

Printout 12-1

```
            TITLE 'SOFTWARE UART ROUTINES FOR 8085'
            ORG 2000H


INPUT:      CALL CI
            MOV B, A
            CALL CO
            JMP INPUT


CI:         DI                  ;DISABLE INTERUPTS
CI05:       RIM                 ;GET INPUT BIT
            RAL                 ;SHIFT IT INTO CARRY BIT
            JC CI05             ;WAIT IF NO START BIT
            LXI D, WAIT         ;HALF BIT TIME
            CALL DELAY
            LXI B, 8            ;B = 0, C = NUMBER OF BITS TO RECEIVE
CI10:       LXI D, TIME         ;FULL BIT TIME
            CALL DELAY          ;WAIT UNTIL MIDDLE OF NEXT BIT
            RIM                 ;GET THE BIT
            RAL                 ;SHIFT IT INTO CARRY
            MOV A, B            ;GET PARTIAL RESULT
            RAR                 ;SHIFT IN NEXT DATA BIT
            MOV B, A            ;REPLACE RESULT
            DCR C               ;DECREMENT BIT COUNT
            JNZ CI10            ;BRANCH IF MORE LEFT
            LXI D, TIME         ;WAIT FOR STOP BIT
            CALL DELAY
            MOV A, B            ;RESULT TO A
            ANI 7FH             ;SET PARITY BIT = 0
            EI                  ;SET UP FOR INTERRUPTS
            RET


CO:         DI                  ;DISABLE INTERRUPTS
            MVI A, 0C0H         ;START BIT MASK
            MVI C, 7            ;C HAS NUMBER OF BITS TO SEND
CO05:       SIM                 ;SEND A BIT
            LXI D, TIME         ;TIME PER BIT
            CALL DELAY
            MOV A, B            ;BITS LEFT TO SEND
            RAR                 ;LOW ORDER BIT TO CARRY
            MOV B, A            ;PUT THE REST BACK
            MVI A, 80H          ;READY FOR ENABLE BIT (SOE)
            RAR                 ;SHIFT IN DATA BIT TOO
            XRI 80H             ;COMPLEMENT DATA BIT
            DCR C               ;DECREMENT BIT COUNTER
            JP CO05             ;SEND ANOTHER BIT
            MVI A, 40H          ;SEND STOP BIT
            SIM                 ;SEND IT
            LXI D, TIME         ;WAIT FOR PARITY BIT
            CALL DELAY
            EI                  ;ENABLE INTERRUPTS
            RET


DELAY:      DCX D               ;DECREMENT DELAY COUNTER
            MOV A, D
            ORA E
            JNZ DELAY
            RET


TIME:       SET 30H             ;FOR 2400 BAUD
WAIT:       SET 15H

            END
```

FIGURE 12-4 The TTL-to-EIA connection.

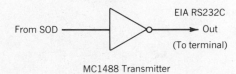

These two routines, with appropriate delay times, constitute a software UART. They are ideal for use with an 8085-based system, and they save the added expense of a hardware version. One important point: since these two separate routines must be run one at a time, you cannot transmit and receive simultaneously as with a hardware UART. Under certain applications (downline loading of data into memory), this may cause problems when attempting to echo received data to the terminal.

5.4 Variable Baud Rate

It may be necessary to be able to vary the baud rate of the software UART. This can be done either by changing the values of TIME and WAIT or by a more interactive procedure. Since TIME is the time per bit and WAIT is a half bit time, different values of WAIT could be stored in a table. TIME can be created by doubling the value of WAIT (RLC). The desired baud rate can be set by an external set of switches (port 0FFH here) and read by the software UART. This can be converted to the correct delay time by consulting a lookup table as in Table 12-2.

The delay times in Table 12-2 depend on two important considerations: first, the speed of the processor used, and second, a concern for the overhead involved in reading the desired baud rate. The second consideration can be eliminated by reading baud rate only once at power-up or system reset. Setting baud rate at startup would probably be a more desirable alternative—it simplifies the program and makes sense because baud rate is not changed frequently. The rate can be determined by storing the quantity in a pair of memory locations and also storing the half time. This is left to the programmer.

TABLE 12-2 Baud Rate to Delay Time Table[a]

Baud Rate	Port 0FF H	Delay Time
110	00	0240 H
300	01	0120 H
600	02	0060 H
1200	03	0030 H
1800	04	0022 H
2400	05	0015 H
4800	06	007 H
9600	07	004 H

[a]Correct only for an 8085 running at 3.125 MHz.

Printout 12-2

```
                       TITLE 'VARIABLE BAUD RATE ROUTINE'
                       ORG 1000H

          DELAY:   LXI H, TABLE              ;WANT TO POINT AT DATA TABLE
                   MVI B, 0                  ;READY FOR TABLE POINTER
          INP:     IN 0FFH                   ;GET DESIRED RATE SETTING
                   ANI 07H                   ;ONLY LOOK AT 3 LSB'S
                   CMP B                     ;CHECK FOR MATCH
                   JZ SETRT                  ;TIME TO DELAY IF MATCH
                   INR B                     ;IF NOT, ONE MORE
                   INX H                     ;INCREMENT TABLE POINTER
                   INX H
                   JMP INP                   ;CONTINUE SEARCH
          SETRT:   MOV E, M                  ;NEED ADDRESS IN HL
                   INX H
                   MOV D, M
                   XCHG                      ;SWAP DE WITH HL
          DEC:     DCX H                     ;START OF DELAY ROUTINE
                   MOV A, L
                   ORA H
                   JNZ DEC                   ;CONTINUE IF HL NOT ZERO
                   RET
          TABLE:   DW 0240H
                   DW 0120H
                   DW 0060H
                   DW 0030H
                   DW 0022H
                   DW 0015H
                   DW 0007H
                   DW 0004H
                   END
```

6 SECURITY SYSTEMS: POLLING MANY LINES

Computer systems are finding increasing use in building security systems. They are called on to monitor many different signal lines that originate from different sources: window and door switches, ultrasonic alarms, heat or smoke detectors, high water warnings, gas detectors, and even simple door bells. Since these sources could result from dozens of signal lines, a computer is an ideal candidate to monitor all these functions. Figure 12-5 is an example of a polling setup.

6.1 Polling

Let us consider a computer system that is required to monitor a number of different functions in a building. Each function is connected to one bit on an input port, and depending on the type of event to be handled, a particular subroutine is called or used. This can be accomplished by using a polling routine, which looks at each of the input bits to determine if any have gone high (indicating an alarm or active condition).

Typically, all the inputs are ORed to an interrupt line. Upon interrupt, the eight lines are scanned or polled to determine which has caused the "emergency." By

FIGURE 12-5 Polling.

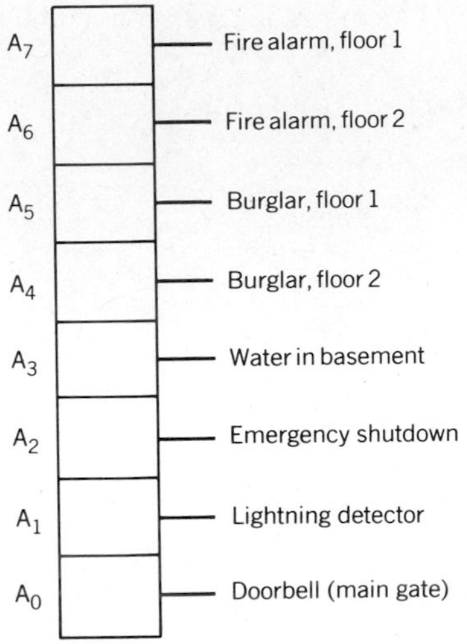

$$A_7$$ — Fire alarm, floor 1

$$A_6$$ — Fire alarm, floor 2

$$A_5$$ — Burglar, floor 1

$$A_4$$ — Burglar, floor 2

$$A_3$$ — Water in basement

$$A_2$$ — Emergency shutdown

$$A_1$$ — Lightning detector

$$A_0$$ — Doorbell (main gate)

scanning from A_7 to A_0, we can ''prioritize'' the signals because A_7, which is observed first, has a higher priority than A_6, and so on.

Once the bit has been found, an address lookup table can get the program to the correct routine. In this version, the table contains a list of addresses for the various routines needed to handle each situation.

6.2 Conclusion

This application is left to you to implement. The basic technique of polling has been described. A full program could include a phone dialer and some type of active alarm system with a readout on a display of the problem area. The system could even activate a speech synthesizer driving a public address system to warn intruders or employees of any impending danger. The possibilities are limited only by the resources of the programmer.

As food for thought, we include a printout of a working building surveillance system that logs all door opening and closing times. It is very hard to sneak around unless you know where the computer is hidden! In addition, there is a speaker driven by a speech synthesizer near each door. Thus every opening or closing of a door is acknowledged.

7 A SOFTWARE CLOCK

The standard digital clock has found its way into clock radios, video tape recorders, microwave ovens, and many other devices. Large electronics suppliers

Printout 12-3

```
        POLL:   IN 40H          ;GET INPUT DATA
                STA TEMP        ;SAVE IT HERE
                MVI B, 80H      ;SET UP MASK REGISTER
                MVI C, 8        ;BIT COUNT
        NEXT:   LDA TEMP        ;GET INPUT DATA BACK
                ANA B           ;LOOK AT SPECIFIC BIT
                JNZ FOUND       ;IF HIGH, DEVICE IS ACTIVE
                MOV A, B        ;ROTATE THE MASK REGISTER
                RRC             ;
                MOV B, A        ;
                DCR C           ;DONE POLLING?
                JNZ NEXT        ;JUMP IF NOT
                JMP POLL        ;OTHERWISE, START OVER
        FOUND:  LXI H, TABLE    ;TOP OF TABLE ADDRESS
                MVI B, 0        ;HERE WE ADD TWO TO
                DCR C           ;HL FOR EVERY BIT
                DAD B           ;TESTED
                DAD B           ;
                MOV A, M        ;NOW WE USE HL TO POINT
                INX H           ;TO A SUBROUTINE ADDRESS
                MOV H, M        ;AND GET THIS ADDRESS
                MOV L, A        ;INTO HL AGAIN. PHEW!
                PCHL            ;
        TABLE:  DW BELL
                DW LIGHT
                DW WATER
                DW BURG1
                DW BURG2
                DW FIRE1
                DW FIRE2
                .
                .
                .
                .                PROGRAM SUBROUTINES
                .
                .
                .
                END
```

now offer LSI chips that do all the normal time-keeping chores in a single chip, and the chip interfaces directly to a computer bus. From a software point of view, it may be desirable to save space and money by implementing a clock function without any hardware external to the computer except for a simple interrupt circuit. Thus, the computer can keep track of the time and at the same time continue to perform a specific function. We simply interrupt the computer 60 times a second and use an interrupt service routine to count the interrupts. When 60 have occurred, we increment the seconds count. If 10 seconds have passed, we increment the tens of seconds, and so on. When we are done, we return to the program that was interrupted. All this happens very fast, so we never miss an interrupt pulse, and the small amount of machine code necessary to perform the clock function does not perceptibly slow down the main program.

Printout 12-4

```
GO SECURITY
ENTER DATE  14-DEC
JLACD BURGLE SYSTEMS

MAIN ENTRANCE EAST --- CLOSED --- 10:59:17   14-DEC
MAIN ENTRANCE EAST --- OPEN ---   10:59:32   14-DEC
TERMINAL ROOM (E105-A) --- OPEN ---  10:59:40   14-DEC
MAIN ENTRANCE EAST --- CLOSED --- 10:59:44   14-DEC
TERMINAL ROOM (E105-A) --- CLOSED --- 10:59:45   14-DEC
TERMINAL ROOM (E105-A) --- OPEN ---  10:59:52   14-DEC
TERMINAL ROOM (E105-A) --- CLOSED --- 11:00:01   14-DEC
MAIN ENTRANCE EAST --- OPEN ---   11:00:07   14-DEC
MAIN ENTRANCE EAST --- CLOSED --- 11:00:37   14-DEC
CONSTRUCTION LAB (E103) --- OPEN ---  11:00:46   14-DEC
MAIN ENTRANCE EAST --- OPEN ---   11:00:53   14-DEC
MAIN ENTRANCE EAST --- CLOSED --- 11:01:25   14-DEC
TERMINAL ROOM (E105-A) --- OPEN ---  11:01:43   14-DEC
MAIN ENTRANCE EAST --- OPEN ---   11:01:43   14-DEC
TERMINAL ROOM (E105-A) --- CLOSED --- 11:01:50   14-DEC
MAIN ENTRANCE EAST --- CLOSED --- 11:01:52   14-DEC
MAIN ENTRANCE EAST --- OPEN ---   11:02:02   14-DEC
MAIN ENTRANCE EAST --- CLOSED --- 11:02:11   14-DEC
MAIN ENTRANCE EAST --- OPEN ---   11:02:15   14-DEC
MAIN ENTRANCE EAST --- CLOSED --- 11:02:33   14-DEC
MAIN ENTRANCE EAST --- OPEN ---   11:03:29   14-DEC
MAIN OFFICE (E101) --- CLOSED --- 11:04:03   14-DEC
MAIN ENTRANCE EAST --- CLOSED --- 11:04:22   14-DEC
MAIN ENTRANCE EAST --- OPEN ---   11:04:43   14-DEC
MAIN ENTRANCE EAST --- CLOSED --- 11:04:51   14-DEC
MACHINES LAB (E105) --- OPEN ---  11:04:55   14-DEC
MACHINES LAB (E105) --- CLOSED --- 11:05:03   14-DEC
MACHINES LAB REAR DOOR (E105) --- OPEN ---   11:05:06   14-DEC
TERMINAL ROOM (E105-A) --- OPEN ---  11:05:08   14-DEC
TERMINAL ROOM (E105-A) --- CLOSED --- 11:05:13   14-DEC
MAIN ENTRANCE EAST --- OPEN ---   11:05:16   14-DEC
MAIN ENTRANCE EAST --- CLOSED --- 11:05:38   14-DEC
TERMINAL ROOM (E105-A) --- OPEN ---   11:05:43   14-DEC
MACHINES LAB REAR DOOR (E105) --- CLOSED --- 11:05:46   14-DEC
MAIN ENTRANCE EAST --- OPEN ---   11:05:50   14-DEC
MACHINES LAB (E105) --- OPEN ---  11:05:54   14-DEC
MACHINES LAB (E105) --- CLOSED --- 11:06:01   14-DEC
MAIN ENTRANCE EAST --- CLOSED --- 11:06:02   14-DEC
CIRCUITS LAB --- CLOSED --- 11:07:00   14-DEC
MAIN ENTRANCE EAST --- OPEN ---   11:07:01   14-DEC
CONSTRUCTION LAB (E103) --- CLOSED --- 11:07:04   14-DEC
MAIN ENTRANCE EAST --- CLOSED --- 11:07:17   14-DEC
MAIN ENTRANCE EAST --- OPEN ---   11:07:18   14-DEC
MICRO-COMPUTER LAB (E201) --- CLOSED --- 11:07:21   14-DEC
MICRO-COMPUTER LAB (E201) --- OPEN ---  11:07:22   14-DEC
CONSTRUCTION LAB (E103) --- OPEN ---  11:07:27   14-DEC
ELECTRONICS LAB (E207) --- CLOSED --- 11:07:29   14-DEC
ELECTRONICS LAB (E207) --- OPEN ---  11:07:30   14-DEC
MAIN ENTRANCE EAST --- CLOSED --- 11:07:31   14-DEC
CONSTRUCTION LAB (E103) --- CLOSED --- 11:07:33   14-DEC
TECHNICIANS LAB (E208) --- CLOSED --- 11:07:35   14-DEC
TECHNICIANS LAB (E208) --- OPEN ---  11:07:36   14-DEC
SECRET DOOR (E207-E208) --- CLOSED --- 11:07:44   14-DEC
SECRET DOOR (E207-E208) --- OPEN ---  11:07:45   14-DEC
TERMINAL ROOM (E105-A) --- CLOSED --- 11:07:56   14-DEC
MAIN ENTRANCE EAST --- OPEN ---   11:08:01   14-DEC
IMSAI UP LABORATORY (E202) --- CLOSED --- 11:08:05   14-DEC
MAIN ENTRANCE EAST --- CLOSED --- 11:08:11   14-DEC
```

7.1 Providing a 60-Hertz Interrupt

The software is set up for a 60-Hz interrupt; if we are to keep accurate time, therefore, we must provide a source of that frequency. Figure 12-6 shows some possible ways of making a 60-Hz interrupt signal.

Now we must condition our 60-Hz signal so that it does not cause more than 60 interrupts a second. Remember, the 8080A/85 will interrupt as long as its interrupt line is high. It does not work on a transition from one state to another. Thus we have at least two ways of conditioning our 60-Hz signal as we see in Fig. 12-7.

FIGURE 12-6 Two ways of producing a 60-Hz interrupt. (*a*) Schmitt trigger interrupt from a-c source. (*b*) Crystal-controlled 60-Hz reference.

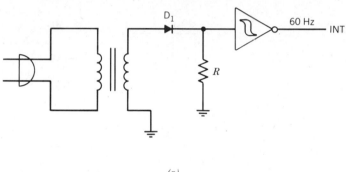

(a)

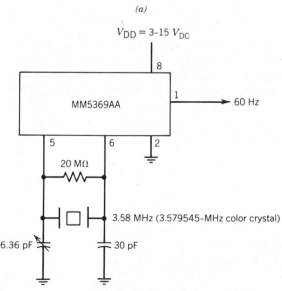

(b)

FIGURE 12-7 Two methods of conditioning interrupt pulses. (*a*) RC
Circuit. (*b*) Flip-flop control.

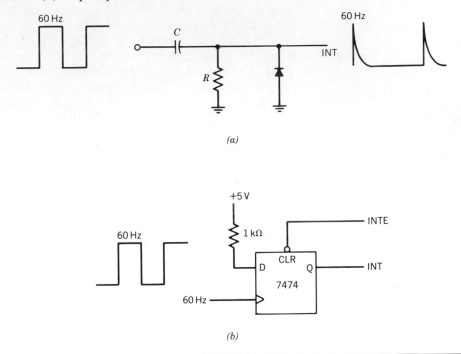

(a)

(b)

The circuit in Fig. 12-7*a* provides a narrow positive spike to the interrupt line.
Values of *R* and *C* can be chosen to provide the correct duration. The diode is
included to eliminate negative spikes. The circuit in Fig. 12-7*b* uses a flip-flop
setup to transfer a logic "1" to the interrupt line on a low-to-high transition. Once
the interrupt has been acknowledged, the INTE line clears the flip-flop, ensuring
that we do not get another interrupt until the next low-to-high transition.

When the 8080A/85 has acknowledged the interrupt request, it looks at the data
bus for an instruction. Although we have the option of forcing a jump instruction
(3 bytes) or any other type of instruction on the bus after an interrupt, we usually
leave the bus alone so that the 8080A/85 sees an FF, which translates to RST 7
in machine code. (This means that the program counter pushed at interrupt time
is actually a return address.) The CPU then goes to address 0038 H and continues
execution. The software CLOCK program takes this into consideration even
though it executes at 1000 H. The first few instructions in CLOCK insert into
memory at 0038 H a JMP instruction that points to the interrupt service routine
(TINT). The interrupt service routine saves all 8080A/85 registers, performs its
time-keeping duties, then restores all registers again. It finds its way back to the
main program by the help of the RET instruction at the end of the routine.

Printout 12-5

```
        CLOCK:  MVI A, 0C3H     ;PLACE THE INTERRUPT SERVICE
                STA 0038H       ;ROUTINE JUMP INTO RST 7 LOCATION
                LXI H, TINT
                SHLD 0039H
                MVI A, 60       ;INIT THE INTERRUPT COUNTER
                STA IBUF
                LXI H, MSG      ;SEND QUESTION TO THE USER
                CALL SEND
                LXI H, TIME     ;TIME ASCII BUFFER
        NEXT:   CALL CIN        ;GET A CHARACTER
                CPI 0DH         ;DONE IF CR
                JZ CRLF
                MOV B, A        ;ECHO CHARACTER TO THE USER
                CALL COUT
                MOV M, B        ;AND PLACE IT IN TIME BUFFER
                INX H
                JMP NEXT
        CRLF:   MVI B, 0DH      ;SEND A CR, LF TO THE USER
                CALL COUT
                MVI B, 0AH
                CALL COUT
                EI              ;ENABLE INTERRUPTS
                JMP 2028H       ;AND RETURN TO DOS
        SEND:   MOV A, M        ;SEND CHARACTERS TO THE USER UNTIL
                ORA A           ;A ZERO (0) IS FOUND
                RZ
                MOV B, A
                CALL COUT
                INX H
                JMP SEND
        MSG:    DB 'ENTER THE TIME: ',0
        IBUF:   DB 0
        TIME:   DB '12:59:59'   ;CORRECT FORMAT FOR TIME
        TINT:   PUSH PSW        ;SAVE ACCUMULATOR
                LDA IBUF        ;GET THE INTERRUPT COUNTER
                DCR A           ;COUNT IT DOWN
                JNZ T1          ;NOT ZERO - NOT 1 SECOND YET
                MVI A, 60       ;ZERO, SO RE-INIT THE COUNTER
        T1:     STA IBUF        ;SAVE THE COUNTER
                JZ T2           ;PROCEED IF COUNTER WAS ZERO
                POP PSW         ;OTHERWISE, GET ACC BACK
                EI              ;ENABLE INTERRUPTS
                RET             ;AND CONTINUE WITH TASK
        T2:     PUSH B          ;SAVE THE OTHER REGISTERS NOW
                PUSH D
                PUSH H
                LXI H, TIME+7   ;INCREMENT SECONDS
                MOV A, M        ;AND CHANGE '9' TO '0'
                CPI 39H         ;IF WE HAVE A '9'
                JZ T3
                INR M
                JMP POPP
        T3:     MVI M, 30H
                DCX H           ;INCREMENT 10'S OF SECONDS
                MOV A, M        ;AND CHANGE '5' TO '0'
                CPI 35H         ;IF WE HAVE A '5'
                JZ T4
                INR M
                JMP POPP
        T4:     MVI M, 30H
                DCX H           ;INCREMENT MINUTES
                DCX H           ;AND CHANGE '9' TO '0'
```

(continued on next page)

Printout 12-5 (continued)

```
                MOV A, M          ;IF WE HAVE A '9'
                CPI 39H
                JZ T5
                INR M
                JMP POPP
T5:             MVI M, 30H
                DCX H             ;INCREMENT 10'S OF MINUTES
                MOV A, M          ;AND CHANGE '5' TO '0'
                CPI 35H           ;IF WE HAVE A '5'
                JZ T6
                INR M
                JMP POPP
T6:             MVI M, 30H
                DCX H
                DCX H
                MOV A, M
                CPI 39H           ;ARE WE AT '9' HOURS?
                JZ T9
                CPI 32H           ;IF NOT A '2' THEN DON'T
                JZ T7             ;CHECK FOR 12 O'CLOCK
                INR M             ;AND JUST INCREMENT HOURS
                JMP POPP
T7:             DCX H             ;NOW IF 10'S OF HOURS ISN'T A '1'
                MOV A, M
                CPI 31H
                JZ T8
                INX H             ;WE JUST INCREMENT THE HOURS
                INR M
                JMP POPP
T8:             MVI M, 20H        ;OTHERWISE, WE SWITCH BACK TO
                INX H             ;1 O'CLOCK
                MVI M, 31H
                JMP POPP
T9:             MVI M, 30H        ;CHANGE HOURS TO '0'
                DCX H             ;AND 10'S OF HOURS TO '1'
                MVI M, 31H
POPP:           POP H             ;RESTORE ALL REGISTERS
                POP D
                POP B
                POP PSW
                EI                ;ENABLE INTERRUPTS
                RET               ;AND CONTINUE WITH TASK
```

7.2 Reading the Time

With the CLOCK program loaded at 1000 H, we execute it and enter the time. The program returns us to a disk-operating system (DOS) after the time has been entered. It could just as well jump to another program in memory. To get the time out of the program, we simply examine the memory buffer area called TIME used by the CLOCK program, and print what we see in it. The following simple BASIC program will read out the time:

```
10   I = 4200
20   FOR J = 0 TO 7
30   PRINT CHR$(EXAM(I + J));
40   NEXT J
50   PRINT
```

A machine language program could just as well display the time, but if we were to show it, you would not get a chance to write it from scratch.

8 HEXDUMP: A PROGRAM FOR MEMORY DUMPS

When debugging a program, it is often desirable to look at a portion of memory to find out whether the program was loaded into memory correctly, whether the data in some locations changed after execution, or possibly whether the stack ran away (too many pushes) and wrote over the program. With HEXDUMP, we are able to look at memory over a specific address range and see the exact hexadecimal data stored in specific memory locations. The process used is to look at all the locations, one at a time. By transferring the data from memory into the accumulator, we are then able to manipulate it into a form (hexadecimal ASCII) that we can recognize. Also included in the hexadecimal dump are the addresses the data came from and the ASCII equivalent of the byte in that location. This last display is useful when a program has ASCII characters stored in it (like the message center of Chapter 11). The user can actually see the message residing in memory. HEXDUMP is a very useful program, and a must for any serious programmer.

We know from Chapter 11 that we can manipulate data in a register into another form, namely, hex to ASCII. In HEXDUMP, we display two ASCII characters for every byte we look at in memory. The process is very simple. Consider Example 12-1.

EXAMPLE 12-1

MOV A, M

This puts a 0B4 H into the accumulator, which in binary looks like this:

1 0 1 1 0 1 0 0

Since the display device we are using prints from left to right (most do), we must figure out how to output the B (1011) first. Simple! We rotate the accumulator right four times, and strip off the bits we want.

1. 0 1 0 1 1 0 1 0 (RRC)

2. 0 0 1 0 1 1 0 1 (RRC)

3. 1 0 0 1 0 1 1 0 (RRC)

4. 0 1 0 0 1 0 1 1 (RRC)

ANI 0F H gives us 0 0 0 0 1 0 1 1 in the accumulator. Now we insert a simple test to decide whether the accumulator is more than 9. If it is, we add 37 H to it; otherwise, we add only 30 H. This puts the correct ASCII code (see Table 12-3) in the accumulator for the upper 4 bits. This we send to the terminal or display device. The lower 4 bits are easier. All we have to do is strip them off and add the correct bias—no rotating.

TABLE 12-3   ASCII Characters and Their
Hexadecimal Equivalent Values

ASCII Characters	Equivalent Values in Hexadecimal
0	30 H
1	31 H
2	32 H
3	33 H
4	34 H
5	35 H
6	36 H
7	37 H
8	38 H
9	39 H
A	41 H
B	42 H
C	43 H
D	44 H
E	45 H
F	46 H

Printout 12-6

```
              ORG 1000H
              LXI SP, STACK
              LXI H, MSG0      ;SEND HELLO MESSAGE
              CALL SEND
    GET1:     LXI H, MSG1      ;ASK FOR THE STARTING ADDR
              CALL SEND
              CALL INPUT
              JC GET1          ;REDO IF ERROR
              SHLD BUF1        ;SAVE IT
    GET2:     LXI H, MSG2      ;ASK FOR THE ENDING ADDR
              CALL SEND
              CALL INPUT
              JC GET2          ;REDO IF ERROR
              INX H            ;FIX THE ENDING ADDRESS
              XCHG             ;PUT ENDING ADDR IN DE
              LHLD BUF1        ;GET STARTING ADDR
              MOV A, L         ;COMPLEMENT L
              CMA
              MOV L, A
              MOV A, H         ;COMPLEMENT H
              CMA
              MOV H, A
              INX H            ;GET THE TWO'S COMPLEMENT
              DAD D            ;SUBTRACT THIS FROM ENDING ADDR
              XCHG             ;BYTES TO DUMP IN DE
              LHLD BUF1        ;GET THE STARTING ADDR
              MOV A, L         ;CHECK THE LOW BYTE FOR 00H
              ANI 0FH
              MOV C, A         ;SKIP COUNT
              MOV A, L         ;ALTER THE LOW BYTE
              ANI 0F0H
              MOV L, A
```

Printout 12-6 (continued)

```
NEX1:   CALL HEAD        ;OUTPUT A HEADER
NEXT:   PUSH B           ;SAVE THE SKIP BYTE
        MOV A, H         ;DISPLAY THE HIGH ADDR BYTE
        CALL DHEX
        MOV A, L         ;AND THE LOW BYTE
        CALL DHEX
        CALL FOUR
        POP B
        CALL DIS16       ;OUTPUT UP TO 16 BYTES IN ASCII
        CALL FOUR
        CALL ASCII       ;AND THEIR ASCII EQUIVALENTS
        LDA DONE         ;CHECK TO SEE IF WE ARE DONE
        ORA A
        JZ EOD
        CALL CRLF
        MVI C, 0         ;FIX THE SKIP COUNT
        CALL CONTC       ;AND LOOK FOR A PAUSE
        MOV A, L         ;CHECK THE LOW BYTE
        CPI 0
        JNZ NEXT
        CALL CRLF        ;DONE WITH A BLOCK
        CALL CRLF
        CALL CRLF
        CALL CRLF
        JMP NEX1
EOD:    CALL CRLF
        LXI H, MSG4      ;SEND GOODBYE MESSAGE
        CALL SEND
        JMP 2028H
CRLF:   MVI B, 13        ;ASCII CR
        CALL COUT
        MVI B, 10        ;ASCII LF
        CALL COUT
        RET
SEND:   MOV A, M         ;SEND ASCII CHARACTERS TO THE USER
        ORA A            ;UNTIL A ZERO (0) IS FOUND
        RZ
        MOV B, A
        CALL COUT
        INX H
        JMP SEND
INPUT:  LXI H, 0         ;GET A HEX NUMBER FROM THE USER
INP2:   CALL CIN         ;GET A CHARACTER
        CPI 13           ;DONE?
        JZ EXIT
        MOV B, A         ;ECHO THE CHARACTER
        CALL COUT
        CPI 30H          ;CHECK FOR HEX DIGIT
        JC ERROR
        CPI 3AH
        JNC ALPHA
        SUI 30H
        JMP DADER
ALPHA:  CPI 41H
        JC ERROR
        SUI 37H
        CPI 10H
        JNC ERROR
DADER:  DAD H            ;ROTATE HL FOUR BITS LEFT
        DAD H
        DAD H
        DAD H
```

(continued on next page)

Printout 12-6 (continued)

```
                ADD L              ;INSERT THE NEW DIGIT
                MOV L, A
                JMP INP2
EXIT:           CALL CRLF
                RET
ERROR:          CALL CRLF
                LXI H, MSG3        ;SAY NUMBER IS NOT HEX
                CALL SEND
                STC                ;CARRY FLAG IS ERROR FLAG
                RET
DHEX:           MOV C, A           ;SAVE THE BYTE
                RRC                ;GET THE UPPER 4 BITS
                RRC
                RRC
                RRC
                ANI 0FH
                CPI 0AH            ;ALPHA CHARACTER?
                JC SKIP
                ADI 7
SKIP:           ADI 30H            ;ADD ASCII BIAS
                MOV B, A           ;SEND THE CHARACTER
                CALL COUT
                MOV A, C           ;GET THE BYTE BACK
                ANI 0FH            ;NOW THE LOWER 4 BITS
                CPI 0AH
                JC SKIP2
                ADI 7
SKIP2:          ADI 30H
                MOV B, A
                CALL COUT
                RET
EIGHT:          CALL FOUR          ;DISPLAY 8 SPACES
FOUR:           PUSH B
                LXI B, 2004H
FOR:            CALL COUT
                DCR C
                JNZ FOR
                POP B
                RET
HEAD:           PUSH H             ;SAVE IMPORTANT REGISTERS
                PUSH B
                CALL EIGHT
                LXI H, HDER        ;HEADER ADDRESS
                CALL SEND
                CALL EIGHT
                LXI H, HDER2
                CALL SEND
                POP B              ;GET THE REGISTERS BACK
                POP H
                RET
CONTC:          CALL CONT          ;CHECK FOR CONTROL C
                RNZ                ;RETURN IF NO CONTROL C HIT
CON2:           CALL CONT          ;ELSE, LOOP UNTIL NEXT CONTROL C
                JNZ CON2
                RET
ASCII:          MOV A, C           ;CHECK THE SKIP BYTE
                ORA A
                JZ CON
                MVI B, 20H         ;SEND A BLANK
                CALL COUT
                INX H
                DCR C
                JMP ASCII
```

Printout 12-6 (continued)

```
        CON:     MOV A, M        ;GET A BYTE
                 CPI 20H         ;FIND OUT IF IT IS PRINTABLE
                 JC BAD
                 CPI 7AH
                 JNC BAD
                 MOV B, A        ;SEND IT
                 CALL COUT
        BACK:    INX H
                 CALL CHECK
                 RZ
                 LDA DONE        ;CHECK END OF DUMP BYTE
                 ORA A
                 JNZ CON         ;CONTINUE IF NOT DONE
                 MVI B, 20H      ;ELSE, MAKE DISPLAY PRETTY
                 CALL COUT
                 JMP BACK
                 RET
        BAD:     MVI B, ','      ;BAD ASCII FILL CHARACTER
                 CALL COUT
                 JMP BACK
        CHECK:   MOV A, D        ;LOOK FOR DE=0
                 ORA E
                 JZ CHK2
                 DCX D           ;IF DE WASN'T 0, DECREMENT IT
                 MOV A, D        ;AND CHECK AGAIN
                 ORA E
        CHK2:    STA DONE        ;SAVE THE VALUE
                 MOV A, L        ;CHECK FOR END OF LINE
                 ANI OFH
                 ORA A
                 RET
        DIS16:   PUSH B          ;SAVE ALL REGISTERS
                 PUSH D
                 PUSH H
        DIS:     MOV A, C        ;CHECK THE SKIP BYTE
                 ORA A
                 JZ DIS2         ;START HEX BYTES IF 0
                 MVI B, 20H      ;ELSE MAKE THE DISPLAY PRETTY
                 CALL COUT
                 CALL COUT
                 CALL COUT
                 INX H
                 DCR C
                 JMP DIS
        DIS2:    MOV A, M        ;GET A BYTE FROM MEMORY
                 CALL DHEX       ;DISPLAY IT
                 MVI B, 20H      ;AND A BLANK
                 CALL COUT
        DIS5:    INX H           ;NEXT LOCATION
        DIS4:    CALL CHECK      ;SEE IF WE ARE DONE
                 JZ DIS3
                 LDA DONE        ;CHECK THE END OF DUMP BYTE
                 ORA A
                 JNZ DIS2
                 MVI B, 20H      ;ELSE, MAKE THE DISPLAY PRETTY
                 CALL COUT
                 CALL COUT
                 CALL COUT
                 JMP DIS5
        DIS3:    POP H           ;RESTORE ALL REGISTERS
                 POP D
                 POP B
                 RET
```

(continued on next page)

Printout 12-6 (continued)

```
COUT:      EQU 200DH
CONT:      EQU 2016H
BUF1:      DS 2
DONE:      DS 1
CIN:       EQU 2010H
MSG0:      ASC 13,10,'*HEX-DUMP*',13,10,0
MSG1:      ASC 'STARTING ADDRESS :',0
MSG2:      ASC 'ENDING ADDRESS :',0
MSG3:      ASC '--NUMBER IS NOT HEX',13,10,0
MSG4:      ASC 'END OF DUMP',13,10,0
HDER:      ASC '-0 -1 -2 -3 -4 -5 -6 -7 '
           ASC '-8 -9 -A -B -C -D -E -F',0
HDER2:     ASC 'ASCII DATA',13,10,0
           DS 50H
STACK:     DS 1
           END
```

9 AN 8085 DISASSEMBLER

A disassembler is a computer program that reads the contents of a range of memory locations and converts the corresponding machine codes (hexadecimal pairs) back to mnemonic form. Very useful in debugging troublesome programming errors, it is also a great aid in looking at programs or subroutines written by another programmer. A disassembler can be written for any of the wide variety of available microcomputers, but each is for a specific instruction set.

9.1 What Does a Disassembler Do?

Given a set of hexadecimal values as obtained from a series of memory locations, a disassembler converts the opcodes back to mnemonic form. Suppose the following bytes are contained in memory locations 2050 H to 2057 H:

AF DB FF D3 FF C3 00 37

These bytes *disassemble* to the following assembly language program:

```
XRA A
IN 0FFH
OUT 0FFH
JMP 3700H
```

A disassembler then performs the reverse process of program assembly. Clearly it is a very important function for the debugging of programs being written for a microcomputer.

The disassembler program must be able to take bytes from a range of desired memory locations and convert the acquired data back to mnemonic form. This is done by comparing data to entries in a large data table containing the mnemonics in sequential order.

Table 12-4 shows the 8085 instructions in ascending numerical order from 00 to FF. The data table then will contain the mnemonics in order, all 256 items.

TABLE 12-4   8085 Mnemonics in Ascending Numeric Order

00	NOP		2B	DCX	H	56	MOV D,M	81	ADD C	AC	XRA H	D7	RST 2
01	LXI	B,dble	2C	INR	L	57	MOV D,A	82	ADD D	AD	XRA L	D8	RC
02	STAX	B	2D	DCR	L	58	MOV E,B	83	ADD E	AE	XRA M	D9	---
03	INX	B	2E	MVI	L,byte	59	MOV E,C	84	ADD H	AF	XRA A	DA	JC adr
04	INR	B	2F	CMA		5A	MOV E,D	85	ADD L	B0	ORA B	DB	IN byte
05	DCR	B	30	SIM*		5B	MOV E,E	86	ADD M	B1	ORA C	DC	CC adr
06	MVI	B,byte	31	LXI	SP,dble	5C	MOV E,H	87	ADD A	B2	ORA D	DD	---
07	RLC		32	STA	adr	5D	MOV E,L	88	ADC B	B3	ORA E	DE	SBI byte
08	---		33	INX	SP	5E	MOV E,M	89	ADC C	B4	ORA H	DF	RST 3
09	DAD	B	34	INR	M	5F	MOV E,A	8A	ADC D	B5	ORA L	E0	RPO
0A	LDAX	B	35	DCR	M	60	MOV H,B	8B	ADC E	B6	ORA M	E1	POP H
0B	DCX	B	36	MVI	M,byte	61	MOV H,C	8C	ADC H	B7	ORA A	E2	JPO adr
0C	INR	C	37	STC		62	MOV H,D	8D	ADC L	B8	CMP B	E3	XTHL
0D	DCR	C	38	---		63	MOV H,E	8E	ADC M	B9	CMP C	E4	CPO adr
0E	MVI	C,byte	39	DAD	SP	64	MOV H,H	8F	ADC A	BA	CMP D	E5	PUSH H
0F	RRC		3A	LDA	adr	65	MOV H,L	90	SUB B	BB	CMP E	E6	ANI byte
10	---		3B	DCX	SP	66	MOV H,M	91	SUB C	BC	CMP H	E7	RST 4
11	LXI	D,dble	3C	INR	A	67	MOV H,A	92	SUB D	BD	CMP L	E8	RPE
12	STAX	D	3D	DCR	A	68	MOV L,B	93	SUB E	BE	CMP M	E9	PCHL
13	INX	D	3E	MVI	A,byte	69	MOV L,C	94	SUB H	BF	CMP A	EA	JPE adr
14	INR	D	3F	CMC		6A	MOV L,D	95	SUB L	C0	RNZ	EB	XCHG
15	DCR	D	40	MOV	B,B	6B	MOV L,E	96	SUB M	C1	POP B	EC	CPE adr
16	MVI	D,byte	41	MOV	B,C	6C	MOV L,H	97	SUB A	C2	JNZ adr	ED	---
17	RAL		42	MOV	B,D	6D	MOV L,L	98	SBB B	C3	JMP adr	EE	XRI byte
18	---		43	MOV	B,E	6E	MOV L,M	99	SBB C	C4	CNZ adr	EF	RST 5
19	DAD	D	44	MOV	B,H	6F	MOV L,A	9A	SBB D	C5	PUSH B	F0	RP
1A	LDAX	D	45	MOV	B,L	70	MOV M,B	9B	SBB E	C6	ADI byte	F1	POP PSW
1B	DCX	D	46	MOV	B,M	71	MOV M,C	9C	SBB H	C7	RST 0	F2	JP adr
1C	INR	E	47	MOV	B,A	72	MOV M,D	9D	SBB L	C8	RZ	F3	DI
1D	DCR	E	48	MOV	C,B	73	MOV M,E	9E	SBB M	C9	RET	F4	CP adr
1E	MVI	E,byte	49	MOV	C,C	74	MOV M,H	9F	SBB A	CA	JZ adr	F5	PUSH PSW
1F	RAR		4A	MOV	C,D	75	MOV M,L	A0	ANA B	CB	---	F6	ORI byte
20	RIM*		4B	MOV	C,E	76	HLT	A1	ANA C	CC	CZ adr	F7	RST 6
21	LXI	H,dble	4C	MOV	C,H	77	MOV M,A	A2	ANA D	CD	CALL adr	F8	RM
22	SHLD	adr	4D	MOV	C,L	78	MOV A,B	A3	ANA E	CE	ACI byte	F9	SPHL
23	INX	H	4E	MOV	C,M	79	MOV A,C	A4	ANA H	CF	RST 1	FA	JM adr
24	INR	H	4F	MOV	C,A	7A	MOV A,D	A5	ANA L	D0	RNC	FB	EI
25	DCR	H	50	MOV	D,B	7B	MOV A,E	A6	ANA M	D1	POP D	FC	CM adr
26	MVI	H,byte	51	MOV	D,C	7C	MOV A,H	A7	ANA A	D2	JNC adr	FD	---
27	DAA		52	MOV	D,D	7D	MOV A,L	A8	XRA B	D3	OUT byte	FE	CPI byte
28	---		53	MOV	D,E	7E	MOV A,M	A9	XRA C	D4	CNC adr	FF	RST 7
29	DAD	H	54	MOV	D,H	7F	MOV A,A	AA	XRA D	D5	PUSH D		
2A	LHLD	adr	55	MOV	D,L	80	ADD B	AB	XRA E	D6	SUI byte		

*8085 Only

Source: All mnemonics copyright Intel Corporation 1976.

Some are not used, but must be included anyway. The basic idea is to get an opcode and then go down the list of mnemonics until the count (represented by the opcode) is reached.

If the opcode is 3E H, we would proceed to the 63rd item in the list, which is "MVI A,". This much could be printed. The remainder of the instruction is contained in the next data byte.

The instructions are basically 1-, 2-, or 3-byte instructions. For example, XRA A is a 1-byte instruction. Once it has been looked up, that instruction is completed. MVI A, is a 2-byte instruction and a second byte of information is required to finish the instruction. LXI H, is a 3-byte instruction.

A method of identifying instructions by length is used in the data table. One-byte instructions are marked with an exclamation point (!). Two-byte instructions are marked with a percent sign (%). Three-byte instructions are marked with a number sign (#).

The data table begins like this:

!NOP#LXI B,!STAX B!INX B

Printout 12-7

```
DB'!NOP#LXI B,!STAX B!INX B'
DB'!INR B!DCR B%MVI B,!RLC'
DB'!??=08!DAD B!LDAX B!DCX B'
DB'!INR C!DCR C%MVI C,!RRC!??=10'
DB'#LXI D,!STAX D!INX D!INR D'
DB'!DCR D%MVI D,!RAL!??=18!DAD D'
DB'!LDAX D!DCX D!INR E!DCR E'
DB'%MVI E,'
DB'!RAR!RIM#LXI H,#SHLD !INX H'
DB'!INR H'
DB'!DCR H%MVI H,!DAA!??=28!DAD H'
DB'#LHLD !DCX H!INR L!DCR L'
DB'%MVI L,'
DB'!CMA!SIM#LXI SP,#STA !INX SP'
DB'!INR M!'
DB'DCR M%MVI M,!STC'
DB'!??=38!DAD SP#LDA !DCX SP'
DB'!INR A'
DB'!DCR A%MVI A,!CMC!'
DB'MOV B,B!MOV B,C'
DB'!MOV B,D!MOV B,E!MOV B,H'
DB'!MOV B,L'
DB'!MOV B,M!MOV B,A!MOV C,B'
DB'!MOV C,C'
DB'!MOV C,D!MOV C,E!MOV C,H'
DB'!MOV C,L'
DB'!MOV C,M!MOV C,A!MOV D,B'
DB'!MOV D,C'
DB'!MOV D,D!MOV D,E!MOV D,H'
DB'!MOV D,L'
DB'!MOV D,M!MOV D,A!MOV E,B'
DB'!MOV E,C'
DB'!MOV E,D!MOV E,E!MOV E,H'
DB'!MOV E,L'
DB'!MOV E,M!MOV E,A!MOV H,B'
DB'!MOV H,C'
DB'!MOV H,D!MOV H,E!MOV H,H'
DB'!MOV H,L'
DB'!MOV H,M!MOV H,A!MOV L,B'
DB'!MOV L,C'
DB'!MOV L,D!MOV L,E!MOV L,H'
DB'!MOV L,L'
DB'!MOV L,M!MOV L,A!MOV M,B'
DB'!MOV M,C'
DB'!MOV M,D!MOV M,E!MOV M,H'
DB'!MOV M,L'
DB'!HLT!MOV M,A!MOV A,B!MOV A,C'
DB'!MOV A,D!MOV A,E!MOV A,H'
DB'!MOV A,L'
DB'!MOV A,M!MOV A,A!ADD B!ADD C'
DB'!ADD D!ADD E!ADD H!ADD L'
DB'!ADD M!ADD A!ADC B'
END
```

The NOP is a 1-byte instruction and begins with the 1-byte marker (!). LXI B, is a 3-byte instruction and begins with the three byte marker (#). This data table is generated using an assembler and multiple DB lines. Printout 12-7 is a sample. This results in the full data table shown in Printout 12-8 (as dumped by the HEXDUMP program).

Printout 12-8

	-0	-1	-2	-3	-4	-5	-6	-7	-8	-9	-A	-B	-C	-D	-E	-F	ASCII DATA
1700	21	4E	4F	50	23	4C	58	49	20	42	2C	21	53	54	41	58	!NOP#LXI B,!STAX
1710	20	42	21	49	4E	58	20	42	21	49	4E	52	20	42	21	44	B!INX B!INR B!D
1720	43	52	20	42	25	4D	56	49	20	42	2C	21	52	4C	43	21	CR B%MVI B,!RLC!
1730	3F	3F	3D	30	38	21	44	41	44	20	42	21	4C	44	41	58	??=08!DAD B!LDAX
1740	20	42	21	44	43	58	20	42	21	49	4E	52	20	43	21	44	B!DCX B!INR C!D
1750	43	52	20	43	25	4D	56	49	20	43	2C	21	52	52	43	21	CR C%MVI C,!RRC!
1760	3F	3F	3D	31	30	23	4C	58	49	20	44	2C	21	53	54	41	??=10#LXI D,!STA
1770	58	20	44	21	49	4E	58	20	44	21	49	4E	52	20	44	21	X D!INX D!INR D!
1780	44	43	52	20	44	25	4D	56	49	20	44	2C	21	52	41	4C	DCR D%MVI D,!RAL
1790	21	3F	3F	3D	31	38	21	44	41	44	20	44	21	4C	44	41	!??=18!DAD D!LDA
17A0	58	20	44	21	44	43	58	20	44	21	49	4E	52	20	45	21	X D!DCX D!INR E!
17B0	44	43	52	20	45	25	4D	56	49	20	45	2C	21	52	41	52	DCR E%MVI E,!RAR
17C0	21	52	49	4D	23	4C	58	49	20	48	2C	23	53	48	4C	44	!RIM#LXI H,#SHLD
17D0	20	21	49	4E	58	20	48	21	49	4E	52	20	48	21	44	43	!INX H!INR H!DC
17E0	52	20	48	25	4D	56	49	20	48	2C	21	44	41	41	21	3F	R H%MVI H,!DAA!?
17F0	3F	3D	32	38	21	44	41	44	20	48	23	4C	48	4C	44	20	?=28!DAD H#LHLD

	-0	-1	-2	-3	-4	-5	-6	-7	-8	-9	-A	-B	-C	-D	-E	-F	ASCII DATA
1800	21	44	43	58	20	48	21	49	4E	52	20	4C	21	44	43	52	!DCX H!INR L!DCR
1810	20	4C	25	4D	56	49	20	4C	2C	21	43	4D	41	21	53	49	L%MVI L,!CMA!SI
1820	4D	23	4C	58	49	20	53	50	2C	23	53	54	41	20	21	49	M#LXI SP,#STA !I
1830	4E	58	20	53	50	21	49	4E	52	20	4D	21	44	43	52	20	NX SP!INR M!DCR
1840	4D	25	4D	56	49	20	4D	2C	21	53	54	43	21	3F	3F	3D	M%MVI M,!STC!??=
1850	33	38	21	44	41	44	20	53	50	23	4C	44	41	20	21	44	38!DAD SP#LDA !D
1860	43	58	20	53	50	21	49	4E	52	20	41	21	44	43	52	20	CX SP!INR A!DCR
1870	41	25	4D	56	49	20	41	2C	21	43	4D	43	21	4D	4F	56	A%MVI A,!CMC!MOV
1880	20	42	2C	42	21	4D	4F	56	20	42	2C	43	21	4D	4F	56	B,B!MOV B,C!MOV
1890	20	42	2C	44	21	4D	4F	56	20	42	2C	45	21	4D	4F	56	B,D!MOV B,E!MOV
18A0	20	42	2C	48	21	4D	4F	56	20	42	2C	4C	21	4D	4F	56	B,H!MOV B,L!MOV
18B0	20	42	2C	4D	21	4D	4F	56	20	42	2C	41	21	4D	4F	56	B,M!MOV B,A!MOV
18C0	20	43	2C	42	21	4D	4F	56	20	43	2C	43	21	4D	4F	56	C,B!MOV C,C!MOV
18D0	20	43	2C	44	21	4D	4F	56	20	43	2C	45	21	4D	4F	56	C,D!MOV C,E!MOV
18E0	20	43	2C	48	21	4D	4F	56	20	43	2C	4C	21	4D	4F	56	C,H!MOV C,L!MOV
18F0	20	43	2C	4D	21	4D	4F	56	20	43	2C	41	21	4D	4F	56	C,M!MOV C,A!MOV

	-0	-1	-2	-3	-4	-5	-6	-7	-8	-9	-A	-B	-C	-D	-E	-F	ASCII DATA
1900	20	44	2C	42	21	4D	4F	56	20	44	2C	43	21	4D	4F	56	D,B!MOV D,C!MOV
1910	20	44	2C	44	21	4D	4F	56	20	44	2C	45	21	4D	4F	56	D,D!MOV D,E!MOV
1920	20	44	2C	48	21	4D	4F	56	20	44	2C	4C	21	4D	4F	56	D,H!MOV D,L!MOV
1930	20	44	2C	4D	21	4D	4F	56	20	44	2C	41	21	4D	4F	56	D,M!MOV D,A!MOV
1940	20	45	2C	42	21	4D	4F	56	20	45	2C	43	21	4D	4F	56	E,B!MOV E,C!MOV
1950	20	45	2C	44	21	4D	4F	56	20	45	2C	45	21	4D	4F	56	E,D!MOV E,E!MOV
1960	20	45	2C	48	21	4D	4F	56	20	45	2C	4C	21	4D	4F	56	E,H!MOV E,L!MOV
1970	20	45	2C	4D	21	4D	4F	56	20	45	2C	41	21	4D	4F	56	E,M!MOV E,A!MOV
1980	20	48	2C	42	21	4D	4F	56	20	48	2C	43	21	4D	4F	56	H,B!MOV H,C!MOV
1990	20	48	2C	44	21	4D	4F	56	20	48	2C	45	21	4D	4F	56	H,D!MOV H,E!MOV
19A0	20	48	2C	48	21	4D	4F	56	20	48	2C	4C	21	4D	4F	56	H,H!MOV H,L!MOV
19B0	20	48	2C	4D	21	4D	4F	56	20	48	2C	41	21	4D	4F	56	H,M!MOV H,A!MOV
19C0	20	4C	2C	42	21	4D	4F	56	20	4C	2C	43	21	4D	4F	56	L,B!MOV L,C!MOV
19D0	20	4C	2C	44	21	4D	4F	56	20	4C	2C	45	21	4D	4F	56	L,D!MOV L,E!MOV
19E0	20	4C	2C	48	21	4D	4F	56	20	4C	2C	4C	21	4D	4F	56	L,H!MOV L,L!MOV
19F0	20	4C	2C	4D	21	4D	4F	56	20	4C	2C	41	21	4D	4F	56	L,M!MOV L,A!MOV

(continued on next page)

Printout 12-8 (continued)

```
        -0  -1  -2  -3  -4  -5  -6  -7  -8  -9  -A  -B  -C  -D  -E  -F        ASCII DATA
1A00    20  4D  2C  42  21  4D  4F  56  20  4D  2C  43  21  4D  4F  56        M,B!MOV M,C!MOV
1A10    20  4D  2C  44  21  4D  4F  56  20  4D  2C  45  21  4D  4F  56        M,D!MOV M,E!MOV
1A20    20  4D  2C  48  21  4D  4F  56  20  4D  2C  4C  21  48  4C  54        M,H!MOV M,L!HLT
1A30    21  4D  4F  56  20  4D  2C  41  21  4D  4F  56  20  41  2C  42        !MOV M,A!MOV A,B
1A40    21  4D  4F  56  20  41  2C  43  21  4D  4F  56  20  41  2C  44        !MOV A,C!MOV A,D
1A50    21  4D  4F  56  20  41  2C  45  21  4D  4F  56  20  41  2C  48        !MOV A,E!MOV A,H
1A60    21  4D  4F  56  20  41  2C  4C  21  4D  4F  56  20  41  2C  4D        !MOV A,L!MOV A,M
1A70    21  4D  4F  56  20  41  2C  41  21  41  44  44  20  42  21  41        !MOV A,A!ADD B!A
1A80    44  44  20  43  21  41  44  44  20  44  21  41  44  44  20  45        DD C!ADD D!ADD E
1A90    21  41  44  44  20  48  21  41  44  44  20  4C  21  41  44  44        !ADD H!ADD L!ADD
1AA0    20  4D  21  41  44  44  20  41  21  41  44  43  20  42  21  41        M!ADD A!ADC B!A
1AB0    44  43  20  43  21  41  44  43  20  44  21  41  44  43  20  45        DC C!ADC D!ADC E
1AC0    21  41  44  43  20  48  21  41  44  43  20  4C  21  41  44  43        !ADC H!ADC L!ADC
1AD0    20  4D  21  41  44  43  20  41  21  53  55  42  20  42  21  53        M!ADC A!SUB B!S
1AE0    55  42  20  43  21  53  55  42  20  44  21  53  55  42  20  45        UB C!SUB D!SUB E
1AF0    21  53  55  42  20  48  21  53  55  42  20  4C  21  53  55  42        !SUB H!SUB L!SUB


        -0  -1  -2  -3  -4  -5  -6  -7  -8  -9  -A  -B  -C  -D  -E  -F        ASCII DATA
1B00    20  4D  21  53  55  42  20  41  21  53  42  42  20  42  21  53        M!SUB A!SBB B!S
1B10    42  42  20  43  21  53  42  42  20  44  21  53  42  42  20  45        BB C!SBB D!SBB E
1B20    21  53  42  42  20  48  21  53  42  42  20  4C  21  53  42  42        !SBB H!SBB L!SBB
1B30    20  4D  21  53  42  42  20  41  21  41  4E  41  20  42  21  41        M!SBB A!ANA B!A
1B40    4E  41  20  43  21  41  4E  41  20  44  21  41  4E  41  20  45        NA C!ANA D!ANA E
1B50    21  41  4E  41  20  48  21  41  4E  41  20  4C  21  41  4E  41        !ANA H!ANA L!ANA
1B60    20  4D  21  41  4E  41  20  41  21  58  52  41  20  42  21  58        M!ANA A!XRA B!X
1B70    52  41  20  43  21  58  52  41  20  44  21  58  52  41  20  45        RA C!XRA D!XRA E
1B80    21  58  52  41  20  48  21  58  52  41  20  4C  21  58  52  41        !XRA H!XRA L!XRA
1B90    20  4D  21  58  52  41  20  41  21  4F  52  41  20  42  21  4F        M!XRA A!ORA B!O
1BA0    52  41  20  43  21  4F  52  41  20  44  21  4F  52  41  20  45        RA C!ORA D!ORA E
1BB0    21  4F  52  41  20  48  21  4F  52  41  20  4C  21  4F  52  41        !ORA H!ORA L!ORA
1BC0    20  4D  21  4F  52  41  20  41  21  43  4D  50  20  42  21  43        M!ORA A!CMP B!C
1BD0    4D  50  20  43  21  43  4D  50  20  44  21  43  4D  50  20  45        MP C!CMP D!CMP E
1BE0    21  43  4D  50  20  48  21  43  4D  50  20  4C  21  43  4D  50        !CMP H!CMP L!CMP
1BF0    20  4D  21  43  4D  50  20  41  21  52  4E  5A  21  50  4F  50        M!CMP A!RNZ!POP


        -0  -1  -2  -3  -4  -5  -6  -7  -8  -9  -A  -B  -C  -D  -E  -F        ASCII DATA
1C00    20  42  23  4A  4E  5A  20  23  4A  4D  50  20  23  43  4E  5A        B#JNZ #JMP #CNZ
1C10    20  21  50  55  53  48  20  42  25  41  44  49  20  21  52  53         !PUSH B%ADI !RS
1C20    54  20  30  21  52  5A  21  52  45  54  23  4A  5A  20  21  3F        T 0!RZ!RET#JZ !?
1C30    3F  3D  43  42  23  43  5A  20  23  43  41  4C  4C  20  25  41        ?=CB#CZ #CALL %A
1C40    43  49  20  21  52  53  54  20  31  21  52  4E  43  21  50  4F        CI !RST 1!RNC!PO
1C50    50  20  44  23  4A  4E  43  20  25  4F  55  54  20  23  43  4E        P D#JNC %OUT #CN
1C60    43  20  21  50  55  53  48  20  44  25  53  55  49  20  21  52        C !PUSH D%SUI !R
1C70    53  54  20  32  21  52  43  21  3F  3F  3D  44  39  23  4A  43        ST 2!RC!??=D9#JC
1C80    20  25  49  4E  20  23  43  43  20  21  3F  3F  3D  44  44  25        %IN #CC !??=DD%
1C90    53  42  49  20  21  52  53  54  20  33  21  52  50  4F  21  50        SBI !RST 3!RPO!P
1CA0    4F  50  20  48  23  4A  50  4F  20  21  58  54  48  4C  23  43        OP H#JPO !XTHL#C
1CB0    50  4F  20  21  50  55  53  48  20  48  25  41  4E  49  20  21        PO !PUSH H%ANI !
1CC0    52  53  54  20  34  21  52  50  45  21  50  43  48  4C  23  4A        RST 4!RPE!PCHL#J
1CD0    50  45  20  21  58  43  48  47  23  43  50  45  20  21  3F  3F        PE !XCHG#CPE !??
1CE0    3D  45  44  25  58  52  49  20  21  52  53  54  20  35  21  52        =ED%XRI !RST 5!R
1CF0    50  21  50  4F  50  20  50  53  57  23  4A  50  20  21  44  49        P!POP PSW#JP !DI


        -0  -1  -2  -3  -4  -5  -6  -7  -8  -9  -A  -B  -C  -D  -E  -F        ASCII DATA
1D00    23  43  50  20  21  50  55  53  48  20  50  53  57  25  4F  52        #CP !PUSH PSW%OR
1D10    49  20  21  52  53  54  20  36  21  52  4D  21  53  50  48  4C        I !RST 6!RM!SPHL
1D20    23  4A  4D  20  21  45  49  23  43  4D  20  21  3F  3F  3D  46        #JM !EI#CM !??=F
1D30    44  25  43  50  49  20  21  52  53  54  20  37  21  B0  DA  00        D%CPI !RST 7!...
END OF DUMP
```

In operation, the program obtains an opcode and proceeds through the data table counting "markers." It does not matter whether the program encounters !, #, or %. Each marker causes the counter to increase. If the opcode is 3E, the 63rd marker will precede the actual mnemonic. In the case of 3E, after the % marker, we should find the data: "MVI A,". The marker indicates that a second byte is required. Printout 12-8 shows the full data table. Each mnemonic is preceded by the appropriate marker.

Two types of data must be printed on the screen: direct ASCII code (MVI A,) or hex data. The ASCII code may be printed directly, whereas the hex data must be converted from hex pairs to ASCII pairs. This is done using a simple conversion routine like those described earlier in Chapter 11.

The program also has the option of printing an address (offset) different from the actual source address. This capability is useful when a printout must appear to be at a particular address. Printout 12-9 shows the program.

Printout 12-9

```
                ;AN 8085 DISASSEMBLER
                ;
                ;
                ;
                ;USER TYPES START ADDRESS AFTER SIGN ON MESSAGE
                ;OPTIONAL OFFSET ADDRESS ACCEPTED AFTER (:)
                ;IE: 2000:3000
                ;
                ;
                ;OPERATES USING NORTHSTAR DOS ROUTINES
                ;       CIN     EQU     2010H
                ;       COUT    EQU     200DH
                ;       CONTC   EQU     2016H
                ;
                ;
                ;DATA TABLE CONTAINING MNEMONICS MUST BE AT 1700H
                ;
                ;       DATA    EQU     1700H
                ;
                ORG 1400H                    ;DISASSEMBLER BEGINS AT 1400H
START:          LXI SP,STACK+30H             ;INITIALIZE STACK POINTER
                CALL MESSG                   ;PRINT SIGNON MESSAGE
MORE:           LXI D, DATA                  ;START OF DATA TABLE
                XRA A                        ;ZERO A REGISTER
                STA FLOP                     ;INITIALIZE OFFSET TOGGLE
                ;
                ;GET A FOUR CHARACTER ADDRESS
HEX:            LXI H, 0000H                 ;
HEX2:           CALL CIN                     ;
NJLA:           CPI ODH                      ;CARRIAGE RETURN TO FINISH
                JZ HEX3                      ;
                CPI ESC                      ;ESCAPE TO MONITOR
                JZ 2028H                     ;
                MOV B, A                     ;WANT TO ECHO CHARACTER
                CALL COUT                    ;
                MOV A, B                     ;
                CPI ':'                      ;COLON MEANS OFFSET DESIRED
                CZ OFFS                      ;
                SUI 30H                      ;
                CPI 10H                      ;
```

(continued on next page)

Printout 12-9 (continued)

```
              JC SKP              ;
              SUI 07H             ;
SKP:          DAD H               ;
              DAD H               ;
              DAD H               ;
              DAD H               ;
              ADD L               ;
              MOV L, A            ;
              JMP HEX2            ;
HEX3:         LDA FLOP            ;CHECK TO SEE IF OFFSET
              ORA A               ;
              CNZ OFFT            ;
              XCHG                ;GET ADDRESS INTO DE!
              MVI A, ODH          ;SEND CARRIAGE RETURN
              CALL COUT           ;
              MVI A, OAH          ;SEND LINE FEED
              CALL COUT           ;
                                  ;
              ;FIND THE MATCHING OP CODE
FIND:         MVI A, 29           ;INITIAL TAB COUNT
              STA TEMP            ;INITIALIZE FOR TAB FUNCTION
              XCHG                ;
              MOV B,M             ;GET FIRST CHAR
              XCHG.               ;HL HAS DATA TABLE
              MVI C,0             ;COUNTER
NXT:          MOV A,M             ;
              CPI '!'             ;ONE BYTE INSTRUCTION
              JNZ SKIP            ;
              MVI A,1             ;IT'S A ONE BYTE INSTRUCTION.
              STA TEMP+5          ;STORE THIS FOR LATER
              JMP SEQ             ;
SKIP:         CPI '%'             ;
              JNZ SKIP2           ;
              MVI A,2             ;IT'S A TWO BYTE INSTRUCTION
              STA TEMP+5          ;SAVE FOR LATER
              JMP SEQ             ;
SKIP2:        CPI '#'             ;
              JNZ CONT2           ;
              MVI A,3             ;IT'S A 3 BYTE INSTRUCTION
              STA TEMP+5          ;SAVE FOR LATER
              JMP SEQ             ;
CONT2:        INX H               ;
              JMP NXT             ;JUST SCANNING MNEMONICS-MUST FIND
                                  ;A TRIGGER CHARACTER - !,%, OR #.
SEQ:          MOV A,B             ;
              CMP C               ;
              JZ PRINT            ;ONLY WAY OUT
                                  ;WE FOUND ONE THAT MATCHES!
              INR C               ;
              INX H               ;HL HAS DATA TABLE OF ALL CODES
              JMP NXT             ;NOT A MATCHING OP CODE YET.
                                  ;FORMAT: ADDRESS - OP CODES - MNEMONIC
                                  ;         - OPERAND - ASCII
                                  ;
              ;PRINT ADDRESS FOUND IN DE REGISTER PAIR
PRINT:        CALL ADOUT          ;
                                  ;
              ;PRINT OP CODES
CODES:        XCHG                ;
              MOV A,M             ;GET FIRST CODE
              XCHG                ;
              CALL HOUT           ;PRINT OP CODES
                                  ;
```

Printout 12-9 (continued)

```
                        ;SEE IF 1, 2, OR 3 BYTE INSTRUCTION
               LDA TEMP+5           ;
               CPI 1                ;
               JZ NXTA              ;
               CPI 2                ;
               JZ BYTE              ;
               INX D                ;
               XCHG                 ;
               MOV A,M              ;
               CALL HOUT            ;
               INX H                ;
               MOV A,M              ;
               CALL HOUT            ;
               XCHG                 ;
               DCX D                ;
               DCX D                ;
               MVI A, 20H           ;FEW SPACES, A 3 BYTE INSTRUCTION.
               CALL SP2             ;
               JMP NXTH             ;
BYTE:          INX D                ;
               XCHG                 ;
               MOV A,M              ;
               XCHG                 ;
               CALL HOUT            ;
               DCX D                ;
               MVI A,20H            ;MORE SPACES - A 2 BYTE INSTGRUCTION.
               CALL SP4             ;
               JMP NXTH             ;
NXTA:          MVI A,20H            ;MORE YET - A 1 BYTE INSTRUCTIO
               CALL SP6             ;
                                    ;
                        ;PRINT MNEMONIC
NXTH:          INX H                ;
               MOV A,M              ;
               CPI '!'              ;
               JZ DONE              ;
               CPI '*'              ;
               JZ DONE              ;
               CPI '%'              ;
               JZ DONE              ;
               CALL COUT            ;
               JMP NXTH             ;
                                    ;
                        ;NOW PRINT OPERAND (DATA OR ADDRESS INFORMATION)
DONE:          MVI A,20H            ;
               CALL SP3             ;
               LDA TEMP+5           ;
               CPI 1                ;
               JZ NEW               ;ALL DONE START OVER
               CPI 2                ;
               JZ PBITE             ;
PADR:          INX D                ;PRINT AN ADDRESS ARGUMENT
               INX D                ;
               XCHG                 ;
               MOV A,M              ;
               CALL HOUT            ;
               DCX H                ;
               MOV A,M              ;
               CALL HOUT            ;
               INX H                ;
               XCHG                 ;
               MVI A,'H'            ;
               CALL COUT            ;
```

(continued on next page)

Printout 12-9 (continued)

```
              JMP NEW                 ;
PBITE:        INX D                   ;PRINT A DATA ARGUMENT
              XCHG                    ;
              MOV A,M                 ;
              XCHG                    ;
              CALL HOUT               ;
              MVI A,'H'               ;
              CALL COUT               ;
              JMP NEW                 ;
                                      ;
              ;PRINT SPACES SUBROUTINES
SP10:         CALL COUT               ;PRINT 10 SPACES
              CALL COUT               ;
              CALL COUT               ;
              CALL COUT               ;
SP6:          CALL COUT               ;PRINT 6 SPACES
SP5:          CALL COUT               ;
SP4:          CALL COUT               ;
SP3:          CALL COUT               ;
SP2:          CALL COUT               ;
              RET                     ;
                                      ;
              ;PRINT ADDRESS
ADOUT:        LDA FLOP                ;REAL ADDRESS OR OFFSET?
              ORA A                   ;
              CNZ OFFSE               ;
              MOV A,D                 ;ADDRESS POINT DE
              CALL HOUT               ;
              MOV A,E                 ;
              CALL HOUT               ;
              MVI A,20H               ;
              CALL SP2                ;
              LDA FLOP                ;
              ORA A                   ;
              CNZ OFFSF               ;
              RET                     ;
              RET                     ;
                                      ;
              ;PRINT BYTE AS ASCII PAIR
HOUT:         PUSH A                  ;
              MVI C,2                 ;HOUT MUST PUT OUT TWO CHARACTERS
              ANI 0F0H                ;
              RRC                     ;
              RRC                     ;
              RRC                     ;
              RRC                     ;
AGA:          SUI 10                  ;
              JM DIGIT                ;
              ADI 41H                 ;
              CALL COUT               ;
              JMP AGAIN               ;
DIGIT:        ADI 3AH                 ;
              CALL COUT               ;
AGAIN:        DCR C                   ;
              RZ                      ;
              POP A                   ;
              ANI 0FH                 ;
              JMP AGA                 ;
                                      ;
              ;START OVER/PRINT MORE/GOTO ASCII
                                      ;PRINT ASCII EQUIVALENT OF OP CODE
NEW:          CALL TAB                ;TAB OVER TO COLUMN 30
              LDA TEMP+5              ;WAS IT 1, 2, OR 3 BYTE INSTRUCTION
```

Printout 12-9 (continued)

```
NEW2:    MOV C,A                    ;SAVE IT
DCR:     DCR A                      ;
         JZ HOT                     ;
         DCX D                      ;MUST GO BACK TO PRINT ASCII
         JMP DCR                    ;
HOT:     XCHG                       ;
HOT2:    MOV A, M                   ;
         CPI 20H                    ;PREPARE TO STAR INVALID CODES
         CM STAR                    ;
         CPI 80H                    ;PREPARE TO STAR INVALID CODES
         CP STAR                    ;
         CALL COUT                  ;
         INX H                      ;
         DCR C                      ;
         JNZ HOT2                   ;
         XCHG                       ;
         MVI A,0DH                  ;CARRIAGE RETURN
         CALL COUT                  ;
         MVI A,0AH                  ;LINE FEED
         CALL COUT                  ;
         LXI H,DATA                 ;NEW START
JLA:     CALL CONTC                 ;START STOP PRINT ROUTINE FOR JAMIE
         JNZ FIND                   ;
         CALL CIN                   ;WE WAIT TO CONTINUE
         CPI 03H                    ;CONTROL C?
         JZ FIND                    ;CONTINUE
         CPI 0DH                    ;CARRIAGE RETURN?
         JZ FIND                    ;CONTINUE
         CPI ESC                    ;ESCAPE?
         JZ 2028H                   ;RETURN TO MONITOR
         LXI D, DATA                ;
         LXI H, 0000H               ;
         PUSH A                     ;
         XRA A                      ;
         STA FLOP                   ;
         POP A                      ;
         JMP NJLA                   ;
                                    ;
         ;OFFSET SUBROUTINE TO SWAP ADDRESSES
OFFS:    PUSH A                     ;
         LDA FLOP                   ;
         XRI 01                     ;
         STA FLOP                   ;
         MOV A, H                   ;
         STA HL0                    ;
         MOV A, L                   ;
         STA HL0+1                  ;
         LXI H, 0                   ;
         POP A                      ;
         CPI ';'                    ;
         JZ FX                      ;
         RET                        ;
FX:      MVI A, 30H                 ;
         RET                        ;
OFFT:    MOV A, H                   ;
         STA OFF                    ;
         MOV A, L                   ;
          STA OFF+1                 ;
         LDA HL0                    ;
         MOV H, A                   ;
         LDA HL0+1                  ;
         MOV L,A                    ;
         RET                        ;
```

(continued on next page)

Printout 12-9 (continued)

```
        OFFSE:  MOV A, D            ;
                STA DE0             ;
                MOV A, E            ;
                STA DE0+1           ;
                LDA OFF             ;
                MOV D, A            ;
                LDA OFF+1           ;
                MOV E, A            ;
                RET                 ;
        OFFSF:  LDA TEMP+5          ;
                CPI 1               ;
                JZ D1               ;
                CPI 2               ;
                JZ D2               ;
                INX D               ;
        D2:     INX D               ;
        D1:     INX D               ;
                MOV A, D            ;
                STA OFF             ;
                MOV A, E            ;
                STA OFF+1           ;
                LDA DE0             ;
                MOV D, A            ;
                LDA DE0+1           ;
                MOV E, A            ;
                RET                 ;
                                    ;
                ;TAB ROUTINE
        TAB:    LDA TEMP            ;
                MOV B,A             ;
                MVI A, 20H          ;
        TAB2:   DCR B               ;
                RZ                  ;
                CALL COUT           ;
                JMP TAB2            ;
                                    ;
                ;FIX UNPRINTABLE ASCII CONTROL CHARACTERS
        STAR:   MVI A, '.'          ; PRINT . FOR NON CHARACTERS
                RET                 ;
                                    ;
                ;PRINT ASCII CHARACTER
        COUT:   PUSH B              ;SAVE BC REGISTERS
                MOV B,A             ;
                LDA TEMP            ;GET NUMBER CHRS ON A LINE
                DCR A               ;ONE LESS
                STA TEMP            ;SAVE IT
                XRA A               ;ZERO A REGISTER
                CALL CCOUT          ;
                POP B               ;
                RET                 ;
                                    ;
                ;SIGN ON MESSAGE SUBROUTINE
        MESSG:  LXI H,MSS           ;SIGN-ON MESSAGE
        XX:     MOV A,M             ;
                CPI '!'             ;END OF MESSAGE DATA TABLE
                RZ                  ;
                CALL COUT           ;
                INX H               ;
                JMP XX              ;
        MSS:    DB'8085 DISASSEMBLER C1982'
                DB ODH,OAH          ;
                DB'!'               ;
        CCOUT:  EQU 200DH           ;NORTHSTAR DOS ROUTINES
```

Printout 12-9 (continued)

```
CIN:      EQU  2010H               ;
CONTC:    EQU  2016H               ;
ESC:      SET  1BH                 ;
DATA:     EQU  1700H               ;
FLOP:     DB  0                    ;
OFF:      DW  0                    ;
HLO:      DW  0                    ;
DEO:      DW  0                    ;
TEMP:     NOP                      ;
          NOP                      ;
          NOP                      ;
          NOP                      ;
          NOP                      ;
          NOP                      ;
          NOP                      ;
          DS 20                    ;
STACK:    NOP                      ;
          END                      ;
```

Printout 12-10 is a sample program disassembly of the disassembler itself! You will notice that the ASCII equivalent of the hex data is also printed, a frequently useful function in examining data tables.

Printout 12-10

```
JLACD 8085 DISASSEMBLER C1982
1400
1400 31CF16 LXI  SP,   16CFH    1..
1403 CD4F16 CALL      164FH     .O.
1406 110017 LXI  D,    1700H    ...
1409 AF     XRA  A              .
140A 327D16 STA       167DH     2}.
140D 210000 LXI  H,    0000H    !..
1410 CD1020 CALL      2010H     ..
1413 FE0D   CPI       0DH       ..
1415 CA3914 JZ        1439H     .9.
1418 FE1B   CPI       1BH       ..
141A CA2820 JZ        2028H     .(
141D 47     MOV  B,A            G
141E CD4016 CALL      1640H     .@.
1421 78     MOV  A,B            x
1422 FE3A   CPI       3AH       .:
1424 CCCE15 CZ        15CEH     ...
1427 D630   SUI       30H       .0
1429 FE10   CPI       10H       ..
142B DA3014 JC        1430H     .0.
142E D607   SUI       07H       ..
1430 29     DAD  H              )
1431 29     DAD  H              )
1432 29     DAD  H              )
1433 29     DAD  H              )
1434 85     ADD  L              .
1435 6F     MOV  L,A            o
1436 C31014 JMP       1410H     ...
1439 3A7D16 LDA       167DH     :}.
143C B7     ORA  A              .
```

(continued on next page)

Printout 12-10 (continued)

```
1433  C4EC15  CNZ     15ECH        ...
1440  EB      XCHG                 .
1441  3E0D    MVI A,  0DH          >.
1443  CD4016  CALL    1640H        .@.
1446  3E0A    MVI A,  0AH          >.
1448  CD4016  CALL    1640H        .@.
144B  3E1D    MVI A,  1DH          >.
144D  328416  STA     1684H        2..
1450  EB      XCHG                 .
1451  46      MOV B,M              F
1452  EB      XCHG                 .
1453  0E00    MVI C,  00H          ..
1455  7E      MOV A,M              ~
1456  FE21    CPI     21H          .!
1458  C26314  JNZ     1463H        .c.
145B  3E01    MVI A,  01H          >.
145D  328916  STA     1689H        2..
1460  C38114  JMP     1481H        ...
1463  FE25    CPI     25H          .%
1465  C27014  JNZ     1470H        .p.
1468  3E02    MVI A,  02H          >.
146A  328916  STA     1689H        2..
146D  C38114  JMP     1481H        ...
1470  FE23    CPI     23H          .#
1472  C27D14  JNZ     147DH        .}.
1475  3E03    MVI A,  03H          >.
1477  328916  STA     1689H        2..
147A  C38114  JMP     1481H        ...
147D  23      INX H                #
147E  C35514  JMP     1455H        .U.
1481  78      MOV A,B              x
1482  B9      CMP C                .
1483  CA8B14  JZ      148BH        ...
1486  0C      INR C                .
1487  23      INX H                #
1488  C35514  JMP     1455H        .U.
148B  CD3615  CALL    1536H        .6.
148E  EB      XCHG                 .
148F  7E      MOV A,M              ~
1490  EB      XCHG                 .
1491  CD5315  CALL    1553H        .S.
1494  3A8916  LDA     1689H        :..
1497  FE01    CPI     01H          ..
1499  CAC714  JZ      14C7H        ...
149C  FE02    CPI     02H          ..
149E  CAB714  JZ      14B7H        ...
14A1  13      INX D                .
14A2  EB      XCHG                 .
14A3  7E      MOV A,M              ~
14A4  CD5315  CALL    1553H        .S.
14A7  23      INX H                #
14A8  7E      MOV A,M              ~
14A9  CD5315  CALL    1553H        .S.
14AC  EB      XCHG                 .
14AD  1B      DCX D                .
14AE  1B      DCX D                .
14AF  3E20    MVI A,  20H          >
14B1  CD3215  CALL    1532H        .2.
14B4  C3CC14  JMP     14CCH        ...
14B7  13      INX D                .
14B8  EB      XCHG                 .
14B9  7E      MOV A,M              ~
```

Printout 12-10 (continued)

```
14BA EB      XCHG                  .
14BB CD5315  CALL     1553H        .S.
14BE 1B      DCX  D                .
14BF 3E20    MVI A,   20H          >
14C1 CD2C15  CALL     152CH        .,.
14C4 C3CC14  JMP      14CCH        ...
14C7 3E20    MVI A,   20H          >
14C9 CD2615  CALL     1526H        .&.
14CC 23      INX  H                #
14CD 7E      MOV A,M               ~
14CE FE21    CPI      21H          .!
14D0 CAE314  JZ       14E3H        ...
14D3 FE23    CPI      23H          .#
14D5 CAE314  JZ       14E3H        ...
14D8 FE25    CPI      25H          .%
14DA CAE314  JZ       14E3H        ...
14DD CD4016  CALL     1640H        .@.
14E0 C3CC14  JMP      14CCH        ...
14E3 3E20    MVI A,   20H          >
14E5 CD2F15  CALL     152FH        ./.
14E8 3A8916  LDA      1689H        :..
14EB FE01    CPI      01H          ..
14ED CA7615  JZ       1576H        .v.
14F0 FE02    CPI      02H          ..
14F2 CA0B15  JZ       150BH        ...
14F5 13      INX  D                .
14F6 13      INX  D                .
14F7 EB      XCHG                  .
14F8 7E      MOV A,M               ~
14F9 CD5315  CALL     1553H        .S.
14FC 2B      DCX  H                +
14FD 7E      MOV A,M               ~
14FE CD5315  CALL     1553H        .S.
1501 23      INX  H                #
```

9.2 A 6502 Disassembler

For variation, we include some information on a 6502 disassembler. The data table changes to reflect a different instruction set. Since the instruction set requires the printing of a number sign (#), this character cannot be used as a marker. In this assembler (6502), the # byte marker is changed to a dollar sign ($) in both the data table and the disassembler.

Here are three tables:

1. Source file for 6502 data (Printout 12-11).

2. Hex dump of 6502 data (Printout 12-12).

3. Sample 6502 disassembly (Printout 12-13).

The 6502 disassembler is presented to illustrate the universal nature of this particular disassembler.

Printout 12-11

```
                                            ;DATA FOR THE 6502 DISASSEMBLER

                                            ;ORG 1700H
          DB        '!BRK$ORA(X)  !NOP!NOP!NOPZORA ZASL !NOP'
          DB        '!PHPZORA #!ASL A!NOP!NOP$ORA $ASL !NOP'
          DB        'ZBPL ZORA(Y) !NOP!NOP!NOP$ORA ZASL(X) !NOP'
          DB        '!CLC$ORA(Y) !NOP!NOP!NOP$ORA(X) $ASL(X) '
          DB        '!NOP$JSR ZAND(X) !NOP!NOPZBIT ZAND ZROL !NOP'
          DB        '!PLPZAND #!ROL A!NOP$BIT $AND $ROL !NOP'
          DB        'ZBMI ZAND(Y) !NOP!NOP!NOPZAND(X) ZROL(X) '
          DB        '!NOP!SEC$AND(Y) !NOP!NOP$AND(X) $ROL(X)'
          DB        '!NOP!RTIZEOR(X) !NOP!NOP!NOPZEOR ZLSR '
          DB        '!NOP!PHAZEOR #!LSR A!NOP$JMP $EOR $LSR '
          DB        '!NOPZBVC ZEOR(Y) !NOP!NOP!NOPZEOR(X) '
          DB        'ZLSR(X) !NOP!CLI$EOR(Y) !NOP!NOP!NOP'
          DB        '$EOR(X) $LSR(X) !NOP!NOP!RTSZADC(X)!NOP!NOP!NOP'
          DB        'ZADC ZROR !NOP!PLAZADC #!ROR A!NOP$JMP-IND '
          DB        '$ADC $ROR !NOPZBVS ZADC(Y) !NOP!NOP!NOP'
          DB        'ZADC(X) ZROR(X) !NOP!SEI$ADC(Y)!NOP!NOP!NOP'
          DB        '$ADC(X) $ROR(X) !NOP!NOPZSTA(X) !NOP!NOP'
          DB        'ZSTY ZSTA ZSTX !NOP!DEY!NOP!TXA!NOP$STY '
          DB        '$STA $STX !NOPZBCC ZSTA(Y) !NOP!NOPZSTY(X) '
          DB        'ZSTA(X) ZSTX(Y) !NOP!TYA$STA(Y) !TXS'
          DB        '!NOP!NOP$STA(X) !NOP!NOPZLDY #ZLDA(X)ZLDX #'
          DB        '!NOPZLDY ZLDA ZLDX !NOP!TAYZLDA #!TAX'
          DB        '!NOP$LDY $LDA $LDX !NOPZBCS ZLDA(Y)!NOP'
          DB        '!NOPZLDY(X) ZLDA(X) ZLDX(Y) !NOP!CLV$LDA(Y) '
          DB        '!TSX!NOP$LDY(X) $LDA(X) $LDX(Y) !NOP'
          DB        'ZCPY #ZCMP(X) !NOP!NOPZCPY ZCMP ZDEC !NOP'
          DB        '!INYZCMP #!DEX!NOP$CPY $CMP $DEC !NOP'
          DB        'ZBNE ZCMP(Y) !NOP!NOP!NOPZCMP(X)ZDEC(X)'
          DB        '!NOP!CLD$CMP(Y) !NOP!NOP!NOP$CMP(X)'
          DB        '$DEC(X) !NOPZCPX #ZSBC(X) !NOP!NOPZCPX '
          DB        'ZSBC ZINC !NOP!INXZSBC #!NOP!NOP$CPX '
          DB        '$SBC $INC !NOPZBEQ ZSBC(Y) !NOP!NOP!NOP'
          DB        'ZSBC(X) ZINC(X) !NOP!SED$SBC(Y) !NOP!NOP!NOP'
          DB        '$SBC(X) $INC(X) !NOP!'
          END
```

Printout 12-12

```
        -0  -1  -2  -3  -4  -5  -6  -7  -8  -9  -A  -B  -C  -D  -E  -F        ASCII DATA
  1700   21  42  52  4B  24  4F  52  41  28  58  29  20  20  21  4E  4F    !BRK$ORA(X)  !NO
  1710   50  21  4E  4F  50  21  4E  4F  50  25  4F  52  41  20  25  41    P!NOP!NOPZORA ZA
  1720   53  4C  20  21  4E  4F  50  21  50  48  50  25  4F  52  41  20    SL !NOP!PHPZORA
  1730   23  21  41  53  4C  20  41  21  4E  4F  50  21  4E  4F  50  24    #!ASL A!NOP!NOP$
  1740   4F  52  41  20  24  41  53  4C  20  21  4E  4F  50  25  42  50    ORA $ASL !NOPZBP
  1750   4C  20  21  4F  52  41  28  59  29  20  21  4E  4F  50  21  4E    L ZORA(Y) !NOP!N
  1760   4F  50  21  4E  4F  50  25  4F  52  41  28  58  29  20  25  41    OP!NOPZORA(X) ZA
  1770   53  4C  28  58  29  20  21  4E  4F  50  21  43  4C  43  24  4F    SL(X) !NOP!CLC$O
  1780   52  41  28  59  29  20  21  4E  4F  50  21  4E  4F  50  21  4E    RA(Y) !NOP!NOP!N
  1790   4F  50  24  4F  52  41  28  58  29  20  24  41  53  4C  28  58    OP$ORA(X) $ASL(X
  17A0   29  20  21  4E  4F  50  24  4A  53  52  20  25  41  4E  44  28    ) !NOP$JSR ZAND(
  17B0   58  29  20  21  4E  4F  50  21  4E  4F  50  25  42  49  54  20    X) !NOP!NOPZBIT
  17C0   25  41  4E  44  20  25  52  4F  4C  20  21  4E  4F  50  21  50    ZAND ZROL !NOP!P
  17D0   4C  50  25  41  4E  44  20  23  21  52  4F  4C  20  41  21  4E    LPZAND #!ROL A!N
  17E0   4F  50  24  42  49  54  20  24  41  4E  44  20  24  52  4F  4C    OP$BIT $AND $ROL
  17F0   20  21  4E  4F  50  25  42  4D  49  20  25  41  4E  44  28  59    !NOPZBMI ZAND(Y
```

Printout 12-12 (continued)

```
        -0 -1 -2 -3 -4 -5 -6 -7 -8 -9 -A -B -C -D -E -F    ASCII DATA
1800    29 20 21 4E 4F 50 21 4E 4F 50 21 4E 4F 50 25 41    ) !NOP!NOP!NOP%A
1810    4E 44 28 58 29 20 25 52 4F 4C 28 58 29 20 21 4E    ND(X) %ROL(X) !N
1820    4F 50 21 53 45 43 24 41 4E 44 28 59 29 20 21 4E    OP!SEC$AND(Y) !N
1830    4F 50 21 4E 4F 50 21 4E 4F 50 24 41 4E 44 28 58    OP!NOP!NOP$AND(X
1840    29 20 24 52 4F 4C 28 58 29 21 4E 4F 50 21 52 54    ) $ROL(X)!NOP!RT
1850    49 25 45 4F 52 28 58 29 20 21 4E 4F 50 21 4E 4F    I%EOR(X) !NOP!NO
1860    50 21 4E 4F 50 25 45 4F 52 20 25 4C 53 52 20 21    P!NOP%EOR %LSR !
1870    4E 4F 50 21 50 48 41 25 45 4F 52 20 23 21 4C 53    NOP!PHA%EOR #!LS
1880    52 20 41 21 4E 4F 50 24 4A 4D 50 20 24 45 4F 52    R A!NOP$JMP $EOR
1890    20 24 4C 53 52 20 21 4E 4F 50 25 42 56 43 20 25     $LSR !NOP%BVC %
18A0    45 4F 52 28 59 29 20 21 4E 4F 50 21 4E 4F 50 21    EOR(Y) !NOP!NOP!
18B0    4E 4F 50 25 45 4F 52 28 58 29 20 25 4C 53 52 28    NOP%EOR(X) %LSR(
18C0    58 29 20 21 4E 4F 50 21 43 4C 49 24 45 4F 52 28    X) !NOP!CLI$EOR(
18D0    59 29 20 21 4E 4F 50 21 4E 4F 50 21 4E 4F 50 24    Y) !NOP!NOP!NOP$
18E0    45 4F 52 28 58 29 20 24 4C 53 52 28 58 29 21       EOR(X) $LSR(X) !
18F0    4E 4F 50 21 52 54 53 25 41 44 43 28 58 29 21 4E    NOP!RTS%ADC(X)!N
```

Printout 12-13

```
2000                                       205C BC2600  LDY(X)   0026H     .&.
2000 060C    ASL    0CH        ..           205F 14      NOP                .
2002 595900  EOR(Y)  0059H      YY.         2060 28      PLP                (
2005 00      BRK                .           2061 203162  JSR     6231H      1b
2006 02      NOP                .           2064 203E59  JSR     593EH      >Y
2007 C900    CMP #   00H        ..          2067 32      NOP                2
2009 00      BRK                .           2068 00      BRK                .
200A C3      NOP                .           2069 203E01  JSR     013EH      >.
200B 62      NOP                b           206C 32      NOP                2
200C 20C33E  JSR     3EC3H      .>          206D 0620    ASL     20H        .
200F 29C3    AND #   C3H        ).          206F 3E0401  ROL(X)   0104H     >..
2011 00      BRK                .           2072 8100    STA(X)   00H       ..
2012 29C3    AND #   C3H        ).          2074 1101    ORA(Y)   01H       ..
2014 0E29C3  ASL     C329H      .).         2076 0521    ORA     21H        .!
2017 8429    STY     29H        .)          2078 00      BRK                .
2019 C3      NOP                .           2079 22      NOP                "
201A 9F      NOP                .           207A CD8320  CMP     2083H      ..
201B 22      NOP                "           207D C2      NOP                .
201C C3      NOP                .           207E 62      NOP                b
201D 2624    ROL     24H        &$          207F 20C32F  JSR     2FC3H      ./
201F C3      NOP                .           2082 27      NOP                '
2020 2023C3  JSR     C323H      #.          2083 F5E5    SBC(X)   E5H       ..
2023 9E      NOP                .           2085 79E6C0  ADC(Y)   C0E6H     y..
2024 23      NOP                #           2088 67      NOP                g
2025 C3      NOP                .           2089 3E14B8  ROL(X)   B814H     >..
2026 5E26C3  LSR(X)   C326H     ^&.         208C 1F      NOP                .
2029 8122    STA(X)   22H       ."          208D 1F      NOP                .
202B 00      BRK                .           208E 1F      NOP                .
202C C3      NOP                .           208F E620    INC     20H        .
202D 792200  ADC(Y)   0022H     y".         2091 B467    LDY(X)   67H       .g
2030 01C626  ORA(X)    26C6H    ..&         2093 79E63F  ADC(Y)   3FE6H     y.?
2033 18      CLC                .           2096 4F      NOP                O
2034 43      NOP                C           2097 FE03DA  INC(X)   DA03H     ...
2035 4F      NOP                O           209A 9F      NOP                .
2036 A421    LDY     21H        .!          209B 2017E6  JSR     E617H      ..
2038 A421    LDY     21H        .!          209E 0C      NOP                .
```

(continued on next page)

Printout 12-13 (continued)

```
203A  A421      LDY     21H          . !        209F  B432      LDY(X)   32H        .2
203C  00        BRK                  .          20A1  FB        NOP                 .
203D  2119      AND(X)  19H          ! .        20A2  26E1      ROL      E1H        & .
203F  2180      AND(X)  80H          ! .        20A4  E57B      SBC      7BH        . {
2041  0F        NOP                  .          20A6  32        NOP                 2
2042  A421      LDY     21H          . !        20A7  FA        NOP                 .
2044  A421      LDY     21H          . !        20A8  26B7      ROL      B7H        & .
2046  A421      LDY     21H          . !        20AA  F2        NOP                 .
2048  00        BRK                  .          20AB  AF        NOP                 .
2049  2119      AND(X)  19H          ! .        20AC  201E00    JSR      001EH      . .
204B  2100      AND(X)  00H          ! .        20AF  D5C5      CMP(X)   C5H        . .
204D  1F        NOP                  .          20B1  CDB621    CMP      21B6H      . . !
204E  010202    ORA(X)  0202H        . . .      20B4  C1D1      CMP(X)   D1H        . .
2051  016323    ORA(X)  2363H        .c‡        20B6  7B        NOP                 {
2054  82        NOP                  .          20B7  B7        NOP                 .
2055  0C        NOP                  .          20B8  C2        NOP                 .
2056  010000    ORA(X)  0000H        . . .      20B9  C520      CMP      20H        .
2059  1A        NOP                  .          20BB  3A        NOP                 :
205A  A421      LDY     21H          . !        +
```

10 THE AUTOMATIC TELEPHONE DIALER

Probably you have read about electronic telephone dialers in one of the current electronics magazines. These devices have been fashioned from pure TTL or CMOS logic gates, calculator keyboards, and even full-scale telephone dialer chips. They work and are relatively inexpensive. Larger electronics stores now carry automatic telephone dialers with number memory and recall and other features. Wouldn't it be a shame to own your own computer if it couldn't even dial the phone for you? The program PHONE (Printout 12-14) takes a number from the ASCII keyboard of a TTY connected to your computer and converts these digits into dial pulses for the telephone.

10.1 Conversion of Bits to Pulses

The execution of the PHONE program begins with a request for a phone number. The user then enters all the digits of the phone number, including area code and

FIGURE 12-8 Relay connections and driver transistor.

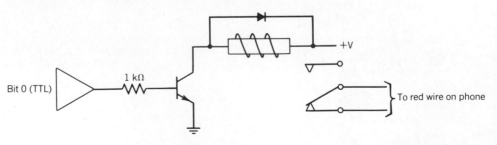

a long distance code if needed (but not more than 10 digits), ending with a carriage return. Then the program dials the phone by alternately setting and resetting a bit in an I/O port. The bit must be connected to a TTL-to-relay driver of some description (see Fig. 12-8), and the relays in turn must be connected to the phone line.

One method of dialing the telephone is by quickly tapping on the hang-up button. The same principle is used inside the phone on the rotary dial, and in the relay driver. As long as the phone is off the hook, the dialer program will dial the phone. The PHONE program is shown in Printout 12-14.

Printout 12-14

```
          ORG 5000H
          LXI SP, STAK              ;SET UP STACK
          MVI B, OCH                ;MAXIMUM OF TWELVE DIGITS
          LXI H, NUMS               ;PHONE NUMBER BUFFER
ZER:      MVI M, 80H                ;THIS IS HOW WE ZERO THE
          INX H                     ;NUMBER BUFFER
          DCR B
          JNZ ZER
          LXI H, MSG                ;PHONE NUMBER MESSAGE
MSGO:     MOV A, M                  ;SEND CHARACTERS FROM MEMORY
          ORA A                     ;TO THE USER UNTIL A ZERO (0)
          JZ GET                    ;IS FOUND
          MOV B, A
          XRA A
          CALL COUT
          INX H
          JMP MSGO
MSG:      DB 'PHONE NUMBER: ',00H
GET:      LXI H, NUMS               ;PHONE NUMBER BUFFER
GET2:     CALL CIN                  ;GET A CHARACTER FROM THE USER
          MOV B, A                  ;SAVE IT IN REGISTER B
          CPI ODH                   ;CARRIAGE RETURN (DONE)?
          JZ STAR
          CALL COUT                 ;ECHO THE CHARACTER TO THE USER
          MOV A, B                  ;AND GET IT BACK
          SUI 30H                   ;REMOVE ASCII BIAS
          ANI OFH                   ;AND LOOK AT 4 LSB'S
          ORA A                     ;IF WE NOW HAVE A ZERO (0)
          JNZ OK
          MVI A, OAH                ;WE MUST COUNT IT AS TEN
OK:       MOV M, A                  ;SAVE THE DIGIT IN THE BUFFER
          INX H
          JMP GET2                  ;AND GET MORE CHARACTERS
STAR:     LXI H, NUMS               ;PHONE NUMBER BUFFER
NXT2:     MOV A, M                  ;GET A DIGIT INTO REGISTER A
          INX H
          ORA A                     ;CHECK FOR THE LAST DIGIT
          JM DONE                   ;JUMP IF 80H IN REGISTER A
          MOV B, A                  ;SAVE THE DIGIT IN REGISTER B
LOP2:     MVI A, 01H                ;TOGGLE THE RELAY BIT IN THE I/O
          OUT PORT                  ;PORT. THIS PULSES THE RELAY
          CALL SDEL
          XRA A
          OUT PORT
          CALL SDEL
          DCR B                     ;REPEAT UNTIL REGISTER B IS ZERO
          JNZ LOP2
          CALL SDEL                 ;DELAY A LITTLE LONGER BETWEEN
```

(continued on next page)

Printout 12-14 (continued)

```
             CALL SDEL                  ;DIGITS
             JMP NXT2                   ;AND GET THE NEXT DIGIT
   SDEL:     LXI D, 3400H               ;SHORT DOUBLE REGISTER DELAY
             XRA A
   DEK:      DCX D
             CMP D
             JNZ DEK
             RET
   DONE:     MVI B, ODH                 ;SEND A CARRIAGE RETURN
             XRA A
             CALL COUT
             MVI B, OAH                 ;AND A LINE FEED TO THE USER
             XRA A
             CALL COUT
             JMP 2028H                  ;BEFORE RETURNING TO DOS
   PORT:     EQU OFFH                   ;RELAY I/O PORT, RELAY DRIVER ON BIT O
   COUT:     SET 200DH
   CIN:      SET 2010H
   NUMS:     DS 12
             DS 10H
   STAK:     DB OOH
             END
```

11 BINARY CODED DECIMAL ARITHMETIC

11.1 Introduction

In this section we will deal exclusively with binary coded decimal (BCD) numbers and the software required to perform arithmetic operations with them. Though we were introduced to BCD addition and subtraction in Chapter 2, we progress much further in this section, where we study multiplication and division, plus methods of inputting and outputting BCD numbers.

11.2 The Floating-Point Number

From the study of physics, we know that the charge on an electron is

$$0.00000000000000000016 \text{ coulomb}$$

and that Avogadro's number is equivalent to

$$602,000,000,000,000,000,000,000 \text{ molecules/mole}$$

These long strings of zeros are hard to work with, and *scientific notation*, whereby $e^- = 1.6 \times 10^{-19}$ C, is certainly preferable. By representing a number in this fashion, we only have two quantities to deal with, a mantissa and an exponent. The mantissa always falls between 1 and 9, and the exponent can be any power of 10. A similar method is employed in computer BCD, or *floating-point* arithmetic. Here, we also deal with the mantissa and exponent, but we also must keep track of the signs of both the mantissa and the exponent.

Table 12-5 shows that we must keep track of four pieces of information to fully

TABLE 12-5 Real Numbers Versus Floating-Point Numbers

Real Number	Floating-Point Representation			
	Sign	Mantissa	Sign	Exponent
483291.7	+	.4832917	+	6
− 214.306	−	.214306	+	3
0.001874	+	.1874	−	2
− 0.000456	−	.456	−	3

represent any real number in floating-point format. The sign, the mantissa, the exponent sign, and the exponent must all be stored in memory, for use by the arithmetic routines.

The following format for representing a real number is not a standard floating-point representation format, but is very similar to methods actually used in floating-point routines. In this format, we use 1 byte to represent the sign of the mantissa. If the number is positive, we set the byte to 00. If the number is negative, we set it to 01. Next, to represent the mantissa, we allocate 5 bytes, which actually hold 10 BCD digits, enough to represent most important numbers. We left-justify the mantissa so that trailing zeros take up the end of the number. Last, we use 1 byte to represent the exponent sign and the exponent. Bit 7 controls the sign (0 = +, 1 = −) and bits 6–0 contain the *binary* representation of the exponent. Therefore, our exponent can go from − 127 to + 127 instead of − 79 to + 79, which would be the case if we used the remaining bits to store a BCD value. So, all told, we use 7 bytes to store any number in floating-point format.

EXAMPLE 12-2

Convert the following numbers into seven-byte floating-point representation:

$$-1035.78 \qquad 238.96 \qquad 0.01478$$
$$1291.35 \qquad -0.00356 \qquad -100015.7$$

SOLUTION

As previously stated, we use 1 byte for sign, 5 for packed BCD data (left-justified), and 1 more for exponent sign and exponent.

The first number (− 1035.78) is negative; therefore, the first byte must be set to 01. The next 3 bytes contain the packed BCD data, followed by 2 bytes of zeros, since there were only 6 digits out of a maximum of 10. Last, we have an 04 for the exponent byte. Thus:

$$-1035.78 \rightarrow 01\ 10\ 35\ 78\ 00\ 00\ 04$$

By the same rules, the rest of the numbers become:

$$238.96 \rightarrow 00\ 23\ 89\ 60\ 00\ 00\ 03$$
$$0.01478 \rightarrow 00\ 14\ 78\ 00\ 00\ 00\ 81 \text{ (remember,}$$
$$\text{bit 7 is the exponent sign bit!)}$$
$$1291.35 \rightarrow 00\ 12\ 91\ 35\ 00\ 00\ 04$$
$$-0.00356 \rightarrow 01\ 35\ 60\ 00\ 00\ 00\ 82$$
$$-100015.7 \rightarrow 01\ 10\ 00\ 15\ 70\ 00\ 06$$

11.3 ASCII to Packed BCD

Before a floating number can be used in a mathematical routine, it must be put into memory, since all 7 bytes will not fit into any of the 8080's registers. It is easier to store the number in memory and then use a register pair to point to the beginning address of the number. Now, however, we must get the number into memory. Since all we know how to do thus far is to enter a decimal (real) number from a keyboard, *not* a packed BCD representation, we now need to learn a means for converting ASCII data into packed BCD data.

The following routine, FPCON, accepts ASCII characters from a keyboard and packs them into the 7-byte format already described. The routine uses a buffer area called BUFF to keep track of the ASCII data entered. When a carriage return is received, FPCON sends a CR,LF and proceeds to pack the input number. The first loop writes seven zeros into the FNUM buffer, which is the floating-point number buffer. This is where the number will be after execution. Next, FPCON checks for a minus sign, "−", and, if it sees one, puts "01" in the first location of FNUM. Then FPCON computes the exponent by: (1) counting the number of digits before the decimal point, or (2) counting the number of zeros before the first nonzero digit is seen. Since no error checking is employed (to simplify the understanding of the routine), the numbers *must* be entered in the following ways:

Positive	*Negative*
141.7 or	−283.2 or
0.018 or	−0.006 or
1007.405	−998.3

This illustrates that decimal numbers must always start with 0. There is a lot of room for improvement by way of error checking for different input formats, but that is left up to you.

The last loop of FPCON does the BCD packing. It simply takes a character from BUFF, rotates it four times, and strips off all other bits. Thus, we are left with the 4 upper bits being 0000 → 1001. These are then ORed with the next character in BUFF (less its ASCII bias) and placed in FNUM. As stated before, FPCON is used to convert numbers into 7-byte packed BCD, but the routine can be modified to accept E+07 characters for setting the exponent. The usefulness is limited only by your imagination. Printout 12-15 shows the FPCON listing.

Printout 12-15

```
              LXI H, BUFF      ;ASCII BUFFER
      LOP2:   CALL 2010H       ;GET A CHARACTER
              CPI 13           ;DONE IF CARRIAGE RETURN
              JZ QUIT
              MOV B, A         ;ECHO IT TO THE USER
              CALL 200DH
              MOV M, B         ;AND PUT IT IN THE BUFFER
              INX H
              JMP LOP2
      QUIT:   MOV M, A         ;SAVE THE CR TOO
              MVI B, 13        ;AND SEND A CR, LF TO THE USER
              CALL 200DH
```

Printout 12-15 (continued)

```
                MVI B, 10
                CALL 200DH
                MVI C, 7        ;CLEAR THE FLOATING POINT BUFFER
                LXI H, FNUM
                XRA A
ZLOP:           MOV M, A
                INX H
                DCR C
                JNZ ZLOP
                LXI H, BUFF     ;ASCII BUFFER
                MOV A, M
                CPI '-'         ;MINUS SIGN?
                JNZ NOMIN
                MVI A, 01H      ;FIX THE SIGN BYTE
                STA FNUM
                INX H           ;NEXT ASCII CHARACTER NOW
NOMIN:          MOV A, M        ;GET ASCII CHARACTER
                CPI '0'         ;IF '0' THEN NUMBER IS < 1
                JNZ PEXP        ;OTHERWISE, POSITIVE EXPONENT
                INX H           ;SKIP OVER '0.'
NXTO:           INX H
                MOV A, M        ;COUNT THE NUMBER OF LEADING '0'
                CPI '0'
                JNZ NEXP
                INR C
                JMP NXTO
NEXP:           MOV A, C        ;CORRECT TO NEGATIVE EXPONENT
                ORI 80H
                STA FNUM+6      ;AND SAVE IT
                DCX H
                JMP ENT2        ;GO PACK THE REST
PEXP:           MOV A, M        ;COUNT THE NUMBER OF DIGITS
                CPI '.'         ;BEFORE THE DECIMAL POINT
                JZ STEXP
                INR C
                INX H
                JMP PEXP
STEXP:          MOV A, C        ;SAVE THE POSITIVE EXPONENT
                STA FNUM+6
                LXI H, BUFF     ;AND SKIP OVER THE MINUS SIGN
                MOV A, M        ;AGAIN IF WE HAVE TO
                CPI '-'
                JZ ENT2
                DCX H
ENT2:           LXI D, FNUM+1   ;PACKED BCD BUFFER
BAK:            INX H
                MOV A, M        ;GET AN ASCII NUM
                CPI '.'         ;IF '.' IGNORE
                JZ BAK
                CPI 13          ;IF CR, THEN DONE
                RZ
                ANI 0FH         ;STRIP OFF ASCII BIAS
                RLC             ;STICK THIS ONE IN UPPER 4 BITS
                RLC
                RLC
                RLC
                MOV C, A        ;SAVE IT HERE
BAK2:           INX H           ;GET THE NEXT CHARACTER
                MOV A, M
                CPI '.'         ;IF '.' IGNORE
                JZ BAK2
                CPI 13          ;IF CR, MUST SAVE LAST DIGIT
                JZ FINAL
```

(continued on next page)

Printout 12-15 (continued)

```
                ANI  0FH            ;OTHERWISE, STRIP OFF BIAS
                ORA  C             ;COMBINE WITH FIRST DIGIT
                STAX D             ;AND SAVE IN BUFFER
                INX  D
                JMP  BAK
        FINAL:  MOV  A, C          ;SAVE THE LAST DIGIT
                STAX D
                RET
        FNUM:   DS 7
        BUFF:   DS 15
```

11.4 Packed BCD to ASCII

We employ one method of programming to convert ASCII to packed BCD, and we use the reverse operation to convert packed BCD to ASCII. FPOUT is a subroutine that takes a packed BCD number stored in memory and outputs a floating-point number in ASCII to the user's console. Remember that because we use a standard format to represent the floating-point number, the program must display the number in that fashion.

EXAMPLE 12-3

Convert the following packed BCD numbers to floating-point notation.

$$00\ 77\ 61\ 38\ 20\ 00\ 09$$
$$01\ 36\ 89\ 71\ 48\ 27\ 04$$
$$00\ 45\ 69\ 23\ 80\ 00\ 87$$
$$01\ 21\ 31\ 55\ 96\ 20\ 81$$

Since these four numbers are packed into the 7-byte format already studied, converting back into floating-point is a simple process. First, we look at the sign byte and display a minus sign if it is an 01. Then we write down "0." to get into our standard format. Next we unpack the five-number bytes so we get 10 numerical digits. Last we write down an E to show that we have an exponent, a plus or minus sign depending on bit 7 of the last byte, and the decimal equivalent of the last 7 bits in the last byte. Thus, the four example numbers unpack in this way:

$$00\ 77\ 61\ 38\ 20\ 00\ 09 \rightarrow \ \ \ 0.7761382000E+09$$
$$01\ 36\ 89\ 71\ 48\ 27\ 04 \rightarrow -0.3689714827E+04$$
$$00\ 45\ 69\ 23\ 80\ 00\ 87 \rightarrow \ \ \ 0.4569238000E-07$$
$$01\ 21\ 31\ 55\ 96\ 20\ 81 \rightarrow -0.2131559620E-01$$

To use FPOUT, the calling program must set the HL register pair to the address pointing to the sign byte (the first of 7 bytes) of the number that is to be unpacked. FPOUT has only one output format, namely, the one used in Example 12-3. You are encouraged to try other ways of displaying (unpacking) the number. For instance, trailing zeros can be eliminated, or the decimal point can move as a function of the exponent. You are encouraged to try other methods because the problem of converting data in this way is an excellent logical programming exercise. Printout 12-16 shows the FPOUT listing.

Printout 12-16

```
         FPOUT:   MOV A, M          ;GET THE SIGN BYTE
                  ORA A
                  JNZ NEGN          ;A 1 MEANS MINUS SIGN
                  MVI B, 20H        ;OTHERWISE, WE PRINT A SPACE
                  JMP SIGN
         NEGN:    MVI B, '-'
         SIGN:    CALL COUT         ;SEND IT TO THE USER
                  MVI B, '0'        ;ALWAYS PRINT THESE FOR
                  CALL COUT         ;SCIENTIFIC NOTATION
                  MVI B, '.'
                  CALL COUT
                  INX H
                  LXI D, FBUF       ;UNPACKED DATA TO GO HERE
                  CALL UNPAK
                  LXI D, FBUF       ;BEGINNING OF UNPACKED DATA
                  MVI C, 10         ;TEN DIGITS
         OUTP:    LDAX D            ;GET A DIGIT
                  ADI 30H           ;ADD ASCII BIAS
                  MOV B, A          ;SEND IT TO THE USER
                  CALL COUT
                  INX D             ;NEXT CHARACTER
                  DCR C             ;DONE?
                  JNZ OUTP
                  MVI B, 20H
                  CALL COUT
                  MVI B, 'E'        ;SEND THIS BEFORE DOING THE
                  CALL COUT         ;EXPONENT
                  MOV A, M          ;GET THE EXPONENT BYTE
                  ORA A
                  JM NEGE           ;BIT 7 HIGH MEANS NEGATIVE
                  MVI B, '+'        ;OTHERWISE WE HAVE A PLUS SIGN
                  JMP EXSGN
         NEGE:    MVI B, '-'
         EXSGN:   CALL COUT
                  MOV A, M          ;GET THE EXPONENT AGAIN
                  ANI 7FH           ;STRIP OFF SIGN BIT
                  CPI 100           ;IS EXP LESS THAN 100?
                  JC NO100          ;YES
                  MVI B, '1'        ;NO, SO PRINT A '1'
                  CALL COUT
                  MOV A, M          ;GET THE EXPONENT AGAIN
                  ANI 7FH           ;STRIP OFF THE SIGN BIT
                  SUI 100           ;AND TAKE AWAY 100
         NO100:   MVI B, '0'        ;INITIALIZE 10'S COUNTER
                  CPI 10            ;IS ACC LESS THAN 10?
                  JC LDIGT          ;YES, LAST DIGIT NOW
                  INR B             ;NO, INCREASE 10'S COUNTER
                  SUI 10            ;TAKE AWAY 10
                  JMP NO100+2       ;AND CONTINUE
         LDIGT:   PUSH PSW          ;SAVE THE LAST DIGIT
                  CALL COUT         ;PRINT THE 10'S DIGIT
                  POP PSW           ;GET THE LAST DIGIT BACK
                  ADI 30H           ;ADD THE ASCII BIAS
                  MOV B, A          ;AND SEND TO THE USER
                  CALL COUT
                  RET
```

11.5 Floating-Point Math

11.5.1 Introduction.   Now that we know how to get numbers into and back out
of memory, we should be able to have a little fun in the process, by adding,

TABLE 12-6   Math Routine Summary

Function	Operation Performed[a]
FPADD	Adds two packed BCD numbers that have been placed in the buffers DIGA and DIGB. The numbers must have the same sign.
FPSUB	Subtracts DIGA from DIGB. Both numbers must have the same sign.
FPMUL	Multiplies the number in the MCAN buffer by the number in the MPLI buffer. No error checking for very large exponents.
FPDIV	Divides number in NUM1 by number in NUM2. Both numbers must have positive exponents. NUM1 must be larger than NUM2.

[a]In all cases, the result is placed in DIGA.

subtracting, multiplying, or dividing them. Sections 11.5.2 through 11.5.5 deal, in succession, with these mathematical operations. The emphasis throughout is on getting across the *method* employed in the subroutine to perform the operation, *not* on time-consuming error checking. It is not our desire to offer an entire floating-point operating system, just the necessary basics. For example, to add -417.3 and 1005.8 would require a sign change and a switch to subtract. This is not our concern and problems of this type are not covered. You may want to add these numbers as a programming exercise, however. Table 12-6 lists the routines and notes the limitations of each.

As before, the programs presented are meant to whet your appetite; they can be improved with additional programming.

11.5.2 Addition.   The subroutine FPADD performs the addition of two BCD packed numbers that reside in the buffers DIGA and DIGB. The result is placed in DIGA. Two numbers must have the same exponent if they are to be added. FPADD makes sure of this by calling a routine called ALIGN, which judiciously changes one of the numbers so that both will have the same exponent. The ALIGN subroutine is explained in Section 11.6, "Auxiliary Subroutines."

When adding numbers in BCD, the programmer must make use of the carry bit and the DAA instructions.

EXAMPLE 12-4

Add 1015 and 728 in BCD arithmetic.

SOLUTION
The first two bytes to be added are 15 and 28. (Remember that a BCD number is represented as hexadecimal digits 0–9.)

$$\begin{array}{r} 15 \\ +28 \\ \hline 3D \text{ H} \end{array}$$

If we now execute a DAA instruction, we will get:

$$\begin{array}{r} 3D \\ +\ 6 \\ \hline 43 \end{array}$$

which is the correct answer. Next we add the next two bytes, 10 and 07.

$$\begin{array}{r} 10 \\ +07 \\ \hline 17 \end{array}$$

A DAA instruction will not affect the result; thus we have 1015 + 728 = 1743.

EXAMPLE 12-5

Add 568 + 284 using BCD math.

SOLUTION
As before, we first add 68 and 84:

$$\begin{array}{r} 68 \\ +84 \\ \hline EC\ H \end{array}$$

Next, we decimal adjust the accumulator:

$$\begin{array}{r} EC \\ +66 \\ \hline (C\ =\ 1)\qquad 52 \end{array}$$

and we notice that the carry flag is equal to one. Now we add the next two bytes, *plus* the carry flag:

$$\begin{array}{r} 05 \\ +02 \\ +01 \\ \hline 08 \end{array}$$

and we get the correct answer, 852.

Examples 12-4 and 12-5 show that we must keep track of the carry flag when performing BCD addition. As you study FPADD, pay particular attention to the ADC M instruction (in the ADDER subroutine). Printout 12-7 shows the FPADD listing.

11.5.3 Subtraction. The BCD subtraction process is more complicated than addition, and there is no easy way to do it in 8080 assembler. However, similar to binary subtraction (where we obtain the 2's complement and then add), the subroutine FPSUB subtracts the number in DIGA from DIGB and places the result in DIGA. It, too, calls ALIGN to line up the exponents in the numbers, and it also calls TCOMP to obtain the 10's complement of the number to be subtracted.

Printout 12-17

```
FPADD:   CALL ALIGN       ;ALIGN EXPONENTS
         CALL ADDER       ;ADD THE TWO NUMBERS
         RNC              ;RETURN IF NO OVERFLOW
         LXI H, DIGA+5    ;OTHERWISE, ROTATE RESULT
         CALL RIGHT
         LDA DIGA+1       ;PLACE OVERFLOW INTO RESULT BUFFER
         ORI 10H
         STA DIGA+1
         LDA DIGA+6       ;GET THE EXPONENT
         CALL INREX       ;INCREMENT THE EXPONENT
         STA DIGA+6       ;AND RESAVE IT
         RET
```

Printout 12-18

```
FPSUB:   CALL ALIGN       ;ALIGN THE EXPONENTS
         LXI H, DIGA+5    ;COMPLEMENT THE SUBTRAHEND
         CALL TCOMP
         CALL ADDER       ;ADD THE TWO NUMBERS
         RC               ;END IF RESULT DID NOT CHANGE SIGNS
         LXI H, DIGA+5    ;OTHERWISE, COMPLEMENT THE RESULT
         CALL TCOMP
         LDA DIGA         ;AND CHANGE THE SIGN
         XRI 01H
         STA DIGA
         RET
```

TCOMP is explained in Section 11.6. Once the 10's complement of one of the numbers has been found, the numbers can be added together just as they are in FPADD. However, the absence of a carry out of the last place when the addition is completed means that the number (result) is negative, and we have to call TCOMP again to get it back into correct form. Printout 12-18 shows the FPSUB listing.

11.5.4 Multiplication.   FPMUL will take the number in MCAN and multiply it by the number in MPLI. It uses a repetitive addition process to call FPADD in a process exactly like the one in Example 12-6. Error checking can be used to speed up the routine by checking for whether one or both numbers are zero, thus making the result zero. Or, if one of the numbers is one, the result is the other number. As usual, this is left up to you to implement.

EXAMPLE 12-6

Multiply 557 by 28.

SOLUTION

First we normalize the numbers into our standard format. Thus, 557 becomes $0.557E+3$ and 28 becomes $0.28E+2$. Now we proceed with the multiple additions.

As step 1, $0.557E + 3$ is added to itself eight times. This gives us $4.456E + 3$, which is automatically converted back to $0.4456E + 4$ by FPADD.

Next, we increase the size of the multiplicand by a power of 10 to $0.557E + 4$ and add this twice to our temporary result, $0.4456E + 4$. Now we have $0.15596E + 5$, which is our correct answer. However, for FPMUL to work for all cases, the actual exponents of both input numbers should be added.

FPMUL does this and replaces the exponent in the result buffer, DIGA. Printout 12-19 shows the FPMUL listing.

Printout 12-19

```
FPMUL:   LXI H, DIGA      ;CLEAR THE DIGA BUFFER
         MVI C, 7
         XRA A
ZERO:    MOV M, A
         INX H
         DCR C
         JNZ ZERO
         LXI H, MCAN      ;MOVE THE MULTIPLICAND INTO
         LXI D, DIGB      ;THE DIGB BUFFER
         MVI C, 7
TRAN:    MOV A, M
         STAX D
         INX H
         INX D
         DCR C
         JNZ TRAN
         LXI H, MPLI+1    ;UNPACK THE MULTIPLIER AND
         LXI D, MDIG      ;STORE IT IN THE MDIG BUFFER
         CALL UNPAK
         LXI H, MDIG+9    ;START WITH RIGHT MOST MULTIPLIER
         MVI B, 10        ;POSSIBLE TEN TIMES
BAK1:    MOV A, M         ;SKIP OVER TRAILING ZEROES
         ORA A
         JNZ MUL1
         DCX H
         DCR B
         JNZ BAK1
         RET              ;RESULT IS ZERO
MUL1:    MOV C, A         ;C IS NUMBER OF TIMES TO ADD
MUL2:    PUSH H           ;SAVE IMPORTANT VALUES
         PUSH B
         CALL FPADD       ;PERFORM ADDITION
         POP B            ;GET IMPORTANT VALUES BACK
         POP H
         DCR C            ;DONE?
         JNZ MUL2
BAK2:    DCR B            ;LAST MULTIPLIER DIGIT?
         JZ EXPO
         DCX H            ;NEXT MULTIPLIER DIGIT
         LDA DIGB+6       ;INCREMENT MULTIPLICAND EXPONENT
         CALL INREX
         STA DIGB+6
         MOV A, M         ;GET THE NEXT MULTIPLIER
         ORA A            ;ZERO?
```

(continued on next page)

Printout 12-19 (continued)

```
                JZ BAK2          ;IF YES, NO ADDITIONS
                JMP MUL1         'OTHERWISE, CONTINUE
        EXPO:   LDA MCAN+6       ;GET THE MULTIPLICAND EXPONENT
                ORA A
                JP EXP1
                XRI 80H          ;IF IT IS NEGATIVE, TAKE THE
                CMA              ;TWO'S COMPLEMENT OF IT
                INR A
        EXP1:   MOV C, A         ;SAVE IT HERE
                LDA MPLI+6       ;GET THE MULTIPLIER'S EXPONENT
                ORA A
                JP EXP2
                XRI 80H          ;IF IT IS NEGATIVE, TAKE THE
                CMA              ;TWO'S COMPLEMENT OF IT
                INR A
        EXP2:   ADD C            ;ADD THEM
                JP EXP3
                CMA              ;FIX EXPONENT
                INR A
                ORI 80H
        EXP3:   STA DIGA+6       ;SAVE THE EXPONENT
                LDA MCAN         ;GET A SIGN
                MOV C, A         ;SAVE IT
                LDA MPLI         ;GET THE OTHER SIGN
                ADD C            ;ADD THEM
                DCR A
                RNZ
                MVI A, 1         ;SAVE A MINUS SIGN
                STA DIGA
                RET
        MDIG:   DS 10
```

11.5.5 Division.   The last mathematical routine presented is FPDIV. FPDIV will divide the number in NUM1 by the number in NUM2. The process used is repetitive subtraction, which is similar to the repetitive addition method in FPMUL. There is no error checking for divide by zero or one and, for simplicity, NUM1 must *always* be larger than NUM2. You may want to modify this aspect of the program, once it is thoroughly understood.

EXAMPLE 12-7

Divide 525.8 by 19 using repetitive subtraction.

SOLUTION

As always, we first convert to floating-point format:

$$525.8 \rightarrow 0.5258E + 3$$
$$19 \rightarrow 0.19E + 2$$

Now we start the repetitive subtraction:

1. Subtract 0.1900 from 0.5258 until we go negative:

$$
\begin{array}{r}
0.5258 \\
-0.1900 \\
\hline
0.3358 \\
-0.1900 \\
\hline
0.1458 \\
-0.1900 \\
\hline
-0.0442
\end{array}
$$

We did this only twice before going negative; thus 2 is the first digit of our result.

2. Subtract 0.01900 from 0.1458 repeatedly:

$$
\begin{array}{r}
0.1458 \\
-0.0190 \\
\hline
0.1268 \\
-0.0190 \\
\hline
0.1078 \\
-0.0190 \\
\hline
0.0888 \\
-0.0190 \\
\hline
0.0698 \\
-0.0190 \\
\hline
0.0508 \\
-0.0190 \\
\hline
0.0318 \\
-0.0190 \\
\hline
0.0128 \\
-0.0190 \\
\hline
-0.0062
\end{array}
$$

We were able to do this operation seven times. Our result so far is 27.

This process continues for a long time, since the two example numbers are not evenly divisible. FPDIV will take the result out to 10 digits to yield 2767368421. FPDIV then finds the difference of the exponents (1 in this example) and adds 1 to it. It does this because the first number is always greater than the second number. So, we get a final answer of 27.67368421, which we know is correct.

FPDIV is the simplest of the four math routines presented here, since NUM1 must always be greater than NUM2.

You may wish to change the routine to accept all exponent types. Printout 12-20 shows the FPDIV subroutine.

Printout 12-20

```
                    MVI A, 8              ;SET UP FOR 8 DIGIT PRECISION
                    STA KNT               ;SAVE IN LOOP COUNTER
                    LXI H, NUM1           ;GET DIVIDEND ADDRESS
                    LXI D, DIGB           ;TRANSFER IT INTO DIGB
                    MVI C, 7
          TRA1:     MOV A, M
                    STAX D
                    INX H
                    INX D
                    DCR C
                    JNZ TRA1
                    LXI H, NUM2           ;GET DIVISOR ADDRESS
                    LXI D, DIGA           ;TRANSFER IT INTO DIGA
                    MVI C, 7
          TRA2:     MOV A, M
                    STAX D
                    INX H
                    INX D
                    DCR C
                    JNZ TRA2
                    LXI H, RES            ;RESULT BUFFER
                    PUSH H                ;SAVE RESULT ADDRESS
                    LDA DIGB              ;GET CURRENT SIGN
                    MOV B, A              ;SAVE IT HERE
          TOP:      MVI C, 0              ;INIT THE LOOP COUNTER
          AGAIN:    PUSH C                ;SAVE THE COUNTER AND SIGN
                    CALL SUB2             ;SUBTRACT DIGA FROM DIGB
                    POP C                 ;GET COUNTER AND SIGN BACK
                    LDA DIGB              ;CHECK FOR A SIGN CHANGE
                    CMP B
                    JNZ FIX               ;CHANGE MEANS WE WENT NEGATIVE
                    INR C                 ;OTHERWISE, WE DO MORE LOOPS
                    JMP AGAIN
          FIX:      POP H                 ;GET BUFFER ADDRESS BACK
                    MOV M, C              ;SAVE THE LOOP COUNT
                    INX H                 ;NEXT BUFFER ADDRESS
                    PUSH H                ;RESAVE IT
                    LXI H, DIGB+5         ;FIX THE DIVISOR
                    CALL TCOMP
                    CALL ADD2
                    XCHG                  ;SAVE THE OLD SIGN
                    MOV M, B
                    LDA DIGB+6            ;AND INCREMENT THE EXPONENT
                    CALL INREX            ;ON THE DIVISOR
                    STA DIGB+6
                    LDA KNT               ;CHECK TO SEE IF WE ARE DONE
                    DCR A
                    STA KNT
                    JNZ TOP
                    LXI H, RES            ;GET BUFFER ADDRESS
                    LXI D, DIGA           ;RESULT IS ALWAYS IN DIGA
                    CALL PACK             ;PUT INTO BCD PAIRS
                    LDA NUM1+6            ;ADD THE EXPONENTS
                    MOV C, A
                    LDA NUM2+6
                    ADD C
                    INR A                 ;PLUS ONE TO BE CORRECT
                    STA DIGA+6            ;PLACE IN RESULT BUFFER
                    RET
          RES:      DS 20
          SUB2:     CALL ALIGN            ;SUB2 IS EXACTLY LIKE FPSUB
                    LXI H, DIGA+5         ;EXCEPT THE RESULT GOES IN DIGB
```

Printout 12-20 (continued)

```
                CALL TCOMP      ;AND DIGA IS SUBTRACTED FROM DIGB
                CALL ADD2
                RC
                LXI H, DIGB+5
                CALL TCOMP
                LDA DIGB
                XRI 01H
                STA DIGB
                RET
ADD2:           LXI H, DIGA+5   ;ADD2 IS EXACTLY LIKE ADDER
                LXI D, DIGB+5   ;EXCEPT THE RESULT IS PLACED
                MVI C, 5        ;IN THE DIGB BUFFER
                LDAX D
                ADC M
                DAA
                MOV M, A
                DCX H
                DCX D
                DCR C
                JNZ ADD2
                RET
```

11.6 Auxiliary Subroutines

11.6.1 Introduction.   All the subroutines that follow are used by the four floating-point mathematical routines. You are encouraged to study them because they can be very useful in modifying any of the floating-point routines.

11.6.2 INREX.   INREX, increment exponent, is used to add 1 to the exponent supplied to the subroutine by the accumulator. The routine first decides whether the exponent is negative, and, if so, decrements it instead. (Example: -5 is larger than -6.) The incremented exponent byte is in the accumulator upon exit. Printout 12-21 shows the INREX listing.

11.6.3 DCREX.   DCREX does the opposite of INREX. That is, it decrements the exponent supplied to it in the accumulator. Printout 12-22 shows the DCREX listing.

Printout 12-21

```
INREX:          ORA A
                JM INR2         ;EXPONENT IS NEGATIVE
                INR A           ;INCREMENT POSITIVE EXPONENT
                RET
INR2:           CPI 80H         ;SWITCH TO POSITIVE EXPONENT?
                JZ INR3
                DCR A           ;DECREMENT NEGATIVE EXPONENT
                RET
INR3:           MVI A, 01H      ;FIRST POSITIVE EXPONENT
                RET
```

Printout 12-22

```
DCREX:   ORA  A
         JZ   DCR2        ;SWITCH TO NEGATIVE EXPONENT?
         JP   DCR3        ;EXPONENT IS POSITIVE
         INR  A           ;INCREMENT NEGATIVE EXPONENT
         RET
DCR2:    MVI  A, 81H      ;FIRST NEGATIVE EXPONENT
         RET
DCR3:    DCR  A           ;DECREMENT POSITIVE EXPONENT
         RET
```

Printout 12-23

```
TCOMP:   PUSH H           ;SAVE THE POINTER
         MVI  C, 5        ;DO 5 BCD PAIRS
         XRA  A           ;CLEAR THE CARRY BIT
TCOM1:   MVI  A, 99H      ;PUT BCD 99 INTO ACCUMULATOR
         SBB  M           ;SUBTRACT MEMORY DIGIT PAIR
         MOV  M, A        ;AND REPLACE IN MEMORY
         DCX  H           ;NEXT PAIR
         DCR  C           ;DONE?
         JNZ  TCOM1
         MVI  C, 5        ;DO 5 PAIRS AGAIN
         STC              ;CARRY IS SET TO ADD 1
         POP  H           ;GET THE POINTER BACK
TCOM2:   MOV  A, M        ;GET 9'S COMPLEMENT PAIR
         ACI  0           ;CORRECT TO TENS COMPLEMENT
         DAA              ;AND PUT INTO BCD
         MOV  M, A        ;REPLACE IT
         DCX  H           ;NEXT PAIR
         DCR  C           ;DONE?
         JNZ  TCOM2
         RET
```

11.6.4 TCOMP.   TCOMP, 10's complement, is a subroutine used to find the 10's complement of 5 bytes stored in memory. Upon entry, HL must point to the last byte of the 5 (i.e., the one with the highest address). TCOMP computes the 10's complement by subtracting the byte in memory from 99 H for all 5 bytes and finishes by adding 1 to the result. The 5 bytes in memory are actually changed to their 10's-complement equivalent. Printout 12-23 shows the TCOMP listing.

11.6.5 ALIGN.   ALIGN is used by both FPADD and FPSUB to align the exponents. ALIGN computes the difference between the exponents and makes up the difference by rotating one of the numbers in DIGA or DIGB to the right.

EXAMPLE 12-8

Align the exponents so that the following numbers can be added: $25E+4$, $7E+3$.

Printout 12-24

```
       ALIGN:   LDA DIGA+6        ;GET FIRST EXPONENT
                MOV C, A          ;SAVE IT HERE
                LDA DIGB+6        ;GET SECOND EXPONENT
                SUB C             ;FIND THE DIFFERENCE
                JM COMP           ;IF MINUS, DIGA > DIGB
                RZ                ;ZERO MEANS ALREADY ALIGNED
                MOV B, A          ;SAVE THE DIFFERENCE
                PUSH B
                MOV D, A          ;INIT THE ROTATE COUNTER
       AGA1:    LXI H, DIGA+5     ;ROTATE DIGA UNTIL D IS ZERO
                CALL RIGHT
                DCR D
                JNZ AGA1
                LDA DIGA+6        ;GET THE OLD EXPONENT
                POP B             ;AND THE DIFFERENCE
                ADD B             ;CORRECT TO NEW EXPONENT
                STA DIGA+6        ;AND SAVE IT
                RET
       COMP:    CMA               ;CORRECT A TO POSITIVE NUMBER
                INR A
                MOV B, A          ;SAVE THE DIFFERENCE
                PUSH B
                MOV D, A          ;INIT THE ROTATE COUNTER
       AGA2:    LXI H, DIGB+5     ;ROTATE DIGB UNTIL D IS ZERO
                CALL RIGHT
                DCR D
                JNZ AGA2
                LDA DIGB+6        ;GET THE OLD EXPONENT
                POP B             ;AND THE DIFFERENCE
                ADD B             ;CORRECT TO NEW EXPONENT
                STA DIGB+6        ;AND SAVE IT
                RET
```

SOLUTION

We first normalize the input numbers to $0.25E+6$ and $0.7E+4$. We must make the smaller number larger by 100 to be able to add the two but at the same time not change the value of the number. To do this, we rotate the 0.7 right twice to get 0.007 and make the exponent larger by 2. Thus, we end up with $0.007E+6$, which we can easily add to $0.25E+6$ to get $0.257E+6$, which is the correct answer.

There is no error checking included in ALIGN to ensure that it changes the correct number. The subroutine always chooses the smaller number, and in some cases this will not be correct. Printout 12-24 shows the ALIGN listing.

11.6.6 UNPAK.   UNPAK is a subroutine used to unpack the 5 packed BCD bytes into 10 separate digits (bytes) and store them in a buffer. Upon entry, HL must point to the first byte of the packed BCD number and DE must point to the first location in the buffer that is to store the unpacked data.

EXAMPLE 12-9

Unpack the number 76 12 45 02 36.

Printout 12-25

```
UNPAK:   MVI C, 5          ;DO 5 BCD PAIRS
NXT:     MOV A, M          ;GET A BCD PAIR
         RRC               ;ROTATE THE LOWER DIGIT OUT
         RRC               ;(PUTS UPPER DIGIT IN LOWER 4 BITS)
         RRC
         RRC
         ANI 0FH
         STAX D            ;PLACE DIGIT INTO BUFFER
         INX D             ;NEXT BUFFER ADDRESS
         MOV A, M          ;GET THE SAME PAIR BACK
         ANI 0FH           ;GET THE LOWER DIGIT
         STAX D            ;PLACE IT INTO THE BUFFER
         INX D             ;NEXT BUFFER ADDRESS
         INX H             ;NEXT DIGIT PAIR
         DCR C             ;DONE?
         JNZ NXT
         RET
```

Printout 12-26

```
RIGHT:   MVI B, 5          ;PREPARE TO ROTATE 5 DIGITS
NXT:     MOV A, M          ;GET A BCD PAIR
         RRC               ;ROTATE OUT LOWER DIGIT
         RRC               ;(PUTS UPPER DIGIT IN LOWER 4 BITS)
         RRC
         RRC
         ANI 0FH
         MOV C, A          ;SAVE IT HERE
         DCR B             ;DONE?
         JZ QUIT
         DCX H             ;NEXT BCD PAIR
         MOV A, M          ;GET IT INTO ACCUMULATOR
         RLC               ;ROTATE OUT UPPER DIGIT
         RLC               ;(PUTS LOWER DIGIT IN UPPER 4 BITS)
         RLC
         RLC
         ANI 0F0H
         ORA C             ;COMBINE UPPER AND LOWER DIGITS
         INX H             ;AND SAVE IT IN THE RIGHT PLACE
         MOV M, A
         DCX H             ;GO TO NEXT DIGIT
         JMP NXT
QUIT:    MOV M, A          ;SAVE THE LAST DIGIT
         RET
```

SOLUTION

The number gets stored in the buffer as follows:

07 06 01 02 04 05 00 02 03 06

UNPAK could be used in a form of output routine, should you wish to write a different one. Printout 12-25 shows the UNPAK listing.

11.6.7 RIGHT. RIGHT is used to rotate a packed BCD number one position to the right and replace the first (left-most) digit with a zero. It is used by both

Printout 12-27

```
ADDER:   LXI  H, DIGA+5      ;FIRST FP NUMBER
         LXI  D, DIGB+5      ;SECOND FP NUMBER
         MVI  C, 5           ;NUMBER OF DIGIT PAIRS
         LDAX D              ;GET 2 BCD DIGITS FROM ONE NUMBER
         ADC  M              ;ADD 2 OTHER BCD DIGITS, PLUS CARRY
         DAA                 ;AND ADJUST BACK INTO BCD
         MOV  M, A           ;SAVE RESULT
         DCX  H              ;ONTO NEXT DIGIT PAIRS
         DCX  D
         DCR  C              ;DONE YET?
         JNZ  ADDER
         RET
```

ALIGN and FPADD; you may find other uses for it, as well. Upon entry, HL must point to the last of the 5 bytes to be rotated. Printout 12-26 shows the RIGHT listing.

11.6.8 ADDER. ADDER is one of the most useful of all the auxiliary subroutines. ADDER is called by FPADD and FPSUB and is used to add the two packed BCD numbers located in DIGA and DIGB. On return, the carry flag represents the overflow out of the left-most position. Printout 12-27 shows the ADDER listing.

11.7 Conclusions

Although Section 8 of Chapter 2 covered basics and fine points of BCD arithmetic, there are other features that you may wish to explore. For instance, a routine called ROUND could be written to round off a packed BCD number if its last digit is greater than or equal to 5. Or, a routine called FPDIS could be written that would decide the format in which a number should be outputted: scientific, or straight decimal with a movable decimal point. You could also modify all the routines to increase accuracy by accepting 6, 7, or more bytes (giving 12, 14, or more digits!) for the number alone (to increase accuracy) and adding 2 bytes at the end for exponent, one for sign and the other for exponent, now as much as 255. Another routine, TRAIL, could eliminate trailing zeros from a floating-point output. All these variations are possible with a little creative thinking, and a lot of time.

12 WAND: A UNIVERSAL PRODUCT CODE LABEL READER

When the supermarket cashier whips your groceries across a glass plate in the counter, causing a bell to ring and numbers to flash on the cash register, you are seeing a bar code reader in action. This device, housed underneath the glass plate, is an electronically controlled laser that constantly scans for an image like the one in Fig. 12-9.

The Universal Product Code (UPC) label, which exists in many forms, appears on most grocery items and on just about everything else that is sold today. The

FIGURE 12-9 Universal Product Code label.

bar code consists of alternate black and white bars of differing widths. The actual width of consecutive black/white bars determines the information contained in the bar code. Every bar code label, like the one in Fig. 12-9, consists of 59 alternating black and white bars. The first three are called the left guard bars. Then come the bars for 6 digits (right, they only display 5!) with each digit represented by two black bars and two white bars. In the middle, there are five more guard bars, three white and two black. Then come the bars for six more digits, and finally the three right guard bars. What a lot of information! This complicated-looking label is actually very simple, however.

12.1 The UPC Label Demystified

As already mentioned, each number in the bar code is made up of four bars, two black ones and two white ones. Though the thickness of each bar can vary, they

FIGURE 12-10 UPC digits broken into modules.

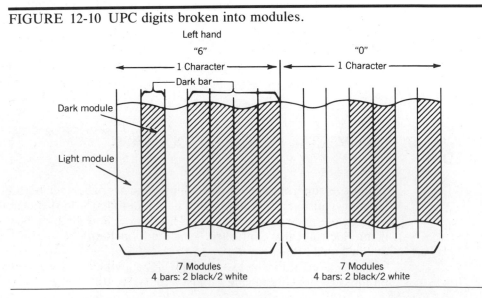

must always occupy a specific space (depending on the size of the label) and this space is always seven modules long (see Fig. 12-10).

A module can be thought of as a bit in UPC label terminology. Therefore, we now know that each digit must, through the use of black and white bars, represent a 7-bit code. Figure 12-10 shows a UPC left-hand 6 and 0, and both numbers occupy seven modules, even though the black bars are of different sizes. In the decoding process, a black bar is read as a one, while a white bar is read as a zero. (This is accomplished by the use of a hand-held bar code reader, such as the reader manufactured by Hewlett-Packard.) Thus, the "6" number decodes as 0 1 0 1 1 1 1 and the "0" decodes as 0 0 0 1 1 0 1. The existence of 7 bits in every number of the UPC label will lend itself well to an 8-bit microcomputer, as we soon shall see. Table 12-7 shows UPC digit binary equivalents.

12.2 Scanning the UPC Label

To read the UPC label, an optical instrument must be used. Since lasers are not in the budget of the hobbyist or experimenter (and certainly not of most students), we must rely on some other form of assistance in reading the UPC label. Hewlett-Packard comes to the rescue with the HEDS-3000 "Digital Wand," consisting of optical sensor, amplifier and digitizer, output transistor, and "push-to-read" switch. As the wand moves across the surface of a UPC label, an LED light source in the optical sensor scans the surface. Black bars absorb the light, while white bars reflect the LED's light onto a photodiode also contained in the optical sensor. The resulting photodiode current is amplified and digitized, and the resulting TTL signal made available as an output. The output remains high as long as the wand "sees" white, or if the push-to-read switch is off. The 1-bit wand signal could be made available to the computer via an 8-bit input port (as in the WAND program) or as an input to SID on the 8085.

12.3 Decoding UPC Data

Gathering the UPC data is a relatively easy task for the microcomputer. We simply look at the wand's input port bit, always looking for a 0–1 or a 1–0 transition. We keep track of the time that the wand "sees" each bar color by incrementing a 16-bit register until the transition to the next level occurs. We then store the

TABLE 12-7   UPC Digit Binary Equivalents

Digit	Left-Hand	Right-Hand
0	0001101	1110010
1	0011001	1100110
2	0010011	1101100
3	0111101	1000010
4	0100011	1011100
5	0110001	1001110
6	0101111	1010000
7	0111011	1000100
8	0110111	1001000
9	0001011	1110100

16-bit time count in a memory buffer area and start over again with a count of zero. When we have accumulated 59 times, we begin processing the UPC data.

Since we are not concerned at this time with vast amounts of error checking (i.e., did you read too fast or too slow?), we skip over the three left guard bar times. Now, because each number in the UPC code is made up of four bars, we look at the next four times stored in the buffer area. By adding them up and dividing by 7, we find the average module time for that number or digit. We then subtract the module time from the stored counts repeatedly. Each time we have to switch to a new bar count, we change the level of the bit that we rotate into the accumulator (see Example 12-10). After seven rotations, we have fully decoded the digit, so we look for it in a table. If we find it, we print the number; otherwise we print an asterisk (*). When all 12 digits have been decoded, we scan another table to see whether it is one of the stored example grocery items in our "store." If it is, we use a message lookup routine and print a description of the product. The same logical sequence occurs in the supermarket, except the computer looks up prices, and you have to give them money.

EXAMPLE 12-10

The following times were recorded for a white/black/white/black sequence:

White	80
Black	245
White	75
Black	170

Decode the UPC code digit (left-hand).

SOLUTION

In accordance with the process described, we must first find the average module time:

$$
\begin{array}{r}
80 \\
245 \\
75 \\
\underline{170} \\
570 \quad \text{total}
\end{array}
\qquad
\begin{array}{r}
81 \\
7\overline{)\,570} \\
\underline{567} \\
3
\end{array}
$$

Since we are not concerned with remainders, we will use 81 as our average module time.

Step 1

Select the first color time, subtract the module time from it, and rotate a zero into the digit register.

$$
\begin{array}{r}
80 \\
\underline{-81} \\
-1
\end{array}
$$

Since we have gone negative, we must be in the second bar; that is, we have observed a white-to-black transition.

Step 2

Select the second color time and subtract the module time from it until done (reaching zero or going negative.)

$$
\begin{array}{cccc}
245 & 164 & 83 & 2 \\
-81 & -81 & -81 & -81 \\
\hline
164 & 83 & 2 & -79
\end{array}
$$

Since we subtracted three times, we rotate three ones (one = black) into the digit register. We now have 0 1 1 1.

Step 3

Select the third color time and subtract the module times from it.

$$
\begin{array}{c}
75 \\
-81 \\
\hline
-6
\end{array}
$$

Since this can be done only once (the software assumes that we *always* have at least one bit per color time; therefore, even though we went negative right away, we get our zero), we rotate one zero into the digit register. We now have 0 1 1 1 0.

Step 4

Fetch the last color time and perform the same operation on it.

$$
\begin{array}{ccc}
170 & 89 & 8 \\
-81 & -81 & -81 \\
\hline
89 & 8 & -73
\end{array}
$$

We were able to do this twice, so we rotate two ones into the digit register. This finally leaves us with 0 1 1 1 0 1 1, which translates to a hexadecimal 3B. A quick look at Table 12-7 reveals we have decoded a UPC "7," and the bar code would look like Fig. 12-11.

FIGURE 12-11 Bar code for a UPC "7."

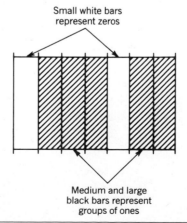

Small white bars represent zeros

Medium and large black bars represent groups of ones

12.4 Forward and Backward

The UPC code is broken into two groups, left-hand digits and right-hand digits. After passing over the three guard bars on the left side (black/white/black), we encounter white/black/white/black digits. When we come up to the middle guard bars (white/black/white/black/white), we switch and encounter black/white/black/white digits; thus we have two separate 10-byte groups for our two different sets of digits. Last, because of the nature of the code, we are able to read forward and even backward getting valuable information either way. The authors have had a WAND program running for more than a year, and our laboratory "store" is becoming quite well stocked with sample label packages. The wand is a most interesting and useful microcomputer application. Printout 12-28 lists the WAND program. Figure 12-12 shows a set of sample items to try on your program!

Printout 12-28

```
                        ;THE 8080 UPC WAND READER


                        ;***VERSION (2) - ADDED ITEM IDENTIFICATION
                        ;***ALL VERSION (2) COMMENTS HAVE 3 STARS(***)



                        ORG 5000H

                        ;IDENTIFY YOURSELF
            START:      LXI SP, STAK+30  ;SET UP THE STACK
                        LXI H, HIMSG     ;GET ADDRESS OF HELLO MESSAGE
                        CALL SEND        ;AND SEND IT

                        ;WAIT FOR WAND ACTIVITY (BLACK TO WHITE)
            BLK:        IN 21H           ;GET WAND STATUS
                        ANI 01H          ;ONLY ON BIT-1
                        JNZ BLK          ;WAND ISN'T ON YET

                        ;DEBOUNCE THE WAND SWITCH (300 MILLI SECONDS)
                        LXI H, 6400H     ;INITIAL COUNT
            DEBO:       DCX H            ;COUNT-DOWN HL
                        MOV A, H
                        ORA L            ;TEST FOR ZERO
                        JNZ DEBO         ;LOOP IF NOT

                        ;WAND IS NOW HOPEFULLY RESTING ON THE
                        ;WHITE GUARD BORDER. WE NOW WAIT 0.5
                        ;SECONDS FOR A WHITE TO BLACK TRANSITION
                        LXI B, 0000H     ;TIMER COUNTER
            WHT:        IN 21H           ;GET THE WAND STATUS
                        INX B            ;INCREMENT TIMER
                        ANI 01H          ;GET THE COLOR BIT
                        JNZ GATHR        ;JUMP IF WE HIT THE BLACK
                                         ;GUARD BAR
                        MOV A, B         ;OTHERWISE, SEE IF THE
                        CPI 0BEH         ;OPERATOR HAS WAITED MORE
                        JNZ WHT          ;THAN 0.5 SECONDS.
                                         ;SLAP HIS WRIST IF HE HAS

                        ;ERROR 1: WHITE DELAY TOO LONG
                        LXI H, MSG1      ;MESSAGE ADDRESS
                        CALL SEND        ;TELL THE OPERATOR
                        JMP START        ;TRY AGAIN
```

Printout 12-28 (continued)

```
                    ;DATA GATHERING SECTION
      GATHR:  MVI C, 3BH          ;THE NUMBER OF BARS TO READ
              LXI H, TIMES        ;STARTING ADDRESS FOR DATA

                    ;BLACK BAR TIMER
      BLKIN:  LXI D, 0000H        ;TIMER COUNTER
      GETB:   IN 21H              ;GET THE WAND STATUS
              INX D               ;INCREMENT TIMER
              ANI 01H             ;GET THE COLOR BIT
              JZ BSAV             ;JUMP IF BLACK TO WHITE
              JMP GETB            ;OTHERWISE CONTINUE
      BSAV:   MOV A, E            ;PUT E INTO ACC
              MOV M, A            ;SAVE IT IN THE DATA TABLE
              INX H               ;PREPARE TO SAVE D
              MOV A, D
              MOV M, A
              INX H
              DCR C               ;ARE WE AT THE LAST BAR?
              JNZ WHTIN           ;NO, CONTINUE

                    ;SEND THE DATA READ MESSAGE
              LXI H, MSG2         ;ADDRESS OF DATA OK MESSAGE
              CALL SEND           ;SEND IT TO THE OPERATOR

                    ;DECODE ALL 12 DIGITS
                    ;FIRST THE LEFT 6
              LXI H, DBUFF        ;DIGIT BUFFER ADDRESS
              SHLD DADR           ;TEMP STORAGE FOR ADDRESS
              LXI H, TIMES+6      ;ADDRESS OF FIRST 4 BAR TIMES
              SHLD NADD           ;SAVE THE BAR ADDRESS
              MVI C, 06H          ;6 LEFT HAND DIGITS
      DEC1:   PUSH B              ;SAVE THE DIGIT COUNT
              CALL DECO           ;DECODE A DIGIT
              LHLD NADD           ;GET THE BAR ADDRESS BACK
              LXI D, 08H          ;PUT 8 INTO DE
              DAD D               ;ADD DE TO HL
              SHLD NADD           ;THIS IS THE NEXT BAR ADDRESS
              LHLD DADR           ;GET THE BUFFER ADDRESS
              MOV C, L            ;TRANSFER INTO BC
              MOV B, H
              LDA DIGX            ;GET THE DECODED DIGIT
              STAX B              ;STORE IT IN THE BUFFER
              INX H               ;NEXT BUFFER ADDRESS
              SHLD DADR           ;SAVE IT
              POP B               ;GET THE DIGIT COUNT
              MOV A, C            ;INTO THE ACCUMULATOR
              DCR A               ;MINUS ONE
              MOV C, A            ;REPLACE C
              JNZ DEC1            ;CONTINUE IF NOT ZERO

                    ;NOW THE RIGHT 6
              LHLD NADD           ;GET THE LAST BAR ADDRESS
              LXI D, 000AH        ;SKIP OVER THE MIDDLE
              DAD D               ;GUARD BARS
              SHLD NADD           ;NEXT BAR ADDRESS
              MVI C, 06H          ;6 MORE DIGITS TO DECODE
      DEC2:   PUSH B              ;SAVE THE DIGIT COUNT
              CALL DECO           ;DECODE ANOTHER DIGIT
              LHLD NADD           ;GET THE BAR ADDRESS BACK
              LXI D, 08H          ;PUT 8 INTO DE
              DAD D               ;ADD DE TO HL
              SHLD NADD           ;NEXT BAR ADDRESS
```

(continued on next page)

Printout 12-28 (continued)

```
                LHLD DADR        ;GET THE BUFFER ADDRESS
                MOV C, L         ;TRANSFER INTO BC
                MOV B, H
                LDA DIGX         ;GET THE DECODED DIGIT
                CMA              ;CORRECT INTO RIGHT HAND CODE
                ANI 7FH          ;CLEAR BIT-8
                STAX B           ;STORE IT IN THE BUFFER
                INX H            ;NEXT BUFFER ADDRESS
                SHLD DADR        ;SAVE IT
                POP B            ;GET THE DIGIT COUNT BACK
                MOV A, C         ;INTO THE ACCUMULATOR
                DCR A            ;COUNT DOWN
                MOV C, A         ;REPLACE C
                JNZ DEC2         ;CONTINUE IF NOT ZERO
                JMP DSCAN        ;SHOW THE RESULTS

                ;DECODE A DIGIT
       DECO:    XRA A            ;CLEAR THE ACCUMULATOR
                STA DIGX         ;CLEAR THE DECODE BYTE
                MVI A, 03H       ;NUMBER OF LOOPS
                STA KNT2         ;SAVE IT
                LHLD NADD        ;GET THE STARTING ADDRESS
                DCX H            ;MINUS TWO
                DCX H
                SHLD ACNT        ;SAVE THIS FOR SUMIT
                CALL NSUM        ;SUM THE BAR TIMES
                SHLD BSUM        ;SAVE IT HERE
                CALL DIV7        ;DIVIDE IT BY 7
                PUSH B           ;SAVE IT ON THE STACK
                POP H            ;GET IT INTO THE HL REG
                PUSH H           ;AND RESAVE IT
                MOV A, L         ;NOW DIVIDE HL BY 2
                RRC              ;THIS IS DONE BY ROTATING
                ANI 7FH
                MOV L, A         ;THE HL REGISTER RIGHT
                MOV A, H         ;ONE BIT.
                STC              ;MAKE SURE THAT THE CARRY
                CMC              ;BIT IS ZERO
                RAR              ;ROTATE H ONE BIT
                MOV H, A         ;AND REPLACE IT
                JNC ROT          ;JUMP IF A 0 WAS ROTATED OUT
                MOV A, L         ;OTHERWISE, STICK A 1 INTO
                ORI 80H          ;THE L REGISTER
                MOV L, A
       ROT:     SHLD SUM         ;THIS IS THE INITIAL SUM
                LXI H, 0000H     ;CLEAR HL
                SHLD SUMR        ;SET THE SUM TO ZERO
       ROT0:    LDA KNT2         ;GET THE COUNT BACK
                DCR A            ;DECREMENT IT
                STA KNT2         ;AND SAVE IT
                JZ ROTE          ;QUIT IF ZERO
                CALL SUMIT       ;LOOK AT THE FIRST BAR TIME
       SUM0:    LHLD SUM         ;GET THE MODULE SUM
                POP D            ;GET THE MODULE TIME BACK
                PUSH D           ;AND RESAVE IT
                DAD D            ;INCREASE THE MODULE TIME
                SHLD SUM         ;AND RESAVE IT
                LDA DIGX         ;GET THE DECODE BYTE BACK
                RLC              ;ROTATE IT LEFT
                STA DIGX         ;RESAVE IT
                XCHG             ;PUT THE SUM IN DE
                LHLD SUMR        ;GET THE BAR SUM
                MOV C, L         ;TRANSFER INTO BC REG
```

Printout 12-28 (continued)

```
                MOV B, H
                MOV A, B        ;PUT THE HIGH BYTE INTO A
                CMP D           ;IS BAR SUM < MODULE SUM
                JC ROT1         ;YES
                JNZ SUMO        ;NO
                MOV A, C        ;MAYBE, CHECK THE LOW BYTE
                CMP E
                JC ROT1
                JNZ SUMO
ROT1:           CALL SUMIT      ;ADD A BAR TIME
SUM1:           LHLD SUM        ;GET THE MODULE SUM
                POP D           ;GET THE MODULE TIME BACK
                PUSH D          ;AND RESAVE IT
                DAD D           ;INCREASE THE MODULE TIME
                SHLD SUM        ;AND RESAVE IT
                LDA DIGX        ;GET THE DECODE BYTE BACK
                ORI 01H         ;PUT A 1 IN
                RLC             ;ROTATE IT LEFT
                STA DIGX        ;RESAVE IT
                XCHG            ;PUT THE SUM IN DE
                LHLD SUMR       ;GET THE BAR SUM
                MOV C, L        ;TRANSFER INTO BC REG
                MOV B, H
                MOV A, B        ;PUT THE HIGH BYTE INTO A
                CMP D           ;IS BAR SUM < MODULE SUM
                JC ROTO         ;YES
                JNZ SUM1        ;NO
                MOV A, C        ;MAYBE, CHECK THE LOW BYTE
                CMP E
                JC ROTO
                JNZ SUM1
                JMP ROTO        ;BACK TO FIRST LOOP
ROTE:           POP PSW         ;FIX THE STACK
                LDA DIGX        ;FIX THE DECODED BYTE
                RRC             ;BY RE-ROTATING IT RIGHT
                ANI 7FH         ;JUST IN CASE
                STA DIGX        ;AND REPLACE IT
                RET

                ;INCREASE THE BAR TIME TOTAL
SUMIT:          LHLD SUMR       ;GET THE SUM BACK
                XCHG            ;PUT IT INTO DE
                LHLD ACNT       ;GET THE ADDRESS BACK
                INX H           ;CORRECT TO THE NEXT
                INX H           ;BAR TIME ADDRESS
                SHLD ACNT       ;AND REPLACE IT
                MOV A, M        ;NOW GET THE BAR TIME
                INX H           ;INTO HL
                MOV H, M
                MOV L, A
                DAD D           ;ADD THIS TO THE TOTAL
                SHLD SUMR       ;THIS IS THE NEW SUM
                RET

                ;WHITE BAR TIMER
WHTIN:          LXI D, 0000H    ;TIMER COUNTER
GETW:           IN 21H          ;GET THE WAND STATUS
                INX D           ;INCREMENT TIMER
                ANI 01H         ;GET THE COLOR BIT
                JNZ WSAV        ;JUMP IF WHITE TO BLACK
                JMP GETW        ;OTHERWISE CONTINUE
WSAV:           MOV A, E        ;PUT E REG INTO ACC
```

(continued on next page)

Printout 12-28 (continued)

```
              MOV M, A          ;SAVE IT IN THE DATA TABLE
              INX H             ;GO TO NEXT ADDRESS
              MOV A, D          ;PUT D REG INTO THE ACC
              MOV M, A          ;AND SAVE IT
              INX H
              DCR C             ;GO TO NEXT BAR
              JMP BLKIN         ;AND CONTINUE

              ;SUM THE BAR TIMES FOR 1 DIGIT
    NSUM:     MVI A, 04H        ;FOUR BARS PER DIGIT
              STA KNT           ;SAVE IT FOR LATER
              LHLD   NADD       ;GET THE DIGIT BASE ADDRESS
              MOV C, L          ;TRANSFER IT INTO BC REG
              MOV B, H
              LXI H, 0000H      ;INITIALIZE THE TIMER SUM TO ZERO
    ADDT:     LDAX B            ;GET THE LOW TIMER VALUE
              MOV E, A          ;TRANSFER IT INTO E REG
              INX B             ;NEXT BYTE
              LDAX B            ;GET THE HIGH TIMER VALUE
              MOV D, A          ;TRANSFER IT INTO D REG
              INX B             ;NEXT LOCATION
              DAD D             ;ADD THIS VALUE TO HL
              LDA KNT           ;GET THE BAR COUNT
              DCR A             ;AND DECREMENT IT
              RZ                ;RETURN IF DONE
              STA KNT           ;OTHERWISE, STORE THE COUNT
              JMP ADDT          ;AND CONTINUE

              ;DIVIDE BY 7
    DIV7:     LXI B, 0000H      ;INITIALIZE RESULT TO ZERO
              LXI D, 0FFF9H     ;2'S COMPLEMENT OF 7
              LHLD BSUM         ;GET THE NUMERATOR INTO HL
    SUB:      DAD D             ;SUBTRACT SEVEN
              INX B             ;COUNT THE NUMBER OF TIMES
              MOV A, H          ;PUT THE HIGH BYTE INTO A
              CPI 00H           ;LESS THAN 256 TO GO?
              JNZ SUB           ;JUMP IF NOT
    SUB2:     MOV A, L          ;PUT THE LOW BYTE INTO A
              CPI 00H           ;EVEN DIVISION?
              RZ                ;RETURN IF YES
              CPI 07H           ;REMAINDER?
              RC                ;RETURN IF #<7
              INX B             ;WE MUST KEEP GOING
              DAD D             ;SUBTRACT 7 MORE
              JMP SUB2          ;HL IS STILL TOO LARGE

              ;DIGIT SEARCH AND DISPLAY
              ;LEFT HAND DIGITS
    DSCAN:    MVI C, 06H        ;6 LEFT HAND DIGITS
              XRA A             ;CLEAR ACCUMULATOR
              STA RERR          ;CLEAR ERROR CONTROL LOCATION
              CALL CRLF         ;SKIP A LINE
              LXI H, DBUFF      ;DIGIT BUFFER ADDRESS
              SHLD TEMP         ;***PUT DBUFF IN TEMP TO
                                ;***PREPARE FOR ITEM DECODE
    NXTD:     PUSH B            ;SAVE THE DIGIT COUNT
              MOV A, M          ;GET AN ENCODED DIGIT
              XCHG              ;SAVE ADDRESS IN DE REG
              LXI H, LS         ;LEFT HAND DIGITS TABLE
              MVI C, 00H        ;START AT DIGIT 0
    DCNT:     CMP M             ;COMPARE ENCODED BYTES
              JZ FND            ;JUMP IF MATCH
              INX H             ;NEXT LOCATION IN TABLE
```

Printout 12-28 (continued)

```
              INR C          ;NEXT DIGIT
              PUSH PSW       ;SAVE THE ACCUMULATOR
              MOV A, C       ;PUT DIGIT NUMBER INTO A
              CPI 0AH        ;GREATER THAN 10?
              JZ DERR        ;GIVE THE DIGIT ERROR
              POP PSW        ;OTHERWISE, CONTINUE
              JMP DCNT       ;TO NEXT DIGIT
      FND:    MOV A, C       ;PUT DIGIT NUMBER INTO ACC
              ADI 30H        ;CORRECT TO ASCII
              MOV B, A       ;GET READY TO SEND
              PUSH H         ;***SAVE HL
              LHLD TEMP      ;***GET DBUFF ADR
              MOV M,B        ;***SAVE ASCII CHARACTER
              INX H          ;***NEXT DBUFF ADR
              SHLD TEMP      ;***RESAVE IT
              POP H          ;***GET HL BACK
              CALL COUT      ;AND SEND TO OPERATOR
      SPC:    MVI B, 20H     ;ASCII SPACE
              CALL COUT      ;SEND IT
              XCHG           ;GET DIGIT ADDRESS BACK
              INX H          ;NEXT DIGIT
              POP B          ;GET THE DIGIT COUNT BACK
              MOV A, C       ;INTO THE ACCUMULATOR
              DCR A          ;NEXT DIGIT
              MOV C, A       ;REPLACE C
              JNZ NXTD       ;CONTINUE IF NOT 6
              JMP DSCN2      ;NOW DO THE RIGHT 6
      DERR:   MVI B, 2AH     ;ERROR SYMBOL
              CALL COUT      ;SEND IT
              MVI A, 0FFH    ;ERROR BYTE
              STA RERR       ;INTO ERROR STATUS LOCATION
              POP PSW        ;FIX THE STACK
              JMP SPC        ;CONTINUE

              ;RIGHT HAND DIGITS
      DSCN2:  MVI C, 06H     ;6 RIGHT HAND DIGITS
              LXI H, DBUFF+6 ;DIGIT BUFFER ADDRESS
              SHLD TEMP      ;***SAVE FOR SECOND HALF
      NXTD2:  PUSH B         ;SAVE THE DIGIT COUNT
              MOV A, M       ;GET AN ENCODED DIGIT
              XCHG           ;SAVE ADDRESS IN DE REG
              LXI H, RS      ;RIGHT HAND DIGITS TABLE
              MVI C, 00H     ;START AT DIGIT 0
      DCNT2:  CMP M          ;COMPARE ENCODED BYTES
              JZ FND2        ;JUMP IF MATCH
              INX H          ;NEXT LOCATION IN TABLE
              INR C          ;NEXT DIGIT
              PUSH PSW       ;SAVE THE ACCUMULATOR
              MOV A, C       ;PUT DIGIT NUMBER INTO A
              CPI 0AH        ;GREATER THAN 10?
              JZ DERR2       ;GIVE THE DIGIT ERROR
              POP PSW        ;OTHERWISE, CONTINUE
              JMP DCNT2      ;TO NEXT DIGIT
      FND2:   MOV A, C       ;PUT DIGIT NUMBER INTO ACC
              ADI 30H        ;CORRECT TO ASCII
              MOV B, A       ;GET READY TO SEND
              PUSH H         ;***SAVE HL
              LHLD TEMP      ;***GET DBUFF+6 ADR
              MOV M,B        ;***SAVE ASCII CHAR
              INX H          ;***NEXT ONE
              SHLD TEMP      ;***RESAVE IT
              POP H          ;***GET HL BACK
```

(continued on next page)

Printout 12-28 (continued)

```
            CALL COUT        ;AND SEND TO OPERATOR
SPC2:       MVI B, 20H       ;ASCII SPACE
            CALL COUT        ;SEND IT
            XCHG             ;GET THE DIGIT ADDRESS BACK
            INX H            ;NEXT DIGIT
            POP B            ;GET THE DIGIT COUNT BACK
            MOV A, C         ;INTO THE ACCUMULATOR
            DCR A            ;NEXT DIGIT
            MOV C, A         ;REPLACE C
            JNZ NXTD2        ;CONTINUE IF NOT 6
            JMP QUIT         ;PREPARE TO END
DERR2:      MVI B, 2AH       ;ERROR SYMBOL
            CALL COUT        ;AND SEND IT
            MVI A, 0FFH      ;ERROR BYTE
            STA RERR         ;INTO ERROR STATUS LOCATION
            POP PSW          ;FIX THE STACK
            JMP SPC2         ;AND CONTINUE
            ;GOODBYE MESSAGE
QUIT:       LDA RERR         ;GET THE ERROR BYTE
            CPI 0FFH         ;WAS THERE A READ ERROR?
            JZ DERR3         ;JUMP IF YES
            CALL CRLF        ;***NEW LINE
            CALL ITEM        ;***HERE WE GO FOR ITEM IDENTIFICATION
            CALL CRLF        ;NEW LINE
            LXI H, BYMSG     ;ADDRESS OF BYE MESSAGE
            CALL SEND        ;SEND IT
            MVI B, 07H       ;ASCII BELL
            CALL COUT        ;BEEP IF A GOOD READ
            JMP START        ;AND DO ANOTHER LABEL


            ;ITEM IDENTIFICATION
ITEM:       MVI C,1          ;***ITEM NUMBER
            LXI H,CODES      ;***REAL CODE DATABLOCK
NXT:        LXI D,DBUFF      ;***ASCII DIGITS READ
            MVI B,12         ;***NUM DIGITS TO MATCH UP
JAMI:       MOV A,M
            CPI '*'          ;***ITEM NOT REGISTERED
            JZ NONE
            XCHG
            CMP M            ;***DIGIT MATCH??
            JNZ ERRX         ;***FAILURE-ADVANCE TO NEXT ITEM
            INX H            ;***NEXT DIGIT REAL CODE
            INX D            ;***NEXT DIGIT ASCII TABLE
            XCHG
            DCR B            ;***OK SO FAR-CHECK ANOTHER DIGIT
            JNZ JAMI         ;***MORE DIGITS IN THIS ITEM
SUCES:      LXI H,GROCS      ;***ADVANCE TO CORRECT  GROC ITEM
CHK:        MOV A,M
            CPI '#'          ;***LAST CHAR?
            JZ FINI
            INX H
            JMP CHK
FINI:       DCR C
            JZ PRINT
            INX H
            JMP CHK
PRINT:      INX H
            CALL SEND
            RET

            ;NOT A STORED ITEM
NONE:       LXI H,OOPS       ;***ITEM NOT REGISTERED
```

Printout 12-28 (continued)

```
                CALL SEND
                RET                 ;***BACK TO WAND
ERRX:           XCHG
NIK:            INX H               ;***NEXT CODE
                DCR B
                JNZ NIK
                INR C               ;***GROCERY COUNT
                JMP NXT
                ;RECOVERY ERROR
DERR3:          CALL CRLF           ;NEW LINE
                LXI H, MSG3         ;LOCATION OF ERROR MESSAGE
                CALL SEND           ;SEND IT OUT
                JMP START           ;AND START OVER

                ;HI MESSAGE
HIMSG:          DB '   *** THE MAGIC WAND ***'

                ;WHITE GUARD DELAY ERROR
MSG1:           DB 'NO ACTIVITY - READ ABORTED#'

                ;OK DATA MESSAGE
MSG2:           DB 'UPC BARS SUCCESSFULLY READ#'

                ;RECOVERY ERROR MESSAGE
MSG3:           DB 'ERROR IN DIGIT RECOVERY#'


                ;BYE MESSAGE
BYMSG:          DB 'RAISIN BRAN --- $1.25#'

                ;MESSAGE OUTPUT ROUTINE
SEND:           CALL CRLF           ;SEND A CR - LF
GET:            MOV A, M            ;PUT CHAR INTO ACC
                CPI 23H             ;IS IT A '#'?
                JZ OUTBK            ;IF YES - JUMP
                MOV B, A            ;MOVE CHAR INTO B REG
                CALL COUT           ;AND SEND IT
                INX H               ;NEXT LETTER
                JMP GET             ;AND CONTINUE
OUTBK:          CALL CRLF           ;SEND A FINAL CR - LF
                RET                 ;AND RETURN

                ;CARRIAGE RETURN - LINE FEED
CRLF:           MVI B, ODH          ;ASCII CR
                CALL COUT           ;SEND IT
                MVI B, OAH          ;ASCII LF
                CALL COUT           ;AND SEND IT
                RET                 ;RETURN TO SEND

                ;COUT DEFINITION
COUT:           SET 200DH           ;NORTHSTAR DOS COUT ADDRESS

                ;COUNT STORAGE
KNT:            DB 00H

KNT2:           DB 00H

                ;DECODED DIGIT STORAGE
DIGX:           DB 00H

                ;ADDRESS STORAGE FOR DECO
ACNT:           DB 00H, 00H
```

(continued on next page)

Printout 12-28 (continued)

```
                ;BUFFER ADDRESS STORAGE
        DADR:   DB 00H, 00H

                ;BAR ADDRESS STORAGE
        NADD:   DB 00H, 00H

                ;ERROR STATUS LOCATION
        RERR:   DB 00H

                ;BAR SUM STORAGE
        BSUM:   DB 00H, 00H

                ;SUM STORAGE
        SUM:    DB 00H, 00H

        SUMR:   DB 00H, 00H

                ;LEFT HAND DIGIT STORAGE
        LS:     DB 0DH, 19H, 13H, 3DH, 23H      ;01234
                DB 31H, 2FH, 3BH, 37H, 0BH      ;56789

                ;RIGHT HAND DIGIT STORAGE
        RS:     DB 72H, 66H, 6CH, 42H, 5CH      ;01234
                DB 4EH, 50H, 44H, 48H, 74H      ;56789

                ;DIGIT BUFFER AREA
        DBUFF:  DS 12               ;ALLOCATE 12 BYTES FOR THE DIGITS

                ;TIMER STORAGE AREA
        TIMES:  DS 32               ;A TOTAL OF 118 LOCATIONS
                DS 32               ;2 FOR EACH BAR IN THE UPC
                DS 32               ;LABEL
                DS 22

        TEMP:   DS 2                ;***STORAGE OF GROCERY ITEM ADR
        STAK:   DS 16               ;SAVE ROOM FOR THE STACK

                ;STACK AREA
                DS 30

                ;STORED UPC LABEL CODES
        CODES:  DB '011160048320'
                DB '050000120727'
                DB '381370037194'
                DB '043000794807'
                DB '071700215612'
                DB '043000793749'
                DB '041000014673'
                DB '012345678905'
                DB '078425701276'
                DB '012000101519'
                DB '041333177052'
                DB '012546923064'
                DB '038000317002'
                DB '071700030215'
                DB '043032004127'
                DB '041192911040'
                DB '071020007119'
                DB '076021001050'
                DB '071009003507'
                DB '071330005621'
                DB '*'              ;***LAST ONE
```

Printout 12-28 (continued)

```
              ;UPC LABEL DESCRIPTIONS
    GROCS:     DB  '#'           ;***MARKS FIRST ITEM IN GROCERY LIST.
               DB  'G UNION ORANGE JUICE - $ 1.69#'
               DB  'CARNATION HOT COCOA - $.89#'
               DB  'JOHNSONS BABY SHAMPOO - $1.49#'
               DB  'MAXWELL HOUSE INST COFFEE '
               DB  '- $3.59#'
               DB  'CROWLEY HALF AND HALF PINT '
               DB  '- $.59#'
               DB  'SANKA INSTANT COFFEE - $2.99#'
               DB  'LIPTON CUP A SOUP - $.59#'
               DB  'PRINTED UPC TEST CODE#'
               DB  'GRAND UNION TEST TAG#'

               DB  'PEPSI LIGHT 6-16OZ - $2.29#'
               DB  'DURACELL ALKALINE BATTERY - $3.20#'
               DB  'GOOD & PLENTY CANDY 1.5 OZ  - $.25#'
               DB  'KELLOG S POP TARTS - $.85#'
               DB  'CROWLEY HOMO MILK - $.79#'
               DB  'BRAND X TOILET TISSUE - $.33#'
               DB  'MURRAY GINGER COOKIES - $.98#'
               DB  'DAIRYLEA ORANGE JUICE - $1.49#'
               DB  'WELDIT CEMENT - $.87#'
               DB  'SHOGUN BY JAMES CLAVELL - $3.50#'
               DB' FREIHOFERS OATMEAL RAISIN DROP'
               DB  ' COOKIES - $1.43#'

              ;NO ITEM MESSAGE
    OOPS:      DB  'NEW ITEM NOT YET REGISTERED#'
               END
```

FIGURE 12-12 Bar codes for sample grocery items.

(continued on next page)

FIGURE 12-12 (continued)

0 50000 12072

0 41000 01467

0 71330 00562

0 41192 91104

13 A DIGITAL VOLTMETER

This section deals with what might be a good example of design overkill, the use of a microprocessor in a fancy analog-to-digital converter scheme. The resulting digital voltmeter illustrates how a microcomputer can be used to read an unknown analog voltage. Many types of transducers produce voltages that must be converted to a set of values based on a data table or a specific algorithm. This method uses an incremental approach to determine the unknown voltage.

13.1 The Analog-to-Digital Converter Circuit

The external circuitry uses an MC1408L8 8-bit D-to-A converter. This allows 256 possible voltages, given the digital input from zero through 255. A comparator is used to determine when the step is equal to the input (unknown) amplitude.

Figure 12-13 shows the circuit using the converter, labeled DAC. (The circuitry is left incomplete to aid you in understanding the principle.) In principle, the eight bit lines to the converter are driven from the 8-bit port of a microcomputer. If a binary count is sent to the D/A converter, the output of the DAC should be an increasing voltage in the form of a staircase waveform. The size of each step is generally the reference voltage divided by the number of steps.

$$\text{step size} = \frac{V_{\text{ref}}}{2^N} = \frac{10}{256} = 0.0390625$$

FIGURE 12-13 Basic diagram of D/A converter circuitry.

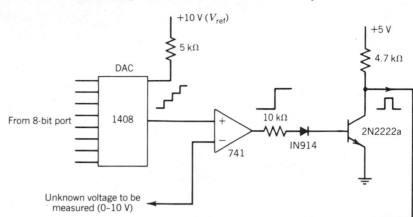

For an 8-bit DAC with a 10-V reference, this would be 39.0625 mV per step. This type of conversion uses a comparator to signal when the staircase has reached a voltage that compares with the unknown voltage. For example, if the unknown voltage is 1.875 V, the D/A is stepped from zero in 39.0625-mV steps until a staircase voltage of 1.875 V is reached or exceeded. At this time the output of the comparator goes high, indicating that the conversion is complete. Where is the digital result? At the 8-bit DAC input. This occurs after 48 steps (i.e., 1.875/0.0390625).

13.2 Using the Resulting Information

After the digital value has been found, it must be made to mean something. After all, 1.875 V has little similarity to an 8-bit binary count of 48 D = 30 H = 00110000 B.

It is desirable to print the result on a terminal or other display. To do this, the quantity 30 H must be converted to a set of ASCII values. One way is to store the constant 0.0390625 in memory as a set of BCD numbers (i.e., 39, 06, 25 in BCD). As the staircase is generated, the quantity 0.0390625 is added to itself until the conversion is complete. Table 12-8 represents the values ending at 1.875.

The trick is to stop incrementing values as soon as the output of the comparator goes high. When the result is reached, it is in packed hexadecimal (i.e., 18, 75, 00, 00 = 0001 1000, 0111 0101, 0000 0000, 0000 0000). This must be converted to a desired form for display (perhaps ASCII). Your programming experience includes the necessary techniques to implement the A-to-D converter, and this is left to you as an exercise.

TABLE 12-8   A/D Accumulated Values

Count	Accumulated Value
0	0
1	0.0390625
2	0.0781250
3	0.1171875
4	0.1562500
5	0.1953125
.	.
.	.
.	.
.	.
47	1.8359375
48	1.8750000

13.3 Analog-to-Digital Conversion Time

It is usually important to convert an analog value rapidly to a digital value. This is the "conversion time" of an A-to-D converter. The incremental technique thus far described requires a length of time equivalent to that required to step the DAC from zero to the unknown value. If the value is near zero, only a few steps are needed and the conversion time is small. If the unknown voltage is near 10 V, then most of the 256 steps will be needed and the conversion time will be great. It is generally desirable to convert in the shortest period of time possible. A better technique uses a process of *halving* to determine the unknown voltage. This technique, as we shall see, requires a maximum of eight outputs to the D/A converter.

If the unknown value is still 1.875 V, the comparator will be used to indicate whether the 8-bit quantity is higher or lower than the unknown voltage. A one on the comparator output means that the unknown is lower, a zero means that the unknown is higher.

We know that the correct result is 48 D = 30 H. The range of outputs is 0 D to 255 D or 0 H to FF H. The procedure is as follows:

1. Output 80 H: Comparator is high, indicating 0–5 V.

2. Output 40 H: Comparator is high, indicating 0–2.5 V.

3. Output 20 H: Comparator is low, indicating 1.25–2.5 V.

4. Output 30 H: Comparator is high. We know they are equal, but we keep going; output is 1.875–2.5 V.

5. Output 35 H: Comparator is high, indicating 1.875–2.020875 V.

6. Output 32.5 H (losing the .5): Comparator is high, indicating 1.875–1.953125 V.

7. Output 31 H: Comparator is high, indicating 1.875–1.941625 V.

8. Output 30 H: Comparator is high; output is 1.875 V.

As you can see, the increment is halved at each step and only eight steps are needed to find the result. This compares with 48 D steps using the incremental technique, clearly an improvement in the conversion process.

Implementing the conversion can be done in several ways. Since a hex result is obtained, the actual result of 1.875 V may be arrived at by doing 48 D repetitive additions of the constant 0.0390625. Since this would require time that may not be available, you may want to compute the result as the conversion process is taking place. This is left to your imagination and as an exercise in programming.

13.4 An Independent Analog-to-Digital Converter

The preceding examples emphasize the use of a microcomputer as an integral part of the A-to-D converter. This may be practical in some applications where the microprocessor is available for such duties. If a lot of A-to-D data is required, however, a separate A-to-D converter circuit may be more desirable.

Figure 12-14a shows a continuous conversion process implemented using an MC1408L8 DAC and a pair of SN74193N TTL counters; a key to the labels appears in Fig. 12-14b. The purpose of this A-to-D is to take a reading from a potentiometer (speed control) and convert the analog quantity to an 8-bit digital value. This digitized value is used in a standard delay loop to vary the speed of a disco light show or other such display.

A rapid conversion is obtained, since the counters are allowed to count either up or down as necessary to "track" any change in potentiometer setting. Points 7–14 are the digital output.

13.5 Commercially Available Converters

There are a wide variety of both D-to-A and A-to-D converters commercially available, with a corresponding wide range of prices. A few are listed here:

D/A
MC1408L8 (also DAC0808LCN): 8 bits
DAC1022LCN: 10 bits
DAC1222LCN: 12 bits

A/D
ADC0800: 8 bits
ADC1210: 12 bits

Figure 12-15, courtesy of National Semiconductor Corporation, defines terms using A/D and D/A converters and presents some sample data sheets.

13.6 Generating Waveforms

It is a relatively simple operation to produce a sinusoidal waveform using a D-to-A converter. The key is to read a table of data values that represent the amplitude of a sine wave. The bottom of the waveform is taken as 00, and the top is FF. The normal center or zero crossing would be at 80 H. Figure 12-16 shows a sine wave.

Printout 12-29 shows the data values derived from a simple basic program that

FIGURE 12-14 (*a*) Circuit for continuous A/D speed control. (*b*) Parts list for circuit.

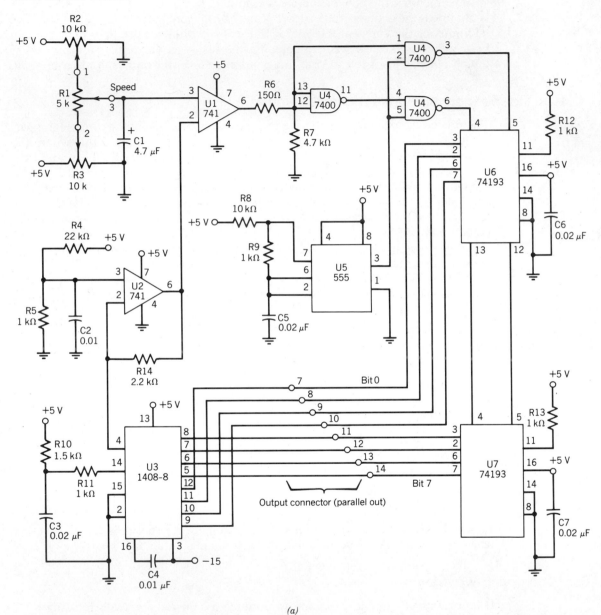

(*a*)

Integrated Circuits

U1, U2	741 Op amp
U3	MC1408-8 D/A converter
U4	SN7400N
U5	555 Timer
U6, U7	SN74193N Counter

Resistors

R1	5-kΩ Linear potentiometer
R2, R3	10-kΩ Linear potentiometer
R4	22-kΩ, 0.25-W Carbon composition
R5, R9, R11,	1-kΩ, 0.25-W Carbon composition
R12, R13	" " "
R6	150 Ω, 0.25 W
R7	4.7 kΩ, 0.25 W
R8	10 kΩ, 0.25 W
R10	1.5 kΩ, 0.25 W
R14	2.2 kΩ, 0.25 W

Capacitors

C1	4.7-μF, 10-VDC electrolytic
C2, C4	0.01-μF, 50-VDC ceramic disc
C3, C5	0.02-μF 50-VDC ceramic disc
C6, C7	0.02-μF 50-VDC ceramic disc

(b)

FIGURE 12-15 National Semiconductor data sheets. (Copyrighted material reprinted with permission of NATIONAL SEMICONDUCTOR CORPORATION.)

 **National Semiconductor**

Definition of Terms

Accuracy: Sum of all errors: non-linearity, zero-scale, full-scale, temperature drift, etc. Careful–this term is sometimes confused with resolution and/or non-linearity.

Conversion Time: The time required for a complete measurement by an A/D converter.

Full-Scale Error: Deviation from true full-scale output when specified reference voltage is applied.

Full-Scale Tempco: Change in scale error due to temperature, usually expressed in parts per million per degree (ppm/°C).

Monotonicity: A DAC whose output always increases for increasing digital input codes is said to be monotonic, i.e., does not decrease at any point.

Non-Linearity: Worst-case deviation from the line between the endpoints (zero and full-scale). Can be expressed as a percentage of full-scale or in fractions of an LSB. ± 1/2 LSB is a desirable specification.

Power-Supply Sensitivity: The sensitivity of a converter to DC changes in power-supply voltages is normally expressed in terms of percentage change in analog input value. Power-supply sensitivity may also be expressed in relation to a specified DC shift of the supply voltage.

Quantizing Error: ± 1/2 LSB error inherent in all A/D conversions. Cannot be eliminated.

Ratiometric Converter: The output of an A/D converter is a digital number proportional to the ratio of (some measure of) the input to a reference. Most requirements for conversions call for an absolute measurement, i.e., against a fixed reference. In some cases, where the measurement is

affected by a changing reference voltage, it is advantageous to use that same reference as the reference for the conversion, to eliminate the effect of variation.

Resolution: The most important converter specification. This is the number of steps the full-scale signal can be divided into, and therefore the size of the steps. May be expressed as the number of bits in the digital word, the size of a least significant bit (smallest step) as a percent of full-scale, or an LSB in millivolts (for a given full-scale).

Bits	Steps (2N)	LSB Size (% of Full-Scale)	LSB Size (10V Full-Scale)
6	64	1.588%	158.8 mV
8	256	0.392%	39.2 mV
10	1,024	0.0978%	9.78 mV
12	4,096	0.0244%	2.44 mV
14	16,384	0.0061%	0.61 mV
16	65,536	0.0015%	0.15 mV

Settling Time: Time from change in input until output remains within ± 1/2 LSB (or some specified percentage) of final output.

3 1/2 Digit BCD: Maximum output count or display is ± 1.999 (± 2000 counts) — approximately 11 binary bits plus sign.

3 3/4 Digit BCD: Maximum output count or display is ± 3.999 (± 4000 counts) — approximately 12 binary bits plus sign.

National Semiconductor

ADC0800 8-Bit A/D Converter

General Description

The ADC0800 is an 8-bit monolithic A/D converter using P-channel ion-implanted MOS technology. It contains a high input impedance comparator, 256 series resistors and analog switches, control logic and output latches. Conversion is performed using a successive approximation technique where the unknown analog voltage is compared to the resistor tie points using analog switches. When the appropriate tie point voltage matches the unknown voltage, conversion is complete and the digital outputs contain an 8-bit complementary binary word corresponding to the unknown. The binary output is TRI-STATE® to permit bussing on common data lines.

The ADC0800PD is specified over −55°C to +125°C and the ADC0800PCD is specified over 0°C to 70°C.

Features

- Low cost
- ±5V, 10V input ranges
- No missing codes
- Ratiometric conversion
- TRI-STATE outputs
- Fast $T_C = 50 \mu s$
- Contains output latches
- TTL compatible
- Supply voltages   5 V_{DC} and −12 V_{DC}
- Resolution   8 bits
- Linearity   ±1 LSB
- Conversion speed   40 clock periods
- Clock range   50 to 800 kHz

Block Diagram

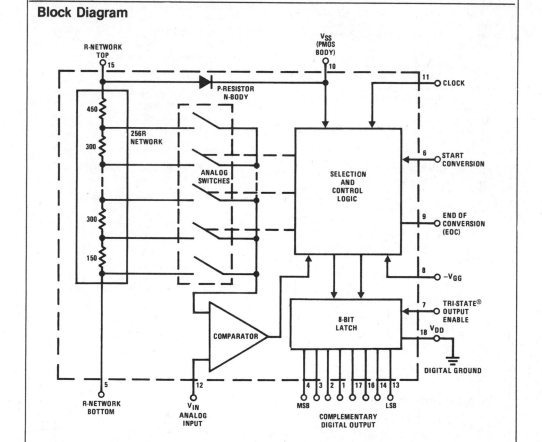

(00000000 = +full-scale)

(continued on next page)

FIGURE 12-15 (continued)

National Semiconductor

A to D, D to A

DAC0800, DAC0801, DAC0802 8-Bit Digital-to-Analog Converters

General Description

The DAC0800 series are monolithic 8-bit high-speed current-output digital-to-analog converters (DAC) featuring typical settling times of 100 ns. When used as a multiplying DAC, monotonic performance over a 40 to 1 reference current range is possible. The DAC0800 series also features high compliance complementary current outputs to allow differential output voltages of 20 Vp-p with simple resistor loads as shown in *Figure 1*. The reference-to-full-scale current matching of better than ±1 LSB eliminates the need for full-scale trims in most applications while the nonlinearities of better than ±0.1% over temperature minimizes system error accumulations.

The noise immune inputs of the DAC0800 series will accept TTL levels with the logic threshold pin, V_{LC}, pin 1 grounded. Simple adjustments of the V_{LC} potential allow direct interface to all logic families. The performance and characteristics of the device are essentially unchanged over the full ±4.5V to ±18V power supply range; power dissipation is only 33 mW with ±5V supplies and is independent of the logic input states.

The DAC0800, DAC0802, DAC0800C, DAC0801C and DAC0802C are a direct replacement for the DAC-08, DAC-08A, DAC-08C, DAC-08E and DAC-08H, respectively.

Features

- Fast settling output current   100 ns
- Full scale error   ±1 LSB
- Nonlinearity over temperature   ±0.1%
- Full scale current drift   ±10 ppm/°C
- High output compliance   −10V to +18V
- Complementary current outputs
- Interface directly with TTL, CMOS, PMOS and others
- 2 quadrant wide range multiplying capability
- Wide power supply range   ±4.5V to ±18V
- Low power consumption   33 mW at ±5V
- Low cost

Typical Applications

FIGURE 1. ±20 Vp-p Output Digital-to-Analog Converter

Connection Diagram

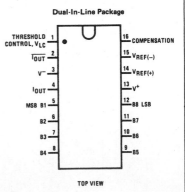

Dual-In-Line Package

TOP VIEW

Ordering Information

NON LINEARITY	TEMPERATURE RANGE	ORDER NUMBERS*					
		D PACKAGE (D16C)		J PACKAGE (J16A)		N PACKAGE (N16A)	
±0.1% FS	−55°C ≤ T_A ≤ +125°C	DAC0802LD	DAC-08AQ				
±0.1% FS	0°C ≤ T_A ≤ +70°C			DAC0802LCJ	DAC-08HQ	DAC0802LCN	DAC-08HP
±0.19% FS	−55°C ≤ T_A ≤ +125°C	DAC0800LD	DAC-08Q				
±0.19% FS	0°C ≤ T_A ≤ +70°C			DAC0800LCJ	DAC-08EQ	DAC0800LCN	DAC-08EP
±0.39% FS	0°C ≤ T_A ≤ +70°C			DAC0801LCJ	DAC-08CQ	DAC0801LCN	DAC-08CP

*Note. Devices may be ordered by using either order number.

FIGURE 12-15 (continued)

Absolute Maximum Ratings

Supply Voltage	±18V or 36V
Power Dissipation (Note 1)	500 mW
Reference Input Differential Voltage (V14 to V15)	V^- to V^+
Reference Input Common-Mode Range (V14, V15)	V^- to V^+
Reference Input Current	5 mA
Logic Inputs	V^- to V^- plus 36V
Analog Current Outputs	Figure 24
Storage Temperature	$-65°C$ to $+150°C$
Lead Temperature (Soldering, 10 seconds)	300°C

Operating Conditions

	MIN	MAX	UNITS
Temperature (T_A)			
DAC0802L	−55	+125	°C
DAC0800L	−55	+125	°C
DAC0800LC	0	+70	°C
DAC0801LC	0	+70	°C
DAC0802LC	0	+70	°C

Electrical Characteristics (V_S = ±15V, I_{REF} = 2 mA, $T_{MIN} \leq T_A \leq T_{MAX}$ unless otherwise specified. Output characteristics refer to both I_{OUT} and $\overline{I_{OUT}}$.)

	PARAMETER	CONDITIONS	DAC0802L/ DAC0802LC MIN	TYP	MAX	DAC0800L/ DAC0800LC MIN	TYP	MAX	DAC0801LC MIN	TYP	MAX	UNITS
	Resolution		8	8	8	8	8	8	8	8	8	Bits
	Monotonicity		8	8	8	8	8	8	8	8	8	Bits
	Nonlinearity				±0.1			±0.19			±0.39	%FS
t_s	Settling Time	To ±1/2 LSB, All Bits Switched "ON" or "OFF", T_A = 25°C		100	135					100	150	ns
		DAC0800L					100	135				ns
		DAC0800LC					100	150				ns
t_{PLH}, t_{PHL}	Propagation Delay	T_A = 25°C										
	Each Bit			35	60		35	60		35	60	ns
	All Bits Switched			35	60		35	60		35	60	ns
TCI_{FS}	Full Scale Tempco			±10	±50		±10	±50		±10	±80	ppm/°C
V_{OC}	Output Voltage Compliance	Full Scale Current Change < 1/2 LSB, R_{OUT} > 20 MΩ Typ	−10		18	−10		18	−10		18	V
I_{FS4}	Full Scale Current	V_{REF} = 10.000V, R14 = 5.000 kΩ R15 = 5.000 kΩ, T_A = 25°C	1.984	1.992	2.000	1.94	1.99	2.04	1.94	1.99	2.04	mA
I_{FSS}	Full Scale Symmetry	$I_{FS4} - I_{FS2}$		±0.5	±4.0		±1	±8.0		±2	±16	μA
I_{ZS}	Zero Scale Current			0.1	1.0		0.2	2.0		0.2	4.0	μA
I_{FSR}	Output Current Range	V^- = −5V	0	2.0	2.1	0	2.0	2.1	0	2.0	2.1	mA
		V^- = −8V to −18V	0	2.0	4.2	0	2.0	4.2	0	2.0	4.2	mA
	Logic Input Levels											
V_{IL}	Logic "0"	V_{LC} = 0V			0.8			0.8			0.8	V
V_{IH}	Logic "1"		2.0			2.0			2.0			V
	Logic Input Current	V_{LC} = 0V										
I_{IL}	Logic "0"	$-10V \leq V_{IN} \leq +0.8V$		−2.0	−10		−2.0	−10		−2.0	−10	μA
I_{IH}	Logic "1"	$2V \leq V_{IN} \leq +18V$		0.002	10		0.002	10		0.002	10	μA
V_{IS}	Logic Input Swing	V^- = −15V	−10		18	−10		18	−10		18	V
V_{THR}	Logic Threshold Range	V_S = ±15V	−10		13.5	−10		13.5	−10		13.5	V
I_{15}	Reference Bias Current			−1.0	−3.0		−1.0	−3.0		−1.0	−3.0	μA
dI/dt	Reference Input Slew Rate	(Figure 24)	4.0	8.0		4.0	8.0		4.0	8.0		mA/μs
$PSSI_{FS+}$	Power Supply Sensitivity	$4.5V \leq V^+ \leq 18V$		0.0001	0.01		0.0001	0.01		0.0001	0.01	%/%
$PSSI_{FS-}$		$-4.5V \leq V^- \leq 18V$ I_{REF} = 1 mA		0.0001	0.01		0.0001	0.01		0.0001	0.01	%/%
	Power Supply Current	V_S = ±5V, I_{REF} = 1 mA										
$I+$				2.3	3.8		2.3	3.8		2.3	3.8	mA
$I-$				−4.3	−5.8		−4.3	−5.8		−4.3	−5.8	mA
		V_S = 5V, −15V, I_{REF} = 2 mA										
$I+$				2.4	3.8		2.4	3.8		2.4	3.8	mA
$I-$				−6.4	−7.8		−6.4	−7.8		−6.4	−7.8	mA
		V_S = ±15V, I_{REF} = 2 mA										
$I+$				2.5	3.8		2.5	3.8		2.5	3.8	mA
$I-$				−6.5	−7.8		−6.5	−7.8		−6.5	−7.8	mA
P_D	Power Dissipation	±5V, I_{REF} = 1 mA		33	48		33	48		33	48	mW
		5V, −15V, I_{REF} = 2 mA		108	136		108	136		108	136	mW
		±15V, I_{REF} = 2 mA		135	174		135	174		135	174	mW

Note 1: The maximum junction temperature of the DAC0800, DAC0801 and DAC0802 is 125°C. For operating at elevated temperatures, devices in the dual-in-line J or D package must be derated based on a thermal resistance of 100°C/W, junction to ambient, 175°C/W for the molded dual-in-line N package.

Printout 12-29

```
> DUMP 0 OFF
        0  1  2  3  4  5  6  7  8  9  A  B  C  D  E  F
0000  82 85 88 8B 8E 91 94 97 9B 9E A1 A4 A7 AA AD AF    ................
0010  B2 B5 B8 BB BE C0 C3 C6 C8 CB CD D0 D2 D4 D7 D9    ................
0020  DB DD DF E1 E3 E5 E7 E9 EB EC EE EF F1 F2 F4 F5    ................
0030  F6 F7 F8 F9 FA FB FB FC FD FD FE FE FE FE FE FF    ................
0040  FE FE FE FE FE FD FD FC FB FB FA F9 F8 F7 F6 F5    ................
0050  F4 F2 F1 EF EE EC EB E9 E7 E5 E3 E1 DF DD DB D9    ................
0060  D7 D4 D2 D0 CD CB C8 C6 C3 C0 BE BB B8 B5 B2 AF    ................
0070  AD AA A7 A4 A1 9E 9B 97 94 91 8E 8B 88 85 82 7E    ................"
0080  7B 78 75 72 6F 6C 69 66 62 5F 5C 59 56 53 50 4E    {xurolifb_\YVSPN
0090  4B 48 45 42 3F 3D 3A 37 35 32 30 2D 2B 29 26 24    KHEB?=:7520-+)&$
00A0  22 20 1E 1C 1A 18 16 14 12 11 0F 0E 0C 0B 09 08    " ..............
00B0  07 06 05 04 03 02 02 01 00 00 00 00 00 00 00 00    ................
00C0  00 00 00 00 00 00 00 01 02 02 03 04 05 06 07 08    ................
00D0  09 0B 0C 0E 0F 11 12 14 16 18 1A 1C 1E 20 22 24    .............. '$
00E0  26 29 2B 2D 30 32 35 37 3A 3D 3F 42 45 48 4B 4E    &)+-0257:=?BEHKN
00F0  50 53 56 59 5C 5F 62 66 69 6C 6F 72 75 78 7B 7F    PSVY\_bfilorux{.
> LOG CON0
```

FIGURE 12-16 Sine wave.

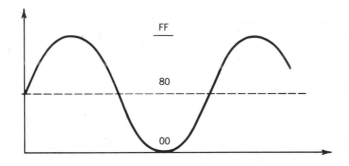

computed sin(*x*) over a range of 360° and split the data into 256 data points. These
were then normalized to range from 0 to FF.

It is simple enough to write a routine to read the data and output it to the D-
to-A converter. The upper frequency is limited by the number of data points and
the speed of the processor. You may be somewhat disturbed to learn that this
produces a waveform in the range of only 200 Hz. Slow!

13.7 Conclusion

It might be said that the only things "digital" in the world are computers. Every
other thing that occurs in nature is analog—temperature, pressure, distance, stress,
light, and sound are all analog signals. The interface between these real signals
and digital computers is the job of A-to-D and D-to-A converters. They occupy
a unique position in the field of digital electronics in their ability to connect the
analog world to the computer.

14 A LARGE MULTIPLEX DISPLAY SYSTEM

We are seeing greatly increased use of large display systems to attract attention and transmit information in the form of advertising for a hotel, entertainment for a discotheque, scores for a sports arena, or basic information purposes. These large displays consist of many hundreds, sometimes thousands, of incandescent lamps. A large multicolor display found on Times Square in New York City has more than 10,000 light bulbs arranged in the familiar red-blue-green triad used in modern color television receivers. Such a display dissipates several hundred kilowatts and is controlled entirely by computer techniques. The images or words or moving displays are purely the manipulation of bits in a large memory. We will examine a small 4-kW display controlled by an 8085 CPU.

14.1 MUX, A Multiplexed Display

This section describes an array of 160 incandescent lamps. These 25-W units are arranged as 10 rows of 16 lamps each, on a 4 ft × 10 ft panel. Each lamp is operated using a solid state relay (SSR) to interface the low-voltage computer levels with the 120-V a-c line. For information purposes, the display is fed from a 230-V, 20-A source. In other words, each half of the display (80 lamps) is fed from a separate 20-A circuit breaker.

The display requires 160 solid state relays, one per lamp. An additional function of the SSR is to completely isolate the computer from the higher voltages for the protection of the computer. The life of the lamps can be greatly extended by switching the lamps on only as the a-c line crosses zero. This occurs 120 times each second. SSRs can be purchased that accomplish zero crossing detect, but it is much cheaper to have the computer sample the line and output data to the display only when the line crosses zero. This both greatly extends the life of the lamps and eliminates any RF noise generated by abrupt switching of the a-c line. Figure 12-17 shows zero cross switching.

Each lamp will be driven by a solid state relay and each relay will be driven by a data latch to hold one bit of data. The data bit will be strobed into the data latch by the microcomputer as in Figure 12-18.

FIGURE 12-17 (*a*) Correct and (*b*) incorrect zero cross switching.

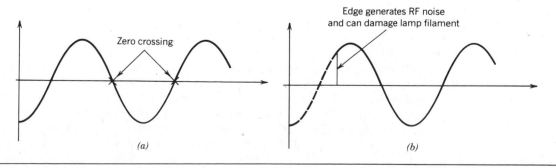

FIGURE 12-18 One-bit memory to drive lamp.

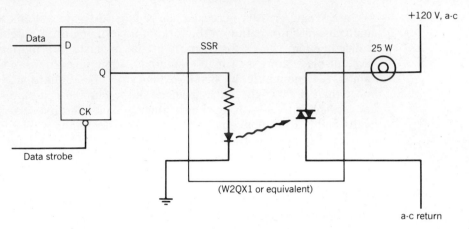

(W2QX1 or equivalent)

The display of 160 lamps, and accompanying SSRs and memory, is driven from two 8-bit ports from a microcomputer. One port contains 8-bit data and the other port selects an 8-bit address. Each address can select either 8 or 16 lamps in a given row: left half, right half, or both halves in parallel.

Since 160 lamps are dynamically driven from two ports having only 16 wires, the display is referred to as a multiplexed display. Typically, in operation, data is sent to the display (8 bits) and then an address is selected to determine which group of eight lamps receives the data. Figure 12-19 shows the multiplex light display system.

The basic circuitry is shown in Figure 12-20. A pair of 7442 BCD-to-ten decoders are used as address decoders. Hex inputs (A–F) are not used and, in fact, are used as deselects on the latch enable inputs. Each data bit must drive 20 data lines because data lines are commoned (i.e., the first and ninth columns are on the same data line); therefore, 8 bits of data are buffered, using a pair of 7475 quad data latches. Only one of 20 lamps gets data, since only one 8-bit segment is normally addressed at a time. A byte is addressed by using peripheral interface adapter (PIA) port B, which drives a pair of 7442 BCD-to-decimal decoders. These select either a byte on the left side of the display or a byte on the right half of the display. Since the display is selected in halves (left or right), it is possible to turn on one or two bulbs at a time.

Figure 12-21 shows a sample programming sheet to determine the necessary data to produce a specific light pattern. For instance, it is desired to produce "8080" in light. The data to do this is shown at the right of the layout. It is a matter of sending the data to the MUX display one byte, one address at a time. A program to output one full "slide" of light follows in Printout 12-30. A program to output many "slides" would be more involved.

A further adaptation involves moving an image on the display. Up, down, sideways, and diagonally are easily implemented through programming. It is suggested that 20 bytes of memory on the computer be manipulated and then trans-

FIGURE 12-19 A multiplex light display system.

MUX is a dynamically selected multiplex light display consisting of 10 rows of 16 lamps each.

```
0    0 0 0 0 0 0 0 0 0 0 0 0 0 0 0 0    0
1    0 0 0 0 0 0 0 0 0 0 0 0 0 0 0 0    1
2    0 0 0 0 0 0 0 0 0 0 0 0 0 0 0 0    2
3    0 0 0 0 0 0 0 0 0 0 0 0 0 0 0 0    3
4    0 0 0 0 0 0 0 0 0 0 0 0 0 0 0 0    4
5    0 0 0 0 0 0 0 0 0 0 0 0 0 0 0 0    5
6    0 0 0 0 0 0 0 0 0 0 0 0 0 0 0 0    6
7    0 0 0 0 0 0 0 0 0 0 0 0 0 0 0 0    7
8    0 0 0 0 0 0 0 0 0 0 0 0 0 0 0 0    8
9    0 0 0 0 0 0 0 0 0 0 0 0 0 0 0 0    9
```

Each row is split into 2 "bytes."

The MUX display system can be thought of as consisting of 20 bytes of 8 lamps each.

Each byte has an address associated with it. The byte at the upper left is byte 0-. The byte at the upper right is -0.

0		0
1		1
2		2
⋮		⋮
9		9

ferred to the MUX memory. This would be done at some convenient rate to allow for the illusion of motion.

Byte addresses range from 00 to 99, with AA to FF ignored or nonselecting: AA is used as an address to select *no* byte; 0A will select the byte in the upper left; A0 will select upper right; 00 will select both bytes, a useful feature.

The displays in this example are driven by a PIA; there are two eight bit data ports (E4, E5, and E7 control). Port A (E4) is data (inverted output). Port B (E5) is byte address. Since there may be a length of cable between the computer and the MUX memory, it is suggested that delays be included to allow settling time between output (about five DCRs).

FIGURE 12-20 A multiplex display memory and decode (insert shows circuitry typical of individual squares).

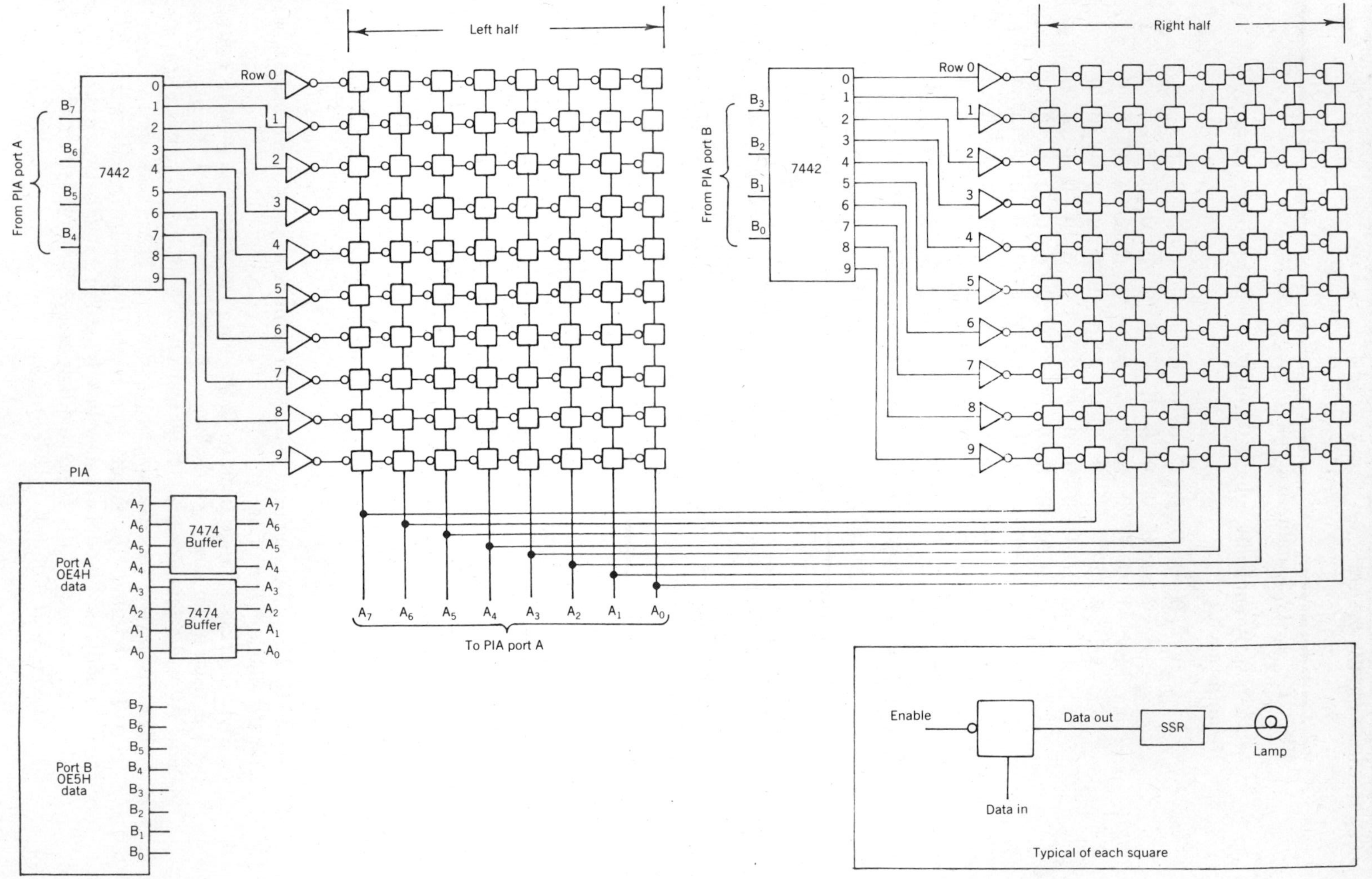

FIGURE 12-21 A multiplex display programming sheet.

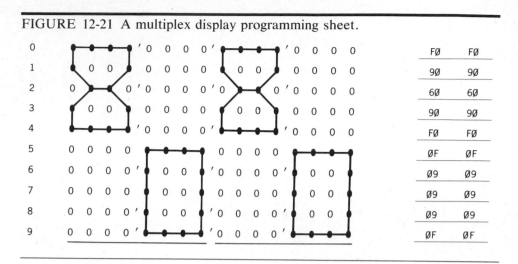

Printout 12-30

```
                        ;THIS IS A DEMONSTRATION PROGRAM TO
                        ;SHOW HOW TO OUTPUT DATA TO THE 160 LAMP
                        ;MULTIPLEXED DISPLAY SYSTEM (MUX).
                        ;
                        ;TWO PARALLEL PORTS ARE USED ON
                        ;A NATIONAL SBC 80/10 MICRO COMPUTER.
                        ;PORT A FOR DATA AT 0E4H
                        ;PORT B FOR BYTE ADDRESS AT 0E5H
                        ;CONTROL PORT = 0E7H
                        ;
                        ;THIS SYSTEM USES AN EXTERNAL DRIVER/
                        ;LATCH MATRIX CONSISTING OF 7475 TTL
                        ;QUAD LATCH CHIPS.
                        ;
                        ;
                        ;
                        ;
                        ORG 3C00H           ;
                        LXI SP,3FEDH        ;
                        MVI A,80H           ;
                        OUT 0E7H            ;PIA SET UP
                        ;
                        MVI B,10            ;1ST TEN BYTES OF DISPLAY
                        LXI H,LOC           ;START OF DATA TO BE DISPLAYED
                        MVI C,0AH           ;LEFT SIDE DISPLAY ADDRESS
                        ;
          MORE:         MOV A,M             ;
                        CMA                 ;OUTPUT ON PORT A IS INVERTED
                        OUT 0E4H            ;DATA TO DISPLAY
                        CALL SLEEP          ;ALLOW TIME DELAY FOR PULSE SETTLING
                        MOV A,C             ;
                        OUT 0E5H            ;
                        ADI 10H             ;NEXT BYTE ADDRESS
                        MOV C,A             ;
                        CALL SLEEP          ;
                        MVI A,0AAH          ;TURN OFF BYTE SELECT
                        OUT 0E5H            ;
                        CALL SLEEP          ;
```

(continued on next page)

Printout 12-30 (continued)

```
                INX H           ;
                DCR B           ;
                JNZ MORE        ;
                                ;
                                ;
                MVI B,10        ;SECOND HALF OF 20 BYTES
                MVI C,0A0H      ;RIGHT SIDE NOW
MORE2:          MOV A,M         ;
                CMA             ;
                OUT 0E4H        ;
                CALL SLEEP      ;
                MOV A,C         ;
                OUT 0E5H        ;
                ADI 1           ;RIGHT HALF
                MOV C,A         ;
                CALL SLEEP      ;
                MVI A,0AAH      ;TURN OFF BYTE SELECT
                OUT 0E5H        ;
                CALL SLEEP      ;
                INX H           ;
                DCR B           ;
                JNZ MORE2       ;
                JMP 8           ;RETURN TO SBC MONITOR
                                ;
SLEEP:          MVI D,5         ;5 DECREMENTS FOR SLEEP TIME
EEP:            DCR D           ;
                JNZ EEP         ;
                RET             ;

LOC:            DB 0FEH,0FEH,0C0H,0C0H,0FEH,0FEH,0C0H,0C0H,0FEH,0FEH
                DB 0FFH,0FFH,18H,18H,18H,18H,18H,18H,18H,18H

                END
```

EXAMPLE 12-11

A typical program sequence for 1 byte.

Output inverted data
    Delay
Output byte address
    Delay
Disable output byte address

Figure 12-22 shows the suggested write timing to the MUX display. To put out the pattern C7 to the byte at bottom left:

```
MVI A, 0C7H   ; Pattern
CMA
OUT 0E4H      ; To port A
CALL SLEEP    ; Delay
MVI A, 9AH    ; Select lower left byte
OUT 0E5H
```

FIGURE 12-22 Write timing.

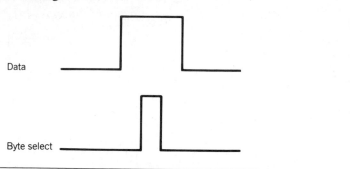

```
        CALL SLEEP
        MVI A, 0AAH   ; Disable byte select
        OUT 0E5H
```

This "sleep" routine (five DCRs) will allow the byte-select pulse to be in the center of the data pulse. External 7475 data latches hold the data to enable the selected lights.

```
        SLEEP:  MVI D,5
        DUD:    DCR D
                JNZ DUD
                RET
```

The series of MUX displays in Fig. 12-23 illustrates the type of display clever programming can create. This affords an ideal example of the usefulness of the microcomputer when combined with clever programming.

14.2 A MUX Controller

The software to actively control a display such as MUX is an endeavor within the realm of users of this text. A three-color display could be constructed, but it is suggested that the monochromatic variety be implemented first. In writing a controller program, the following features may be considered.

1. The program should feature a way to return to the controlling monitor.
2. There should be a way to restart the program with initial or default parameters.
3. The program could include a way to change the speed of the display while it is operating.
4. An option should be included to alter the basic stored patterns.
5. The programmer could consider changing the number of times that a pattern is repeated before proceeding to the next pattern.
6. The program could run a series of routines when unattended but respond to specific requests for routines when desired.
7. A routine that simply tests all of the lamps should always be included.

FIGURE 12-23 Sample multiplex displays. (*a*) 8080. (*b*) USA*. (*c*) ET. (*d*) Sine wave. (*e*) μp. (*f*) Snoopy. (*g*) Arrow. (*h*) Spider.

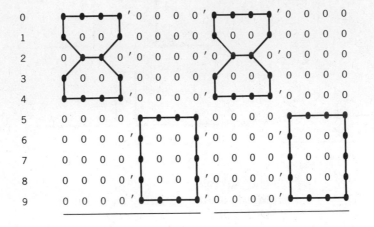

(*a*)

(*b*)

FIGURE 12-23 (continued)

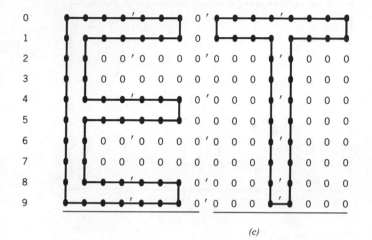

(c)

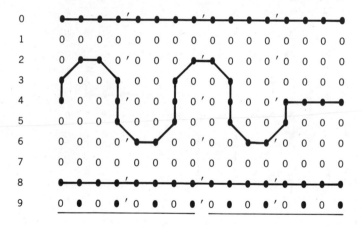

(d)

(continued on next page)

FIGURE 12-23 (continued)

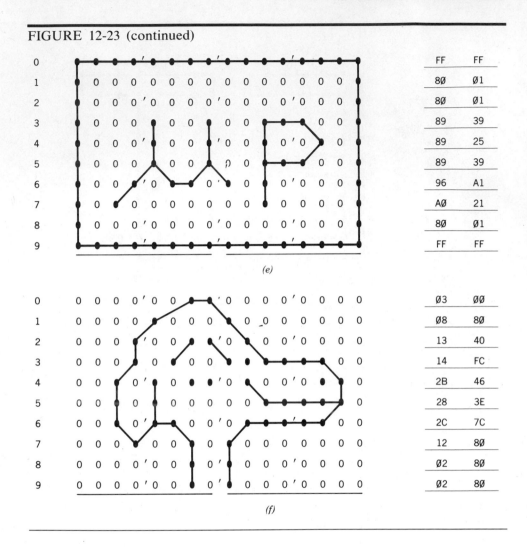

(e)

(f)

8. A single keystroke on the keyboard could complement the pattern, thus providing an inverse image.

9. A routine could be included that sends hex data from the keyboard to the display.

10. An option to set a motion bit to move the images around on the "slide show" is an interesting feature.

11. A routine to transfer keyboard data to the lamps in ASCII code could be written.

12. The addition of a character generator would allow sequential ASCII data to be sent to the display in "Times Square" fashion.

13. A number of interesting displays could be created using box-type shapes.

FIGURE 12-23 (continued)

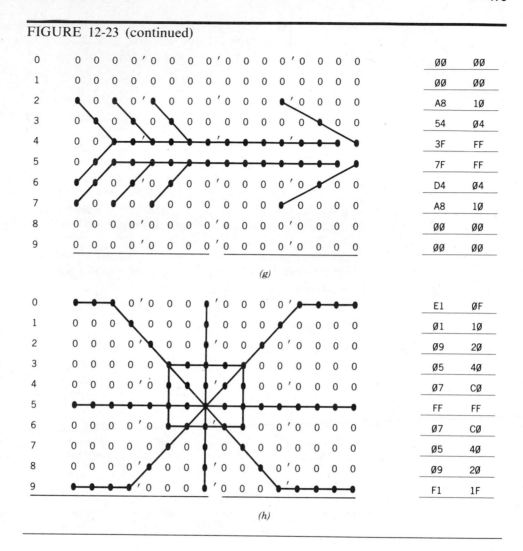

(g)

(h)

14. An opportunity to adjust the "sleep" time would allow the programmer to study the effects of timing on the latch circuitry.

15. A routine to constantly send pulses to the latches would allow the programmer to use an oscilloscope to observe the multiplexing operations.

16. A routine to allow for manual change of patterns and data should be included.

17. The addition of a HELP (H) command is always useful to an operator of a multiplex display. It should summarize the commands to which the display responds.

As you might surmise, this type of program has already been implemented by the authors on a large working version of MUX. The program is omitted here, but the suggestions should aid in creation of software to drive your own multiplexed display. To build and program such a display is really a team effort and makes an ideal student/faculty project. It's also a great application of microprocessors, a most interesting recruiting tool, and a lot of fun!

14.3 Variable Speed

The rate at which motion occurs on such a display depends directly on the rate at which 160 bits of data are sent to the MUX memory. The illusion of motion occurs at fewer than 10 "slides" or "frames" per second. A software delay loop will easily do the task, but variable delay is always a desirable feature. We further suggest the addition of an A-to-D converter for controlling the speed at which the display changes its pattern and thus appears to move—the user adjusts a linear potentiometer, and the resistance is changed from a linear signal to an 8-bit digital signal. The software reads the data from a third port and uses this binary quantity to establish a delay.

The circuit in Figure 12-14a uses an 8-bit up/down counter and an MC1408L8 D/A converter to continuously track the position of the potentiometer. Hence, it converts an analog level to a digital quantity.

14.4 Using a Character Generator

Since this application will inevitably occur to the serious student, we now discuss turning the MUX display into a Times Square display. In this versatile type of display, words (letters) are continuously sent to the display and the characters are slid sideways from right to left. In this way, words, sentences, and messages may be seen on the display.

With 16 bits across, it is possible to show 2.5 characters at a time when each character is represented using a 5×7 dot matrix. It is desirable to enter ASCII data via a terminal and to have the characters appear on the display in a 5×7

FIGURE 12-24 2513 Character generator.

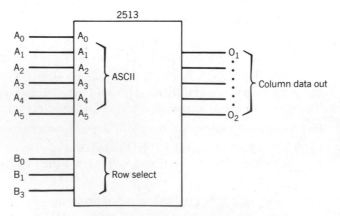

dot matrix format. While the data for each character could be stored in a large (huge, $5 \times 7 \times 64$ characters) data table, it makes more sense to utilize a standard character generator ROM. This example uses a readily available 2513 5×7 character generator. This ROM has uppercase letters only and works well within the format of MUX. A pair of I/O ports are used to read the ROM and a third I/O port accepts column data (i.e., row 1, columns 1–5). Figure 12-24 shows the 2513 character generator.

A subroutine can be used to send the ASCII code to the 2513 (A_0–A_5). Then by applying row data (0–6) to the row select lines, obtain the column data a row at a time, thus reading the character data from the ROM. The data is then placed in memory to be manipulated and output to the MUX display.

15 DOWNLINE LOADING

15.1 Transfer of Program Data from a Host Computer to a User Computer

When a microcomputer program is written for use by a microcomputer, the program in mnemonic form is prepared on another computer referred to as the host computer. This is usually a large, general-purpose machine, perhaps capable of servicing many users at one time. The host computer can be used to prepare a source file, which is then assembled to produce a file of hexadecimal information. This "hex" file contains the machine language (opcodes) required by the microcomputer. This string of hexadecimal information must next be loaded into the memory of the microcomputer on which the program is to be run.

This transfer of program data can be accomplished in several ways. One way is to put the program on an EPROM or PROM and carry the program to the microcomputer. Another method might be the manual entry of each instruction into the microcomputer. This section describes the electronic transfer of program data from the host computer to the microcomputer. This is done using serial data transmission techniques and can be quite efficient and fast.

For example, the following program generates a string of hexadecimal data when assembled:

```
          ORG 2000 H
CIN:      XRA A
          IN OFF H
          ANI 02
          JZ CIN
          IN OFF H
          OUT OFF H
          RET
          END
```

The hexadecimal string may look like this:

AF DB FF E6 02 CA 00 20 DB FF D3 FF C9

Figure 12-25 shows two computers in a downline load.

The data is generated as a string of ASCII characters. We may refer to this as a hex file or a hex string, but that is wishful thinking. An ASCII A contains 8 bits

FIGURE 12-25 Two computers in a downline load.

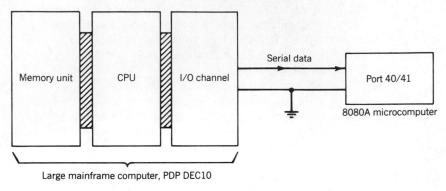

(41 H) and an ASCII F contains 8 bits (46 H). However, the opcode for XRA A is AF. These are clearly dissimilar types of data:

$$ASCII \text{ "A"} = 01000001 \text{ B}$$
$$ASCII \text{ "F"} = 01000110 \text{ B}$$
$$XRA \text{ A} = 10101111 \text{ B}$$

These two types of data, ASCII pairs and packed hexadecimal, are very different. The data must be converted from one form to another before data can be exchanged between computers.

15.2   The Loader Program

The conversion of data to packed hexadecimal (opcode form ready for use) can be done either at the transmission end or at the reception end of the data line. The computer that assembled the source program and produced the ASCII hex file (AF DB FF . . . , etc.) can do the conversion before transmitting any data. If this is the case, each byte of data coming down the line can be loaded sequentially directly into the microcomputer's memory.

If the data is not converted at the sending end and arrives instead as a string of ASCII characters, it must be converted before it is loaded into sequential memory locations.

Let us examine the "simple" case first, in which the real work has already been done by the computer that originally assembled the hex file. In this case, the bytes of program data (not ASCII) will be coming down the line in correct form (i.e., AF H) and can be loaded directly into sequential memory locations on the microcomputer.

The actual electronic load does not take place automatically. It requires a simple machine language program to receive the data and place it into sequential memory locations. We will refer to this program as the "loader." It resides in the microcomputer at the receiving end of the line.

```
             TITLE 'DOWNLINE LOADER'
             ORG 1000 H
             LXI H, 2000 H    ; Data load address
      CIN:   IN 40 H          ; Get USART status
             ANI 02           ; Mask RRDY bit
             JZ CIN           ; Wait
             IN 41 H          ; Get data byte
             MOV M, A         ; Store in HL address
             INX H            ; Next location
             JMP CIN          ; Continue
             END
```

This simple "loader" program is entered on the microcomputer, and data will be taken from the serial port as it comes down the line and will be stored in sequential memory locations. This "loader" program must be running as data is sent down the line from the "host" computer.

This data is received in machine language form. It represents the opcodes required by the microcomputer, and two addresses are involved:

1000 H, address of loader itself

2000 H, address of program being loaded (PROGRAM ORG)

This program can be loaded at any desired address by modifying the LXI H, 2000 H instruction.

The hex file for loader is as follows:

21 00 20 DB 40 E6 02 CA 03 10 DB 41 77 23 C3 03 10

This is a short program and does no data conversion. Its sole purpose is to place received data bytes into sequential memory locations. (*Note*: This type of data is not ASCII and as such will not be displayed correctly on a CRT display.)

15.3 An Improved Loader

A slightly better loader program accomplishes two more interesting functions:

1. The location at which data is to be loaded will be read from the port FF switch port. Since this is an eight-switch port, only the high-order 8 bits of the address will be read.

2. As data is received, a counter will increment and be displayed on the port FF lights. This will serve to measure the size of the file being loaded and also will act as an activity indicator.

```
             TITLE 'BETTER LOADER'
             ORG 1000 H
             MVI B, 0      ;Counter
             MOV L, B      ;Zero L register
             IN 0FF H      ;Get MSB of load address
             MOV H, A      ;Save in H
      CIN:   IN 40 H
             ANI 02
```

```
        JZ CIN
        IN 41 H
        MOV M, A
        INX H
        INR B        ;Increment counter
        MOV A, B     ;Transfer to A
        OUT 0FF H    ;Out to light port
        JMP CIN
        END
```

15.4 An Even Better Loader

Loaders can get fancier and fancier. For instance, as soon as the program is loaded, control can be immediately transferred to the starting address of the program just loaded. To do this, the loader must detect a sequence of characters to indicate that the data has been completely transmitted. Here are two ways of doing this. Both involve placing a "trigger" character or sequence of characters at the end of the hex file.

1. Send an unused opcode (e.g., 0DD H).
2. Send a sequence (e.g., 'STOP', 53 H, 54 H, 4F H, 50 H).

Method 1 has the advantage of simplicity. It is easy to implement in the loader program:

```
        CPI 0DD H
        JZ 2000H     ;Start address
```

However, any DD in the file can cause premature (and disastrous) program execution. We are taking our chances with this method! But it does work 99.8% of the time!

Method 2 is almost foolproof. How often does the sequence 'STOP' come up in a program? Not very often! However, it lengthens the loader considerably. This can be a problem if the loader is being entered manually, but not if it is coming from disk storage.

These "trigger" characters are entered in the main program (host computer) at the end of the program. They must appear at the end of the hex file:

```
        DB 0DD H              DB 'STOP'
                        or
        END                   END
```

To implement detection of the sequence of characters 'STOP', it is necessary to test first for an S. If the next character is not a T, we retest for an S. If the first character is S and the second is T, we test for O. If we don't get an O, we start over! 'STOP' must be received in exact sequence to start running the program. It is left to you to implement this.

Another feature that could be added to a loader program is information at the start of the hex file. A good thing to send is the starting address of the program! Then each program is loaded automatically at its current starting address. The source program might look like this:

```
                    TITLE 'IC TESTER PROGRAM'
                    DW 1000 H
                    ORG 1000 H
            TOP:    LXI H, DATA
                     ⋮
```

The resulting hex file would begin:

00 10 21 2B 14 ...

The first 2 bytes are the starting address of the program to be loaded. (*Note:* The DW 1000 H goes before the ORG statement. This keeps all the program addresses correct.)

The loader would have to be modified to use the first two bytes received for the HL register.

```
            ORG 1000H
            LXI SP, STACK
            CALL GTBYT      ;Get a byte
            MOV L, A
            CALL GTBYT
            MOV H, A         ;HL now has start address
    GET:    CALL GTBYT
            MOV M, A
            INX H
            JMP GET
GTBYT:      IN 40H
            ANI 02
            JZ GTBYT
            IN 41H
            RET
            DS 10
STACK:      DB 0
            END
```

This version is purposely simple for illustrative purposes. A program combining the following features is left to the student to write and debug:

1. Auto start.

2. Auto increment of port FF lights.

3. Auto run of loaded program.

All these loaders assume that the data coming from the host computer is in packed hexadecimal format. If this is not the case, if the host is not converting the data from pair to packed hex form, that conversion must also be done by the loader before transfer into sequential memory locations.

15.5 Converting from ASCII to Packed Hexadecimal Form

If the hex file is coming down the line in ASCII form, with each pair of ASCII characters forming the machine language code, conversion at the micro end by the load is required. Each ASCII character (8 bits) must first be converted to 4-

TABLE 12-9   ASCII Character to Four-Bit
Hexadecimal Conversion

ASCII Character	4-Bit Hexadecimal
0 = 30 H	0 = 0000 B
1 = 31 H	1 = 0001 B
2 = 32 H	2 = 0010 B
3 = 33 H	3 = 0011 B
4 = 34 H	4 = 0100 B
5 = 35 H	5 = 0101 B
6 = 36 H	6 = 0110 B
7 = 37 H	7 = 0111 B
8 = 38 H	8 = 1000 B
9 = 39 H	9 = 1001 B
A = 41 H	A = 1010 B
B = 42 H	B = 1011 B
C = 43 H	C = 1100 B
D = 44 H	D = 1101 B
E = 45 H	E = 1110 B
F = 46 H	F = 1111 B

bit hexadecimal as in Table 12-9. Next, the 4-bit hexadecimal characters must be put together in pairs.

Received ASCII characters are first tested to determine whether they fall in the 0–9 range (digits) or the A–F range (letters). Since an ASCII "zero" is 30 H, we subtract the bias of 30 H from each character. If the ASCII character is a digit, what is left over after removal of the bias is the correct hex form. Otherwise, subtracting 7 provides the correct hex value.

EXAMPLE 12-12

An ASCII C equals:

$$\begin{array}{r} 43\text{ H} \\ -30\text{ H} \\ \hline 13\text{ H} \\ -7\text{ H} \\ \hline 0\text{C H} \end{array}$$

Let's look at the program sequence:

```
CONVRT:   CALL CIN     ;Get character to A
          SUI 30H      ;Remove ASCII bias
          CPI 10
          CM LTRS      ;Letter if minus
          RLC          ;Put in MSB
          RLC
          RLC
          RLC
```

```
                              MOV C, A    ;Save in C register
                              CALL CIN    ;Get second character
                              SUI 30H
                              CPI 10
                              CM LTRS
                              ORA C       ;Result in A
                              RET
                    LTRS:     SUI 7       ;Remove letter bias
                              RET
                    CIN:      IN 40H      ;Get USART status
                              ANI 02      ;RRDY = one?
                              JZ CIN      ;No, wait
                              IN 41H      ;Get character
                              ANI 7FH     ;Set parity bit = zero
                              RET
```

Combining this conversion routine with the loader routine presented previously allows downline loading of a microcomputer when the host is transmitting serial ASCII data. Since CONVRT gets data as needed via the CIN routine, a complete loader would minimally look like this:

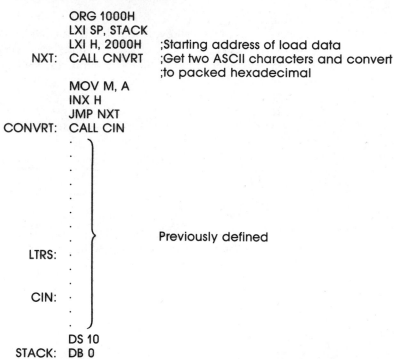

```
                    ORG 1000H
                    LXI SP, STACK
                    LXI H, 2000H     ;Starting address of load data
          NXT:      CALL CNVRT       ;Get two ASCII characters and convert
                                     ;to packed hexadecimal

                    MOV M, A
                    INX H
                    JMP NXT
          CONVRT:   CALL CIN
                      .
                      .
                      .
                      .
                      .
                      .
                      .                    Previously defined
                      .
          LTRS:       .
                      .
                      .
          CIN:        .
                      .
                      .
                    DS 10
          STACK:    DB 0
                    END
```

15.6 Loading: A Final Word

The concept of sending data or programs in machine language form from one computer to another is an important one. Often, a program that has been assembled on a large general-purpose computer must be loaded into the memory of a remote microcomputer. Seventy-five percent of the programs in this book were done in this manner! The form that the data takes may either be ASCII pair or packed hexadecimal. The conversion from ASCII to packed hex can be done at either end of the line. However, only true machine language (packed hex) can be loaded into microcomputer memory and executed.

When a program in machine language form is downline loaded, it must start in the correct (intended) location, and subsequent data must go into sequential locations. This starting location may be specified at the head of the data for the convenience of loading. In addition, the technique can be made more complete by detecting the end of the file and automatically starting execution of the program just loaded.

All this is accomplished by first placing a "DOWNLINE LOADER" routine into the microcomputer. This "loader" supervises the reception, conversion, and loading of data into the microcomputer. An important concept indeed!

16 SORTING

It often is necessary to sort a group of numbers into ascending (increasing) or descending order. In the following application an 8080A/85 routine is used to sort 16 positive (0–255) numbers from lowest to highest. The important aspect of a sort routine is keeping track of how the numbers being sorted are compared. In this way, negative numbers can be included in the group, and they will end up in the proper place.

The procedure used in SORT is to check two consecutive numbers in the group. If the second number is less than the first, the numbers are swapped in memory.

Printout 12-31

```
          ORG     1000H
  SORT:   MVI     E,15        ;prepare for 15 passes
  CONT1:  MOV     D,E         ;init loop counter
          LXI     H,NUMS      ;HL points to data
  CONT2:  MOV     A,M         ;get a data value
          INX     H
          CMP     M           ;compare it with the next value
          JC      SKIP        ;jump if second value is larger
          MOV     B,M         ;otherwise, swap the two numbers
          DCX     H
          MOV     M,B
          INX     H
          MOV     M,A
  SKIP:   DCR     D           ;done with pass?
          JNZ     CONT2
          DCR     E           ;done with sort?
          JNZ     CONT1
          RET
  NUMS:   DS      16          ;save room for 16 values
          END
```

Thus after each pass through the group, the highest, second highest, third highest move to the end of the group.

17 SAVING BYTES

Oftentimes a machine code program must be shortened to fit into a memory space of a certain size. This is extremely important if the program is to be burned into a ROM. How then can a program that is 1035 bytes long fit into a 2708 (1K $\times$ 8)? When faced with this problem, you review the entire program and look for ways to save bytes. Examples 12-13, 12-14, and 12-15 represent basic ideas a good programmer should keep in mind.

EXAMPLE 12-13

The following loop is used to clear a block of memory:

```
CLEAR:  LXI H, 2000H   ;Starting address
        LXI D, 1000H   ;Number of locations to clear
LOOP:   MVI M, 0
        INX H
        DCX D
        MOV A, D
        ORA E          ;DE = 0?
        JNZ LOOP
```

By rewriting the loop slightly, we can accomplish the same task:

```
CLEAR:  LXI H, 2000H   ;Starting address
        MVI A, 30H     ;Ending high-order address
LOOP:   MVI M, 0
        INX H
        CMP H          ;H = 30 H?
        JNZ LOOP
```

Notice that we have saved 3 bytes by rewriting the loop (the first method took 15, the second 12).

EXAMPLE 12-14

There are 37 CALL COUT instructions in a program. In what way can an RST be used to save bytes?

SOLUTION
Instead of 37 CALL COUT instructions, we merely use RST 2. We save 37 $\times$ 2 = 74 bytes. If we then place a JMP COUT at the RST 2 location (10 H), we lose 3 bytes but still save 71 bytes. That is enough for a good 8080 subroutine.

EXAMPLE 12-15

A programmer has a TTY status port set up as follows:

Port 7

7   0

R	X	X	X	X	X	X	T

If bit 7 is high, a character has been received. If bit 0 is high, it is OK to transmit. The programmer codes the port in this way:

```
IN 7
ANI 80H
JNZ REC
IN 7
ANI 01H
JNZ TRAN
```

Making use of the 8080A/85 flags, the programmer can save 4 bytes by rewriting as follows:

```
IN 7
ORA A
JM REC
RAR
JC TRAN
```

The three examples just given suggest that you should always examine your programming procedure. Work for ways to save bytes and make full use of the flags and registers. Make *them* work for *you*.

18 SUMMARY

This chapter has looked at a variety of programming examples dealing with interesting applications of a microcomputer system. The techniques presented apply to unlimited other uses for a microcomputer. All the examples are from real situations, and a thorough study of them will benefit any reader. They are food for thought and should stir the imagination.

Use of the software UART can save an extra package in some printed circuit board designs. Often a computer can be tested with no more than a UART and a data terminal. Under normal conditions, no terminal is required. It would be costly to put a UART on every board if needed only for troubleshooting. When the requirement arises, a monitor ROM can be inserted containing the necessary routines.

The security system, software clock, and phone dialer can be combined to produce a sophisticated surveillance system. The clock would allow the logging of the date and time of any event detected by the polling routines. In practice, we have a 16-door alarm system in operation complete with a data logger. Each coming and going is logged, along with the time of the event.

HEXDUMP and the disassembler are invaluable tools in debugging programs under development. They can show program errors due to poor loading or just plain incorrect thinking.

The sections on BCD arithmetic and sorting are very useful. Frequently a microcomputer in a control application must do a bit of data crunching. A full-blown math package including trig functions is unnecessary, undesirable, and perhaps expensive. We are forced to write some short routines like those presented to do the math required.

The math done may be related to interpreting digital data derived from an analog-to-digital converter. The section on the digital voltmeter introduced A/D and also D/A techniques. Frequently the D/A converter is used to generate analog waveforms of specific types. For example, we might want to generate a ramp to drive a digital plotter or the drive signals for a servo motor. Not only do these motors drive plotters, but they play a very important part in "positioners" such as X–Y positioning tables for drafting purposes, wire wrap tables, or even component insertion equipment. D/A converters are also called on to generate waveforms to approximate human speech. The application of bandpass filters can shape the resulting waveforms as required. The data sent to the D/A converter is subject to certain optimization algorithms determined by the microprocessor software.

A/D and D/A converters figure into many other exciting areas including the following:

1. *Speech Recognition.* Not an easy application; however, an ideal use for a dedicated microcomputer. The recognition of words can free the hands of a test console operator. Recognition of a key word ("execute," "load," "number one," "test sequence," "abort," etc.) causes a specific subroutine or sequence to be activated. Speech recognizers have been around for half a decade and are commonly available. They use microprocessors.

2. *Digitizing Video Waveforms.* The standard television signal is generated in a camera by sweeping an electron beam over a phosphor surface. The typical TV scan of 512 lines in $\frac{1}{30}$ sec can be digitized easily so that an image can be stored in a computer for later recall. The A-to-D converter does this. On remote missions like photographing the surface of the planet Mars, slow-scan television is used to give the A-to-D time to digitize the data, which is transmitted when the scan is completed.

3. *Video Games.* Entirely microprocessor driven. Many use X–Y type displays where the X and the Y axes are separately driven by high-speed, 12-bit D-to-A converters. Everything that you see is under the control of a microprocessor performing the functions specified by a cleverly written program. The sounds are generally the product of the computer and a bit of analog circuitry. Game inputs are probably potentiometers or switches. The pots drive A-to-D converters so that the microprocessor can determine the position of the control.

Returning to the examples presented in the chapter, the UPC bar code reader is an item of immense interest. Nearly all consumer goods in grocery stores are now identified using the familiar symbols of UPC. The wave of a wand or the flash of a laser beam can quickly and accurately measure the width of each light and dark bar, convert this information to a set of ASCII numbers, and consult a lookup table for information about the product. A running inventory can be made for restocking and inventory purposes as the sale of each item is registered.

The section on downline loading details the intricacies of sending data from one place to another. In this application the data is for a microcomputer program that has been written and assembled on a host computer. The resulting object file (opcodes) is sent via a wire into the intended microcomputer. This is a very common industrial occurrence where many computers around a plant control small operations. A large central computer communicates with the remote processors when needed. This type of computer control is referred to as "distributed" processing. Its advantage is that if the main computer or one of the remote computers goes down or malfunctions, the plant is not shut down. Each of the smaller computers continues to perform its task and the company remains in operation. The section on downline loading introduces you to these necessary techniques and includes data conversion when necessary.

This type of data transfer is also used in telemetering. There may be a set of temperature, pressure, radiation, and wind velocity gauges on the top of a mountain or in a satellite orbiting the earth. The necessary data can be transmitted in serial form via either wire (a neat trick with a satellite, we admit) or via radio. In any event, the computer receives the data, processes it, and presents the information.

The section on writing shorter programs responds to a concern of all dedicated programmers. Shorter programs run faster and occupy less memory space.

The multiplex display is an idea that has found great use in advertising signs, in ball stadiums, in discos, and on television. Large displays can have thousands of incandescent lamps controlled by a microcomputer to produce the illusions of color and motion. A new innovation allows a user to place a finger on a TV screen to select an option. Again, a computer can be used to determine the position of the finger on the screen, freeing the operator from finding buttons to press, and so on.

Two areas will find a great increase in programming applications. These are color graphics displays for the human eye to see and speech synthesis for the human ear to hear. Speech synthesis can be associated with speech recognition, and all these areas can be applied in the burgeoning field of study called "robotics." Computers are becoming "humanized" in many ways. They are being designed and programmed to make their use by humans easier, faster, and more comfortable. User-friendly computers appear to be emulating humans, dispelling the fears of many beginners. Ease and speed of operation are both important. Color displays are used in emergency situations in power distribution systems, nuclear plants, and satellite receiving stations. Red indicators on color picture tubes indicate danger, green stands for normal, and yellow for warning. And everywhere, the simple computer is at work waiting for the next instruction or word from the programmer. Are the machines in control? Do they make the decisions? Can they handle any situation that might arise? The answer is not an answer. They do exactly as they are programmed.

Remember: You are the programmer, the engineer, and the power behind the machine.

19 GLOSSARY

A-to-D converter. A circuit that converts an analog signal to a binary (digital) representation.

BCD. Binary coded decimal. Also the 8421 code used for doing decimal arithmetic in binary.

Bar code. See Universal Product Code.

Character generator. Generally a ROM that given an ASCII input produces the shape of a specified character in dot matrix form ($5 \times 7, 7 \times 9, 11 \times 13$, etc.).

Conversion time. The time required for an A-to-D converter to change an analog level to a digital equivalent.

D-to-A converter. A circuit that converts a set of bits to an analog level.

DI. The Disable Interrupt instruction.

Data logger. A computer that records or "logs" events as they happen, for future reference.

Digitizing. Changing an analog signal to a digital signal.

Disassembler. A program that converts object code (opcodes) back to source code (mnemonic form).

Downline Loading. The process by which one computer may serially communicate with another computer.

EI. The Enable Interrupt instruction.

Interrupt. A signal that interrupts the normal operation of a program for some infrequent purpose.

Mantissa. The significant digits of a number in scientific notation.

Multiplex (MUX). To time-share a set of wires.

Packed BCD. The representation as a byte of two 4-bit BCD numbers.

Polling. The process of checking many bits on a series of input ports to determine which has caused an interrupt.

Resolution. The step size (often in percent) of a D/A converter. For an 8-bit DAC, the resolution is $\frac{1}{2^8} = \frac{1}{256} \approx .0039$.

SID. The serial input data line in an 8085 CPU.

SOD. The serial output data line in an 8085 CPU.

Settling time. The time from a change in the analog input level until the digital output is stable within $\pm\frac{1}{2}$ LSB.

Software clock. An interrupt-driven program that keeps accurate time for use by a microcomputer. The program requires a very small percentage of the CPU's time, freeing the CPU to handle major tasks.

Speech synthesizer. A processor that can simulate the waveforms of human speech.

UART. A universal asynchronous receiver/transmitter used to send or receive serial data between computers or terminals.

Universal Product Code (UPC). The standard bar code used in consumer items.

Zero crossing. Refers to the point on the 60-Hz a-c line at which the a-c waveform crosses zero.

PROBLEMS

12-1 In the PHONE program, an 8-bit output port was used to control a relay, but only 1 bit was used out of 8. Can you modify the program to use SOD (8085) to control the relay and free up the I/O port?

12-2 Can you write a machine language routine to read the characters in the TIME buffer in the CLOCK program, and send them to the user's terminal with COUT?

12-3 Consider the delay used in the software UART program. A delay equal to 15 H is used for 2400 baud. Determine the error that this causes because of rounding of the actual required figure.

12-4 Complete the program that was started in Section 6, "Security Systems." Assign subroutines to the various "emergencies" and show how the necessary subroutine addresses are loaded and implemented.

12-5 Consider a security system with the main job of monitoring 16 doors. Write a machine language program to log each opening and closing of any door. Be sure to consider that a door may open during the time required to print a line of data, and you would not want to miss such an event.

12-6 In the HEXDUMP program, explain how the total number of bytes to be printed is computed.

12-7 Explain the operation of a disassembler and identify the two different ways in which the data stored in memory can be interpreted.

12-8 The key to any scientific programming language lies in its ability to perform floating-point arithmetic (add, subtract, multiply, and divide). The use of BCD eliminates the rounding errors associated with using pure binary. Write a subroutine to compute $\sin(x)$ where x is in radians based on the formula:

$$\sin(x) = x - \frac{x^3}{3!} + \frac{x^5}{5!} - \frac{x^7}{7!} + \cdots$$

Does this give you any hint of the complexity of producing a full scientific library of functions?

12-9 Modify the bar code reader program to correctly read the shortened UPC found on many magazines.

12-10 Determine the resolution in percent of a 10- and 12-bit D-to-A converter.

12-11 Write the program to read a voltage using the circuitry presented in Section 13. Assume that the comparator goes high when a match in voltages occurs. Let this be connected to the LSB of an input port.

12-12 Assign 20 bytes in memory to represent the 160 lamps of display in the multiplex display (Section 14). Write a subroutine to output the 20 bytes to the display. Then write four subroutines to shift the data in memory one place to the left; that is, when displayed, the data appears to shift left. Then shift right, shift up, shift down.

12-13 In the SORT routine, only positive numbers were allowed. Modify the routine to include negative numbers, i.e., 0 to 7F positive, 80 to FF negative.

12-14 Modify the SORT routine to return immediately if a single pass through the group fails to produce a swap (meaning that the numbers are completely sorted).

12-15 The 60-Hz interrupt service routine in a software clock takes 20 msec to execute. Why is this a problem?

12-16 Calculate and normalize 256 data points for one cycle of a 10-V peak triangle wave. How can the data be reduced?

12-17 Modify the clock program to keep track of AM/PM and the date.

CHAPTER 13

A HARDWARE LOOK AT
THE 8080A/85

1 INSTRUCTIONAL OBJECTIVES

As a result of reading this chapter, you should gain familiarity with the architecture of the 8080A and 8085 microprocessors and of the timing and control signals required. The various systems in use in industry are illustrated by means of a presentation of standard busses.

2 SELF-EVALUATION QUESTIONS

Keep the following questions in mind and try to answer them when you have completed the chapter:

1. How do the internal organizations of the 8080A and 8085 microprocessors differ?
2. What is a processor cycle, and how does it relate to the sync pulse?
3. What is a machine cycle?
4. What is the status word, and what is its function in microcomputer control?
5. What occurs at RESET, HOLD, and INTerrupt?
6. How is a multiplexed address/data bus separated (decoded)?
7. What is a TRAP?

3 INTRODUCTION

In our study of the architecture of two central processing units, the 8080A and the 8085, we will see exactly how the CPU addresses memory, performs input and output to and from data ports, executes instructions, and handles interrupts. This chapter is divided into two parts, dealing with the 8080A CPU and the architecture of the 8085, respectively.

4 THE 8080A CENTRAL PROCESSING UNIT

4.1 Introduction

Figure 13-1 shows a photomicrograph of the 8080A central processing unit chip (magnified, of course) and a pinout of the DIP package. The 8080A is fabricated with large-scale integration and Intel's *n*-channel silicon gate MOS technology. The 8080A uses an 8-bit architecture (D0–D7) and has address lines (A0–A15). The 8080A also has six timing and control outputs (SYNC, DBIN, WAIT, $\overline{WR}$, HLDA, and INTE) and four control inputs (READY, HOLD, INT, and RESET). The most important (and necessary) inputs remaining are the two clock inputs (Ø1 and Ø2) and the four power inputs ($+5$, -5, $+12$, GND). In the next few sections, we will study each of the control/timing lines in detail, to see how they affect and control the operation of the 8080A CPU.

4.2 Internal Organization of the 8080A

The block diagram of the internal organization of the 8080A CPU (Fig. 13-2) shows eight 8-bit registers, B, C, D, E, H, L, W, and Z, but only the first six are available to the programmer—the W and Z registers are reserved for internal 8080A operations.

The registers are all accessed through the register multiplexer, which controls the data transfer to and from the internal 8-bit data bus. Sixteen-bit register transfers or operations are handled by the incrementer/decrementer address latch circuit. The 16-bit stack pointer and the 16-bit program counter are also located in the register array. Data transfer to any register (except the accumulator) is controlled by the register multiplexer and by the register select circuit, both of which are controlled by the instruction register, the instruction decoder and machine cycle timing circuit, and the timing and control circuit.

The 8-bit accumulator, an 8-bit temporary accumulator (ACT), a 5-bit flag register (zero, sign, carry, parity, and auxiliary carry), and an 8-bit temporary register (TMP) are all contained in the arithmetic and logic unit (ALU), which executes all arithmetic, logical, and rotate operations. Both the TMP and ACT registers plus the carry flip-flop feed data to the ALU, which transfers the result either to the accumulator (ACC) or to the internal 8-bit data bus via signals from the timing and control unit.

The 5-bit flag register can be fed either by the ALU (as the result of an instruction) or by the TMP register.

The accumulator can receive data from the internal 8-bit data bus or from the ALU. The accumulator can be tested by the decimal adjust accumulator (DAA) circuit, and its contents can be altered (decimal corrected) during the execution of a DAA instruction.

All data transfer into and out of the 8080A CPU takes place in the 8-bit data bus buffer/latch, which contains input and output buffers connected to the D0 $\rightarrow$ D7 data I/O pins on the chip and the 8-bit internal data bus.

4.3 The Processor Cycle

The processor cycle, or instruction cycle, is defined as the amount of time required to fetch and execute an instruction. Instruction cycles are composed of between

FIGURE 13-1 Photomicrograph and DIP pinout of the 8080 CPU.
(Reprinted by permission of Intel Corporation, copyright 1980.)

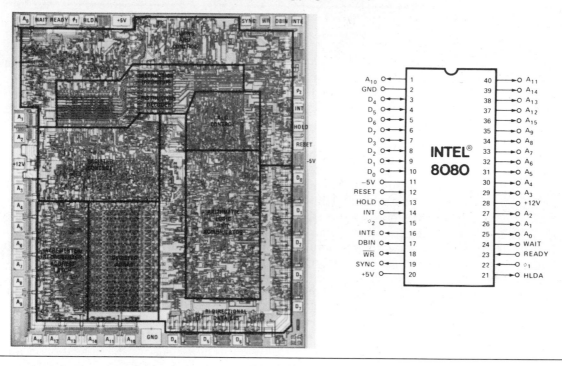

FIGURE 13-2 Functional block diagram of the 8080 CPU.
(Reprinted by permission of Intel Corporation, copyright 1980.)

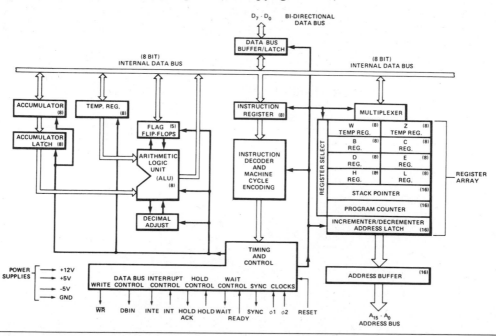

TABLE 13-1   Duties of Machine Cycle States

State	Operation
1	Checks the HALT line and halts the CPU if it is set.
2	Checks the hold line and enters a wait state until ready to continue.
3	Performs opcode fetch or memory/IO read or write.
4	If used, performs the opcode decode. Most instruction fetch machine cycles are four states long.

one and five machine cycles. A machine cycle is needed whenever the 8080A accesses memory or an input/output port. A machine cycle is further divided into three to five states. A state is the amount of time between successive positive-going transitions on the Ø1 clock line—thus, the Ø1 clock line directly controls the timing for all instruction cycles. You can see at this time that 4 to 18 states may be necessary to complete an instruction cycle. The length of the instruction cycle naturally depends on the type of instruction being executed. The different states of a machine cycle have the duties listed in Table 13-1.

In Fig. 13-3 we see that a state is the exact amount of time in one Ø1 clock period. The synchronizing signal produced by the 8080A (i.e., SYNC) indicates the beginning of a new machine cycle. The Ø2 clock provides the timing for the SYNC pulse.

The SYNC pulse serves a dual purpose in the 8080A: the indication of a new machine cycle, and the signification that status information is available on the data bus D0–D7. Figure 13-4 shows how the SYNC pulse is used to gate data on the 8080A's data bus into an 8212 latch.

The timing waveforms in Fig. 13-4*b* show that when SYNC goes high, it sets up the DS_2 pin on the 8212 (in Fig. 13-4*a*). At the onset of the next transition to high of the Ø1 clock line, the DS_1 pin of the 8212 is pulled high, which causes the 8212 to latch onto the status word previously placed on the data bus by the 8080A. The status word identifies the present machine cycle. Table 13-2 shows a bit-by-bit representation of the status word, and Table 13-3 shows how each word is decoded to one of 10 different 8080A machine cycles.

FIGURE 13-3 Clocks Ø1 and Ø2, and sync timing. (Reprinted by permission of Intel Corporation, copyright 1980.)

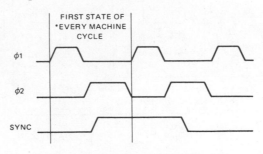

FIGURE 13-4 (*a*) Status word latch for the 8080A and (*b*) its timing diagram. (Reprinted by permission of Intel Corporation, copyright 1980.)

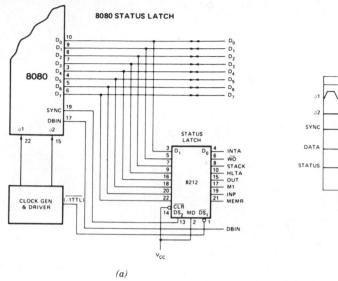

(*a*)

TABLE 13-2   Status Word Definition

STATUS INFORMATION DEFINITION

Symbols	Data Bus Bit	Definition
INTA*	D_0	Acknowledge signal for INTERRUPT request. Signal should be used to gate a restart instruction onto the data bus when DBIN is active.
$\overline{WO}$	D_1	Indicates that the operation in the current machine cycle will be a WRITE memory or OUTPUT function ($\overline{WO} = 0$). Otherwise, a READ memory or INPUT operation will be executed.
STACK	D_2	Indicates that the address bus holds the pushdown stack address from the Stack Pointer.
HLTA	D_3	Acknowledge signal for HALT instruction.
OUT	D_4	Indicates that the address bus contains the address of an output device and the data bus will contain the output data when $\overline{WR}$ is active.
M_1	D_5	Provides a signal to indicate that the CPU is in the fetch cycle for the first byte of an instruction.
INP*	D_6	Indicates that the address bus contains the address of an input device and the input data should be placed on the data bus when DBIN is active.
MEMR*	D_7	Designates that the data bus will be used for memory read data.

*These three status bits can be used to control the flow of data onto the 8080 data bus.

Source: Reprinted by permission of Intel Corporation, copyright 1980.

TABLE 13-3 Status Word Chart

DATA BUS BIT	STATUS INFORMATION	INSTRUCTION FETCH ①	MEMORY READ ②	MEMORY WRITE ③	STACK READ ④	STACK WRITE ⑤	INPUT READ ⑥	OUTPUT WRITE ⑦	INTERRUPT ACKNOWLEDGE ⑧	HALT ACKNOWLEDGE ⑨	INTERRUPT ACKNOWLEDGE WHILE HALT ⑩
D_0	INTA	0	0	0	0	0	0	0	1	0	1
D_1	$\overline{WO}$	1	1	0	1	0	1	0	1	1	1
D_2	STACK	0	0	0	1	1	0	0	0	0	0
D_3	HLTA	0	0	0	0	0	0	0	0	1	1
D_4	OUT	0	0	0	0	0	0	1	0	0	0
D_5	M_1	1	0	0	0	0	0	0	1	0	1
D_6	INP	0	0	0	0	0	1	0	0	0	0
D_7	MEMR	1	1	0	1	0	0	0	0	1	0

TYPE OF MACHINE CYCLE

Ⓝ STATUS WORD

Source: Reprinted by permission of Intel Corporation, copyright 1980.

To understand the status word chart, you need to know that a machine cycle type is decoded whenever a bit or bits change in the status word. For instance, whenever bit 4 goes high, we have an OUTPUT WRITE machine cycle, which means that the 8080A will place data on the data bus and a port address on the address lines to select an output port. In practice, we decode and use only 8 of the 10 machine cycles. We do not have to decode STACK READ or STACK WRITE, since the stack resides in RAM, and the machine cycles for STACK are the same as those for MEMORY READ/WRITE except for the STACK (D_2) bit. Although it is not essential, some machines use the INSTRUCTION FETCH machine cycle status word to implement single-step functions on a front panel. Circuitry is designed to hold the 8080A from running at the onset of each new instruction fetch. However, the remaining seven machine cycle types are *very* useful and are all decoded to make a very powerful microcomputing system.

4.4 8080A Control Inputs

As mentioned in Section 4.1, the 8080A has four control inputs, READY, HOLD, INT, and RESET. Let's take a look at each one and see how it affects the operation of the 8080A.

4.4.1 RESET. The RESET input to the 8080A performs several functions, as indicated by the timing diagram (Fig. 13-5). While it is high, it holds the 16 address lines and 8 data lines high. When it returns to a low state, the 8080A will start executing data from address 0000 H. In other words, the RESET line serves to

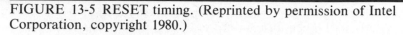

FIGURE 13-5 RESET timing. (Reprinted by permission of Intel Corporation, copyright 1980.)

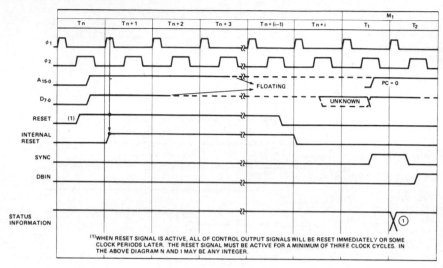

(1)WHEN RESET SIGNAL IS ACTIVE, ALL OF CONTROL OUTPUT SIGNALS WILL BE RESET IMMEDIATELY OR SOME CLOCK PERIODS LATER. THE RESET SIGNAL MUST BE ACTIVE FOR A MINIMUM OF THREE CLOCK CYCLES. IN THE ABOVE DIAGRAM N AND I MAY BE ANY INTEGER.

reset the program counter (PC) to zero. It also clears the interrupt enable flip-flop (see Section 4.5.5, "INTE") and the hold enable flip-flop (see Section 4.5.4, "HLDA"), which are internal 8080A flip-flops. No registers or flags are affected.

4.4.2 READY. The READY input is used to synchronize the 8080A with slow memory devices. During program execution, the 8080A places an address on the address bus and waits for valid data to be placed on the data bus. The READY line indicates that this has occurred. If, for some reason, the memory or device addressed has not placed the data on the bus before the next state, the device will indicate this by pulling READY low, and the 8080A will go into a WAIT state until the READY line goes high again.

4.4.3 HOLD. The HOLD control input tells the 8080A that an external device wishes to use its address and data bus. Figure 13-6 shows the read and write modes. A logic "1" on the HOLD line will cause the 8080A to stop executing at the end of the current instruction and place the address and data lines in a high impedance state. As long as the HOLD line is high, the 8080A will remain suspended. When the HOLD line is brought low again, the 8080A will resume program execution where it left off. No registers or flags are affected. The 8080A will acknowledge that it is in a "hold" state with the HLDA line (see Section 4.5.4, "HLDA").

4.4.4 INT. The INT (interrupt) control input is a very important and useful control line. The 8080A will recognize an interrupt request on this control input at the end of the current instruction cycle (see Figure 13-7) or if it is halted (i.e., if the HLT instruction had been executed previously—see Chapter 10 for an explanation

FIGURE 13-6 (*a*) Read mode and (*b*) write mode for HOLD timing. (Reprinted by permission of Intel Corporation, copyright 1980.)

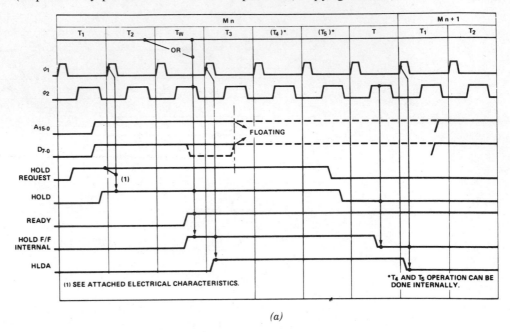

(*a*)

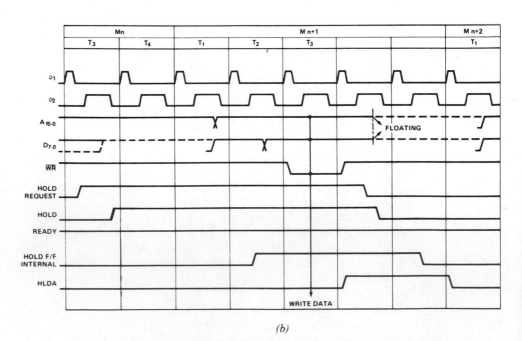

(*b*)

FIGURE 13-7 INT timing. (Reprinted by permission of Intel Corporation, copyright 1980.)

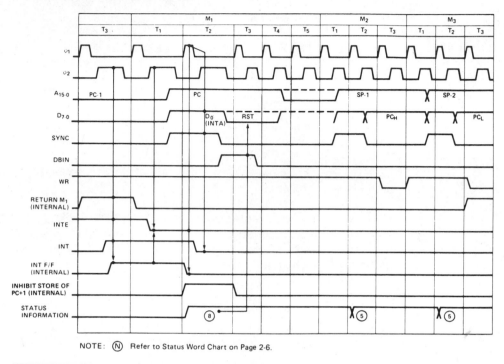

NOTE: (N) Refer to Status Word Chart on Page 2-6.

of the HLT instruction). The 8080A will not recognize an INT request if it is in a HOLD mode or if the INTE (see Section 4.5.5, "INTE") flip-flop is cleared. The INTE flip-flop is set by the EI instruction (see Chapter 10).

When the interrupt is acknowledged, many things happen: the INTE flip-flop is cleared, to prevent the occurrence of any more interrupts until the next EI instruction has been executed; the INTA (interrupt acknowledge) status bit is set, indicating that the 8080A has acknowledged the interrupt; and the 8080A looks to the data bus for an instruction (usually an RST instruction).

The INT request does not alter any flags or registers, but if program execution is to be resumed after an interrupt routine has completed its task, the routine will have to save all 8080A registers (PSW, BC, DE, HL) that it destroys (alters).

4.5 8080A Control Outputs

The last 8080A signals we are going to learn about are the six control outputs: SYNC, DBIN, WAIT, $\overline{WR}$, HLDA, and INTE. We are already familiar with SYNC, having seen how it is used to gate the status word into an 8212 (Section 4.3). Now we will look at the other five control outputs.

4.5.1 DBIN.   The data bus input (DBIN) line is perhaps one of the most important signals in the 8080A CPU. The DBIN line is used to indicate that the data bus is in the input mode. When DBIN is high, data from memory or an input port can be placed on the data bus.

4.5.2 WAIT.   The WAIT output indicates that the 8080A is in a WAIT mode (i.e., the READY line has been pulled low). No program execution takes place during WAIT mode.

4.5.3 $\overline{WR}$.   The $\overline{WR}$ output, or WRITE, is used to write data into a memory location or to send data to an output port. The data for either operation is on the data bus when $\overline{WR}$ is low (0).

4.5.4 HLDA.   The hold-acknowledge output is active when the 8080A's HOLD line has been activated.

4.5.5 INTE.   The interrupt-enable output is used to indicate that the 8080A will honor an interrupt request on the INT line. The INTE line reflects the state of the internal interrupt-enable flip-flop, which can be set or cleared by the EI/DI instruction, or cleared by the RESET control input.

4.6 Using the 8080A

Now that we know what all the control lines on the 8080A are used for, we will see how they are used in actual digital circuits to make the 8080A operate.

4.6.1 Providing a Clock for the 8080A.   The 8080A needs a two-phase clock to prevent the microprocessor from attempting to write into memory at the same time it wishes to read from memory, as well as for other reasons. The two-phase clock is most easily generated by the use of an 18-MHz crystal and an 8224 clock generator and driver IC.

In Fig. 13-8, we see that the 8224 directly drives the Ø1 and Ø2 lines on the 8080A. The 8224 has an internal divide-by-9 counter to divide the 18-MHz crystal down to 2 MHz, as well as some flip-flops and other gates. In addition to supplying the clocks, the 8224 controls the READY and RESET inputs to the 8080A. The RESET input is pulled high momentarily on power-up due to the *RC* network on pin 2 of the 8224. An external RESET switch is also provided. The 8080A will enter a wait state whenever the RDYIN line is pulled low. This is the input that must be chosen if slow memories are being used with the 8080A. Last, the 8224 also generates a STSTB (status-strobe) signal, which is used to latch the status word into an 8212 during the SYNC output from the 8080A. This line can be used directly, or the circuit of Fig. 13-4*a* may be used.

4.6.2 Generating an Eight-Bit Data Bus.   We already know that all data passes into and out of the 8080A CPU on an 8-bit data bus. This type of data bus is called a bidirectional data bus because it passes data in both directions.

Figure 13-9 shows how two 8216s can be used to implement an 8-bit bidirectional data bus. The eight DB lines coming from the 8216s are the bidirectional data

FIGURE 13-8 Generating a two-phase clock.

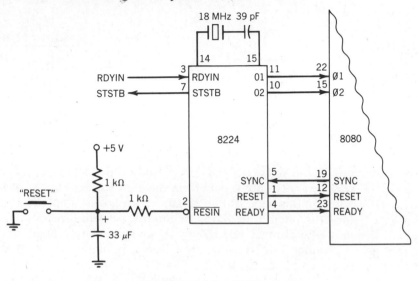

FIGURE 13-9 An 8-bit bidirectional data bus; all resistors are 4.7 KΩ.

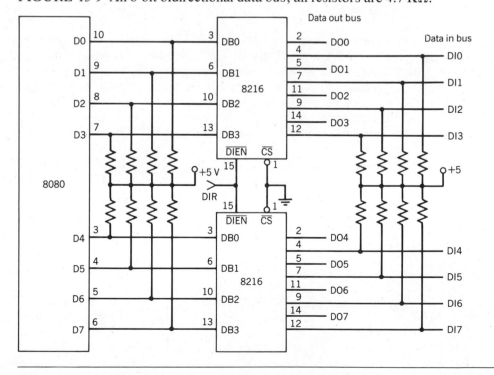

FIGURE 13-10 Direction control for the 8216. (*a*) NOR gate. (*b*) Truth table.

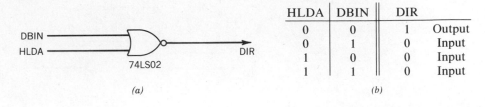

HLDA	DBIN	DIR	
0	0	1	Output
0	1	0	Input
1	0	0	Input
1	1	0	Input

(a)   *(b)*

bus. Input data (DI0–DI7) is placed on the 8080A's data bus when DIR is low; consequently output data from the 8080A is sent to the data out bus (DO0–DO7) when DIR is high. In Fig. 13-10, we see how a simple NOR gate is used to control the DIR line on the 8216s to control the data direction. Whenever the 8080A is in a HOLD mode (HLDA = 1), we place the 8216s in input mode. If the 8080A is running (HLDA = 0), the DIR line is simply an inverted version of the DBIN line.

Remember that we can disable the 8080A's busses through the HOLD line. When we do this, we must also disable the 8216s so that they will have no effect on the data busses. A simple circuit used in the HOLD process could also control the $\overline{CS}$ lines on the 8216s. When the 8080A is using the data bus, the $\overline{CS}$ lines must be low, but when an external device wishes to use the data busses, it must be taken high.

4.6.3 Addressing RAM and ROM.

The process for reading or writing memory in the 8080A is relatively simple. The address is placed on the address bus, and then the data is either read on the data input bus with a memory read pulse or placed on the data output bus, to be followed by a memory write pulse.

The memory read pulse is generated directly. It is bit 7 of the status word. The write pulse has to be generated with simple logic like that in Fig. 13-11. We can see that we only get a positive memory WRITE pulse when both $\overline{WR}$ and OUT are low.

Once we are able to generate the MEMR/MEMW pulses, our next step is to *partition* memory. That is, certain areas of memory must be set aside for RAM,

FIGURE 13-11 Generating a memory write pulse. (*a*) NOR gate. (*b*) Truth table.

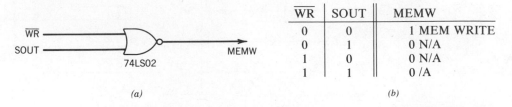

$\overline{WR}$	SOUT	MEMW	
0	0	1	MEM WRITE
0	1	0	N/A
1	0	0	N/A
1	1	0	/A

(a)   *(b)*

and certain areas for ROM. It is important to understand why RAM and ROM may not reside in the same area of memory. So, once we have decided which areas of memory will contain the desired type of memory, we must design select circuitry for each.

EXAMPLE 13-1

Determine the addressing circuitry needed for a 4K block of RAM with first byte at 3000 H.

SOLUTION

First refer to Chapter 7 to refresh your understanding of RAMs and the circuitry needed for a 4K block of RAM. You will realize that 10 address lines (A0–A9) are needed to address 1K of RAM, and two additional address lines (A10, A11) to determine which of the four 1K blocks we wish to look at. We are then left with four address lines, A12, A13, A14, and A15 (Fig. 13-12).

If we make a diagram of the address for 3000 H, we have:

$$
\begin{array}{ccccccc}
A15 & A14 & A13 & A12 & A11 & A10 & A9 \rightarrow A0 \\
0 & 0 & 1 & 1 & X & X & X \ldots X
\end{array}
$$

where X = don't care, which means that we are looking at the 3000 H address range whenever A14 and A15 are low, and A12 and A13 are high. We do not care about A11 through A0, since these lines are used by the 4K RAM circuitry itself. By convention, we will need a low ($\overline{CS}$ or $\overline{EN}$) on the master decoder in the 4K RAM circuit to select it; therefore, we need a simple circuit to produce a low when the upper four address bits are at 3H. The four-input NAND gate in Fig. 13-12 is set up to produce a low when it sees four high inputs. Inverters are added to the A14 and A15 address lines, to produce ones when they are low.

FIGURE 13-12 Address select for 3000 H.

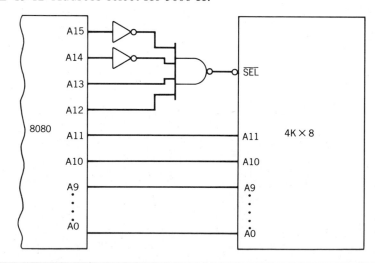

FIGURE 13-13 Address partitioning for 4K blocks (*a*) using exclusive-NOR and (*b*) using a comparator.

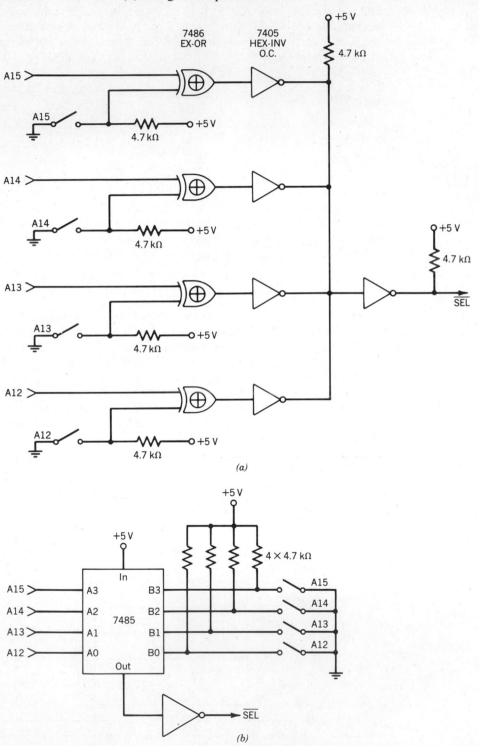

FIGURE 13-14 (*a*) Input and (*b*) output timing strobes. (Reprinted
by permission of Intel Corporation, copyright 1980.)

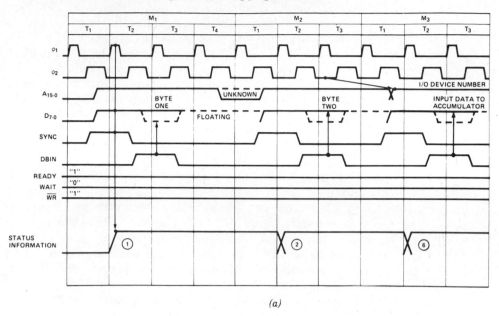

(a)

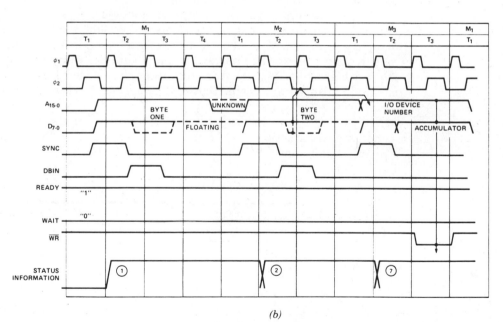

(b)

It may become desirable to select any of the sixteen 4K blocks to suit your particular memory requirements. In this case, four switches can be used to select the 4K boundary that you wish to use as RAM. Figure 13-13 illustrates two methods of selecting a 4K memory boundary. The process used on commercial memory boards (8K, 16K, 32K, and up) is not so different from either of these. In both approaches four switches are used to select the address. When a switch is closed, the associated 8080A address line must be a zero to match it. If the switch is open, the 8080A address line must be a one. In both cases, if all four 8080A address lines match the switches, the $\overline{\text{SEL}}$ output will go low from its normal high state.

4.6.4 Providing I/O Ports for the 8080A.   The 8080A is capable of sending data to or inputting data from any one of 256 data ports. To do this, the 8080A places the port address on the address bus and then does an IN or an OUT instruction (refer to the status word description section). The data port is decoded in the same way that we decoded a memory boundary, namely, by using circuitry to look at the address bits and send out a low when they match some predetermined number. The easiest example would be decoding port 0FF H, which would simply require an eight-input NAND gate (74LS30) connected to address lines A0–A7. Thus, when they are all high, the output goes low, selecting port 0FF H (255 D). See Fig. 13-14 for input/output timing.

The second step is generating either an input strobe to gate external data onto the 8080A's data bus, or an output strobe to send data to an external device. The circuitry in Fig. 13-15 can be used for generating both. We see that we will get a positive input pulse when INP and $\overline{\text{WR}}$ are high, and a positive output pulse when OUT is high and $\overline{\text{WR}}$ is low. As a precaution, the gate used for the INPUT function could be changed to a three-input AND gate, the third input being DBIN.

4.6.5 Providing an Interrupt for the 8080A.   Now we are ready to add interrupt capability to the 8080A. In Fig. 13-16, we see how a flip-flop is used to control the INT line on the CPU.

FIGURE 13-15 Generating input/output strobes.

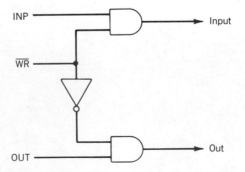

FIGURE 13-16 Interrupt control circuit using the 7474.

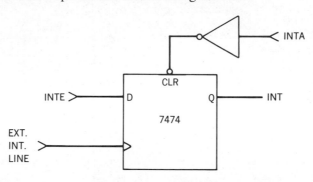

The INT line of the 8080A is fed by the Q output of a D flip-flop (74LS74). We know that whenever the INT line is high, the 8080A will acknowledge an interrupt (assuming that the interrupts are enabled). Note that we are not talking about a low-to-high transition on the INT line, just a high level; thus, the flip-flop becomes important. Assume that the flip-flop is cleared (Q = 0) and that interrupts are enabled. Then INTE is one and INTA is zero. When we get a low-to-high transition on the clock line of the D flip-flop, the logic "1" from INTE will get clocked through to the Q output and interrupt the 8080A. When the 8080A processes the interrupt, it will disable all future interrupts (INTE = 0) and also clear the flip-flop (INTA = 1). Now additional transitions on the clock line of the flip-flop will be ignored until an EI instruction is executed in the 8080A to reset INTA and set INTE.

5 THE 8085 CENTRAL PROCESSING UNIT

5.1 Introduction

Unlike the 8080A, the 8085 CPU (see Fig. 13-17) is a much more simplified (or integrated) central processing unit. While remaining 100% compatible with the software set of the 8080A, the 8085 includes two more instructions (RIM, SIM), an additional four interrupt lines (TRAP, RST7.5, RST6.5, and RST5.5), an internal clock generator requiring only an external crystal, direct status output (the status word does not have to be latched), and increased speed. The 8085 also has two input/output pins for serial data (SID, SOD) and best of all, the chip operates on a single 5-V supply.

5.2 Internal Organization of the 8085

A quick glance at the internal functional block diagram of the 8085 (Fig. 13-18) indicates that this microprocessor is much less complex than the 8080A, which does not seem right because we know that the 8085 is by far the more powerful

FIGURE 13-17 Pinout for the 8085 CPU. (Reprinted by permission of Intel Corporation, copyright 1982.)

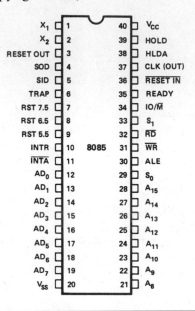

FIGURE 13-18 Functional block diagram for the 8085 CPU. (Reprinted by permission of Intel Corporation, copyright 1982.)

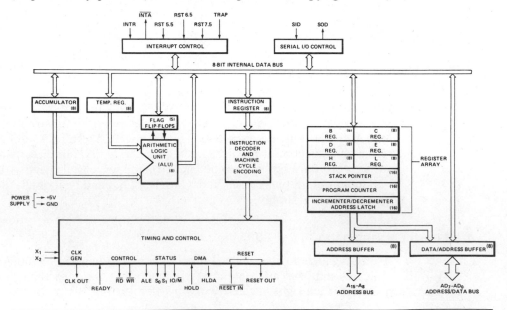

processor. The major difference between the functional block diagram in Figs. 13-2 and 13-18 is the addition of the SERIAL I/O and INTERRUPT CONTROL blocks. We remember that the 8085 has increased interrupt-handling power as well as separate serial input and output pins (refer to Chapter 12 for an understanding of SID and SOD).

5.3 Basic System Timing in the 8085

Since we are already familiar with processor cycles from Section 4, this material is not repeated here. The 8085, however, has a multiplexed data/lower address bus, which needs to be studied. The 8085 has a control signal, ALE, which stands for address latch enable. ALE is actually a strobe pulse that the 8085 sends out to indicate to external components that the lower 8 bits of the current program counter are available on the data bus. In the basic system timing for the 8085 (Fig. 13-19), we see that ALE is high whenever we need to look at the lower 8 address bits. The ALE signal also pulses when an I/O port address is outputted, since the port address ($0 \leq$ port ≤ 255) is duplicated on the upper *and* lower address bus. The ALE signal always takes place in the T_1 state of any machine cycle, which is indicated in Fig. 13-19 and also in Table 13-4. Special integrated circuits that employ the addressing and timing schemes of the 8085 and do not

FIGURE 13-19 Basic system timing. (Reprinted by permission of Intel Corporation, copyright 1982.)

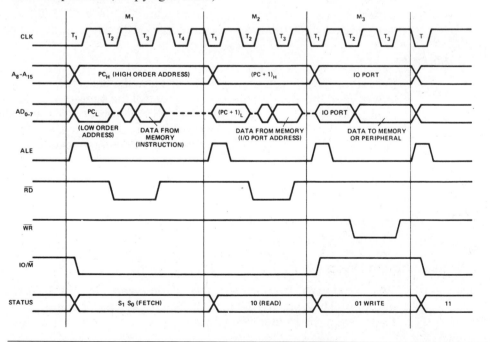

TABLE 13-4   8085 Machine State Chart

Machine State	Status & Buses				Control		
	S1,S0	IO/$\overline{\text{M}}$	A_8-A_{15}	AD_0-AD_7	$\overline{\text{RD}}$,$\overline{\text{WR}}$	$\overline{\text{INTA}}$	ALE
T_1	X	X	X	X	1	1	1†
T_2	X	X	X	X	X	X	0
T_{WAIT}	X	X	X	X	X	X	0
T_3	X	X	X	X	X	X	0
T_4	1	0*	X	TS	1	1	0
T_5	1	0*	X	TS	1	1	0
T_6	1	0*	X	TS	1	1	0
T_{RESET}	X	TS	TS	TS	TS	1	0
T_{HALT}	0	TS	TS	TS	TS	1	0
T_{HOLD}	X	TS	TS	TS	TS	1	0

0 = Logic "0"   1 = Logic "1"   TS = High Impedance   X = Unspecified

†ALE not generated during 2nd and 3rd machine cycles of DAD instruction.

*IO/$\overline{\text{M}}$ = 1 during T_4-T_6 states of RST and INA cycles.

Source: Reprinted by permission of Intel Corporation, copyright 1982.

require a latched lower address byte are covered later in Section 6. However, the lower address byte can be latched into an 8212 in Fig. 13-20. It is clear that the 8212 latches onto the lower order address byte when ALE is high.

5.4 8085 Status Lines

Instead of having to latch a status word like the 8080A, the 8085 has three status outputs and three control outputs that control the operation of the external devices used to implement a system with the 8085 microprocessor. In Table 13-5 we see how the 8085 indicates its current machine cycle by use of the six outputs. The three status outputs indicate one of seven machine cycles at the onset of the T_1 machine cycle state (which was the case with the 8080A except that SYNC is not required to latch the status—i.e., it is *always* present). The three control outputs become active shortly afterward when data must be transferred. The READ signal $\overline{\text{RD}}$ indicates that a read operation (from memory or an I/O port) is in progress. The WRITE signal $\overline{\text{WR}}$ indicates the opposite: when low, data is written into memory or to an I/O port.

5.5 Remaining Pin Descriptions

We are already familiar with five of the remaining 8085 CPU pins (inputs or outputs). They are READY, HOLD, HLDA, INTR, and $\overline{\text{INTA}}$. In addition to these, we have 11 more signals to look at.

FIGURE 13-20 Lower address byte latch for the 8085. (Reprinted by permission of Intel Corporation, copyright 1982.)

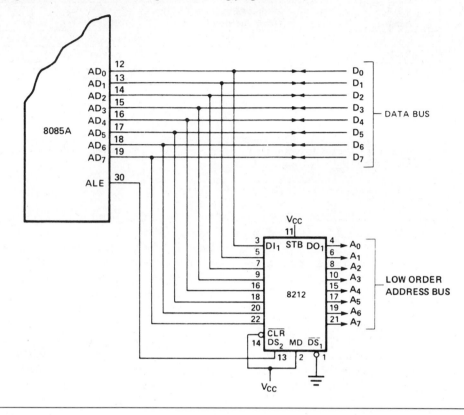

TABLE 13-5 8085 Machine Cycle Chart

MACHINE CYCLE			STATUS			CONTROL		
			IO/M̄	S1	S0	R̄D̄	W̄R̄	ĪNTA
OPCODE FETCH	(OF)		0	1	1	0	1	1
MEMORY READ	(MR)		0	1	0	0	1	1
MEMORY WRITE	(MW)		0	0	1	1	0	1
I/O READ	(IOR)		1	1	0	0	1	1
I/O WRITE	(IOW)		1	0	1	1	0	1
INTR ACKNOWLEDGE	(INA)		1	1	1	1	1	0
BUS IDLE	(BI)	DAD	0	1	0	1	1	1
		INA(RST/TRAP)	1	1	1	1	1	1
		HALT	TS	0	0	TS	TS	1

0 = Logic "0"   1 = Logic "1"   TS = High Impedance   X = Unspecified

Source: Reprinted by permission of Intel Corporation, copyright 1982.

TABLE 13-6   8085 Interrupt Priority

Name	Priority	Address Branched To (1) When Interrupt Occurs	Type Trigger
TRAP	1	24H	Rising edge AND high level until sampled.
RST 7.5	2	3CH	Rising edge (latched).
RST 6.5	3	34H	High level until sampled.
RST 5.5	4	2CH	High level until sampled.
INTR	5	See Note (2).	High level until sampled.

NOTES:

(1) The processor pushes the PC on the stack before branching to the indicated address.

(2) The address branched to depends on the instruction provided to the cpu when the interrupt is acknowledged.

Source: Reprinted by permission of Intel Corporation, copyright 1981.

5.5.1 TRAP.   The TRAP signal is a nonmaskable interrupt. That is, the interrupt cannot be enabled or disabled by software (as is the case with INT). The TRAP *always* interrupts the 8085. It has the highest priority of all the interrupts available (see Table 13-6), which is to say that TRAP will be acknowledged if it is received at the same time as another interrupt. TRAP saves the program counter on the stack when acknowledged, causing a jump to address 0024H. This is the address at which the service routine for TRAP should be located.

5.5.2 RST5.5, RST6.5, and RST7.5.   The restart interrupts 5.5, 6.5, and 7.5 are all similar to the INTR interrupt, with the exception that they have different priorities (see Table 13-6). Like TRAP, they all have a specific address that is jumped to when the interrupt is acknowledged.

5.5.3 $\overline{\text{RESET IN}}$.   Like the 8080A, the 8085's $\overline{\text{RESET IN}}$ causes the program counter to be cleared and the interrupts to be disabled (with the exception of TRAP). Unlike the 8080A, the $\overline{\text{RESET IN}}$ pulse may unpredictably alter the contents of the 8085's registers.

5.5.4 RESET OUT.   The RESET OUT output indicates that the 8085 is being reset and can be used to reset external devices in the 8085 system.

5.5.5 CLK.   The CLK output has a period twice that of the input to the X_1, X_2 inputs (see Section 5.5.8) and can be used as a system clock for synchronization if so desired.

5.5.6 SID.   The serial input data (SID) line is the input line to the 8085 for serial data. The state of the SID line (high or low) is placed in bit 7 of the accumulator whenever the RIM instruction is executed.

5.5.7 SOD.   The serial output data line (SOD) is the output line from the 8085 containing serial output data. The state of the line (set, reset) is specified by the SIM instruction.

FIGURE 13-21 Five clock driver circuits. (Reprinted by permission of Intel Corporation, copyright 1982.)

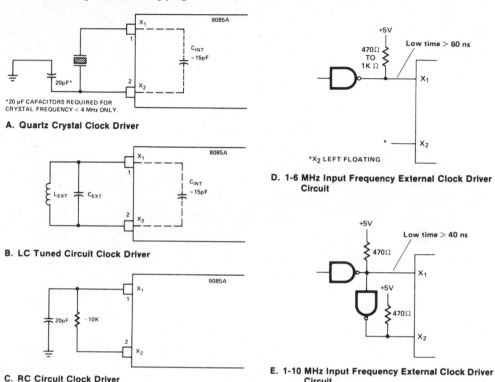

A. Quartz Crystal Clock Driver

B. LC Tuned Circuit Clock Driver

C. RC Circuit Clock Driver

D. 1-6 MHz Input Frequency External Clock Driver Circuit

E. 1-10 MHz Input Frequency External Clock Driver Circuit

5.5.8 X_1 and X_2. Although the 8085 does *not* require a clock generator/driver to make it operate, Fig. 13-21 presents five ways in which the 8085 can be clock driven. We can see that a simple crystal can be used, or *RC*, *LC* combinations. This type of clock circuitry provides the designer with much freedom in the design process and is a great advantage of the 8085 microprocessor, although a crystal would be the best choice for a stable clock for the CPU.

6 USING THE 8085

You should realize that we have two choices open to us in operating the 8085 in a basic system. We can choose to operate with a standard 16-bit address bus (demultiplexed) like that seen in Fig. 13-22. In this system, memories (RAM and ROM) and their associated support circuitry (like those seen in Chapter 7) must be used, as well as an 8212 to demultiplex the lower address byte.

Figure 13-23 shows another basic system setup; this one employs the multiplexed bus scheme and is implemented with standard Intel support chips. In this

FIGURE 13-22 Basic 8085 system using standard memories.
(Reprinted by permission of Intel Corporation, copyright 1982.)

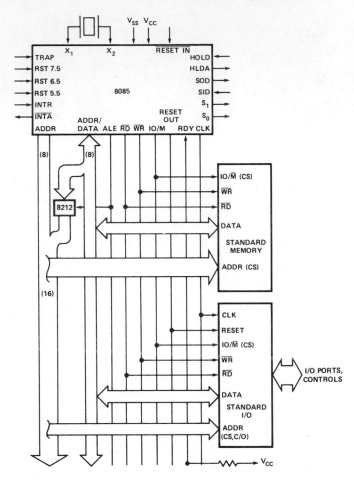

system, utilizing an 8355 or 8755 (ROM I/O or EPROM I/O) and an 8156
(RAM–I/O–Counter–Timer), we have a complete working 8085 microcomputer,
with 2K × 8 bits of ROM, 256 × 8 bits of RAM, and five parallel I/O ports. In
addition, we have the SID and SOD serial I/O lines, a timer control, and, best of
all, the support chips do all the fancy address/data/status decoding internally.

If the system in Figure 13-23 is not large enough for your needs, the system in
Fig. 13-24 can be used. The expanded system features a greatly increased number
of parallel I/O ports (there are 12 now!) and a significantly increased amount of
ROM (6K × 8!). The RAM has also been doubled to 512 bytes, but, in the case
of a dedicated system, it is more important to have more ROM to implement an
operating system.

FIGURE 13-23 The basic 8085 system (multiplexed address).
(Reprinted by permission of Intel Corporation, copyright 1982.)

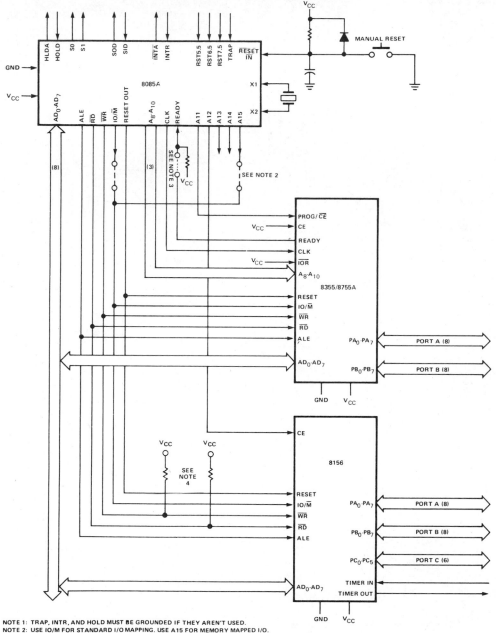

NOTE 1: TRAP, INTR, AND HOLD MUST BE GROUNDED IF THEY AREN'T USED.
NOTE 2: USE IO/M FOR STANDARD I/O MAPPING. USE A15 FOR MEMORY MAPPED I/O.
NOTE 3: CONNECTION IS NECESSARY ONLY IF ONE T_{WAIT} STATE IS DESIRED.
NOTE 4: PULL-UP RESISTORS RECOMMENDED TO AVOID SPURIOUS SELECTION WHEN $\overline{RD}$ AND $\overline{WR}$ ARE
          3-STATED. THESE RESISTORS ARE NOT INCLUDED ON THE PC BOARD LAYOUT OF FIGURE 3-7.

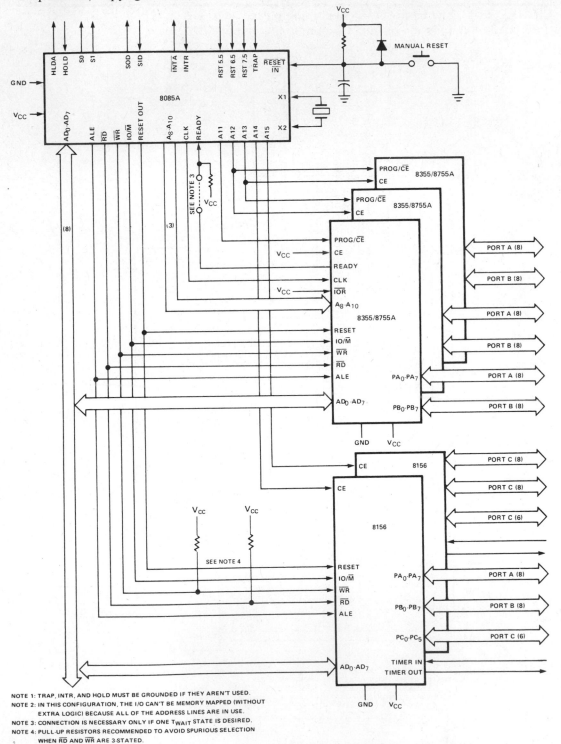

FIGURE 13-24 An expanded system. (Reprinted by permission of Intel Corporation, copyright 1982.)

NOTE 1: TRAP, INTR, AND HOLD MUST BE GROUNDED IF THEY AREN'T USED.
NOTE 2: IN THIS CONFIGURATION, THE I/O CAN'T BE MEMORY MAPPED (WITHOUT EXTRA LOGIC) BECAUSE ALL OF THE ADDRESS LINES ARE IN USE.
NOTE 3: CONNECTION IS NECESSARY ONLY IF ONE T_{WAIT} STATE IS DESIRED.
NOTE 4: PULL-UP RESISTORS RECOMMENDED TO AVOID SPURIOUS SELECTION WHEN $\overline{RD}$ AND $\overline{WR}$ ARE 3-STATED.

7 INTERNAL AND EXTERNAL BUSSES

We have already learned that data is moved about within a computer using parallel paths called busses. Every computer has an address bus, a data bus, and a control bus, but the number of wires required depends on the type of processor involved. Generally, 8-bit microprocessors (e.g., 6800, 8080A, 8085, 6502, 1802, and Z80 models) require an 8-bit data bus and a 16-bit address bus.

To contribute to product development among manufacturers and at the same time to standardize the interconnection between microprocessor components, standards have been established. There are many standard busses around, and your computer probably uses one of them. This section examines eight of them:

1. S100 bus.

2. STD bus.

3. Multibus.

4. VME bus and VERSA bus.

5. Apple® II bus.

6. IBM PC bus.

7. PDP-11 and LSI-11 bus.

8. IEEE 488 bus.

For example, numerous companies provide computer products that attach easily to the S100 bus. Typically, a bus standard is desirable whenever a wide variety of circuit cards must be used together. There may be a processor (CPU) card, an I/O card, a disk controller card, a video display card (perhaps in color), a digitizer card, an A-to-D or D-to-A card, a speech synthesizer card, a speech recognizer card, a real-time clock, power control, assorted memory cards (RAM, EPROM, etc.), music boards, and even game cards. Often a front panel card is used to help write and debug programs for the chosen microprocessor.

These cards must be connected to communicate with one another and that is where a common bus enters the picture. Usually each card plugs into an edge connector mounted in a mother board. The mother board may consist of 100 parallel lines that go to every connector, hence to every card inserted into the mother board. Although all boards are connected to the bus lines, each board is active only when selected. The particular arrangement of data lines, address lines, control lines, and powerlines is the subject of each bus standard. This allows different manufacturers to specialize in an area and produce boards compatible with a particular bus. One company might produce A/D–D/A boards, and another may focus on speech synthesis. Several (many) produce memory boards. Such variety is desirable and possible only when bus standards exist.

7.1 The S100 Bus (IEEE 696)

The S100 bus is a system that uses 100 pin connectors as shown in Fig. 13-25. These connectors are all interconnected on a 100-conductor mother board. The S100 bus is a very popular industrial bus that can adapt to use by both 8- and 16-

FIGURE 13-25 Mother board for the S100. (Courtesy of
Imsai/Fulcrum.)

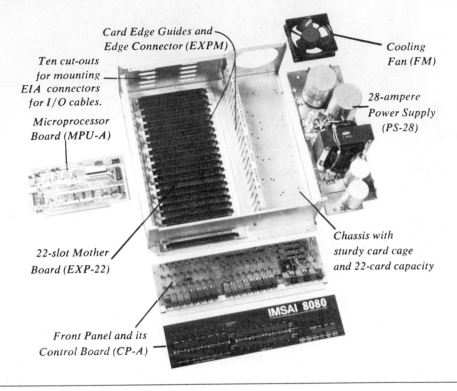

bit microprocessors. This bus is assured a secure future by its adoption through
Standard 696 of the Institute of Electrical and Electronics Engineers, Inc. (IEEE
696) and also a diverse selection of compatible boards manufactured for it. The
100 lines are used for power distribution, address, data, and control signals.
Provision is made for up to 24 address lines, 16 data lines, 8 vectored interrupts,
and numerous status and control signals.

The S100 bus was first used with 8-bit microprocessor-based systems and used
16 address lines (A0–A15) and 16 data lines. The data lines were eight unidirectional data input lines (DI0–DI7) and eight unidirectional output lines (DO0–DO7).
To accommodate a 16-bit processor, these data lines become bidirectional. Table
13-7 shows the IEEE 696 (S100) standard pin assignments. Power is distributed
from three separate power supplies (+8, +18, −18 V d-c) to S100 boards, which
have individual voltage regulators as required. All other signals on the bus lines
are TTL levels (0–5 V). This bus promises to continue to grow and to be a very
popular method of interconnecting circuit cards. Figure 13-26 shows an 8080A
CPU card.

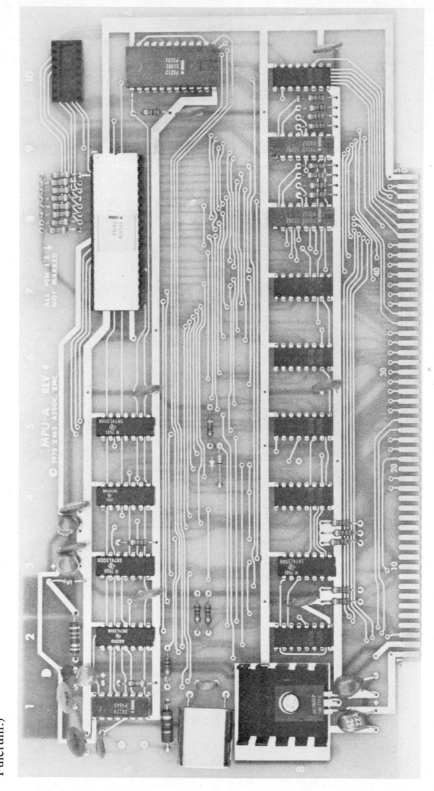

FIGURE 13-26 Card for the 8080A MPU. (Courtesy of Imsai/Fulcrum.)

TABLE 13-7   IEEE 696 (S100) Standard Pin Assignments

Pin No.	Signal and Type	Active Level	Description
1	+8 VOLTS (B)		Instantaneous minimum greater than 7 volts, instantaneous maximum less than 25 volts, average maximum less than 11 volts.
2	+16 VOLTS (B)		Instantaneous minimum greater than 14.5 volts, instantaneous maximum less than 35 volts, average maximum less than 21.5 volts.
3	XRDY (S)	H	One of two ready inputs to the current bus master. The bus is ready when both these ready inputs are true. See pin 72.
4	VI0*(S)	L O.C.	Vectored interrupt line 0.
5	VI1*(S)	L O.C.	Vectored interrupt line 1.
6	VI2*(S)	L O.C.	Vectored interrupt line 2.
7	VI3*(S)	L O.C.	Vectored interrupt line 3.
8	VI4*(S)	L O.C.	Vectored interrupt line 4.
9	VI5*(S)	L O.C.	Vectored interrupt line 5.
10	VI6*(S)	L O.C.	Vectored interrupt line 6.
11	VI7*(S)	L O.C.	Vectored interrupt line 7.
12	NMI*(S)	L O.C.	Nonmaskable interrupt.
13	PWRFAIL*(B)	L	Power fail bus signal.
14	DMA3*(M)	L O.C.	Temporary master priority bit 3.
15	A18 (M)	H	Extended address bit 18.
16	A16 (M)	H	Extended address bit 16.
17	A17 (M)	H	Extended address bit 17.
18	SDSB* (M)	L O.C.	The control signal to disable the 8 status signals.
19	CDSB* (M)	L O.C.	The control signal to disable the 5 control output signals.
20	GND (B)		Common with pin 100.
21	NDEF		Not to be defined. Manufacturer must specify any use in detail.
22	ADSB* (M)	L O.C.	The control signal to disable the 16 address signals.
23	DODSB* (M)	L O.C.	The control signal to disable the 8 data output signals.
24	ϕ (B)	H	The master timing signal for the bus.
25	pSTVAL*(M)	L	Status valid strobe.
26	pHLDA (M)	H	A control signal used in conjunction with HOLD* to coordinate bus master transfer operations.
27	RFU		Reserved for future use.
28	RFU		Reserved for future use.
29	A5 (M)	H	Address bit 5.
30	A4 (M)	H	Address bit 4.
31	A3 (M)	H	Address bit 3.
32	A15 (M)	H	Address bit 15 (most significant for non-extended addressing.)
33	A12 (M)	H	Address bit 12.
34	A9 (M)	H	Address bit 9.
35	DO1 (M)/DATA1 (M/S)	H	Data out bit 1, bidirectional data bit 1.
36	DO0 (M)/DATA0 (M/S)	H	Data out bit 0, bidirectional data bit 0.
37	A10 (M)	H	Address bit 10.
38	DO4 (M)/DATA4 (M/S)	H	Data out bit 4, bidirectional data bit 4.
39	DO5 (M)/DATA5 (M/S)	H	Data out bit 5, bidirectional data bit 5.

TABLE 13-7 (continued)

Pin No.	Signal and Type	Active Level	Description
40	DO6 (M)/DATA6 (M/S)	H	Data out bit 6, bidirectional data bit 6.
41	DI2 (S)/DATA10 (M/S)	H	Data in bit 2, bidirectional data bit 10.
42	DI3 (S)/DATA11 (M/S)	H	Data in bit 3, bidirectional data bit 11.
43	DI7 (S)/DATA15 (M/S)	H	Data in bit 7, bidirectional data bit 15.
44	sM1 (M)	H	The status signal which indicates that the current cycle is an opcode fetch.
45	sOUT (M)	H	The status signal identifying the data transfer bus cycle to an output device.
46	sINP (M)	H	The status signal identifying the data transfer bus cycle from an input device.
47	sMEMR (M)	H	The status signal identifying bus cycles which transfer data from memory to a bus master, which are not interrupt acknowledge instruction fetch cycle(s).
48	sHLTA (M)	H	The status signal which acknowledges that a HLT instruction has been executed.
49	CLOCK(B)		2 MHz (0.5%) 40-60% duty cycle. Not required to be synchronous with any other bus signal.
50	GND (B)		Common with pin 100.
51	+8 VOLTS (B)		Common with pin 1.
52	−16 VOLTS (B)		Instantaneous maximum less than −14.5 volts, instantaneous minimum greater than −35 volts, average minimum greater than −21.5 volts.
53	GND (B)		Common with pin 100.
54	SLAVE CLR* (B)	L O.C.	A reset signal to reset bus slaves. Must be active with POC* and may also be generated by external means.
55	DMAO* (M)	L O.C.	Temporary master priority bit 0.
56	DMA1* (M)	L O.C.	Temporary master priority bit 1.
57	DMA2* (M)	L O.C.	Temporary master priority bit 2.
58	sXTRQ* (M)	L	The status signal which requests 16-bit slaves to assert SIXTN*.
59	A19 (M)	H	Extended address bit 19.
60	SIXTN* (S)	L O.C.	The signal generated by 16-bit slaves in response to the 16-bit request signal sXTRQ*.
61	A20 (M)	H	Extended address bit 20.
62	A21 (M)	H	Extended address bit 21.
63	A22 (M)	H	Extended address bit 22.
64	A23 (M)	H	Extended address bit 23.
65	NDEF		Not to be defined signal.
66	NDEF		Not to be defined signal.
67	PHANTOM* (M/S)	L O.C.	A bus signal which disables normal slave devices and enables phantom slaves—primarily used for bootstrapping systems without hardware front panels.
68	MWRT (B)	H	pWR—sOUT (logic equation). This signal must follow pWR* by not more than 30 ns.

(continued on next page)

TABLE 13-7   (continued)

Pin No.	Signal and Type	Active Level	Description
69	RFU		Reserved for future use.
70	GND (B)		Common with pin 100.
71	RFU		Reserved for future use.
72	RDY (S)	O.C.	See comments for pin 3.
73	INT* (S)	L O.C.	The primary interrupt request bus signal.
74	HOLD* (M)	L O.C.	The control signal used in conjunction with pHLDA to coordinate bus master transfer operations.
75	RESET*(B)	L O.C.	The reset signal to reset bus master devices. This signal must be active with POC* and may also be generated by external means.
76	pSYNC (M)	H	The control signal identifying BS_1.
77	pWR* (M)	L	The control signal signifying the presence of valid data on DO bus or data bus.
78	pDBIN (M)	H	The control signal that requests data on the DI bus or data bus from the currently addressed slave.
79	A0 (M)	H	Address bit 0 (least significant).
80	A1 (M)	H	Address bit 1.
81	A2 (M)	H	Address bit 2.
82	A6 (M)	H	Address bit 6.
83	A7 (M)	H	Address bit 7.
84	A8 (M)	H	Address bit 8.
85	A13 (M)	H	Address bit 13.
86	A14 (M)	H	Address bit 14.
87	A11 (M)	H	Address bit 11.
88	DO2 (M)/DATA2 (M/S)	H	Data out bit 2, bidirectional data bit 2.
89	DO3 (M)/DATA3 (M/S)	H	Data out bit 3, bidirectional data bit 3.
90	DO7 (M)/DATA7 (M/S)	H	Data out bit 7, bidirectional data bit 7.
91	DI4 (S)/DATA12 (M/S)	H	Data in bit 4 and bidirectional data bit 12.
92	DI5 (S)/DATA13 (M/S)	H	Data in bit 5 and bidirectional data bit 13.
93	DI6 (S)/DATA14 (M/S)	H	Data in bit 6 and bidirectional data bit 14.
94	DI1 (S)/DATA9 (M/S)	H	Data in bit 1 and bidirectional data bit 9.
95	DI0 (S)/DATA8 (M/S)	H	Data in bit 0 (least significant for 8-bit data) and bidirectional data bit 8.
96	sINTA (M)	H	The status signal identifying the bus input cycle(s) that may follow an accepted interrupt request presented on INT*.
97	sWO* (M)	L	The status signal identifying a bus cycle which transfers data from a bus master to a slave.
98	ERROR* (S)	L O.C.	The bus status signal signifying an error condition during present bus cycle.
99	POC* (B)	L	The power-on clear signal for all bus devices; when this signal goes low, it must stay low for at least 10 ms.
100	GND (B)		System ground.

Source: Courtesy of the Institute of Electrical and Electronics Engineers.

7.2 The STD Bus (IEEE 961)

The STD bus is another popular bus that has been adopted by a number of industries. It was first proposed by Pro-Log Corporation, and other companies have joined Pro-Log with a wide range of STD bus-related products. The card for the STD bus is roughly half the size of the card for the S100 bus and is compact for industrial control applications. The STD bus is 56 lines wide and uses a card $4\frac{1}{2}$ in. $\times$ $6\frac{1}{2}$ in. The 8-bit data bus is bidirectional (unlike S100) and includes a 16-bit address bus and control bus. All address and data lines are tri-state buffered. The power lines are +5 for logic and +12 for peripheral chips. The individual circuit cards are not intended to have their own regulators. For devices requiring substrate levels, −5 V is provided.

Pro-Log makes available three different STD bus processor cards. These allow the designer the option of using the 6800, 8085, or Z80 CPUs. The bus is in the public domain; it is neither patented nor copyrighted. The pin assignments of the STD bus and a Pro-Log 8085 processor card using the STD bus are shown in Table 13-8 and Fig. 13-27, respectively.

TABLE 13-8   STD Bus Pin Assignments

		COMPONENT SIDE				CIRCUIT SIDE		
	PIN	MNEMONIC	SIGNAL FLOW	DESCRIPTION	PIN	MNEMONIC	SIGNAL FLOW	DESCRIPTION
LOGIC POWER BUS	1	+5VDC	In	Logic Power (bussed)	2	+5VDC	In	Logic Power (bussed)
	3	GND	In	Logic Ground (bussed)	4	GND	In	Logic Ground (bussed)
	5	VBB #1	In	Logic Bias #1 (-5V)	6	VBB #2	In	Logic Bias #2 (-5V)
DATA BUS	7	D3	In/Out	Low-Order Data Bus	8	D7	In/Out	High-Order Data Bus
	9	D2	In/Out	Low-Order Data Bus	10	D6	In/Out	High-Order Data Bus
	11	D1	In/Out	Low-Order Data Bus	12	D5	In/Out	High-Order Data Bus
	13	D0	In/Out	Low-Order Data Bus	14	D4	In/Out	High-Order Data Bus
ADDRESS BUS	15	A7	Out	Low-Order Address Bus	16	A15	Out	High-Order Address Bus
	17	A6	Out	Low-Order Address Bus	18	A14	Out	High-Order Address Bus
	19	A5	Out	Low-Order Address Bus	20	A13	Out	High-Order Address Bus
	21	A4	Out	Low-Order Address Bus	22	A12	Out	High-Order Address Bus
	23	A3	Out	Low-Order Address Bus	24	A11	Out	High-Order Address Bus
	25	A2	Out	Low-Order Address Bus	26	A10	Out	High-Order Address Bus
	27	A1	Out	Low-Order Address Bus	28	A9	Out	High-Order Address Bus
	29	A0	Out	Low-Order Address Bus	30	A8	Out	High-Order Address Bus
CONTROL BUS	31	WR*	Out	Write to Memory or I/O	32	RD*	Out	Read Memory or I/O
	33	IORQ*	Out	I/O Address Select	34	MEMRQ*	Out	Memory Address Select
	35	IOEXP	In/Out	I/O Expansion	36	MEMEX	In/Out	Memory Expansion
	37	REFRESH*	Out	Refresh Timing	38	MCSYNC*	Out	CPU Machine Cycle Sync.
	39	STATUS 1*	Out	CPU Status	40	STATUS 0*	Out	CPU Status
	41	BUSAK*	Out	Bus Acknowledge	42	BUSRQ*	In	Bus Request
	43	INTAK*	Out	Interrupt Acknowledge	44	INTRQ*	In	Interrupt Request
	45	WAITRQ*	In	Wait Request	46	NMIRQ*	In	Nonmaskable Interrupt
	47	SYSRESET*	Out	System Reset	48	PBRESET*	In	Push-Button Reset
	49	CLOCK*	Out	Clock from Processor	50	CNTRL*	In	AUX Timing
	51	PCO	Out	Priority Chain Out	52	PCI	In	Priority Chain In
AUXILIARY POWER BUS	53	AUX GND	In	AUX Ground (bussed)	54	AUXGND	In	AUX Ground (bussed)
	55	AUX +V	In	AUX Positive (+12V DC)	56	AUX -V	In	AUX Negative (-12V DC)

*Low-level active indicator

Source: Prolog Corporation.

FIGURE 13-27 Processor card for the Pro-Log 8085 using the STD
bus. (Courtesy of Pro-Log Corporation.)

7.3 The Multibus (IEEE 796).

The multibus, a product of Intel Corporation, has been widely adopted by industry.
Many products are available that are compatible with this bus. The multibus
contains 20 address lines, a 16-bit bidirectional data bus, control lines, and power
distribution lines. Four voltages are provided for: ± 5 V and ± 12 V.

This bus uses two connectors. The primary connector (P1) has power, address,
data, and control information. The auxiliary connector (P2) uses optional signals
involving battery backup and memory protection.

Unlike the S100 bus or the STD bus, the multibus uses inverted signals. Table
13-9 shows the signals on the multibus. Notice that the bus transfers inverted
address, data, and control information (negative logic). Products manufactured
for the multibus include single-board computers like the Intel 80/10 (Fig. 13-28),
memory products including bubble memory and disk cards, I/O cards, A-to-D
and D-to-A cards. A card cage is available to make the use of several cards easy.

TABLE 13-9 Multibus Signals

Name	Symbol	Source	Purpose
P1 signals:			
Initialization	$\overline{\text{INIT}}$	Master/external switch	Resets system.
Address line	$\overline{\text{ADR0-ADR13}}$ (numbered, 0-13$_{16}$)	Master	Memory and I/O address.

TABLE 13-9 (continued)

Name	Symbol	Source	Purpose
Inhibit RAM signal	$\overline{\text{INH1}}$	Master	Prevents RAM response.
Inhibit ROM signal	$\overline{\text{INH2}}$	Master	Prevents ROM response.
Data lines	DAT0-DATF	Master/slave	Bidirectional data to or from memory/I/O port.
Byte high enable	$\overline{\text{BHEN}}$	Master	Used with 16-bit memory and I/O transfers.
Bus clock	$\overline{\text{BCLK}}$	Bus control	Negative edge is used to sync bus priority resolution circuits.
Constant clock	$\overline{\text{CCLK}}$	Bus control	General-purpose clock.
Bus priority in	$\overline{\text{BPRN}}$	Bus control	Indicates to a particular master that it has highest priority.
Bus priority out	$\overline{\text{BPRO}}$	Master	Used in daisy chain priority resolution schemes.
Bus busy	$\overline{\text{BUSY}}$	Master	Indicates bus is in use.
Bus request	$\overline{\text{BREQ}}$	Master	Indicates that a master requires use of the bus.
Common request	$\overline{\text{CREQ}}$	Master	Informs current bus master that another master wants to use the bus.
Memory read	$\overline{\text{MRDC}}$	Master	Address of a memory location to be read is on the address bus.
Memory write	$\overline{\text{MWTC}}$	Master	Address of a memory location to be written is on the address bus.
I/O read command	$\overline{\text{IORC}}$	Master	Address of an input port to be read has been placed on the address bus.
I/O write command	$\overline{\text{IOWC}}$	Master	Address of an output port to be written has been placed on the address bus.
Transfer acknowledge	$\overline{\text{XACK}}$	Slave	Response when specified read/write operation has been completed.
Interrupt request	$\overline{\text{INT0-INT7}}$	Slave	Multilevel parallel interrupt request lines.
Interrupt acknowledge	$\overline{\text{INTA}}$	Master	Requests transfer of interrupt information.

P2 signals:

Name	Symbol	Source	Purpose
AC low	$\overline{\text{ACLO}}$	Power supply	AC input voltage too low.
Power fail interrupt	$\overline{\text{PFIN}}$	External power fail circuit	Power failure.
Power fail sense	$\overline{\text{PFSN}}$	External power fail circuit	Output of a latch indicating that a power failure has occurred.
Power fail reset	$\overline{\text{PFSR}}$	External power fail circuit	Resets the power failure sense latch.
Address latch enable	ALE	Master	From 8085 or 8086 as an auxiliary address latch.
Halt	$\overline{\text{HALT}}$	Master	MPU has halted.
Wait state	$\overline{\text{WAIT}}$	Master	Master processor is in the wait state.
Auxiliary reset	$\overline{\text{AUXRESET}}$	External	Initiates power up sequence.
Memory protect	MPRO	External	Prevents memory operations when power is uncertain.

Source: Reprinted by permission of Intel Corporation, copyright 1982.

FIGURE 13-28 The Intel 80/10B single-board computer. (Reprinted by permission of Intel Corporation, copyright 1982.)

7.4 The VME Bus and the VERSA Bus

The VME and VERSA busses are generally associated with Motorola. The VERSA bus uses a card size of $14\frac{1}{2}$ in. $\times$ $9\frac{1}{4}$ in. and is larger than the VME bus card, which is approximately 9 in. $\times$ 6 in. The VERSA bus is intended to support 6800-based products, whereas the VME bus is specifically a 68000-based system. Each bus is designed to handle, 8-, 16-, or 32-bit formats and establishes an arbitration scheme when multiple processors share the bus. Table 13-10 shows comparable features of these two busses.

7.5 The Apple II Bus

The very popular Apple II home computer comes with a wide variety of peripheral cards available. The peripheral connector allows various devices to communicate with the 6502 processor. This bus contains 50 parallel lines and includes 16 address lines, 8 bidirectional data lines, power distribution, and control lines including interrupts. It is a simple bus structure, as indicated by its peripheral connector pinout (Fig. 13-29) and its pin assignments (Table 13-11).

TABLE 13-10   Comparable Features of the VME Bus and the VERSA Bus

Feature	VERSA Bus	VME Bus
CPU	6800 family	68000 family
Data width	8, 16, 32 bits	8, 16, 32 bits
Address field	32 bits	16, 24, or 32 bits
Number of cards	Up to 15	Up to 20
Signal line length	18 in. maximum	19 in. maximum
Power supply	+5, ±12, ±15 V	+5, ±12 V

Source: Courtesy of Motorola, Inc.

FIGURE 13-29 Peripheral connector pinout for the Apple® II.
(Courtesy of Apple.)

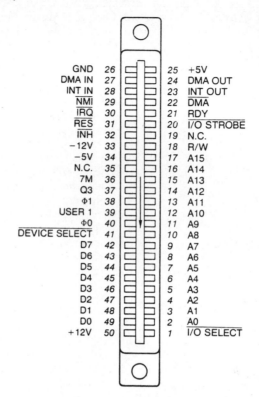

TABLE 13-11   Peripheral Connector Pin Assignments

Pin	Name	Description
1	I/O SELECT	This line, normally high, will become low when the microprocessor references page SC$_n$, where n is the individual slot number. This signal becomes active during Φ0 and will drive 10 LSTTL loads*. This signal is not present on peripheral connector 0.
2-17	A0-A15	The buffered address bus. The address on these lines becomes valid during Φ1 and remains valid through Φ0. These lines will each drive 5 LSTTL loads*.
18	R/$\overline{W}$	Buffered Read/Write signal. This becomes valid at the same time the address bus does, and goes high during a read cycle and low during a write. This line can drive up to 2 LSTTL loads*.
19	SYNC	On peripheral connector 7 *only*, this pin is connected to the video timing generator's SYNC signal.

(continued on next page)

TABLE 13-11   (continued)

Pin	Name	Description
20	I/O STROBE	This line goes low during Φ0 when the address bus contains an address between SC800 and $CFFF. This line will drive 4 LSTTL loads*.
21	RDY	The 6502's RDY input. Pulling this line low during Φ1 will halt the microprocessor, with the address bus holding the address of the current location being fetched.
22	DMA	Pulling this line low disables the 6502's address bus and halts the microprocessor. This line is held high by a 3KΩ resistor to +5v.
23	INT OUT	Daisy-chained interrupt output to lower priority devices. This pin is usually connected to pin 28 (INT IN).
24	DMA OUT	Daisy-chained DMA output to lower priority devices. This pin is usually connected to pin 22 (DMA IN).
25	+5v	+5 volt power supply. 500mA current is available for *all* peripheral cards.
26	GND	System electrical ground.
27	DMA IN	Daisy-chained DMA input from higher priority devices. Usually connected to pin 24 (DMA OUT).
26	INT IN	Daisy-chained interrupt input from higher priority devices. Usually connected to pin 23 (INT OUT).
29	NMI	Non-Maskable Interrupt. When this line is pulled low the Apple begins an interrupt cycle and jumps to the interrupt handling routine at location $3FB.
30	IRQ	Interrupt ReQuest. When this line is pulled low the Apple begins an interrupt cycle only if the 6502's I (Interrupt disable) flag is not set. If so, the 6502 will jump to the interrupt handling subroutine whose address is stored in locations $3FE and $3FF.
31	RES	When this line is pulled low the microprocessor begins a RESET cycle.
32	INH	When this line is pulled low, all ROMs on the Apple board are disabled. This line is held high by a 3KΩ resistor to +5v.
33	−12v	−12 volt power supply. Maximum current is 200mA for all peripheral boards.
34	−5v	−5 volt power supply. Maximum current is 200mA for all peripheral boards.
35	COLOR REF	On peripheral connector 7 *only*, this pin is connected to the 3.5MHz COLOR REFerence signal of the video generator.
36	7M	7MHz clock. This line will drive 2 LSTTL loads*.
37	Q3	2MHz asymmetrical clock. This line will drive 2 LSTTL loads*.
38	Φ1	Microprocessor's phase one clock. This line will drive 2 LSTTL loads*.
39	USER 1	This line, when pulled low, disables *all* internal I/O address decoding*.
40	Φ0	Microprocessor's phase zero clock. This line will drive 2 LSTTL loads*.

TABLE 13-11 (continued)

Pin	Name	Description
41	DEVICE SELECT	This line becomes active (low) on each peripheral connector when the address bus is holding an address between SC0n0 and SC0nF, where n is the slot number plus $8. This line will drive 10 LSTTL loads*.
42-49	D0-D7	Buffered bidirectional data bus. The data on this line becomes valid 300nS into Φ0 on a write cycle, and should be stable no less than 100ns before the end of Φ0 on a read cycle. Each data line can drive one LSTTL load.
50	+12v	+12 volt power supply. This can supply up to 250mA total for all peripheral cards.

*Loading limits are for each peripheral card.
Source: Apple Computer.

7.6 The IBM Personal Computer I/O Bus

The IBM PC has five connectors on the expansion backplane and represents access to the 8088 CPU resident within the machine. There are 20 address lines and 8 data lines. All signals are TTL. The power lines supply ±5 and ±12 V. The IBM PC uses a 62-pin edge connector (Fig. 13-30), and many companies offer peripheral cards for this microprocessor.

7.7 The PDP-11 and LSI-11 Busses

Digital Equipment Corporation's PDP-11 has been a very popular 16-bit mini-computer. It has retained its popularity in a microchip version, the LSI-11. These computers are covered in Chapter 15. The PDP-11 bus is called the UNIBUS, whereas the LSI-11 uses the LSI-11 bus. The UNIBUS and the LSI-11 bus are similar. However the UNIBUS has separate address and bidirectional data lines, whereas the LSI-11 uses multiplexed (shared) address and data lines. Both busses operate on a strict master/slave relationship. That is, a device must request and receive permission to use the bus. Therefore an arbitration scheme has been established by which contenders for use of the bus are assigned priorities. For further information on these busses, see the manufacturer's literature (DEC PDP-11 Architecture Handbook).

7.8 The IEEE 488 bus (GPIB)

This very popular bus is NOT an internal bus within processors. Its purpose is entirely different. It is essentially a cable (or many cables) allowing the interconnection of different types of test equipment to a central controller. It is known as the general-purpose information bus (GPIB) and was adopted from the Hewlett-Packard information bus (HPIB). The GPIB is now standardized by the IEEE (Standard 488).

IEEE 488 is used to interface remotely programmable and nonprogrammable test equipment. Such equipment may include calculators, computers, signal gen-

FIGURE 13-30 Backplane connector for the IBM PC. (Courtesy of IBM.)

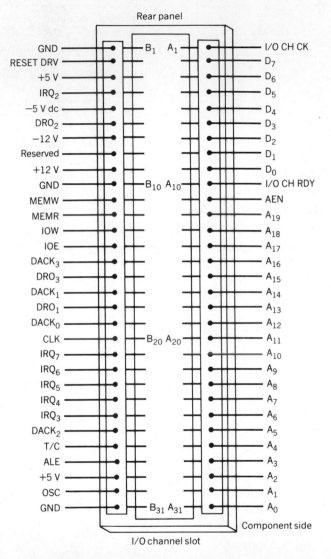

erators, multimeters, counters, oscilloscopes, computer terminals, and controllers. Figure 13-31 shows examples of IEEE 488 compatible equipment. A controller may be programmed to perform a series of test measurements through instruments connected to the bus. Data from measurements can then be evaluated or logged appropriately. Hewlett-Packard, Tektronix, and other manufacturers supply instrumentation that can communicate via the IEEE 488 interface bus, which is used mainly in automated testing of electronic circuitry. Figure 13-32 and Table 13-12 give details of the 488 bus.

FIGURE 13-31 Example of 488-compatible equipment. (Courtesy of Tektronix, Inc.)

FIGURE 13-32 Bus details for the 488.

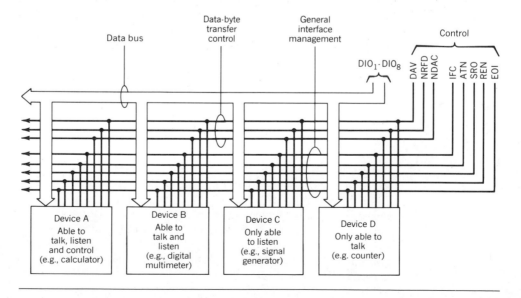

TABLE 13-12   The IEEE-488 Lines

Designation	Description
DIO_1–DIO_8	*Data Input/Output:* Eight data transfer lines; also called the data bus.
ATN	*Attention:* Issued only by the controller, to gain the attention of bus devices before beginning a handshake sequence and to denote address or control information on the data bus.
DAV	*Data Valid:* Issued by a talker to notify the listener(s) that data has been placed on the DIO lines.
EOI	*End or Identify:* Issued by a talker to notify the listener(s) that the data byte currently on the DIO lines is the last one. The controller issues this together with ATN to initiate a parallel poll sequence.
IFC	*Interface Clear:* Issued only by the controller to bring all active bus devices to a known state.
NDAC	*Not Data Accepted:* Issued by a listener while fetching data from the DIO lines.
NRFD	*Not Ready for Data:* Issued by all listeners and released by each listener as it becomes ready to accept data.
REN	*Remote Enable:* Grounded to the maintain control over the system.
SRQ	*Service Request:* Issued by any device needing service from the controller.

Source: IEEE Standard 488.

8 SUMMARY

We have examined the hardware side of two microprocessors and have observed the various relationships between timing and control signals on the respective busses. The internal architecture was explored for the 8080A and 8085 microprocessors. The chapter included a description of various bus standards used by both industry and consumers and examined an instrumentation bus that is very popular in the area of automatic testing.

9 GLOSSARY

Bus.   A system of parallel lines used to transfer address, data, control, and power from one component to another.

Interrupt.   Hardware interrupt; used to divert the processor's attention.

Machine cycle.   The individual components of a processor cycle; subdivided into three to five states.

Processor cycle or instruction cycle.   The time required to fetch and execute one instruction; consists of one to five machine cycles.

Restart.   An interrupt. May be hardware or software in nature.

Standard.   A document that establishes electrical specifications and conventions to be followed by manufacturers.

State. Subdivision of a machine cycle; the shortest time in a microprocessor at the clock frequency.

Sync pulse. Issued at the beginning of each machine cycle (8080A).

TRAP. A nonmaskable hardware interrupt line.

PROBLEMS

13-1 For the 8080A microprocessor running at 2 MHz, compute the state time and instruction cycle time for the MVI A instruction.

13-2 Repeat Problem 13-1 for the 8085 microprocessor.

13-3 On the 8080A, how many sync pulses are issued for the MVI A instruction?

13-4 Look up the 8224 clock chip and determine how an 18-MHz signal becomes a 2-MHz two-phase clock.

13-5 How is status information decoded (captured) from the 8080A?

13-6 What are the differences between the 8080A and 8085 MPUs?

13-7 Show NAND circuitry to generate an 8080A memory write pulse (MEMW).

13-8 Design address select circuitry for address E800 H to E8FF H.

13-9 How much RAM and how much EPROM are provided in the basic 8085 system of Fig. 13-23?

13-10 Explain why it is desirable to have standards for busses.

13-11 Design a decoder to operate an input/output port at port address 37 H.

13-12 Design a multiple port decoder to operate four input ports at addresses 80, 81, 82, and 83 H.

13-13 In a "ghost area" in RAM, the CPU can address two different areas of memory and get the same data. For example, a 256-byte RAM is fully decoded to base address 0000, but A8 is ignored in the address decoder; therefore the processor can get the same data from the RAM by addressing 0–FF or 100–1FF because A8 is ignored. In the decoder shown in Fig. 13-33, ROM is from 0 to 7FFF and RAM is from 8000 to FFFF. If a 2K × 8 RAM is used and a 4K × 8 ROM is used, how many ghosting areas exist?

FIGURE 13-33 Decoder circuitry for Problem 13-13.

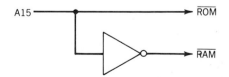

CHAPTER 14

PERIPHERAL CONTROLLERS

1 INSTRUCTIONAL OBJECTIVES

As a result of reading this chapter you should be able to:

1. Recognize that certain functions are better done by an LSI special-purpose chip than by a microprocessor.
2. Be familiar with a few of the available peripheral chips and their general function.
3. Infer the need for other types of peripheral controller chips.
4. Learn the part numbers of the following LSI peripheral chips:

   8741 Universal peripheral interface

   8202 DRAM controller

   8232 Floating-point processor

   8231 Arithmetic processor

   8253 Programmable interval timer

   8271/72 Floppy disk controller

   8275 Programmable CRT controller

   8279 Programmable keyboard/display interface

2 SELF-EVALUATION QUESTIONS

Keep the following questions in mind and try to answer them when you have completed the chapter:

1. What is the function of each of the peripheral chips listed in Section 1?
2. How does a controller chip like the 8741 vary from an 8085 microprocessor?
3. What is the main advantage of a math processor chip over a set of software subroutines?

4. How does a programmable interval timer free up a microprocessor?

5. What is the general idea behind any of the peripheral interface chips?

6. Do you suppose that other manufacturers produce LSI chips of similar function? (Affirmative.)

3 INTRODUCTION

A peripheral, generally speaking, is any device connected to a computer that performs a function for the computer. We have already learned about a few peripherals: the TV typewriter (video display with keyboard), floppy disk drives, and memories. In this chapter, we take a look at some advanced peripheral integrated circuits, all of which are very useful when designing computer systems or adding to the power of existing systems.

We are not going to get deeply involved with the theory of operation of each peripheral/controller we intend to study. Rather than investigate timing waveforms, we merely find out what kind of peripheral controllers are available and what they can do for us. For further interest, a deep look into an Intel catalog should satisfy the inquisitive mind.

4 THE 8741A UNIVERSAL PERIPHERAL INTERFACE

The 8041A/8641A/8741A universal peripheral interface is an 8-bit microcomputer with 1K bytes of ROM (or EPROM), 64 bytes of RAM, more than 90 instructions, and 18 programmable I/O pins on a single 40-pin DIP. Figure 14-1 presents the pin designations for the interface. The 8041 and 8641 are ROM versions; 8741, an EPROM version, can be erased with ultraviolet light and reprogrammed if necessary.

If the internal block diagram of the interface in Fig. 14-2 reminds you of the one for the 8080A/85, that is because the 8741 is actually an 8-bit microcomputer. The 8741 has more than 90 instructions that enable it to become a dedicated I/O controller, eliminating the need for peripheral routines on the host system. For instance, a memory-mapped video display is being used as output for a computer system. Instead of writing a COUT routine on the host system, the 8741 could take care of all scrolling, cursor, and other functions. All the main computer would have to do is send out character data to the 8741, like an output port. The instruction set presented in Table 14-1 gives you an idea of the 8741's range.

In addition, the 8741 has 18 I/O lines and interrupt capability—clearly it is virtually a complete microcomputing system in itself. The 8741 is ideal for controlling printers, displays, scanning keyboards, and so on. A peripheral like the 8741 should be used whenever program execution speed must be kept at a maximum.

FIGURE 14-1 Pin assignments for the 8741A. (Reprinted by permission of Intel Corporation, copyright 1982.)

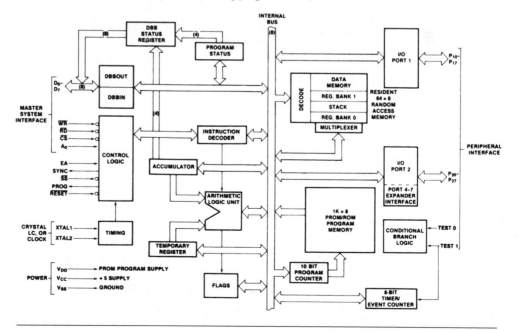

```
TEST 0 [ 1        40 ] Vcc
 XTAL1 [ 2        39 ] TEST 1
 XTAL2 [ 3        38 ] P27/DACK
 RESET [ 4        37 ] P26/DRQ
    SS [ 5        36 ] P25/IBF
    CS [ 6        35 ] P24/OBF
    EA [ 7        34 ] P17
    RD [ 8        33 ] P16
    A0 [ 9        32 ] P15
    WR [ 10   8041A/   31 ] P14
  SYNC [ 11   8741A   30 ] P13
    D0 [ 12       29 ] P12
    D1 [ 13       28 ] P11
    D2 [ 14       27 ] P10
    D3 [ 15       26 ] VDD
    D4 [ 16       25 ] PROG
    D5 [ 17       24 ] P23
    D6 [ 18       23 ] P22
    D7 [ 19       22 ] P21
   VSS [ 20       21 ] P20
```

FIGURE 14-2 Internal block diagram for the 8741A. (Reprinted by permission of Intel Corporation, copyright 1982.)

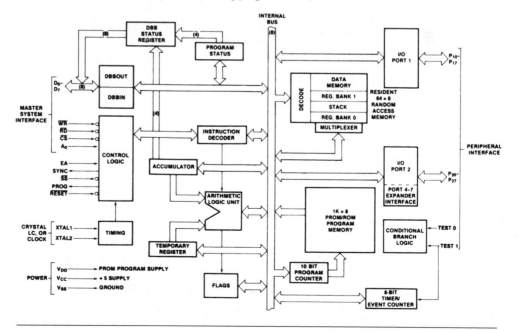

TABLE 14-1   8741 Instruction Set

Mnemonic	Description	Bytes	Cycles
ACCUMULATOR			
INC A	Increment A	1	1
DEC A	Decrement A	1	1
CLR A	Clear A	1	1
CPL A	Complement A	1	1
DA A	Decimal Adjust A	1	1
SWAP A	Swap nibbles of A	1	1
RL A	Rotate A left	1	1
RLC A	Rotate A left through carry	1	1
RR A	Rotate A right	1	1
RRC A	Rotate A right through carry	1	1
INPUT/OUTPUT			
IN A, Pp	Input port to A	1	2
OUTL Pp, A	Output A to port	1	2
ANL Pp, #data	AND immediate to port	2	2
ORL Pp, #data	OR immediate to port	2	2
IN A, DBB	Input DBB to A, clear IBF	1	1
OUT DBB, A	Output A to DBB, set OBF	1	1
MOV STS, A	A_4–A_7 to Bits 4–7 of Status	1	1
MOVD A, Pp	Input Expander port to A	1	2
MOVD Pp, A	Output A to Expander port	1	2
ANLD Pp, A	AND A to Expander port	1	2
ORLD Pp, A	OR A to Expander port	1	2
TIMER/COUNTER			
MOV A, T	Read Timer/Counter	1	1
MOV T, A	Load Timer/Counter	1	1
STRT T	Start Timer	1	1
STRT CNT	start Counter	1	1
STOP TCNT	Stop Timer/Counter	1	1
EN TCNTI	Enable Timer/ Counter Interrupt	1	1
DIS TCNTI	Disable Timer/ Counter Interrupt	1	1
CONTROL			
EN DMA	Enable DMA Hand- shake Lines	1	1
EN I	Enable IBF Interrupt	1	1
DIS I	Disable IBF Inter- rupt	1	1
EN FLAGS	Enable Master Interrupts	1	1
SEL RB0	Select register bank 0	1	1
SEL RB1	Select register bank 1	1	1
NOP	No Operation	1	1

Mnemonic	Description	Bytes	Cycles
REGISTERS			
INC Rr	Increment register	1	1
INC @Rr	Increment data memory	1	1
DEC Rr	Decrement register	1	1
SUBROUTINE			
CALL addr	Jump to subroutine	2	2
RET	Return	1	2
RETR	Return and restore status	1	2
FLAGS			
CLR C	Clear Carry	1	1
CPL C	Complement Carry	1	1
CLR F0	Clear Flag 0	1	1
CPL F0	Complement Flag 0	1	1
CLR F1	Clear F1 Flag	1	1
CPL F1	Complement F1 Flag	1	1
BRANCH			
JMP addr	Jump unconditional	2	2
JMPP @A	Jump indirect	1	2
DJNZ Rr, addr	Decrement register and jump	2	2
JC addr	Jump on Carry=1	2	2
JNC addr	Jump on Carry=0	2	2
JZ addr	Jump on A Zero	2	2
JNZ addr	Jump on A not Zero	2	2
JT0 addr	Jump on T0=1	2	2
JNT0 addr	Jump on T0=0	2	2
JT1 addr	Jump on T1=1	2	2
JNT1 addr	Jump on T1=0	2	2
JF0 addr	Jump on F0 Flag=1	2	2
JF1 addr	Jump on F1 Flag=1	2	2
JTF addr	Jump on Timer Flag =1, Clear Flag	2	2
JNIBF addr	Jump on IBF Flag =0	2	2
JOBF addr	Jump on OBF Flag =1	2	2
JBb addr	Jump on Accumula- tor Bit	2	2

TABLE 14-1 (continued)

Mnemonic	Description	Bytes	Cycles
ACCUMULATOR			
ADD A, Rr	Add register to A	1	1
ADD A, @Rr	Add data memory to A	1	1
ADD A, #data	Add immediate to A	2	2
ADDC A, Rr	Add register to A with carry	1	1
ADDC A, @Rr	Add data memory to A with carry	1	1
ADDC A, #data	Add immediate to A with carry	2	2
ANL A, Rr	AND register to A	1	1
ANL A, @Rr	AND data memory to A	1	1
ANL A, #data	AND immediate to A	2	2
ORL A, Rr	OR register to A	1	1
ORL A, @Rr	OR data memory to A	1	1
ORL A, #data	OR immediate to A	2	2
XRL A, Rr	Exclusive OR register to A	1	1
XRL A, @Rr	Exclusive OR data memory to A	1	1
XRL A, #data	Exclusive OR immediate to A	2	2

Mnemonic	Description	Bytes	Cycles
DATA MOVES			
MOV A, Rr	Move register to A	1	1
MOV A, @Rr	Move data memory to A	1	1
MOV A, #data	Move immediate TO A	2	2
MOV Rr, A	Move A to register	1	1
MOV @Rr, A	Move A to data memory	1	1
MOV Rr, #data	Move immediate to register	2	2
MOV @Rr, #data	Move immediate to data memory	2	2
MOV A, PSW	Move PSW to A	1	1
MOV PSW, A	Move A to PSW	1	1
XCH A, Rr	Exchange A and register	1	1
XCH A, @Rr	Exchange A and data memory	1	1
XCHD A, @Rr	Exchange digit of A and register	1	1
MOVP A, @A	Move to A from current page	1	2
MOVP3, A, @A	Move to A from page 3	1	2

Source: Reprinted by permission of Intel Corporation, copyright 1981.

5 THE 8202 DYNAMIC RAM CONTROLLER

Remember from Chapter 7 that there are two kinds of memories, static and dynamic. In a dynamic RAM, the address lines are multiplexed, enabling a single dynamic RAM chip to have a large amount of bit storage, with a small number of address lines. For instance, for a 4K × 1 bit dynamic RAM (DRAM), only six address lines would be needed [$2^{(2 \times 6)} = 2^{12} = 4096$)], plus a Row Address Select (RAS) and Column Address Select (CAS) line. When we start to design the decoding/multiplexing logic needed to implement a large memory system, we begin to realize the value of a stock dynamic RAM controller chip. With the seven $\overline{OUT}$ outputs and the four $\overline{RAS}$ outputs, the 8202 (Fig. 14-3) has the capability of driving up to 128K bytes of DRAM without the use of external decoding. The 8202 takes care of all timing, multiplexing, and refresh for the dynamic RAMs. Figure 14-4 shows the internal block diagram for the 8202 which, not surprisingly, contains a multiplexer, a timing and control block, and an arbiter. The arbiter controls which of the three cycles the 8202 is capable of operating in: read, write, and refresh.

Figure 14-5 shows the flowchart the 8202 uses in arbitration. The 8202 has control signal inputs designed for ease of use and interfacing with an 8085 microprocessor. No address bus latching is required. In Fig. 14-6 we see the 8202 in a

FIGURE 14-3 Pinout for the 8202 dynamic RAM controller.
(Reprinted by permission of Intel Corporation, copyright 1982.)

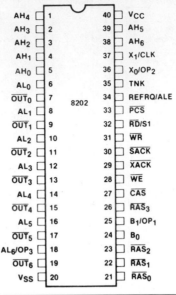

FIGURE 14-4 Block diagram for the 8202 controller. (Reprinted by
permission of Intel Corporation, copyright 1982.)

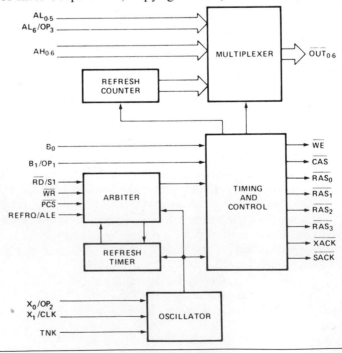

FIGURE 14-5 Arbitration flowchart for the 8202. (Reprinted by permission of Intel Corporation, copyright 1982.)

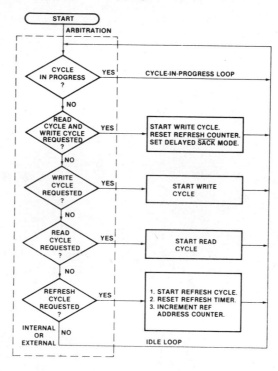

typical application. This setup uses 2117 DRAMs, which are all 16K × 1 bit memories. Thus, the 8202 is used to implement a 64K memory for an 8085 system. The 8202 can be used with other dynamic RAMs, such as 2104s or 2118s, just as easily. It is all a matter of selecting the right logic levels on control pins.

6 THE 8232 FLOATING-POINT PROCESSOR

In Chapter 12 we saw how to implement a simple floating-point package on the 8080A/85 microcomputer. It is not difficult to see that the accuracy and power of a floating-point routine (hence the speed) depends on the software used to implement it. An alternate method, however, would be to use an external hardware device to do the floating-point math for us. The 8232 does just this. The 8232 floating-point processor will add, subtract, multiply, and divide 32- or 64-bit binary numbers. The 8232 communicates via an 8-bit data bus, which makes it ideal for use as a peripheral.

Figures 14-7 and 14-8 give the pinout and description and the block diagram, respectively, of the 8232. The 8232 has an internal 768 × 16 bit ROM containing microinstructions that enable the 8232 to perform its math operations. The 8232

FIGURE 14-6 Typical application of the 8202. (Reprinted by permission of Intel Corporation, copyright 1982.)

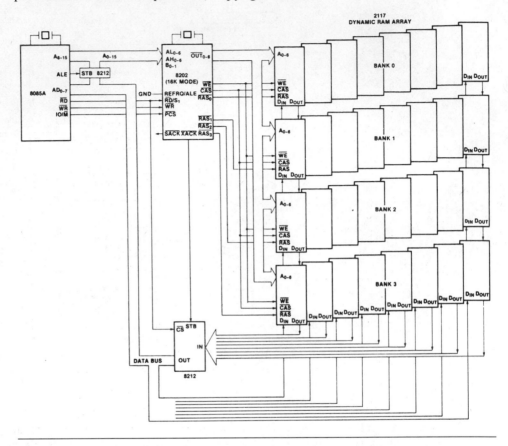

is controlled by a command word, sent to it over the 8-bit data bus. By using the 8232, the programmer simply writes routines to convert the decimal input to binary, control the 8232, and reconvert the results back into decimal. Table 14-2 summarizes the functions available in the 8232. Keep in mind that the 8232 is a binary processor, which explains the data of Table 14-3 (some sample math operations and their execution times).

The 8232 has its own internal stack and registers for keeping track of temporary values. The 8232 uses a simple encoding function, which the programmer must implement, to encode decimal numbers (floating point) into their binary representation. The expression

$$N = (-1)^S \, 2^{E-(2^7-1)} \, (1.M)$$

with annotations: "Bias" pointing to the (2^7-1) term and "Binary point" pointing to the $1.M$ term.

FIGURE 14-7 Pinout and description for the 8232. (Reprinted by permission of Intel Corporation, copyright 1982.)

PIN CONFIGURATION

Vss	1		24	END
Vcc	2		23	CLK
$\overline{EACK}$	3		22	RESET
$\overline{SVACK}$	4		21	A₀
SVREQ	5		20	$\overline{RD}$
ERR	6	8232	19	$\overline{WR}$
DO NOT USE	7		18	$\overline{CS}$
DB0	8		17	READY
DB1	9		16	VDD
DB2	10		15	DB7
DB3	11		14	DB6
DB4	12		13	DB5

PIN NAMES

RESET	RESET
DB0–DB7	DATA BUS
$\overline{CS}$	CHIP SELECT
$\overline{RD}$	READ DATA REGISTER
$\overline{WR}$	WRITE DATA OR COMMAND
A₀	COMMAND/DATA, INPUT
READY	READY, OUTPUT
END	END EXECUTION, OUTPUT
$\overline{EACK}$	END ACKNOWLEDGE, INPUT
SVREQ	SERVICE REQUEST, OUTPUT
$\overline{SVACK}$	SERVICE ACKNOWLEDGE, INPUT
CLK	CLOCK, INPUT
ERR	ERROR, OUTPUT

FIGURE 14-8 Block diagram for the 8232. (Reprinted by permission of Intel Corporation, copyright 1982.)

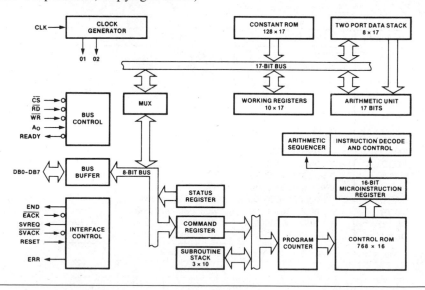

TABLE 14-2   8232 Command Survey

Commands Bits								Mnemonic	Description
7	6	5	4	3	2	1	0		
X	0	0	0	0	0	0	1	SADD	Add TOS to NOS single precision and result to NOS. Pop stack.
X	0	0	0	0	0	1	0	SSUB	Subtract TOS from NOS single precision and result to NOS. Pop stack.
X	0	0	0	0	0	1	1	SMUL	Multiply NOS by TOS single precision and result to NOS. Pop stack.
X	0	0	0	0	1	0	0	SDIV	Divide NOS by TOS single precision and result to NOS. Pop stack.
X	0	0	0	0	1	0	1	CHSS	Change sign of TOS single precision operand.
X	0	0	0	0	1	1	0	PTOS	Push single precision operand on TOS to NOS.
X	0	0	0	0	1	1	1	POPS	Pop single precision operand from TOS. NOS becomes TOS.
X	0	0	0	1	0	0	0	XCHS	Exchange TOS with NOS single precision.
X	0	1	0	1	1	0	1	CHSD	Change sign of TOS double precision operand.
X	0	1	0	1	1	1	0	PTOD	Push double precision operand on TOS to NOS.
X	0	1	0	1	1	1	1	POPD	Pop double precision operand from TOS. NOS becomes TOS.
X	0	0	0	0	0	0	0	CLR	CLR status.
X	0	1	0	1	0	0	1	DADD	Add TOS to NOS double precision and result to NOS. Pop stack.
X	0	1	0	1	0	1	0	DSUB	Subtract TOS from NOS double precision and result to NOS. Pop stack.
X	0	1	0	1	0	1	1	DMUL	Multiply NOS by TOS double precision and result to NOS. Pop stack.
X	0	1	0	1	1	0	0	DDIV	Divide NOS by TOS double precision and result to NOS. Pop stack.

Notes: X = Don't care.   Operation for bit combinations not listed above is undefined.

Source: Reprinted by permission of Intel Corporation, copyright 1982.

TABLE 14-3   8232 Execution Times

Command	TOS	NOS	Result	Clock Periods
SADD	3F800000	3F800000	40000000	58
SSUB	3F800000	3F800000	00000000	56
SMUL	40400000	3FC00000	40900000	198
SDIV	3F800000	40000000	3F000000	228
CHSS	3F800000	—	BF800000	10
PTOS	3F800000	—	—	16
POPS	3F800000	—	—	14
XCHS	3F800000	40000000	—	26
CHSD	3FF0000000000000	—	BFF0000000000000	24
PTOD	3FF0000000000000	—	—	40
POPD	3FF0000000000000	—	—	26
CLR	3FF0000000000000	—	—	4
DADD	3FF000000A000000	800000000000000	3FF00000A0000000	578
DSUB	3FF00000A0000000	800000000000000	3FF00000A0000000	578
DMUL	BFF800000000000	3FF800000000000	C002000000000000	1748
DDIV	BFF800000000000	3FF800000000000	BFF0000000000000	4560

Note: TOS, NOS and result are in hexadecimal; clock period is in decimal.

Source: Reprinted by permission of Intel Corporation, copyright 1982.

is used to represent a 32-bit, single-precision, floating-point number. In this method bit 31 is a sign bit, where zero means a positive number, and a one means a negative number, hence the $(-1)^S$ of the equation. Bits 23 through 30 represent the exponent. With these 8 bits, we can represent powers of 2 from $+128$ to -127. The last 23 bits (0–22) stand for the mantissa. There is a similar equation for double-precision numbers (64 bits). It is not difficult to see that it is easier to write two conversion routines and use the 8232 instead of developing all the math routines.

7 THE 8231 ARITHMETIC PROCESSOR

The 8232 allows us to use an integrated circuit to perform addition, subtraction, multiplication, and division. The 8231 provides us with these functions plus a great deal more. Figures 14-9 and 14-10 indicate that the pinout is basically the same, but there is a big difference in the block diagram. In the 8232 we need a microinstruction ROM to perform the math operations. In the 8231 we now see an algorithm controller, due to the large variety of operations that the 8231 will

FIGURE 14-9 Pinout and description for the 8231. (Reprinted by permission of Intel Corporation, copyright 1982.)

PIN CONFIGURATION

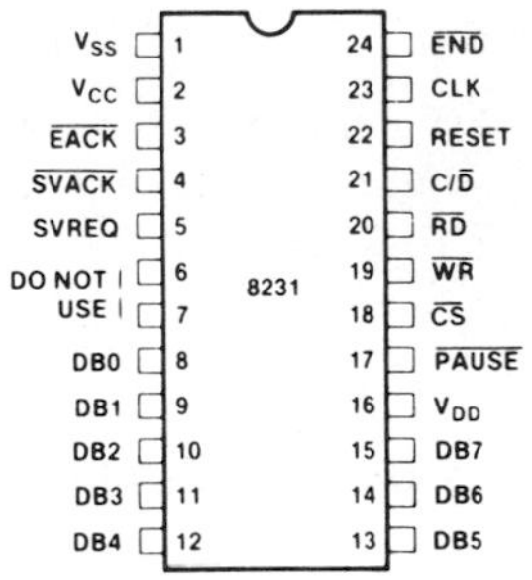

		8231		
V_{SS}	1		24	$\overline{END}$
V_{CC}	2		23	CLK
$\overline{EACK}$	3		22	RESET
$\overline{SVACK}$	4		21	$C/\overline{D}$
SVREQ	5		20	$\overline{RD}$
DO NOT	6		19	$\overline{WR}$
USE	7		18	$\overline{CS}$
DB0	8		17	$\overline{PAUSE}$
DB1	9		16	V_{DD}
DB2	10		15	DB7
DB3	11		14	DB6
DB4	12		13	DB5

PIN NAMES

RESET	RESET
DB0–DB7	DATA BUS
$\overline{CS}$	CHIP SELECT
$\overline{RD}$	READ, DATA OR STATUS
$\overline{WR}$	WRITE, DATA OR COMMAND
$C/\overline{D}$	COMMAND/DATA, INPUT
$\overline{PAUSE}$	PAUSE, OUTPUT
$\overline{END}$	END EXECUTION, OUTPUT
$\overline{EACK}$	END ACKNOWLEDGE, INPUT
SVREQ	SERVICE REQUEST, OUTPUT
$\overline{SVACK}$	SERVICE ACKNOWLEDGE, INPUT
CLK	CLOCK, INPUT

FIGURE 14-10 Block diagram for the 8231. (Reprinted by permission of Intel Corporation, copyright 1982.)

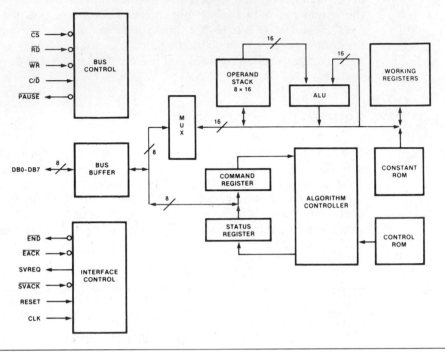

perform. Table 14-4 reveals that the 8231 performs trig functions, logarithmic functions, and single- and double-precision math. Like the 8232, the 8231 uses an algorithm for representing the floating-point number. The function, which is simply:

$$N = M2^E$$

is less complex than the preceding one.

EXAMPLE 14-1

Use the conversion equation to represent the decimal number 184 in binary.

SOLUTION

Since the mantissa must be between 0.5 and 1, we must use an exponent of 8. Thus, $2^8 = 256$. So, the mantissa we must convert into binary is 184 divided by 256, or 0.71875. Refer to Chapter 1 if you cannot convert 0.71875 into binary by yourself, but remember that all we have to do is keep multiplying by 2 until (hopefully) we get to zero. With luck, 0.71875 will convert to .10111 binary. Now it remains to put these bits into a format that the 8231 will understand.

With the addition of the 8231 to an existing operating system (BASIC) or to an operating system designed to implement the 8231, we can greatly increase the

TABLE 14-4 8231 Command Survey

Command Code							Command Mnemonic	Command Description[1]
7	6	5	4	3	2	1	0		

<table>
<tr><td colspan="10" align="center">FIXED POINT SINGLE PRECISION</td></tr>
<tr><td>R</td><td>1</td><td>1</td><td>0</td><td>1</td><td>1</td><td>0</td><td>0</td><td>SADD</td><td>Adds TOS to NOS. Result to NOS. Pop Stack.</td></tr>
<tr><td>R</td><td>1</td><td>1</td><td>0</td><td>1</td><td>1</td><td>0</td><td>1</td><td>SSUB</td><td>Subtracts TOS from NOS. Result to NOS Pop Stack.</td></tr>
<tr><td>R</td><td>1</td><td>1</td><td>0</td><td>1</td><td>1</td><td>1</td><td>0</td><td>SMUL</td><td>Multiplies NOS by TOS. Result to NOS. Pop Stack.</td></tr>
<tr><td>R</td><td>1</td><td>1</td><td>0</td><td>1</td><td>1</td><td>1</td><td>1</td><td>SDIV</td><td>Divides NOS by TOS. Result to NOS. Pop Stack.</td></tr>
<tr><td colspan="10" align="center">FIXED POINT DOUBLE PRECISION</td></tr>
<tr><td>R</td><td>0</td><td>1</td><td>0</td><td>1</td><td>1</td><td>0</td><td>0</td><td>DADD</td><td>Adds TOS to NOS. Result to NOS. Pop Stack.</td></tr>
<tr><td>R</td><td>0</td><td>1</td><td>0</td><td>1</td><td>1</td><td>0</td><td>1</td><td>DSUB</td><td>Subtracts TOS from NOS. Result to NOS. Pop Stack.</td></tr>
<tr><td>R</td><td>0</td><td>1</td><td>0</td><td>1</td><td>1</td><td>1</td><td>0</td><td>DMUL</td><td>Multiplies NOS by TOS. Result to NOS. Pop Stack.</td></tr>
<tr><td>R</td><td>0</td><td>1</td><td>0</td><td>1</td><td>1</td><td>1</td><td>1</td><td>DDIV</td><td>Divides NOS by TOS. Result to NOS. Pop Stack.</td></tr>
<tr><td colspan="10" align="center">FLOATING POINT</td></tr>
<tr><td>R</td><td>0</td><td>0</td><td>1</td><td>0</td><td>0</td><td>0</td><td>0</td><td>FADD</td><td>Adds TOS to NOS. Result to NOS. Pop Stack.</td></tr>
<tr><td>R</td><td>0</td><td>0</td><td>1</td><td>0</td><td>0</td><td>0</td><td>1</td><td>FSUB</td><td>Subtracts TOS from NOS. Result to NOS. Pop Stack.</td></tr>
<tr><td>R</td><td>0</td><td>0</td><td>1</td><td>0</td><td>0</td><td>1</td><td>0</td><td>FMUL</td><td>Multiplies NOS by TOS. Result to NOS. Pop Stack.</td></tr>
<tr><td>R</td><td>0</td><td>0</td><td>1</td><td>0</td><td>0</td><td>1</td><td>1</td><td>FDIV</td><td>Divides NOS by TOS. Result to NOS. Pop Stack.</td></tr>
<tr><td colspan="10" align="center">DERIVED FLOATING POINT FUNCTIONS[2]</td></tr>
<tr><td>R</td><td>0</td><td>0</td><td>0</td><td>0</td><td>0</td><td>0</td><td>1</td><td>SQRT</td><td>Square Root of TOS. Result in TOS.</td></tr>
<tr><td>R</td><td>0</td><td>0</td><td>0</td><td>0</td><td>0</td><td>1</td><td>0</td><td>SIN</td><td>Sine of TOS. Result in TOS.</td></tr>
<tr><td>R</td><td>0</td><td>0</td><td>0</td><td>0</td><td>0</td><td>1</td><td>1</td><td>COS</td><td>Cosine of TOS. Result in TOS.</td></tr>
<tr><td>R</td><td>0</td><td>0</td><td>0</td><td>0</td><td>1</td><td>0</td><td>0</td><td>TAN</td><td>Tangent of TOS. Result in TOS.</td></tr>
<tr><td>R</td><td>0</td><td>0</td><td>0</td><td>0</td><td>1</td><td>0</td><td>1</td><td>ASIN</td><td>Inverse Sine of TOS. Result in TOS.</td></tr>
<tr><td>R</td><td>0</td><td>0</td><td>0</td><td>0</td><td>1</td><td>1</td><td>0</td><td>ACOS</td><td>Inverse Cosine of TOS. Result in TOS.</td></tr>
<tr><td>R</td><td>0</td><td>0</td><td>0</td><td>0</td><td>1</td><td>1</td><td>1</td><td>ATAN</td><td>Inverse Tangent of TOS. Result in TOS.</td></tr>
<tr><td>R</td><td>0</td><td>0</td><td>0</td><td>1</td><td>0</td><td>0</td><td>0</td><td>LOG</td><td>Common Logarithm (base 10) of TOS. Result in TOS.</td></tr>
<tr><td>R</td><td>0</td><td>0</td><td>0</td><td>1</td><td>0</td><td>0</td><td>1</td><td>LN</td><td>Natural Logarithm (base e) of TOS. Result in TOS.</td></tr>
<tr><td>R</td><td>0</td><td>0</td><td>0</td><td>1</td><td>0</td><td>1</td><td>0</td><td>EXP</td><td>Exponential (e^x) of TOS. Result in TOS.</td></tr>
<tr><td>R</td><td>0</td><td>0</td><td>0</td><td>1</td><td>0</td><td>1</td><td>1</td><td>PWR</td><td>NOS raised to the power in TOS. Result to NOS. Pop Stack.</td></tr>
<tr><td colspan="10" align="center">DATA MANIPULATION COMMANDS[3]</td></tr>
<tr><td>R</td><td>0</td><td>0</td><td>0</td><td>0</td><td>0</td><td>0</td><td>0</td><td>NOP</td><td>No Operation.</td></tr>
<tr><td>R</td><td>0</td><td>0</td><td>1</td><td>1</td><td>1</td><td>1</td><td>1</td><td>FIXS</td><td>Converts TOS from floating point to single precision fixed point format.</td></tr>
<tr><td>R</td><td>0</td><td>0</td><td>1</td><td>1</td><td>1</td><td>1</td><td>0</td><td>FIXD</td><td>Converts TOS from floating point to double precision fixed point format.</td></tr>
<tr><td>R</td><td>0</td><td>0</td><td>1</td><td>1</td><td>1</td><td>0</td><td>1</td><td>FLTS</td><td>Converts TOS from single precision fixed point to floating point format.</td></tr>
<tr><td>R</td><td>0</td><td>0</td><td>1</td><td>1</td><td>1</td><td>0</td><td>0</td><td>FLTD</td><td>Converts TOS from double precision fixed point to floating point format.</td></tr>
<tr><td>R</td><td>1</td><td>1</td><td>1</td><td>0</td><td>1</td><td>0</td><td>0</td><td>CHSS</td><td>Changes sign of single precision fixed point operand on TOS.</td></tr>
<tr><td>R</td><td>0</td><td>1</td><td>1</td><td>0</td><td>1</td><td>0</td><td>0</td><td>CHSD</td><td>Changes sign of double precision fixed point operand on TOS.</td></tr>
<tr><td>R</td><td>0</td><td>0</td><td>1</td><td>0</td><td>1</td><td>0</td><td>1</td><td>CHSF</td><td>Changes sign of floating point operand on TOS.</td></tr>
<tr><td>R</td><td>1</td><td>1</td><td>1</td><td>0</td><td>1</td><td>1</td><td>1</td><td>PTOS</td><td>Push single precision fixed point operand on TOS to NOS.</td></tr>
<tr><td>R</td><td>0</td><td>1</td><td>1</td><td>0</td><td>1</td><td>1</td><td>1</td><td>PTOD</td><td>Push double precision fixed point operand on TOS to NOS.</td></tr>
<tr><td>R</td><td>0</td><td>0</td><td>1</td><td>0</td><td>1</td><td>1</td><td>1</td><td>PTOF</td><td>Push floating point operand on TOS to NOS.</td></tr>
<tr><td>R</td><td>1</td><td>1</td><td>1</td><td>1</td><td>0</td><td>0</td><td>0</td><td>POPS</td><td>Pop single precision fixed point operand from TOS. NOS becomes TOS.</td></tr>
<tr><td>R</td><td>0</td><td>1</td><td>1</td><td>1</td><td>0</td><td>0</td><td>0</td><td>POPD</td><td>Pop double precision fixed point operand from TOS. NOS becomes TOS.</td></tr>
<tr><td>R</td><td>0</td><td>0</td><td>1</td><td>1</td><td>0</td><td>0</td><td>0</td><td>POPF</td><td>Pop floating point operand from TOS. NOS becomes TOS.</td></tr>
<tr><td>R</td><td>1</td><td>1</td><td>1</td><td>1</td><td>0</td><td>0</td><td>1</td><td>XCHS</td><td>Exchange single precision fixed point operands TOS and NOS.</td></tr>
<tr><td>R</td><td>0</td><td>1</td><td>1</td><td>1</td><td>0</td><td>0</td><td>1</td><td>XCHD</td><td>Exchange double precision fixed point operands TOS and NOS.</td></tr>
<tr><td>R</td><td>0</td><td>0</td><td>1</td><td>1</td><td>0</td><td>0</td><td>1</td><td>XCHF</td><td>Exchange floating point operands TOS and NOS.</td></tr>
<tr><td>R</td><td>0</td><td>0</td><td>1</td><td>1</td><td>0</td><td>1</td><td>0</td><td>PUPI</td><td>Push floating point constant "π" onto TOS. Previous TOS becomes NOS.</td></tr>
</table>

Notes: 1. NOMENCLATURE: TOS is Top Of Stack. NOS is Next On Stack.
 2. All derived floating point functions destroy the contents of the stack. Only the result can be counted on to be valid upon command completion.
 3. Format conversion commands (FIXS, FIXD, FLTS, FLTD) require that floating point data format be specified (command bits 5 and 6 must be 0).

Source: Reprinted by permission of Intel Corporation, copyright 1982.

TABLE 14-5   8231 Execution Times

Command Mnemonic	Clock Cycles	Command Mnemonic	Clock Cycles	Command Mnemonic	Clock Cycles	Command Mnemonic	Clock Cycles
SADD	17	FADD	56-350	LN	4478	POPF	12
SSUB	30	FSUB	58-352	EXP	4616	XCHS	18
SMUL	92	FMUL	168	PWR	9292	XCHD	26
SDIV	92	FDIV	171	NOP	4	XCHF	26
DADD	21	SORT	800	CHSS	26	PUPI	16
DSUB	38	SIN	4464	CHSD	34		
DMUL	208	COS	4118	CHSF	16		
DDIV	208	TAN	5754	PTOS	16		
FIXS	92-216	ASIN	7668	PTOD	20		
FIXD	100-346	ACOS	7734	PTOF	20		
FLTS	98-186	ATAN	6006	POPS	10		
FLTD	98-378	LOG	4490	POPD	12		

Source: Reprinted by permission of Intel Corporation, copyright 1982.

FIGURE 14-11 Pinout and description for the 8253. (Reprinted by permission of Intel Corporation, copyright 1982.)

PIN CONFIGURATION

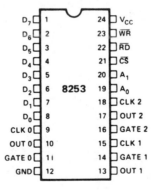

PIN NAMES

$D_7 \, D_0$	DATA BUS (8-BIT)
CLK N	COUNTER CLOCK INPUTS
GATE N	COUNTER GATE INPUTS
OUT N	COUNTER OUTPUTS
RD	READ COUNTER
WR	WRITE COMMAND OR DATA
CS	CHIP SELECT
$A_0 \, A_1$	COUNTER SELECT
V_{CC}	+5 VOLTS
GND	GROUND

operating speed of the language. Table 14-5 lists execution speeds of the 8231 as a representation of clock cycles. Notice that the longest operation is PWR, which takes 9292 clock cycles to perform. The fastest math operation is SADD (17 cycles), which translates into a large savings in time over software routines that accomplish the same task.

8 THE 8253 PROGRAMMABLE INTERVAL TIMER

The 8253 programmable interval timer is composed of three separate 16-bit counters, which can be programmed to count in binary or BCD. The 8253 would be a valuable aid in the process of providing software time delays. The 8253 has a maximum clock rate of 2 MHz. If the clock input is known, it is easy to compute binary constants to represent a certain time delay, and, through software, to use the 8253 instead of the microprocessor to time out the delay. Figures 14-11 and 14-12 show a pinout and description of the 8253 and its internal block diagram, respectively. The three independent counters are loaded or preset (programmed) via a control word. Through software, the programmer can read/write to any of the three

FIGURE 14-12 Internal block diagram for the 8253. (Reprinted by permission of Intel Corporation, copyright 1982.)

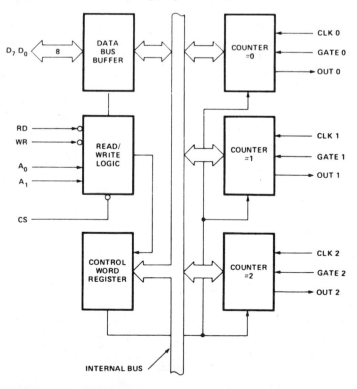

registers. The 8253 can be configured to interrupt the microprocessor when a preset count is reached. This enables the microprocessor to perform other tasks while the 8253 is timing out. The 8253 has many applications; for example, it can be used as a programmable one-shot, a rate/square-wave generator, or a software/hardware-triggered strobe.

9 FLOPPY DISK CONTROLLERS

In Chapter 7 we briefly introduced the idea of a floppy diskette for mass permanent storage. In a floppy disk drive, a floppy diskette is mounted inside a protective packet and coated with a magnetic material. The floppy diskette spins on a spindle at a synchronous rate of 360 rpm. On the smaller disk drives ($5\frac{1}{4}$ in.), each diskette contains 35 circular tracks, with each track containing 10 sectors. In a certain disk operating system (DOS), each sector contains 256 bytes (single density) or 512 bytes (double density). A quick calculation will show that (for double density) a single diskette can hold $35 \times 10 \times 512 = 179K$ bytes of data. Strictly speaking, this figure is for an unformatted diskette. In a formatted diskette, the first track, track 0, contains the diskette directory (catalog of programs) on the first few sectors. You generally lose a few kilobytes of storage capacity when you format a diskette.

Now, when the diskette is placed in a drive, it starts spinning on its spindle. When the disk controller issues a read or write command, the read/write head is

FIGURE 14-13 Pinouts for the 8271 and 8272 floppy disk controllers. (Reprinted by permission of Intel Corporation, copyright 1982.)

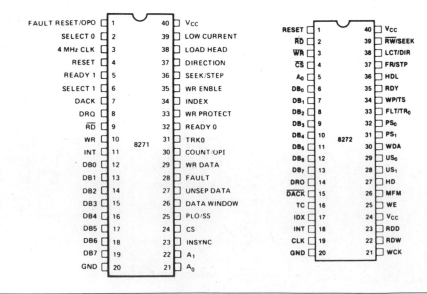

FIGURE 14-14 Controller connections for the 8271. (Reprinted by permission of Intel Corporation, copyright 1982.)

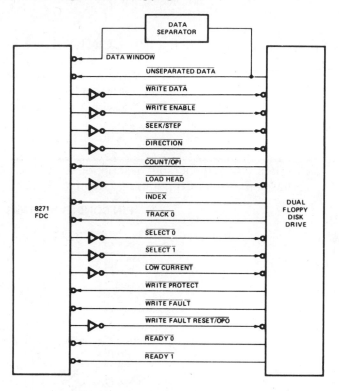

NOTE: INPUTS TO CHIP MAY REQUIRE RECEIVERS
         (AT LEAST PULL UP/DOWN PAIRS).

lowered close to the diskette surface to enable the transfer of data, much like the operation of a tape recorder. The read/write head is mounted on a movable assembly connected to a stepper motor that can move in and out in small steps to select each track on the diskette. When a disk operating system boots up, the software first seeks out track 0 by stepping the read/write head in. When the disk is at track 0, the DOS keeps track of the read/write head and never loses track of it. Because the operation of a disk drive is so complicated, two chips, the 8271 and the 8272, have been developed to aid in this process (see Fig. 14-13). These peripheral chips are designed to work with the IBM recording formats for floppy disks. The 8271 is programmable to accommodate sectors of different length (128, 256, or 512 bytes), among other things.

Regarding the theory of operation in the 8271 and the 8272, we note simply that these chips control *all* functions of the disk drive. Figure 14-14 shows a standard interface for the 8271. As you can see, the 8271 is capable of driving two floppy disk drives. The requirement for electronics is greatly reduced when a floppy disk controller chip is used because such arrangements as a port system (one for data, one for status, one for commands) or a memory-mapped method for implementing a disk system are not necessary.

10 THE 8275 PROGRAMMABLE CRT CONTROLLER

The Intel 8275 programmable CRT controller is a peripheral controller that greatly reduces the electronics required to implement a raster-scan (TV) display. Figure 14-15 gives a pinout of the 8275 with a description of the pins. All the sync and

FIGURE 14-15 Pinout and description for the 8275. (Reprinted by permission of Intel Corporation, copyright 1982.)

PIN CONFIGURATION

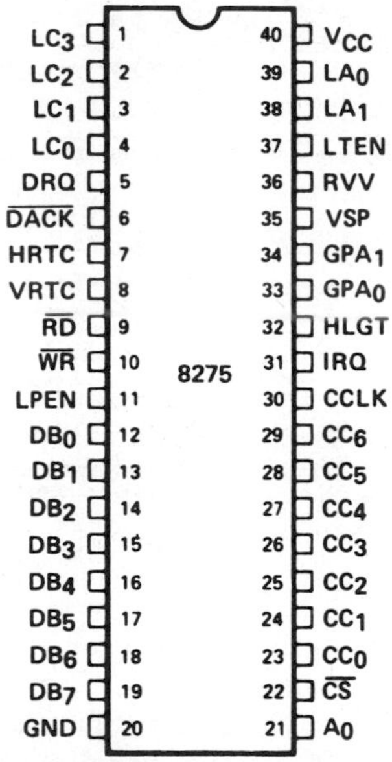

	8275	
LC_3 ☐ 1		40 ☐ V_{CC}
LC_2 ☐ 2		39 ☐ LA_0
LC_1 ☐ 3		38 ☐ LA_1
LC_0 ☐ 4		37 ☐ LTEN
DRQ ☐ 5		36 ☐ RVV
$\overline{DACK}$ ☐ 6		35 ☐ VSP
HRTC ☐ 7		34 ☐ GPA_1
VRTC ☐ 8		33 ☐ GPA_0
$\overline{RD}$ ☐ 9		32 ☐ HLGT
$\overline{WR}$ ☐ 10		31 ☐ IRQ
LPEN ☐ 11		30 ☐ CCLK
DB_0 ☐ 12		29 ☐ CC_6
DB_1 ☐ 13		28 ☐ CC_5
DB_2 ☐ 14		27 ☐ CC_4
DB_3 ☐ 15		26 ☐ CC_3
DB_4 ☐ 16		25 ☐ CC_2
DB_5 ☐ 17		24 ☐ CC_1
DB_6 ☐ 18		23 ☐ CC_0
DB_7 ☐ 19		22 ☐ $\overline{CS}$
GND ☐ 20		21 ☐ A_0

PIN NAMES

DB_{0-1}	B1–DIRECTIONAL DATA BUS	LC_{0-3}	LINE COUNTER OUTPUTS
DRQ	DMA REQUEST OUTPUT	LA_{0-1}	LINE ATTRIBUTE OUTPUTS
$\overline{DACK}$	DMA ACKNOWLEDGE INPUT	HRTC	HORIZONTAL RETRACE OUTPUT
IRQ	INTERRUPT REQUEST OUTPUT	VRTC	VERTICAL RETRACE OUTPUT
$\overline{RD}$	READ STROBE INPUT	HLGT	HIGHLIGHT OUTPUT
$\overline{WR}$	WRITE STROBE INPUT	RVV	REVERSE VIDEO OUTPUT
A_0	REGISTER ADDRESS INPUT	LTEN	LIGHT ENABLE OUTPUT
CS	CHIP SELECT INPUT	VSP	VIDEO SUPPRESS OUTPUT
CCLK	CHARACTER CLOCK INPUT	GPA_{0-1}	GENERAL PURPOSE ATTRIBUTE OUTPUTS
CC_{0-6}	CHARACTER CODE OUTPUTS	LPEN	LIGHT PEN INPUT

FIGURE 14-16 Internal block diagram for the 8275. (Reprinted by permission of Intel Corporation, copyright 1982.)

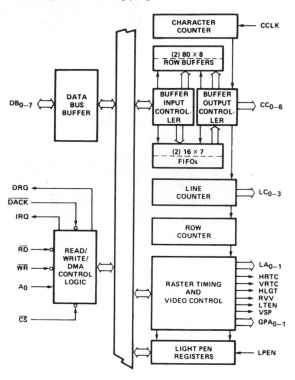

video signals are generated. In addition, since the 8275 is user programmable, the programmer can select a wide variety of display formats, including graphics. Cursor control is implemented, and there are connections for a light pen. Figure 14-16 shows the internal block diagram for the 8275. The line and row counters keep track of where data belongs (or is being placed) on the screen. In Fig. 14-17 we see how the 8275 forms letters by placing line after line of data onto the screen to form the characters. To implement the fancier functions of the 8275 (black on white), the 8275 must be provided with minimal electronics to decode the character attribute functions. Figure 14-18 suggests a circuit to generate actual raster-scan video. The 8275 is an excellent cost-effective circuit for use in a microcomputer system for which video, graphics, or both are a must.

11 THE 8279 PROGRAMMABLE KEYBOARD/DISPLAY INTERFACE

The 8279 is an extremely versatile I/O device. It can control the scanning of a keyboard, and it can also display data on a multiplexed display. The pinout and description of the 8279 and a logic symbol for the peripheral are shown in Figs.

FIGURE 14-17 Display of a character row. (Reprinted by permission of Intel Corporation, copyright 1982.)

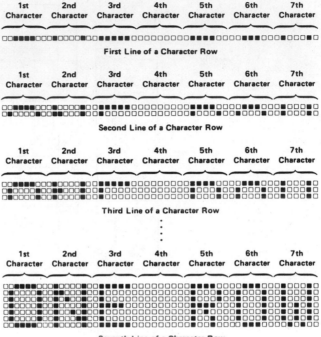

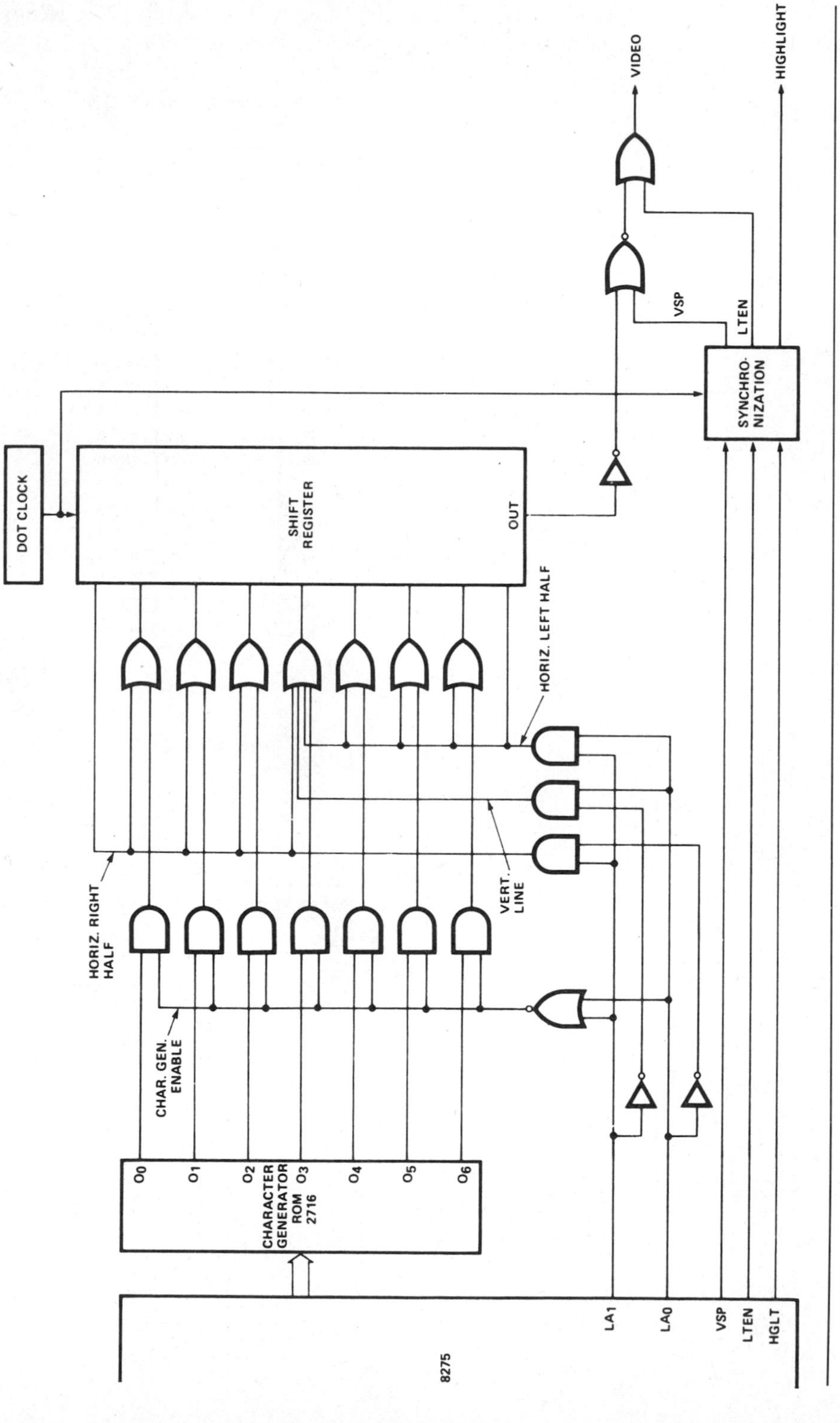

FIGURE 14-18 Circuitry for video generation. (Reprinted by permission of Intel Corporation, copyright 1982.)

FIGURE 14-19 Pinout and description for the 8279. (Reprinted by permission of Intel Corporation, copyright 1982.)

PIN CONFIGURATION

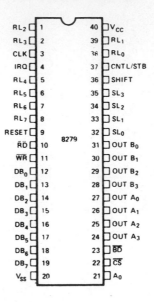

PIN NAMES

$DB_0 /$	I,O	DATA BUS (BI DIRECTIONAL)
CLK	I	CLOCK INPUT
RESET	I	RESET INPUT
$\overline{CS}$	I	CHIP SELECT
$\overline{RD}$	I	READ INPUT
$\overline{WR}$	I	WRITE INPUT
A_0	I	BUFFER ADDRESS
IRQ	O	INTERRUPT REQUEST OUTPUT
SL_{0-3}	O	SCAN LINES
RL_{0-7}	I	RETURN LINES
SHIFT	I	SHIFT INPUT
CNTL/STB	I	CONTROL/STROBE INPUT
OUT A_{0-3}	O	DISPLAY (A) OUTPUTS
OUT B_{0-3}	O	DISPLAY (B) OUTPUTS
$\overline{BD}$	O	BLANK DISPLAY OUTPUT

FIGURE 14-20 The 8279 logic symbol. (Reprinted by permission of Intel Corporation, copyright 1982.)

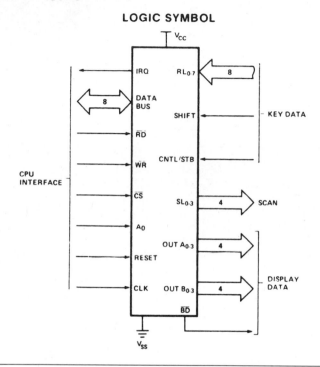

14-19 and 14-20, respectively. The 8279 was designed for 8-bit microcomputer systems to lighten the burden on the microprocessor in performing basic I/O functions.

From Fig. 14-21, the internal block diagram of the 8279, we can see that the 8279 has many RAM areas, plus a timing and control block. The 8279 can be used to scan (and debounce) an 8×8 key keyboard. The 8279 employs two-key lockout or N-key rollover (i.e., the last key hit is entered, even if N keys are already held down from previous hits) and has an 8-key buffer for the keyboard. The 8279 will also drive an 8- or 16-digit display (see Fig. 14-22), and right or left entry can be selected. All in all, the 8279 is a versatile display/input device that would make a good addition to any microcomputing system.

12 CONCLUSION

We have just studied eight peripheral controllers. Intel makes others—a dot matrix printer controller (8295), a host of GPIB (general-purpose interface bus) peripherals (8291, 8292, and 8293), a data encryption unit (8294), and an industrial digital processor (iSBC941). All these peripherals have useful functions, but they are not

FIGURE 14-21 Internal block diagram for the 8279. (Reprinted by permission of Intel Corporation, copyright 1982.)

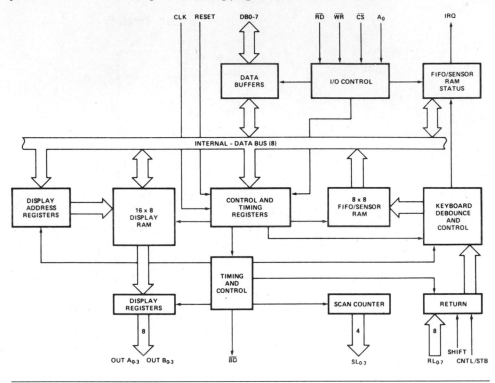

within the realm we are studying. For more information on these other peripherals, plus more detailed information on those just examined, a good look through an Intel catalog would be helpful.

13 GLOSSARY

Controller.   A microprocessor and its software used to control an external piece of equipment.

CRT.   A computer terminal using a cathode ray tube for display.

DRAM.   Dynamic random access memory.

GPIB.   General-purpose interface bus (now called IEEE-488).

Math processor.   A chip that performs math operations on data.

Peripheral.   A functional device attached to a computer (e.g., terminal, printer, memory, modem, math processor).

Timer.   A circuit that produces a delay by event count.

FIGURE 14-22 An application of the 8279. (Reprinted by
permission of Intel Corporation, copyright 1982.)

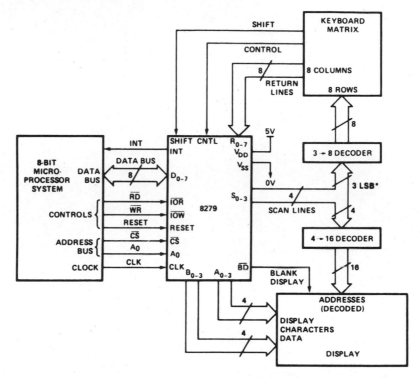

*Do not drive the keyboard decoder with the MSB of the scan lines.

CHAPTER 15
OTHER MICROPROCESSORS

1 INSTRUCTIONAL OBJECTIVES

This chapter introduces 8-bit and 16-bit microprocessors, a 32-bit microprocessor, single-chip microcomputers, a bit slice approach, and the microprocessor version of a minicomputer. A study of these powerful industrial processors will make you more comfortable in a world that is increasingly complex and controlled by different microprocessors. After reading this chapter you should:

1. Know the similarities between all microprocessors including interrupt structures, register architecture, speeds, and addressing capabilities.
2. Know the differences between the instruction sets and addressing modes of the given microprocessors.
3. Understand the difference between hardware and software interrupts and the various restart locations in memory.
4. Know the architecture of the single-chip microcomputers.
5. Be able to compare characteristics of various processors.

2 SELF-EVALUATION QUESTIONS

Keep the following questions in mind and try to answer them when you have completed the chapter.

1. What is a vectored interrupt and where does each CPU vector to?
2. How is the Z80 an improvement over the 8080A, 8085, 6800, and 6502 microprocessors?
3. What advantages do 16-bit CPUs have over the 8-bit variety?
4. Why use a single-chip microcomputer?
5. What is the LSI-11?
6. How are the 68000 and the LSI-11 similar?
7. What is the importance of indexed addressing?
8. Which processors are set up to handle *arrays* of data?

583

9. How do clock rates really affect processor speed?

10. How are many address lines and many data lines connected on the various processors?

3 INTRODUCTION

Thus far we have made an extensive study of the 8080A and the 8085 microprocessors. Their internal register set is well known to us, and extensive programming examples have been studied with a wide variety of applications in mind. We have seen the power of these two 8-bit microprocessors and have studied the many ways of using them. Having established a firm base in the use of these two similar microprocessors, let us now look at other available microprocessors. They vary in size, speed, and architecture and yet maintain a certain similarity. You are encouraged to re-read Chapter 9 on the organization of computers, for this chapter is the basis for our continuing study.

We will, in particular, study three very popular 8-bit microprocessors: the Z80, the 6800, and the 6502. We will then study the organization of two 16-bit microprocessors: the 8086 and the 68000. We will also look at the 432, a 32-bit machine intended to support the new Defense Department language, Ada (modeled after ''C'' and Pascal).

We will then turn to the concept of the single-chip microprocessor. Such a device can contain CPU, RAM, ROM, and I/O (including D/A or A/D) on one chip. Single-chip CPUs include the 8048 and the 8051.

At the end, to acquire an appreciation of the expanse of computing power available to the designer today, we examine a substantial list of many available microprocessors. Their differing architectures, speeds, instruction sets, and memory sizes make these microprocessors impressive to compare. Let us begin with three units with which we are somewhat familiar.

4 THE 8080A, THE 8085, AND THE Z80

We have already studied the 8080A and the 8085. We have learned that although they are not pin-for-pin compatible, they are architecturally identical. The register set is shown again for your review (Fig. 15-1).

The Z80 by Zilog contains twice the registers of the 8080A and the 8085 and features better speed (4 or 8 MHz). It is very similar to the 8080A and 8085 but has an expanded set of available instructions. This is due in part to the addition of a set of index registers.

Figure 15-2 shows the architecture of the Z80, with its duplicate set of registers (two As, Bs, Cs, etc.). How many times have we all wished for more registers to use as counters or for storing constants? In addition, the Z80 has two index registers (IX and IY), which allow for another way to refer to memory locations. Thus, we have three ways (IX, IY, and HL) to point to a memory location (M). Table 15-1 compares the 8080A, the 8085, and the Z80.

FIGURE 15-1 Architecture for the 8080A/85. (Reprinted by permission of Intel Corporation, copyright 1980.)

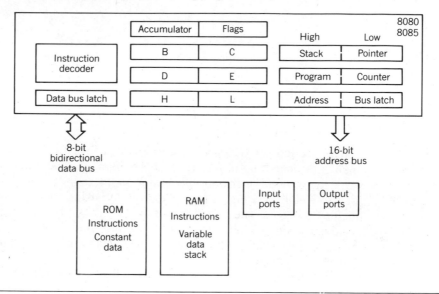

FIGURE 15-2 Architecture for the Z80. [Reproduced by permission, © 1983 Zilog, Inc. This material shall not be reproduced without the written consent of Zilog, Inc. Zilog, Z80®, and/or Z8000™ (whichever applicable) are/is a trademark(s) of Zilog, Inc., with whom John Wiley & Sons, Inc., is not associated.]

TABLE 15-1 Comparison of the 8080A, the 8085, and the Z80

CPU	8-Bit Registers	Maximum Clock (MHz)
8080A	7	2
8085	7	3.125[a]
Z80	14, plus 2 index registers	4 MHz[b]

[a]This is the maximum *internal* clock rate. A 6.25-MHz crystal must be used to achieve this frequency. In addition, the 8085 is available with a maximum internal clock of 5 MHz.
[b]6 and 8 MHz available.

The second set of registers in the Z80, called the "prime" set (A', B', etc.), makes the handling of interrupts more efficient. The index registers (IX and IY) are used to point to memory locations and, like the HL register pair, both IX and IY are 16-bit registers. A typical instruction using either index register also might include a displacement, d: IX + d or IY + d. In this case, the effective address used is the sum of the contents of the appropriate index register and the displacement, d. The trick is that d is a signed 2's-complement number (8 bits) in the range of −128 to +127.

For example, if IX = 728 H and the displacement is 40 H bytes, the effective address (EA) is:

$$
\begin{array}{ll}
IX & = 728 \text{ H} \\
d & = \underline{40 \text{ H}} \\
EA & = 768 \text{ H}
\end{array}
$$

If IX = 1268 H and the displacement is −40 H bytes, we first compute the signed 2's-complement quantity:

$$
\begin{array}{ll}
-40 \text{ H} & = -01000000 \\
1\text{'s complement} & = 10111111 \\
2\text{'s complement} & = 11000000 \text{ or C0 H}
\end{array}
$$

If

$$
\begin{array}{ll}
IX & = 1268 \text{ H} \\
d & = \underline{FFC0 \text{ H}} \quad \text{or} \quad -40 \text{ H} \\
EA & = 1228 \text{ H}
\end{array}
$$

In other words, the processor computes an effective address less than that specified by the index register if −128 ≤ d ≤ −1 and greater if 0 ≤ d ≤ 127.

5 THE Z80 MICROPROCESSOR

The Z80 can be purchased in a variety of speeds:

$$
\begin{array}{ll}
\text{Z80} & 2.5 \text{ MHz} \\
\text{Z80A} & 4 \text{ MHz} \\
\text{Z80B} & 6 \text{ MHz} \\
\text{Z80C} & 8 \text{ MHz}
\end{array}
$$

FIGURE 15-3 Pinout for the Z80 MPU. (Reproduced by permission, © 1983 Zilog, Inc. This material shall not be reproduced without the written consent of Zilog, Inc.)

Package Outline

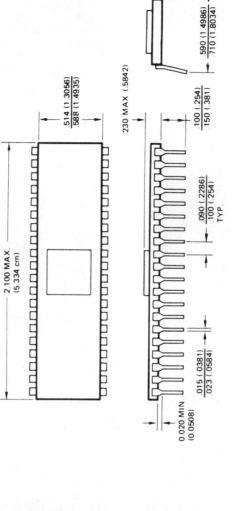

*Dimensions for metric system are in parentheses

Package Configuration

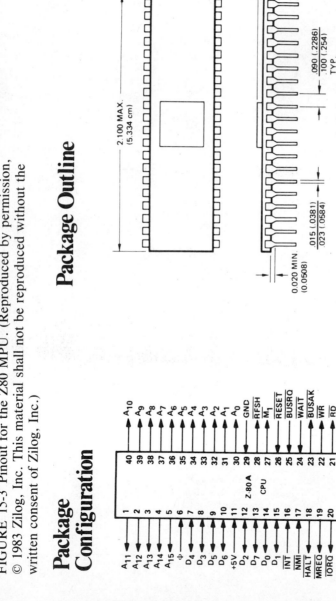

TABLE 15-2 Pin Description for the Z80 MPU

Pin Descriptions

A₀-A₁₅. *Address Bus* (output, active High, 3-state). A_0–A_{15} form a 16-bit address bus. The Address Bus provides the address for memory data bus exchanges (up to 64K bytes) and for I/O device exchanges.

BUSACK. *Bus Acknowledge* (output, active Low). Bus Acknowledge indicates to the requesting device that the CPU address bus, data bus, and control signals MREQ, IORQ, RD, and WR have entered their high-impedance states. The external circuitry can now control these lines.

BUSREQ. *Bus Request* (input, active Low). Bus Request has a higher priority than NMI and is always recognized at the end of the current machine cycle. BUSREQ forces the CPU address bus, data bus, and control signals MREQ, IORQ, RD, and WR to go to a high-impedance state so that other devices can control these lines. BUSREQ is normally wire-ORed and requires an external pullup for these applications. Extended BUSREQ periods due to extensive DMA operations can prevent the CPU from properly refreshing dynamic RAMs.

D₀-D₇. *Data Bus* (input/output, active High, 3-state). D_0–D_7 constitute an 8-bit bidirectional data bus, used for data exchanges with memory and I/O.

HALT. *Halt State* (output, active Low). HALT indicates that the CPU has executed a Halt instruction and is awaiting either a non-maskable or a maskable interrupt (with the mask enabled) before operation can resume. While halted, the CPU executes NOPs to maintain memory refresh.

INT. *Interrupt Request* (input, active Low). Interrupt Request is generated by I/O devices. The CPU honors a request at the end of the current instruction if the internal software-controlled interrupt enable flip-flop (IFF) is enabled. INT is normally wire-ORed and requires an external pullup for these applications.

IORQ. *Input/Output Request* (output, active Low, 3-state). IORQ indicates that the lower half of the address bus holds a valid I/O address for an I/O read or write operation. IORQ is also generated concurrently with M1 during an interrupt acknowledge cycle to indicate that an interrupt response vector can be placed on the data bus.

M1. *Machine Cycle One* (output, active Low). M1, together with MREQ, indicates that the current machine cycle is the opcode fetch cycle of an instruction execution. M1, together with IORQ, indicates an interrupt acknowledge cycle.

MREQ. *Memory Request* (output, active Low, 3-state). MREQ indicates that the address bus holds a valid address for a memory read or memory write operation.

NMI. *Non-Maskable Interrupt* (input, negative edge-triggered). NMI has a higher priority than INT. NMI is always recognized at the end of the current instruction, independent of the status of the interrupt enable flip-flop, and automatically forces the CPU to restart at location 0066H.

RD. *Read* (output, active Low, 3-state). RD indicates that the CPU wants to read data from memory or an I/O device. The addressed I/O device or memory should use this signal to gate data onto the CPU data bus.

RESET. *Reset* (input, active Low). RESET initializes the CPU as follows: it resets the interrupt enable flip-flop, clears the PC and Registers I and R, and sets the interrupt status to Mode 0. During reset time, the address and data bus go to a high-impedance state, and all control output signals go to the inactive state. Note that RESET must be active for a minimum of three full clock cycles before the reset operation is complete.

RFSH. *Refresh* (output, active Low). RFSH, together with MREQ, indicates that the lower seven bits of the system's address bus can be used as a refresh address to the system's dynamic memories.

WAIT. *Wait* (input, active Low). WAIT indicates to the CPU that the addressed memory or I/O devices are not ready for a data transfer. The CPU continues to enter a Wait state as long as this signal is active. Extended WAIT periods can prevent the CPU from refreshing dynamic memory properly.

WR. *Write* (output, active Low, 3-state). WR indicates that the CPU data bus holds valid data to be stored at the addressed memory or I/O location.

The Z80 features an on-chip dynamic memory refresh counter to make the use of dynamic memories less complicated. Only RAS and CAS timing must be supplied. The Z80 CPU pinout and the Z80 CPU pin description are shown in Fig. 15-3 and Table 15-2, respectively.

The internal registers may be looked at as either 14 8-bit registers or as a pair of accumulators (A, A′) and six 16-bit registers (BC, DE, HL, BC′, DE′, HL′). A group of "exchange" instructions is used to refer to the main or alternate (prime) set of registers accessible to the user.

The instructions for the Z80 are divided into the following categories, some identical to the 8080A and 8085 CPUs. The manufacturer's summary of instructions for the Z80 and the Z80A is presented in Fig. 15-4.

	8080A/8085	Z80
8-bit register load	X	X
16-bit register load	X	X
Exchange, block transfer, search		X
8-bit arithmetic	X	X
Logic operations	X	X
CPU control	X	X
16-bit arithmetic	X	X
Rotate and shift	X	X*
Bit set, reset, and test		X
Jumps	X	X
Jump relative		X
Calls, returns, restarts	X	X
Input and output	X	X**

*Enhanced: all registers plus M.
**Enhanced: port may be specified by the C register.

5.1 The Z80 Load Group

Considering the 8-bit load group (LD), let us compare the Z80 and 8080A/8085 mnemonics.

8080A/8085	Z80	Opcode (8080A/Z80)	
MOV r, r′	LD r, r′	A, B	: 78H
MVI r, n	LD r, n	A, data	: 3EH
MOV r, m	LD r, (HL)	A, M	: 7EH
—	LD r, (IX + d)		: DD . . . .
LDAX B	LD A, (BC)		: 0AH
STAX D	LD (DE), A		: 12H
LXI B, nn	LD BC, nn	B, nn	: 01 . . . .

A few points:

1. The mnemonics appear to be different between 8080A/85 and Z80, but the opcodes are the same. This means that *any* 8080A program (and any 8085 except for RIM and SIM!) will run on a Z80 without modification (but *not* the

FIGURE 15-4 The Z80 instruction set. (Reproduced by permission, © 1983 Zilog, Inc. This material shall not be reproduced without the written consent of Zilog, Inc.)

Instruction Set

The following is a summary of the Z80, Z80A instruction set showing the assembly language mnemonic and the symbolic operation performed by the instruction. A more detailed listing appears in the Z80-CPU technical manual, and assembly language programming manual. The instructions are divided into the following categories:

8-bit loads	Miscellaneous Group
16-bit loads	Rotates and Shifts
Exchanges	Bit Set, Reset and Test
Memory Block Moves	Input and Output
Memory Block Searches	Jumps
8-bit arithmetic and logic	Calls
16-bit arithmetic	Restarts
General purpose Accumulator & Flag Operations	Returns

In the table the following terminology is used.

b ≡ a bit number in any 8-bit register or memory location

cc ≡ flag condition code
 NZ ≡ non zero
 Z ≡ zero
 NC ≡ non carry
 C ≡ carry
 PO ≡ Parity odd or no over flow
 PE ≡ Parity even or over flow
 P ≡ Positive
 M ≡ Negative (minus)

d ≡ any 8-bit destination register or memory location
dd ≡ any 16-bit destination register or memory location
e ≡ 8-bit signed 2's complement displacement used in relative jumps and indexed addressing
L ≡ 8 special call locations in page zero. In decimal notation these are 0, 8, 16, 24, 32, 40, 48 and 56
n ≡ any 8-bit binary number
nn ≡ any 16-bit binary number
r ≡ any 8-bit general purpose register (A, B, C, D, E, H, or L)
s ≡ any 8-bit source register or memory location
s_b ≡ a bit in a specific 8-bit register or memory location
ss ≡ any 16-bit source register or memory location
subscript "L" ≡ the low order 8 bits of a 16-bit register
subscript "H" ≡ the high order 8 bits of a 16-bit register
( ) ≡ the contents within the ( ) are to be used as a pointer to a memory location or I/O port number

8-bit registers are A, B, C, D, E, H, L, I and R
16-bit register pairs are AF, BC, DE and HL
16-bit registers are SP, PC, IX and IY

Addressing Modes implemented include combinations of the following:

Immediate	Indexed
Immediate extended	Register
Modified Page Zero	Implied
Relative	Register Indirect
Extended	Bit

	Mnemonic	Symbolic Operation	Comments
8-BIT LOADS	LD r, s	r ← s	s ≡ r, n, (HL), (IX+e), (IY+e)
	LD d, r	d ← r	d ≡ (HL), r (IX+e), (IY+e)
	LD d, n	d ← n	d ≡ (HL), (IX+e), (IY+e)
	LD A, s	A ← s	s ≡ (BC), (DE), (nn), I, R
	LD d, A	d ← A	d ≡ (BC), (DE), (nn), I, R
16-BIT LOADS	LD dd, nn	dd ← nn	dd ≡ BC, DE, HL, SP, IX, IY
	LD dd, (nn)	dd ← (nn)	dd ≡ BC, DE, HL, SP, IX, IY
	LD (nn), ss	(nn) ← ss	ss ≡ BC, DE, HL, SP, IX, IY
	LD SP, ss	SP ← ss	ss ≡ HL, IX, IY
	PUSH ss	(SP-1) ← ss_H; (SP-2) ← ss_L	ss ≡ BC, DE, HL, AF, IX, IY
	POP dd	dd_L ← (SP); dd_H ← (SP+1)	dd ≡ BC, DE, HL, AF, IX, IY
EXCHANGES	EX DE, HL	DE ↔ HL	
	EX AF, AF'	AF ↔ AF'	
	EXX	$\begin{pmatrix} BC \\ DE \\ HL \end{pmatrix} ↔ \begin{pmatrix} BC' \\ DE' \\ HL' \end{pmatrix}$	
	EX (SP), ss	(SP) ↔ ss_L, (SP+1) ↔ ss_H	ss ≡ HL, IX, IY

	Mnemonic	Symbolic Operation	Comments
MEMORY BLOCK MOVES	LDI	(DE) ← (HL), DE ← DE+1 HL ← HL+1, BC ← BC-1	
	LDIR	(DE) ← (HL), DE ← DE+1 HL ← HL+1, BC ← BC-1 Repeat until BC = 0	
	LDD	(DE) ← (HL), DE ← DE-1 HL ← HL-1, BC ← BC-1	
	LDDR	(DE) ← (HL), DE ← DE-1 HL ← HL-1, BC ← BC-1 Repeat until BC = 0	
MEMORY BLOCK SEARCHES	CPI	A-(HL), HL ← HL+1 BC ← BC-1	
	CPIR	A-(HL), HL ← HL+1 BC ← BC-1, Repeat until BC = 0 or A = (HL)	A-(HL) sets the flags only. A is not affected
	CPD	A-(HL), HL ← HL-1 BC ← BC-1	
	CPDR	A-(HL), HL ← HL-1 BC ← BC-1, Repeat until BC= 0 or A = (HL)	
8-BIT ALU	ADD s	A ← A + s	CY is the carry flag
	ADC s	A ← A + s + CY	
	SUB s	A ← A - s	
	SBC s	A ← A - s - CY	s ≡ r, n, (HL) (IX+e), (IY+e)
	AND s	A ← A ∧ s	
	OR s	A ← A ∨ s	
	XOR s	A ← A ⊕ s	

FIGURE 15-4 (continued)

	Mnemonic	Symbolic Operation	Comments
8-BIT ALU	CP s	A − s	s = r, n (HL) (IX+e), (IY+e)
	INC d	d ← d + 1	d = r, (HL) (IX+e), (IY+e)
	DEC d	d ← d − 1	
16-BIT ARITHMETIC	ADD HL, ss	HL ← HL + ss	ss ≡ BC, DE HL, SP
	ADC HL, ss	HL ← HL + ss + CY	
	SBC HL, ss	HL ← HL − ss − CY	
	ADD IX, ss	IX ← IX + ss	ss ≡ BC, DE, IX, SP
	ADD IY, ss	IY ← IY + ss	ss ≡ BC, DE, IY, SP
	INC dd	dd ← dd + 1	dd ≡ BC, DE, HL, SP, IX, IY
	DEC dd	dd ← dd − 1	dd ≡ BC, DE, HL, SP, IX, IY
GP ACC. & FLAG	DAA	Converts A contents into packed BCD following add or subtract.	Operands must be in packed BCD format
	CPL	A ← $\overline{A}$	
	NEG	A ← 00 − A	
	CCF	CY ← $\overline{CY}$	
	SCF	CY ← 1	
MISCELLANEOUS	NOP	No operation	
	HALT	Halt CPU	
	DI	Disable Interrupts	
	EI	Enable Interrupts	
	IM 0	Set interrupt mode 0	8080A mode
	IM 1	Set interrupt mode 1	Call to 0038H
	IM 2	Set interrupt mode 2	Indirect Call
ROTATES AND SHIFTS	RLC s		s ≡ r, (HL) (IX+e), (IY+e)
	RL s		
	RRC s		
	RR s		
	SLA s		
	SRA s		
	SRL s		
	RLD		
	RRD		

	Mnemonic	Symbolic Operation	Comments
BIT S., R. & T	BIT b, s	Z ← $\overline{s_b}$	Z is zero flag
	SET b, s	s_b ← 1	s ≡ r, (HL)
	RES b, s	s_b ← 0	(IX+e), (IY+e)
INPUT AND OUTPUT	IN A, (n)	A ← (n)	
	IN r, (C)	r ← (C)	Set flags
	INI	(HL) ← (C), HL ← HL + 1 B ← B − 1	
	INIR	(HL) ← (C), HL ← HL + 1 B ← B − 1 Repeat until B = 0	
	IND	(HL) ← (C), HL ← HL − 1 B ← B − 1	
	INDR	(HL) ← (C), HL ← HL − 1 B ← B − 1 Repeat until B = 0	
	OUT(n), A	(n) ← A	
	OUT(C), r	(C) ← r	
	OUTI	(C) ← (HL), HL ← HL + 1 B ← B − 1	
	OTIR	(C) ← (HL), HL ← HL + 1 B ← B − 1 Repeat until B = 0	
	OUTD	(C) ← (HL), HL ← HL − 1 B ← B − 1	
	OTDR	(C) ← (HL), HL ← HL − 1 B ← B − 1 Repeat until B = 0	
JUMPS	JP nn	PC ← nn	cc { NZ PO / Z PE / NC P / C M
	JP cc, nn	If condition cc is true PC ← nn, else continue	
	JR e	PC ← PC + e	
	JR kk, e	If condition kk is true PC ← PC + e, else continue	kk { NZ NC / Z C
	JP (ss)	PC ← ss	ss ≡ HL, IX, IY
	DJNZ e	B ← B − 1, if B = 0 continue, else PC ← PC + e	
CALLS	CALL nn	(SP−1) ← PC_H (SP−2) ← PC_L, PC ← nn	cc { NZ PO / Z PE / NC P / C M
	CALL cc, nn	If condition cc is false continue, else same as CALL nn	
RESTARTS	RST L	(SP−1) ← PC_H (SP−2) ← PC_L, PC_H ← 0 PC_L ← L	
RETURNS	RET	PC_L ← (SP), PC_H ← (SP+1)	
	RET cc	If condition cc is false continue, else same as RET	cc { NZ PO / Z PE / NC P / C M
	RETI	Return from interrupt, same as RET	
	RETN	Return from non-maskable interrupt	

reverse!). Therefore, since the Z80A CPU is twice as fast, *any* 8080A program will run faster on a Z80!

2. All the mnemonics that load data to a register or are register-to-register data MOVEs are called register loads (LD). This makes program origination somewhat simpler.

3. Load instructions (LD) involving the index registers are prefixed with FDH (IY) or DDH (IX).

5.2 Exchange, Block Move, and Search for the Z80

The exchange instructions are used to handle the additional operations due to the double (prime) register set. These instructions, which are unique to the Z80, include a very useful set of instructions that can be used to move *blocks* of memory:

8080A/8085	Z80	
XCHG	EX DE, HL	DE $\rightleftarrows$ HL
—	EX AF, AF'	AF' $\rightleftarrows$ AF'
—	EXX	BC $\rightleftarrows$ BC'
		DE $\rightleftarrows$ DE'
		HL $\rightleftarrows$ HL'
XTHL	EX (SP), HL	
—	LDI	(DE) $\leftarrow$ (HL)
		DE $\leftarrow$ DE + 1
		HL $\leftarrow$ HL + 1
		BC $\leftarrow$ BC $-$ 1
BLOCK MOVE:	LDIR	(DE) $\leftarrow$ (HL)
		DE $\leftarrow$ DE + 1
		HL $\leftarrow$ HL + 1
		BC $\leftarrow$ BC $-$ 1
		Repeat until BC = 0
	LDD	(DE) $\leftarrow$ (HL)
		DE $\leftarrow$ DE $-$ 1
		HL $\leftarrow$ HL $-$ 1
		BC $\leftarrow$ BC $-$ 1
	LDDR	(DE) $\leftarrow$ (HL)
		DE $\leftarrow$ DE $-$ 1
		HL $\leftarrow$ HL $-$ 1
		BC $\leftarrow$ BC $-$ 1
		Repeat until BC = 0
BLOCK SEARCH:	CPI	A $-$ (HL)
		HL $\leftarrow$ HL + 1
		BC $\leftarrow$ BC $-$ 1
	CPIR	A $-$ (HL)
		HL $\leftarrow$ HL + 1
		BC $\leftarrow$ BC $-$ 1
		Repeat until A = (HL)
		or BC = 0

CPD	A − (HL)
	HL ← HL − 1
	BC ← BC − 1
CPDR	A − (HL)
	HL ← HL − 1
	BC ← BC − 1
	Repeat until A = (HL)
	or BC = 0

As an example, in the memory-mapped video display program (Chapter 11), it was necessary to move 1024 (400 H) bytes of ASCII data to screen memory. This was a transfer of a block of data to a range of memory locations starting at CC00 H. The corresponding program in 8080A/85 assembly language is as follows:

8080A/85

```
              LXI D, 0CC00H   ;Screen memory
              LXI H, DATA     ;Data table
              LXI B, 400H     ;# data items
MORE:         MOV A, M        ;Get data
              XCHG
              MOV M, A        ;Send to screen
              XCHG
              INX H           ;Next locations
              INX D
              DCX B           ;One less item
              MOV A, B
              ORA C
              JNZ MORE        ;Continue if not
                              zero
```

The corresponding sequence in Z80 assembly language is much shorter due to the block transfer instruction:

Z80

```
LD HL, DATA      ;Data table
LD DE, 0CC00H    ;Screen memory
LD BC, 400H      ;# data items
LDIR             ;Block move
```

As you can see, with LDIR, HL and DE are memory pointers and BC is the counter.

5.3 The Z80 Rotate and Shift Group

Another interesting capability that the Z80 has over the 8080A/85 instruction set is the ability to shift data left or right in any register or memory location and not be limited to accumulator operations. Some of these are listed to give you an idea of the added power of the Z80.

Instruction (Z80)	Description
RLC r RRC r	Rotate register left or right (eight bits).
RLC (IX + d) RLC (IY + d) RRC (IY + d) RRC (IY + d)	Rotate memory location left or right. Address equals index register X or Y plus displacement. The displacement ''d'' is a signed 2's-complement quantity in the range −128 to +127.

An interesting instruction that aids in BCD arithmetic is RLD or RRD.

RLD   Rotate digit left between
                                        the accumulator and
                                        location (HL).

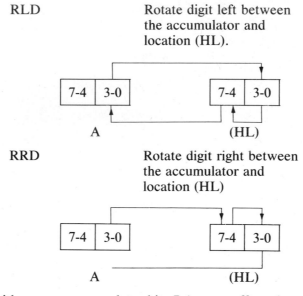

RRD   Rotate digit right between
                                        the accumulator and
                                        location (HL)

In either case, accumulator bits 7-4 are unaffected.

5.4 The Z80 Bit Set, Reset, and Test Group

The following instructions are unique to the Z80 and allow the programmer to affect bits in most CPU registers and all the memory locations.

SET 6, D	Will set bit 6 of register D (bits are MSB = 7, LSB = 0).
RES 4, B	Will reset bit 4 of register B.
SET 4, (HL)	Will set bit 4 of the memory location specified by HL.

The bit test instruction will either set or reset the carry bit to indicate whether a bit is high or low as specified in the instruction. For example:

BIT 6, C　　　Will test bit 6 of register C.
　　　　　　　The carry flag will match
　　　　　　　the bit tested.

Additional examples are as follows:

BIT 4, (HL)
BIT 7, (IX + 45H)
SET 6, A
RES 4, (IY + F7H)

5.5 The Z80 Input and Output Group

The Z80 I/O instructions have been augmented with a variety of instructions besides the basic into A and out from A. These are listed and described below:

Register I/O

IN r, (C)　　　Into register r from the port number specified by register C.
　　　　　　　r ← (C)

Block I/O

INI　　　Into memory (HL) from the port number specified by register C. B is decremented, HL is incremented.
　　　　　(HL) ← (C)
　　　　　B ← B − 1
　　　　　HL ← HL + 1

INIR　　　(HL) ← C
　　　　　B ← B − 1
　　　　　HL ← HL + 1
　　　　　Repeat until B = 0

IND　　　(HL) ← C
　　　　　B ← B − 1
　　　　　HL ← HL − 1

IND R　　　(HL) ← (C)
　　　　　B ← B − 1
　　　　　HL ← HL − 1
　　　　　Repeat until B = 0

The output instructions work the same way, using register C to specify a port number and HL as a pointer to a range of memory locations. Register B is used to count the operations. These block I/O instructions are used to read or write a block of memory via the I/O port specified in register C.

5.6 The Z80 Interrupts

The Z80 responds to two hardware interrupts: the $\overline{\text{INT}}$ and the $\overline{\text{NMI}}$. The former is a maskable interrupt line that can be either enabled (EI) or disabled (DI) at the programmer's discretion and causes a response by the CPU that depends on one of three modes selected by the programmer.

Mode 0: Identical to the 8080A response. In this mode the interrupting device can place an instruction on the data bus and the CPU will execute it. This may be a single-byte restart instruction causing the CPU to vector to a specific location.

Mode 1: In this mode the CPU loads 0038 H into the program counter and the effect is the same as an 8080A RST7.

Mode 2: This mode, the most powerful interrupt on the Z80, uses the I register (interrupt vector); it facilitates responses to many different sources of interrupt. The programmer sets up a table of addresses, which are the starting addresses of the various interrupt service routines. The appropriate element in this service routine table is pointed to by both the I register and the interrupting device. The I register must contain the upper 8 bits of the address of the table. The lower 8 bits of the address of the table are supplied by the interrupting device. Since the lower 8-bit information is placed on the data bus by the interrupting device, it is obvious that additional circuitry is required to use Z80 interrupt mode 2.

The three maskable interrupt modes are set by the use of one of three Z80 instructions:

Mnemonic	Opcode	Instruction
IM 0	ED46	Set mode 0
IM 1	ED56	Set mode 1
IM 2	ED5E	Set mode 2

The simplest use of the maskable interrupt is to set interrupt mode 1 and enable interrupts:

$$\text{IM 1}$$
$$\text{EI}$$

The nonmaskable interrupt $\overline{\text{NMI}}$ is unaffected by the EI and DI instructions. This negative edge-triggered interrupt causes the CPU to vector to location 0066 H. At this time, the program counter is automatically saved on the user's stack so that a return will be possible in the manner of the 8080A and 8085 CPUs.

The Z80 has the same software restart instructions as the 8080A (RST0–RST7); hence Z80 has a wide range of interrupt options.

5.7 Conclusion: The Z80 CPU

We have discussed most of the instructions that the Z80 has in addition to those of the 8080A/85 CPUs. They combine with the higher speed of the Z80 (4, 6, or 8 MHz) to make a CPU even more powerful than the 8080A or 8085. Keep in mind that any 8080A program is compatible with the Z80. You can double your speed if you just go to the Z80 CPU! Clever use of the additional instructions can increase the value of the Z80 CPU to the user. But the measure of that added value is in the ability of the programmer, as usual.

Let's review some of the Z80's attributes:

1. 8080A software compatibility is maintained. (*Note:* RIM and SIM are not implemented on the Z80.)

2. 8-, 6-, 4-MHz versions are available.

3. The instruction set includes string, bit, byte, and word operations with block searches and block transfers. Indexed and relative addressing are powerful features.

4. There is a vectored interrupt capability.

5. There is a duplicate set of registers.

6. The on-chip dynamic memory refresh counter reduces the amount of off-chip circuitry required for dynamic RAM.

And best of all, you already know the 8080A/85, so a shift to this new processor is relatively easy. It also introduces you to the idea of using an 8-bit signed 2's-complement displacement.

6 THE 6800 MICROPROCESSOR

Motorola, Inc., brought out the 6800 MPU at about the same time Intel introduced the 8080A. The 6800 is a 1-MHz, 8-bit microprocessor that, like the 8080A, 8085, and Z80, can address up to 65,536 memory locations but operates from a single 5-V power supply. The 6800 has two 8-bit accumulators (A and B) and an index register (X). The register architecture of the 6800 MPU, its instruction set, its pinout, and its pin description are shown in Figs. 15-5 to 15-7 and Table 15-3.

FIGURE 15-5 The 6800 MPU register set. (Courtesy of Motorola, Inc.)

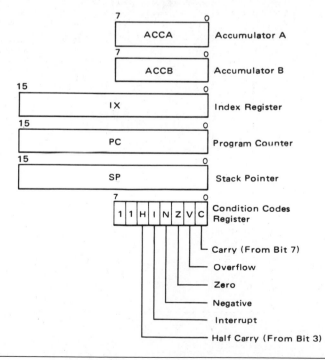

FIGURE 15-6 The 6800 MPU instruction set. (Courtesy of Motorola, Inc.)

OPERATIONS	MNEMONIC	IMMED OP	~	=	DIRECT OP	~	=	INDEX OP	~	=	EXTND OP	~	=	IMPLIED OP	~	=	BOOLEAN/ARITHMETIC OPERATION (All register labels refer to contents)	H (5)	I (4)	N (3)	Z (2)	V (1)	C (0)
Add	ADDA	8B	2	2	9B	3	2	AB	5	2	BB	4	3				A + M → A	↕	●	↕	↕	↕	↕
	ADDB	CB	2	2	DB	3	2	EB	5	2	FB	4	3				B + M → B	↕	●	↕	↕	↕	↕
Add Acmltrs	ABA													1B	2	1	A + B → A	↕	●	↕	↕	↕	↕
Add with Carry	ADCA	89	2	2	99	3	2	A9	5	2	B9	4	3				A + M + C → A	↕	●	↕	↕	↕	↕
	ADCB	C9	2	2	D9	3	2	E9	5	2	F9	4	3				B + M + C → B	↕	●	↕	↕	↕	↕
And	ANDA	84	2	2	94	3	2	A4	5	2	B4	4	3				A · M → A	●	●	↕	↕	R	●
	ANDB	C4	2	2	D4	3	2	E4	5	2	F4	4	3				B · M → B	●	●	↕	↕	R	●
Bit Test	BITA	85	2	2	95	3	2	A5	5	2	B5	4	3				A · M	●	●	↕	↕	R	●
	BITB	C5	2	2	D5	3	2	E5	5	2	F5	4	3				B · M	●	●	↕	↕	R	●
Clear	CLR							6F	7	2	7F	6	3				00 → M	●	●	R	S	R	R
	CLRA													4F	2	1	00 → A	●	●	R	S	R	R
	CLRB													5F	2	1	00 → B	●	●	R	S	R	R
Compare	CMPA	81	2	2	91	3	2	A1	5	2	B1	4	3				A − M	●	●	↕	↕	↕	↕
	CMPB	C1	2	2	D1	3	2	E1	5	2	F1	4	3				B − M	●	●	↕	↕	↕	↕
Compare Acmltrs	CBA													11	2	1	A − B	●	●	↕	↕	↕	↕
Complement, 1's	COM							63	7	2	73	6	3				$\overline{M}$ → M	●	●	↕	↕	R	S
	COMA													43	2	1	$\overline{A}$ → A	●	●	↕	↕	R	S
	COMB													53	2	1	$\overline{B}$ → B	●	●	↕	↕	R	S
Complement, 2's	NEG							60	7	2	70	6	3				00 − M → M	●	●	↕	↕	①	②
(Negate)	NEGA													40	2	1	00 − A → A	●	●	↕	↕	①	②
	NEGB													50	2	1	00 − B → B	●	●	↕	↕	①	②
Decimal Adjust, A	DAA													19	2	1	Converts Binary Add. of BCD Characters into BCD Format	●	●	↕	↕	↕	③
Decrement	DEC							6A	7	2	7A	6	3				M − 1 → M	●	●	↕	↕	④	●
	DECA													4A	2	1	A − 1 → A	●	●	↕	↕	④	●
	DECB													5A	2	1	B − 1 → B	●	●	↕	↕	④	●
Exclusive OR	EORA	88	2	2	98	3	2	A8	5	2	B8	4	3				A ⊕ M → A	●	●	↕	↕	R	●
	EORB	C8	2	2	D8	3	2	E8	5	2	F8	4	3				B ⊕ M → B	●	●	↕	↕	R	●
Increment	INC							6C	7	2	7C	6	3				M + 1 → M	●	●	↕	↕	⑤	●
	INCA													4C	2	1	A + 1 → A	●	●	↕	↕	⑤	●
	INCB													5C	2	1	B + 1 → B	●	●	↕	↕	⑤	●
Load Acmltr	LDAA	86	2	2	96	3	2	A6	5	2	B6	4	3				M → A	●	●	↕	↕	R	●
	LDAB	C6	2	2	D6	3	2	E6	5	2	F6	4	3				M → B	●	●	↕	↕	R	●
Or, Inclusive	ORAA	8A	2	2	9A	3	2	AA	5	2	BA	4	3				A + M → A	●	●	↕	↕	R	●
	ORAB	CA	2	2	DA	3	2	EA	5	2	FA	4	3				B + M → B	●	●	↕	↕	R	●
Push Data	PSHA													36	4	1	A → M$_{SP}$, SP − 1 → SP	●	●	●	●	●	●
	PSHB													37	4	1	B → M$_{SP}$, SP − 1 → SP	●	●	●	●	●	●
Pull Data	PULA													32	4	1	SP + 1 → SP, M$_{SP}$ → A	●	●	●	●	●	●
	PULB													33	4	1	SP + 1 → SP, M$_{SP}$ → B	●	●	●	●	●	●
Rotate Left	ROL							69	7	2	79	6	3				M	●	●	↕	↕	⑥	↕
	ROLA													49	2	1	A	●	●	↕	↕	⑥	↕
	ROLB													59	2	1	B	●	●	↕	↕	⑥	↕
Rotate Right	ROR							66	7	2	76	6	3				M	●	●	↕	↕	⑥	↕
	RORA													46	2	1	A	●	●	↕	↕	⑥	↕
	RORB													56	2	1	B	●	●	↕	↕	⑥	↕
Shift Left, Arithmetic	ASL							68	7	2	78	6	3				M	●	●	↕	↕	⑥	↕
	ASLA													48	2	1	A	●	●	↕	↕	⑥	↕
	ASLB													58	2	1	B	●	●	↕	↕	⑥	↕
Shift Right, Arithmetic	ASR							67	7	2	77	6	3				M	●	●	↕	↕	⑥	↕
	ASRA													47	2	1	A	●	●	↕	↕	⑥	↕
	ASRB													57	2	1	B	●	●	↕	↕	⑥	↕
Shift Right, Logic	LSR							64	7	2	74	6	3				M	●	●	R	↕	⑥	↕
	LSRA													44	2	1	A	●	●	R	↕	⑥	↕
	LSRB													54	2	1	B	●	●	R	↕	⑥	↕
Store Acmltr.	STAA				97	4	2	A7	6	2	B7	5	3				A → M	●	●	↕	↕	R	●
	STAB				D7	4	2	E7	6	2	F7	5	3				B → M	●	●	↕	↕	R	●
Subtract	SUBA	80	2	2	90	3	2	A0	5	2	B0	4	3				A − M → A	●	●	↕	↕	↕	↕
	SUBB	C0	2	2	D0	3	2	E0	5	2	F0	4	3				B − M → B	●	●	↕	↕	↕	↕
Subtract Acmltrs.	SBA													10	2	1	A − B → A	●	●	↕	↕	↕	↕
Subtr. with Carry	SBCA	82	2	2	92	3	2	A2	5	2	B2	4	3				A − M − C → A	●	●	↕	↕	↕	↕
	SBCB	C2	2	2	D2	3	2	E2	5	2	F2	4	3				B − M − C → B	●	●	↕	↕	↕	↕
Transfer Acmltrs	TAB													16	2	1	A → B	●	●	↕	↕	R	●
	TBA													17	2	1	B → A	●	●	↕	↕	R	●
Test, Zero or Minus	TST							6D	7	2	7D	6	3				M − 00	●	●	↕	↕	R	R
	TSTA													4D	2	1	A − 00	●	●	↕	↕	R	R
	TSTB													5D	2	1	B − 00	●	●	↕	↕	R	R
																		H	I	N	Z	V	C

FIGURE 15-6 (continued)

LEGEND:

OP	Operation Code (Hexadecimal);
~	Number of MPU Cycles;
=	Number of Program Bytes;
+	Arithmetic Plus;
−	Arithmetic Minus;
·	Boolean AND;
M_{SP}	Contents of memory location pointed to be Stack Pointer;

+	Boolean Inclusive OR;
⊙	Boolean Exclusive OR;
$\overline{M}$	Complement of M;
→	Transfer Into;
0	Bit = Zero;
00	Byte = Zero;

Note – Accumulator addressing mode instructions are included in the column for IMPLIED addressing

CONDITION CODE SYMBOLS:

H	Half-carry from bit 3;
I	Interrupt mask
N	Negative (sign bit)
Z	Zero (byte)
V	Overflow, 2's complement
C	Carry from bit 7
R	Reset Always
S	Set Always
↕	Test and set if true, cleared otherwise
●	Not Affected

FIGURE 15-7 The 6800 MPU pinout. (Courtesy of Motorola, Inc.)

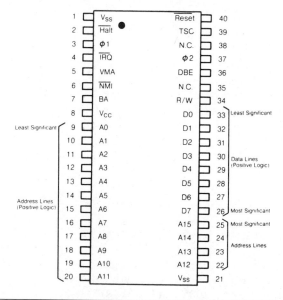

TABLE 15-3   Pin Description for the 6800 MPU

SUMMARY OF CONTROL LINES

	LOW "0" V_{ss} ≤0.4 V	HIGH "1" ≥2.4 V
HALT	All machine activity halted	Machine will fetch and execute instructions
IRQ	Interrupt request pending	No Interrupt Request
VMA	Address Bus data is INVALID	Address Bus data is VALID
NMI	A nonmaskable Interrupt Request is pending	Nonmaskable interrupt is not pending
BA	Address bus is not available	Address bus is available and microprocessor activity has stopped
RESET	All microprocessor registers cleared, all machine activity halted	On ⌐ microprocessor executes initial start-up sequence
TSC	Address lines and R/W line are not HI-IMPEDANCE	Address lines and R/W in HI-IMPEDANCE
DBE	Data bus lines are in HI-IMPEDANCE	Data bus drivers are enabled
R/W	MPU "WRITE" operation-data from microprocessor to memory or peripheral	MPU "READ" operation-data from memory or peripheral into microprocessor

Source: Courtesy of Motorola, Inc.

(*Note:* The 68A00 is a 1.5-MHz version of the 6800 and the 68B00 is a 2.0-MHz version.)

Since the 6800 runs at 1 MHz, you might initially assume that it is slower than the 8080A. However, the 6800 requires fewer machine cycles for most instructions, with the overall effect that the two microprocessors are comparable in speed.

The two accumulators of the 6800 are handy to have, and instructions typically specify which of the two (A or B) is to be used.

LDAA	Load accumulator A	$M \rightarrow A$
LDAB	Load accumulator B	$M \rightarrow B$
ADDA	Add to A	$A + M \rightarrow A$
ADDB	Add to B	$B + M \rightarrow B$

(The arrow is read "transferred to." Thus $M \rightarrow A$ is read "Memory is transferred to A.")

The stack pointer (SP) register and the index register are both 16-bit registers used to point to a memory location. The SP is used in the same way that it is in the 8080A, 8085, and Z80 MPUs. Since the stack area is used for interrupts, subroutines, push operations, and so on, the stack builds toward lesser locations (i.e., toward zero from FFFF).

The 6800 does *not* use byte swapping, so the high-order byte is loaded ahead of the low-order byte. This is reversed from the microprocessors studied earlier.

The index register is extremely useful in working with data tables and is used to contain the base address of a table. It can be incremented or decremented and is also used with instructions that specify an offset to be added to the base address contained in the index register. This can be used to point to the *next* data table.

This offset is an *unsigned* 8-bit number; hence a negative offset is not allowed when working with the index register on the 6800. (It was OK on the Z80.)

Suppose there are three data tables, each 50 hex bytes long, starting at 2000 H: Table 1 is at 2000 H, Table 2 is at 2050 H, and Table 3 begins at 20A0 H. The index register is loaded with the base address of 2000 H or LDX #$2000 depending on assembler (LDX #2000 H = CE2000). Any instruction dealing with a data table then may specify Table 1, 2, or 3 by including an offset (or displacement) of zero, 50 H, or 0A0 H, respectively. To point to successive elements of a table, the index register is incremented or decremented (INX, DEX). The index register is the significant architectural difference between microprocessors, and this section summarizes its operation.

The 6800 instruction set is divided into the following groups:

Transfer (load and store): Accumulators, index and stack register, memory

Logical: AND, OR, XOR accumulators and memory

Rotate and Shift: Accumulators and memory

Arithmetic: Add, subtract, increment, decrement, complement

Test, Compare, Clear: Test with AND

Jumps: Branch, jump, call

I/O: The 6800 uses memory-mapped I/O

Note: The 6800 has six flags (N = negative, I = interrupt, H = half carry, Z = zero, V = overflow, C = carry). Many of the instructions affect the flags. Consult the manufacturer's data sheets on the 6800.

The 6800 assembler terminology is summarized here to assist in understanding the description of instruction types:

$	Hexadecimal number
#	Immediate data
#$	Immediate hex data

Instruction	*Comment*
ADDA #28	Add 28 D to accumulator A
ADDB #$28	Add 28 H to accumulator B
ADDA $128	Add contents of location 128 H to A
ADDA $20, X	Add contents of address specified by 20H plus index register to accumulator A

The 6800 uses six modes of addressing:

	Mode	Example	Comment
1.	Immediate	LDAA #30	Put 30 D in A
2.	Direct	LDAA $20	Put contents of location 20 H in A
3.	Indexed	LDAA 0, X	Put contents of location X + 0 in A
4.	Extended	LDAA $2000	Put contents of location 2000 H in A
5.	Inherent (rotates)	ROLA	Rotate accumulator A left through carry
6.	Accumulator	COMA	Complement accumulator A

6.1 The 6800 Load and Transfer Group

The 6800 uses a variety of methods for moving data from register to register or memory and from memory to register. Any of the addressing modes may be utilized. Immediate addressing allows the basic registers to be loaded with initial data.

Example	Comment
LDAA #25	Load A with 25 D
LDAB #$50	Load B with 50 H
LDX #$2040	Load index register with 2040 H
LDS #1000	Load stack pointer register with 1000 D

Direct and extended* addressing deals with memory locations.

Example	Comment
*LDAB $1000	$(1000\ H) \rightarrow B$
*STAA $2020	$A \rightarrow (2020\ H)$
LDX $50	$(50\ H) \rightarrow X_H, (51\ H) \rightarrow X_L$
*STX $1008	$X_H \rightarrow (1008\ H), X_L \rightarrow (1009\ H)$
STS 200	$SP_H \rightarrow (200\ D), SP_L \rightarrow (201\ D)$

*Extended addressing requires 2 bytes for the address and can address 0–FFFF H. Direct addressing requires 1 byte (zero page) and can address only 00–00FF H. Indexed addressing uses the index register to determine which memory address is involved in an operation. In indexed addressing, an effective address is calculated from the sum of the contents of the index (X) register and the displacement (d).

The displacement is an *unsigned* quantity from 0 to 255 D.

Example	*Comment*
LDAA $20, X	Load A from the address X + 20 H
STAB 10, X	Store B in address 10 D + X
LDX $20, X	Load index register with 2 bytes from X + 20 H and X + 21 H

Also included are transfer instructions that move one register to another.

Example	*Comment*
TAB	Transfer A to B
TBA	Transfer B to A
TXS	X − 1 → SP
TSX	SP + 1 → X

6.2 The 6800 Logical and Shift Instructions

Both accumulators (A and B) and any memory location of the 6800 may be affected by the logical and shift instructions, which include rotates and shifts, and the logical AND, OR, XOR, and invert instructions.

Instruction	*Comment*
ANDA #$80	AND A with 80 H; result in A
ANDB $500	AND B with contents of location 500 H; result in B
NEG $40, X	2's complement of location (X + 40 H)
NEGA	2's complement of A
NEGB	2's complement of B
COM $4000	1's complement of contents of location 4000 H
COMA	1's complement of A
COMB	1's complement of B
EORA #01	XOR A with 01
EORB $20, X	XOR B with contents of location (X + 20 H)
ORAA $1000	OR A with contents of location 1000 H
ORAB #0	OR B with 0
ROL 20, X	Rotate contents of location (X + 20) left (9 bit)
ROLA	Rotate A left (9 bit)
ROR $1000	Rotate contents of location 1000 H right (9 bit)
RORB	Rotate B right (9 bit)
ASL $10, X*	Shift contents of location (X + 10 H) left (8 bit)
ASR $1000**	Shift contents of location 1000 H right (8 bit)
ASLA*	Shift A left (8 bit)
ASLB**	Shift B left (8 bit)

*$B_0 = 0$
**$B_7 = B_7$

Note: See "Programming the 6800 Microprocessor" (Fig. 15-6) for flags affected.

6.3 The 6800 Arithmetic Group

The arithmetic group includes add, subtract, increment, and decrement and deals with both registers and memory locations.

Example	Comment
ADDA #41	Add 41 D to A
ADDB $30, X	Add contents of location (X + 30 H) to B
ABA	Add B to A
ADCA #100	Add 100 D to A with carry
DAA	Decimal adjust A (A only)
DEC $20, X	Decrement contents of location (X + 20 H)
DECA	Decrement A (A − 1 → A)
DECB	Decrement B (B − 1 → B)
INC $100	Add 1 to contents of location 100 H
INCA	Increment A (A + 1 → A)
INCB	Increment B (B + 1 → B)
SUBA #40	Subtract 40 D from A
SUBB $20	Subtract contents of location 20 H from B
SBA	Subtract B from A (A − B → A)
SBCA $40, X	Subtract contents of location (X + 40 H) and carry from A
SBCB $2000	Subtract contents of location 2000 H and carry from B
DEX	X − 1 → X
DES	SP − 1 → SP
INX	X + 1 → X
INS	SP + 1 → SP

6.4 Tests, Compares, and Clears for the 6800

The 6800's testing group for comparing quantities with one another is represented by the following examples.

Example	Comment
CLR 40, X	00 → M, 00 → (X + 40 D)
CLRA	00 → A
CLRB	00 → B
CMPA $20	Compare A to contents of location 20 H (A − M)
CMPB 0, X	Compare B to contents of location (X) (B − M)
CBA	Compare accumulators (A − B)
BITA #$F0	Logical AND of A with F0; sets flags
BITB $2000	AND of B with location (2000 H); sets flags
TST	Compare M with 0, sets N, Z appropriately

TSTA	Compare A with 0, sets N, Z appropriately
TSTB	Compare B with 0, sets N, Z appropriately
CLC	Clear carry bit
CLI	Clear interrupt mask
CLV	Clear overflow (V)
SEI	Set interrupt mask
SEV	Set overflow bit (B)
CPX $200	Compare index register to M (200 H)

6.5 Jump and Branch Instructions for the 6800

The 6800 uses jump instructions to locate to a 2-byte address within the range 0–FFFF H. The branch instructions are to an address equal to 2 plus the 2's complement of the byte following the branch instruction. The left-most bit indicates whether the branch is forward ($b_7 = 0$) or backward ($b_7 = 1$). The displacement quantity is limited to $+127$ D in the foward direction (7FH) and by -128 in the reverse direction (80 H). The unconditional jump (JMP) is a 3-byte instruction that can transfer control to any of 65,536 locations. The unconditional branch (BRA) is a 2-byte instruction that can transfer control forward to $127 + 2$ (or 129) locations. BRA can transfer control backward ($2 - 128$) locations for 126 locations back. This calculation can be tricky. Remember, the CPU will branch to the current location plus 2 plus d.

EXAMPLE 15-1

Branching forward: if BRA is at 2020 H and it is desired to branch to 2040 H, calculate d.

$$2020 \text{ H} + 2 = 2022 \text{ H}$$

$$\begin{array}{r} 2040 \text{ H} \\ -2022 \text{ H} \\ \hline 1\text{E H} \end{array}$$

Answer: BRA $1E

EXAMPLE 15-2

Branching backward: if BRA is at 2020 H and it is desired to branch to 1FF0 H, calculate d.

$$2020 \text{ H} + 2 = 2022 \text{ H}$$

$$\begin{array}{r} 2022 \text{ H} \\ -1\text{FF0 H} \\ \hline 32 \text{ H} \end{array}$$

Since this is a reverse branch, we calculate the 2's complement of 32 H.

$$00110010 \text{ H}$$

1's complement: 11001101 H

2's complement: 11001110 H $= $ CE H

Let's check this:

$$2020 \text{ H} + 2 = 2022 \text{ H}$$

$$\begin{array}{r} 2022 \text{ H} \\ + \text{FFCE H} \\ \hline 1\ 1\text{FF0 H} \end{array}$$

Ignored $\curvearrowleft$

1FF0 H is the correct address.

Subroutine calls are handled by either jumping to the subroutine (JSR) or branching to the subroutine (BSR). The 6800 pushes a return address onto the user stack so that upon return (RTS) a correct return will be performed to the main program. A separate return from interrupt (RTI) instruction is used when an interrupt is involved.

6.6 A 6800 Program Example

One of our earlier program examples dealt with clearing a memory-mapped video display by writing 400 H space codes (20 H) into a range of memory locations beginning at 0CC00 H. Since this routine is familiar to us in 8080A/85 code, we'll do the same thing using the 6800.

```
        NAME 'CLEARSCREEN'
        ORG $1000            ;Set up origin
CLEAR:  LDX #$400            ;Set up byte counter
        STX $1100            ;Save X at 1100H, 1101H
        LDX #$0CC00          ;Set up display pointer
        LDAA #$20            ;Put code for space in A
LOOP:   STAA 0, X            ;Send space to screen
        INX                  ;Next screen location
        STX $1102            ;Save next screen location
        LDX $1100            ;Get byte counter
        DEX                  ;Decrement count
        BEQ RET              ;Done if zero
        STX $1100            ;Otherwise, save count
        LDX $1102            ;Get screen address
        BRA LOOP             ;And continue
RET:    RTS                  ;Return to main program
        END
```

6.7 Interrupt Handling for the 6800

The 6800 responds to four types of interrupt: a general reset (hardware reset = $\overline{\text{RST}}$), a hardware nonmaskable interrupt ($\overline{\text{NMI}}$), hardware maskable interrupt request ($\overline{\text{IRQ}}$), and a software interrupt (SWI). These four interrupts cause the

TABLE 15-4 Address Table of Service Routines

| Interrupt | Locations of Service Routine Addresses | |
	Address Hi	Address Lo
Restart	FFFE	FFFF
NMI	FFFC	FFFD
SWI	FFFA	FFFB
IRQ	FFF8	FFF9

CPU to vector to an address table of service routines to handle the interrupts. These are shown in Table 15-4.

The general reset causes the CPU to look for the address of an initialization routine at addresses FFFE and FFFF. This means that user ROM will be in place at these locations and that it will contain the 2-byte address of the start of a program.

EXAMPLE 15-3

On restart, the CPU vectors to address 2000 H.

FFFE	20
FFFF	00

The nonmaskable interrupt is not affected by the interrupt mask bit in the condition code (flag) register. The CPU automatically saves the index register, the PC, the accumulators, and the CC register on the user stack. Then it retrieves a service routine address from FFFC and FFFD. The RTI (return from interrupt) instruction automatically restores all the registers from the stack area.

Both the maskable interrupt (service routine address found at FFF8 and FFF9) and the software interrupt (service routine address found at FFFA and FFFB) can be disabled by the CPU by setting the interrupt mask bit to 1. This prohibits the interruption of any interrupt service routine in progress. Again, the CPU automatically saves the registers in the stack area and RTI automatically restores them before returning. The programmer has control of the interrupt mask bit via the clear interrupt mask (CLI) and the set interrupt mask (SEI) instructions.

The wait for interrupt (WAI) instruction causes the CPU to halt until interrupted, then the registers and return address is pushed onto the stack.

6.8 Other 6800 Series CPUs

There is an expanded series of 6800-based CPUs featuring improved CPU clock rates, on-board clock circuitry, on-board RAM and ROM types of memory, I/O, and timers (see Table 15-5). Other 8-bit microprocessors based on the 6800 are as follows:

TABLE 15-5   Expanded Series of 6800 MPUs

CPU	Clock-Rate (MHz)	Instruction Set
MC6800	1	6800
MC68A00	1.5	6800
MC68B00	2.0	6800
MC6801	1	6800[a]
MC68A01	1.5	6800[a]
MC68B01	2.0	6800[a]
MC68701	1	6800[a]
MC68A701	1.5	6800[a]
MC68B701	2.0	6800[a]
MC6802	1	6800
MC68A02	1.5	6800
MC68B02	2.0	6800
MC6808	1	6800
MC68A08	1.5	6800
MC68B08	2.0	6800
MC6805P2	4.2	6805
MC6805R2	4.2	6805[a]
MC68705	4.2	6805
MC6809	1	6809
MC68A09	1.5	6809
MC68B09	2.0	6809

[a]A/D converter.

6801 Series.   In these single-chip MPUs, A and B registers can be combined to form the 16-bit D accumulator. They have the following additional features:

2048-Byte ROM

128-Byte RAM

8×8 Multiply instruction

Serial interface

29 Parallel I/O lines

Three-function programmable timer

Internal clock generator

Enhanced 6800 instruction set

This series includes the following MPUs:

68701: EPROM version

6803: A 6801 without ROM

6802: General 6800 upgrade
On-board clock (just add a crystal)
128-Byte RAM (0–7FH)
Software compatible with 6800

6808: A 6802 without RAM (i.e., a 6800 with an on-board clock circuit).

6805: Microcomputer unit (MCU), single-chip MPU
Similar to 6800
64-Byte RAM
1100-Byte user ROM
20 Parallel I/O lines
On-chip clock generator
8-Bit timer is internal
One accumulator
Note: No external memory (11 internal address bits)

68705: EPROM version

6809 Series.   Features of these architecturally improved 6800 MPUs include the following:

Three added registers: second index register, direct page register, and user stack pointer register

On-board clock circuitry

A + B accumulators can be combined to form 16-bit D accumulator

For further information on these 8-bit MPUs, see the Motorola Microprocessor Data Manual.

6.9 Conclusion: The 6800 MPU

Since the 6800 has a pair of accumulators, but no other 8-bit registers, the programmer has to think much more in terms of using memory locations for data, counters, pointers, and other constants. From this standpoint, there are many memory locations available for this type of storage. The 6800 is approximately equal to the 8080A in its operations and shares many similarities. The two accumulators are a real convenience, and extensive use of the index register makes the use of memory fairly straightforward.

7 THE 6502 MICROPROCESSOR

The 6502 is an 8-bit microprocessor developed by Rockwell International and based on the 6800. It operates from a single 5-V supply and has on-board clock circuitry. Its characteristics are similar to the 6800, but it has only one accumulator. However, there are two 8-bit index registers. Figure 15-8 shows the internal register structure of the 6502. Programming the 1-MHz 6502 is basically the same as using the 6800. It is a memory-oriented MPU with memory-mapped I/O standard.

Two major differences between the 6502 and the 6800 that considerably alter the two instruction sets are the one accumulator of the 6502 and the added index register (Y). In addition, the operation of the 6502's index register is quite different from that of the 6800. There are 13 different types of addressing used with the 6502 (Table 15-6), and this accounts for the different uses of the index registers.

FIGURE 15-8 The 6502 microprocessor register set. (Used with the permission of Rockwell International Corporation.)

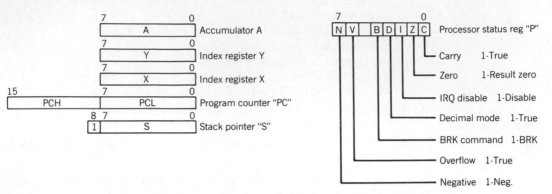

TABLE 15-6   6502 Addressing Modes

Code	Mode	Description
IMM	Immediate addressing	The operand is contained in the second byte of the instruction.
ABS	Absolute addressing	The second byte of the instruction contains the 8 low-order bits of the effective address (EA). The third byte contains the 8 high-order bits of the EA.
Z PAGE	Zero-page addressing	Second byte contains the 8 low-order bits of the EA. The 8 high-order bits are zeros.
A	Accumulator	One-byte instruction operating on the accumulator.
Z PAGE, X-Z PAGE, Y	Zero-page indexed	The second byte of the instruction is added to the index (carry is dropped) to form the low-order byte of the EA. The high-order byte of the EA is zero.
ABS, X-ABS, Y	Absolute indexed	The EA is formed by adding the index to the second and third byte of the instruction.
(IND, X)	Indexed indirect	The second byte of the instruction is added to the X index, discarding the carry. The result points to a location on page zero, which contains the 8 low-order bits of the EA. The next byte contains the 8 high-order bits.
(IND), Y	Indirect indexed	The second byte of the instruction points to a location in page zero. The contents of this memory location is added to the Y index, the result being the low-order 8 bits of the EA. The carry from this operation is added to the contents of the next page-zero location, the result being the 8 high-order bits of the EA.

7.1 The 6502 Instruction Set

Table 15-7 lists the 6502 instructions in alphabetical order. As you can see, they are similar to those of the 6800 but reflect the absence of accumulator B and the addition of the index register.

Note: The 6502 assembler syntax is different from the 6800. Immediate data is assumed, and memory references are in parentheses. For example:

	6800		*6502*
Immediate:	LDAA #$40	=	LDA $40
Direct:	LDAA $28	=	LDA ($28)

In addition, the 6502 uses byte swapping in the manner of the 8080A, 8085, and Z80 microprocessors. In an unconditional jump, the code forms as follows:

JMP $2041   4C4120

7.2 The 6502 Interrupt Modes

Like the 6800, the 6502 has both a nonmaskable interrupt ($\overline{NMI}$) and a maskable interrupt ($\overline{IRQ}$). Using either of these 6502 hardware interrupts causes the CPU to vector to an address as shown in the tabulation that follows:

NMI	FFFA (address low)
	FFFB (address high)
Reset	FFFC (address low)
	FFFD (address high)
IRQ	FFFE (address low)
	FFFF (address high)

Byte swapping causes the addresses to be stored as in the 8080A, 8085, and Z80 microprocessors and reversed for the 6800. The reset vector is located at FFFC and FFFD.

EXAMPLE 15-4

On reset, the 6502 vectors to address 2000 H.

FFFC	00
FFFD	20

The software interrupt is accomplished on the 6502 by the break instruction (BRK = $00). This is treated as a maskable hardware interrupt (IRQ) and, therefore, the MPU vectors to locations FFFE and FFFF for an address.

TABLE 15-7   6502 Instruction Set

Processor status codes column headers: 7 6 5 4 3 2 1 0 → N V • B D I Z C

MNEMONIC	OPERATION	IMM OP	n	#	ABS OP	n	#	ZP OP	n	#	ACC OP	n	#	IMPL OP	n	#	(IND,X) OP	n	#	(IND),Y OP	n	#	ZP,X OP	n	#	ABS,X OP	n	#	ABS,Y OP	n	#	REL OP	n	#	IND OP	n	#	ZP,Y OP	n	#	STATUS (N V • B D I Z C)	MNEMONIC
ADC	A + M + C → A (4)(1)	69	2	2	6D	4	3	65	3	2							61	6	2	71	5	2	75	4	2	7D	4	3	79	4	3										N V • • • • Z C	ADC
AND	A ∧ M → A (1)	29	2	2	2D	4	3	25	3	2							21	6	2	31.	5	2	35	4	2	3D	4	3	39	4	3										N • • • • • Z •	AND
ASL	C ← [7 ... 0] ← 0				0E	6	3	06	5	2	0A	2	1											16	6	2	1E	7	3												N • • • • • Z C	ASL
BCC	BRANCH ON C = 0 (2)																																90	2	2							• • • • • • • •	BCC
BCS	BRANCH ON C = 1 (2)																																B0	2	2							• • • • • • • •	BCS
BEQ	BRANCH ON Z = 1 (2)																																F0	2	2							• • • • • • • •	BEQ
BIT	A ∧ M				2C	4	3	24	3	2																															M_7 M_6 • • • • Z •	BIT
BMI	BRANCH ON N = 1 (2)																																30	2	2							• • • • • • • •	BMI
BNE	BRANCH ON Z = 0 (2)																																D0	2	2							• • • • • • • •	BNE
BPL	BRANCH ON N = 0 (2)																																10	2	2							• • • • • • • •	BPL
BRK	BREAK													00	7	1																									• • • • • 1 • •	BRK
BVC	BRANCH ON V = 0 (2)																																50	2	2							• • • • • • • •	BVC
BVS	BRANCH ON V = 1 (2)																																70	2	2							• • • • • • • •	BVS
CLC	0 → C													18	2	1																									• • • • • • • 0	CLC
CLD	0 → D													D8	2	1																									• • • • 0 • • •	CLD
CLI	0 → I													58	2	1																									• • • • • 0 • •	CLI
CLV	0 → V													B8	2	1																									• 0 • • • • • •	CLV
CMP	A − M (1)	C9	2	2	CD	4	3	C5	3	2							C1	6	2	D1	5	2	D5	4	2	DD	4	3	D9	4	3										N • • • • • Z C	CMP
CPX	X − M	E0	2	2	EC	4	3	E4	3	2																															N • • • • • Z C	CPX
CPY	Y − M	C0	2	2	CC	4	3	C4	3	2																															N • • • • • Z C	CPY
DEC	M − 1 → M				CE	6	3	C6	5	2																D6	6	2	DE	7	3										N • • • • • Z •	DEC
DEX	X − 1 → X													CA	2	1																									N • • • • • Z •	DEX
DEY	Y − 1 → Y													88	2	1																									N • • • • • Z •	DEY
EOR	A ⊻ M → A (1)	49	2	2	4D	4	3	45	3	2							41	6	2	51	5	2	55	4	2	5D	4	3	59	4	3										N • • • • • Z •	EOR
INC	M + 1 → M				EE	6	3	E6	5	2																F6	6	2	FE	7	3										N • • • • • Z •	INC
INX	X + 1 → X													E8	2	1																									N • • • • • Z •	INX
INY	Y + 1 → Y													C8	2	1																									N • • • • • Z •	INY
JMP	JUMP TO NEW LOC				4C	3	3																												6C	5	3				• • • • • • • •	JMP
JSR	JUMP SUB				20	6	3																																		• • • • • • • •	JSR
LDA	M → A (1)	A9	2	2	AD	4	3	A5	3	2							A1	6	2	B1	5	2	B5	4	2	BD	4	3	B9	4	3										N • • • • • Z •	LDA
LDX	M → X (1)	A2	2	2	AE	4	3	A6	3	2																			BE	4	3							B6	4	2	N • • • • • Z •	LDX
LDY	M → Y (1)	A0	2	2	AC	4	3	A4	3	2													B4	4	2	BC	4	3													N • • • • • Z •	LDY
LSR	0 → [7 ... 0] → C				4E	6	3	46	5	2	4A	2	1											56	6	2	5E	7	3												0 • • • • • Z C	LSR
NOP	NO OPERATION													EA	2	1																									• • • • • • • •	NOP
ORA	A ∨ M → A (1)	09	2	2	0D	4	3	05	3	2							01	6	2	11	5	2	15	4	2	1D	4	3	19	4	3										N • • • • • Z •	ORA
PHA	A → Ms S − 1 → S													48	3	1																									• • • • • • • •	PHA
PHP	P → Ms S − 1 → S													08	3	1																									• • • • • • • •	PHP
PLA	S + 1 → S Ms → A													68	4	1																									N • • • • • Z •	PLA
PLP	S + 1 → S Ms → P													28	4	1																									(RESTORED)	PLP
ROL	[7 ... 0] ← C				2E	6	3	26	5	2	2A	2	1											36	6	2	3E	7	3												N • • • • • Z C	ROL
ROR	C → [7 ... 0]				6E	6	3	66	5	2	6A	2	1											76	6	2	7E	7	3												N • • • • • Z C	ROR
RTI	RTRN INT													40	6	1																									(RESTORED)	RTI
RTS	RTRN SUB													60	6	1																									• • • • • • • •	RTS
SBC	A − M − C̄ → A (1)	E9	2	2	ED	4	3	E5	3	2							E1	6	2	F1	5	2	F5	4	2	FD	4	3	F9	4	3										N V • • • • Z C (3)	SBC
SEC	1 → C													38	2	1																									• • • • • • • 1	SEC
SED	1 → D													F8	2	1																									• • • • 1 • • •	SED
SEI	1 → I													78	2	1																									• • • • • 1 • •	SEI
STA	A → M				8D	4	3	85	3	2							81	6	2	91	6	2	95	4	2	9D	5	3	99	5	3										• • • • • • • •	STA
STX	X → M				8E	4	3	86	3	2																												96	4	2	• • • • • • • •	STX
STY	Y → M				8C	4	3	84	3	2													94	4	2																• • • • • • • •	STY
TAX	A → X													AA	2	1																									N • • • • • Z •	TAX
TAY	A → Y													A8	2	1																									N • • • • • Z •	TAY
TSX	S → X													BA	2	1																									N • • • • • Z •	TSX
TXA	X → A													8A	2	1																									N • • • • • Z •	TXA
TXS	X → S													9A	2	1																									• • • • • • • •	TXS
TYA	Y → A													98	2	1																									N • • • • • Z •	TYA

(1) ADD 1 to 'N' IF PAGE BOUNDARY IS CROSSED
(2) ADD 1 TO 'N' IF BRANCH OCCURS TO SAME PAGE
    ADD 2 TO 'N' IF BRANCH OCCURS TO DIFFERENT PAGE
(3) CARRY NOT = BORROW
(4) IF IN DECIMAL MODE, Z FLAG IS INVALID
    ACCUMULATOR MUST BE CHECKED FOR ZERO RESULT

X	INDEX X	+	ADD
Y	INDEX Y	−	SUBTRACT
A	ACCUMULATOR	∧	AND
M	MEMORY PER EFFECTIVE ADDRESS	∨	OR
Ms	MEMORY PER STACK POINTER	⊻	EXCLUSIVE OR
M_7	MEMORY BIT 7		
M_6	MEMORY BIT 6		
n	NO. CYCLES		
#	NO. BYTES		

Source: Used with permission of Rockwell International Corporation.

Masking of the interrupt bit is automatic or by the set interrupt disable status bit instruction (SEI). It is reset by the clear interrupt disable bit instruction (CLI).

A branch to a subroutine and the various interrupts automatically cause the program counter and status registers to be saved on the stack. The location of the stack area is limited to a range of 256 memory locations from 0100 H to 01FF H. This is because the 6502 has a 9-bit stack pointer register with $b_8 = 1$. The stack pointer is always initialized to 1FF when a reset occurs, saving the programmer this little chore.

7.3 Conclusion: The 6502

Although the architecture of the 6502 is similar to that of the 6800, we see that there are certain differences between the two microprocessors, such as the addition of a second index register in the 6502. Other examples of these differences are the byte swapping, on-board oscillator, and differences in interrupt handling techniques. The 6502 is included here mainly because of its popularity in the Apple II home and business computer systems and because it is used in certain control applications due to its economical pricing. Its versatility is well illustrated in its use on the Apple computer, and its similarities to the 6800 MPU make it an easy microprocessor to learn.

8 SIXTEEN-BIT MICROPROCESSORS

The 16-bit microprocessor has an internal 16-bit data path, which is really why we call them 16-bit microprocessors. The use of a 16-bit data path essentially can halve the time required to perform an operation. Consider how an 8-bit MPU loads data into a register:

LDA $40

is a basic 2-byte instruction. The MPU fetches the load instruction and decodes it. It *then* determines that a piece of data is required and examines memory for the missing information. Simplifying somewhat, we can say that in a 16-bit machine it would be possible to get both the instruction and the data at the same time, with a resulting increase in program speed. A further benefit of a 16-bit data path is that pieces of data are not limited to 8 bits. If BCD arithmetic is involved, the choice is between getting a pair of BCD numbers at a time (1 byte) versus getting two pair (2 bytes).

Another reason for the popularity of the 16-bit MPUs is that the newer fabrication processes yield devices that are inherently faster. The new MPUs operate at higher clock frequencies, which contribute to faster programs. These new MPUs also include larger register sets and advanced instructions that are more efficient in addition/subtraction routines and most include a (software) multiply function to assist in applications that require number crunching. The use of these MPUs with high-level languages also implies that more memory would be of advantage, and all these 16-bit MPUs have added address lines to address many more memory locations as shown in Table 15-8, which extends to 36 address lines only because large mainframe computers can have this many address lines.

TABLE 15-8   Additional Address Lines

Address Lines (N)	Limit of Memory Locations
16	65,536
20	1,048,576
24	16,777,216
28	268,435,456
32	4,294,967,296
36	68,719,476,736

8.1 Data Size for Sixteen-Bit MPUs

The 8-bit MPUs were generally concerned with storing 8-bit quantities or perhaps a pair of BCD digits. The 16-bit registers commonly found in the newer MPUs allow a new set of data types (see Table 15-9).

Many of the 16-bit MPUs can load a register with words of varying lengths, and this must be specified in the mnemonic to aid in assembly. For example:

```
MOVE.B    #40H, $2000    ;Move 40H to 2000H
MOVE.W    #40H, $2000    ;Move 0040 to 2000, 2001
MOVE.L    #40H, $2000    ;Move 00000040 to 2000–2003
```

8.2 Organization of Memory

The 16-bit MPUs we will be studying typically still use 8-bit (byte-oriented) memory locations. This means that a 16-bit instruction will require two consecutive memory locations. Figure 15-9 shows this organization.

In addition to the possibility of very large active memories, we may have to consider the possibility of using portions or *segments* of the memory at any one time. The ability of a computer to address a very large memory does not necessarily imply that it can all be addressed at any one time. We will find that certain sections of memory may be portioned off from our use for use by the MPU (supervisory functions). Since programs are frequently written in sections or *modules*, it may be neither desirable nor necessary to have an infinitely large memory available at any given moment. This introduces us to the concept of a segmented memory.

Figure 15-10 shows a memory addressed by 20 address lines and, therefore, containing 1,048,576 possible memory locations. The decision to install that much memory on a given computer is at the discretion of the designer. Three discrete segments that have been set aside for particular uses are shown.

TABLE 15-9   Data Lengths in 16-Bit MPUs

Data Type	Length
One-half byte	Nibble = 4 bits
Byte (B)	8 bits
Word (W)	2 bytes = 16 bits
Long word (L)	4 bytes = 32 bits

FIGURE 15-9 Data formats in memory.

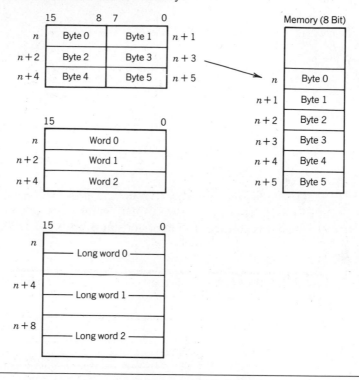

FIGURE 15-10 Memory addressed by 20 address lines.

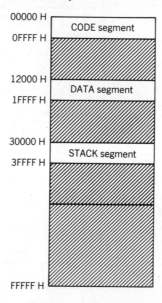

FIGURE 15-11 64K Byte-overlapping segments.

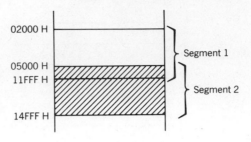

The code segment is a 65,536-byte area beginning at zero and extending to 0FFFF H. This area, called CODE, is for the actual opcodes of the program. The segment for storage of data or constants begins at 12000 H and extends to address

FIGURE 15-12 The internal register set for the 8086/8088.
(Reprinted by permission of Intel Corporation, copyright 1982.)

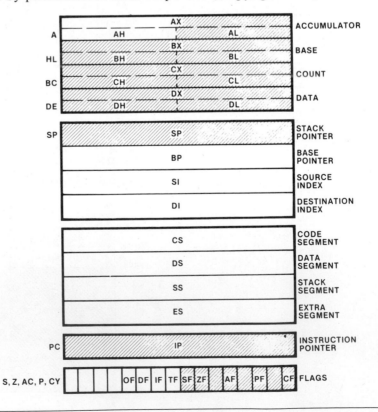

1FFFF H. A third area of memory in this example has been set aside as a user stack area, beginning at address 3FFFF H and working toward lesser memory locations.

The actual size of a segment would depend on the microprocessor involved and the available memory. It is also conceivable that one memory segment might overlap another memory segment without harm, as shown in Fig. 15-11.

9 THE 8086/8088

The Intel 8086 microprocessor is a 16-bit MPU both internally and externally. It has a full 16-bit data bus that can transfer 16 bits at a time. Internal registers are all 16 bits and the 8086 can handle $2^{20} = 1,048,576$ memory locations in 65,536-byte segments. The 8088 microprocessor is identical to the 8086 except that the 8088 has an 8-bit external data bus and can transfer data 8 bits at a time. The 8088 was produced for designers already comfortable with 8-bit MPUs who wanted to take advantage of the expanded 8086 architecture and at the same time address additional memory locations ($16 \times 65,536$). The 8086 and 8088 are software compatible and share the same instruction set. Figure 15-12 shows the internal register set of the 8086 and 8088. From this point on we will refer only to the 8086.

The 8086 consists of a dozen 16-bit registers available to the user. The first group consists of a 16-bit accumulator (AX), which can be used separately as two 8-bit accumulators (AH) and (AL). Registers B, C, and D function in the same manner. That is, they can be considered either as 16-bit registers (BX, CX, DX) or as a group of individual 8-bit registers (BH, BL, CH, CL, DH, DL). See Table 15-10 for register operations.

The data register (DX) is used to specify ports for I/O operations as well as for general data manipulation. The count register (CX) is used with certain instructions for automatic incrementing or decrementing.

The 8086 has four 16-bit registers to be used as address pointers: the stack pointer, the base pointer (for table manipulations), and two index registers.

The 8086 is segment oriented and works with 65,536 bytes at a time. The four segment registers are used to determine the upper 16 bits of a 20-bit address.

TABLE 15-10   Register Operations

Register	Operations
AX	Word multiply, word divide, word I/O
AL	Byte multiply, byte divide, byte I/O, translate, decimal arithmetic
AH	Byte multiply, byte divide
BX	Translate
CX	String operations, loops
CL	Variable shift and rotate
DX	Word multiply, word divide, indirect I/O
SP	Stack operations
SI	String operations
DI	String operations

These are the code segment registers (used with the instruction pointer), the data segment register (used with the base pointer), the stack segment, and an extra segment register for those really tough applications.

The 16-bit instruction pointer (program counter) keeps track of where current code is stored. The actual 20-bit address is determined by adding the contents of the instruction pointer (bits 15–0) to the contents of the code segment register (bits 19–4).

If the code segment register contains 2500 H and the instruction pointer register contains 3208 H, the effective 20-bit address at which the process is operating is found as follows:

$$
\begin{aligned}
\text{Code segment} &= 2500- \\
\text{Instruction pointer} &= \underline{3208} \\
\text{Current address} &= 28208 \text{ H}
\end{aligned}
$$

To operate at address 0 (ORG 0), the code segment register must have 0000 in it and the instruction pointer be at 0 also.

Initialization of the stack would require the initialization of both the stack segment register and the stack pointer. The effective address of the stack area is computed in the same manner as the program counter. The stack segment register contains the upper 16 of 20 address bits (19–4) and the stack pointer contains the lower 16 of 20 address bits (15–0).

The 8086 uses an assembly language that is similar to that of the 8080Â/85 MPUs. It has 24 addressing modes and can perform 8- and 16-bit signed or unsigned

FIGURE 15-13 Pinouts for the 8086/8088. (Reprinted by permission of Intel Corporation, copyright 1982.)

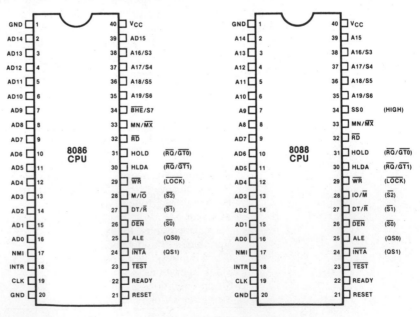

arithmetic in binary or decimal, including multiplication and division. It is available to 8 MHz.

CPU	Clock
8086	5 MHz
8086-2	8 MHz
8086-4	4 MHz

The address and data lines are multiplexed to fit this powerful processor into a 48-pin package. Decoding is accomplished using the address latch enable (ALE) signal. Figure 15-13 gives the 8086/8088 pinouts.

Contributing to the high speed of the 8086 is its design incorporating two separately functioning units: the execution unit (EU) and the bus interface unit (BIU). Previously, the MPU fetched an instruction from memory, and the address and data bus lines were idle during the decoding process. In the 8086, however, the bus interface unit is busy getting the next value from memory while the execution unit is working on the current instruction. This improved efficiency makes the 8086 a very fast MPU.

9.1 The 8086 Instruction Set

Instructions on the 8086 may consist of 1 to 6 bytes, depending on the data lengths involved and the type of instruction. Generally, data transfers are either byte (B) or word (W) lengths, and this is part of the mnemonic specification. For example, to load an immediate byte of data into the accumulator, the format of the instruction is as follows:

MOV: Immediate to Register

Byte 1			Byte 2	Byte 3
1011	W	Register	Data	Data if W = 1

To make the opcode, the missing bits must be supplied: W = 0 if a byte is involved, W = 1 if a double byte (word) is involved. The register information is a 3-bit quantity based on Table 15-11. *Note on Syntax:* Brackets are used to distinguish stored from immediate data; thus 47 H means immediate data, and [28 H] means data at address 28 H.

TABLE 15-11   The 8086 Register Designators

16-Bit (W = 1)		8-Bit (W = 0)	
000	AX	000	AL
001	CX	001	CL
010	DX	010	DL
011	BX	011	BL
100	SP	100	AH
101	BP	101	CH
110	SI	110	DH
111	DI	111	BH

To move a byte of data equal to 47 H into the lower 8 bits of the A register, the instruction becomes:

MOV AL, 47H

| 10110000 | 01000111 | = B047 |

To move a double byte or word equal to 2061 H into the B register:

MOV BX, 2061H

| 10111011 | 01100001 | 00100000 | = BB6120 |

Notice that the MOV in this case was either a 2- or a 3-byte instruction in this *mode*, depending on the size of the data involved. You should also get the idea that an *assembler* will be a definite requirement to do any serious programming as is the case for any 16-bit microprocessor. Table 15-12 shows the 8086 instruction set summary, and we will examine a few sets to establish a base for you for the use of the 8086 (and also the 8088).

TABLE 15-12 8086 Instruction Set

DATA TRANSFER

MOV = Move:

	7 6 5 4 3 2 1 0	7 6 5 4 3 2 1 0	7 6 5 4 3 2 1 0	7 6 5 4 3 2 1 0	7 6 5 4 3 2 1 0	7 6 5 4 3 2 1 0
Register/memory to/from register	1 0 0 0 1 0 d w	mod reg r/m	(DISP-LO)	(DISP-HI)		
Immediate to register/memory	1 1 0 0 0 1 1 w	mod 0 0 0 r/m	(DISP-LO)	(DISP-HI)	data	data if w = 1
Immediate to register	1 0 1 1 w reg	data	data if w = 1			
Memory to accumulator	1 0 1 0 0 0 0 w	addr-lo	addr-hi			
Accumulator to memory	1 0 1 0 0 0 1 w	addr-lo	addr-hi			
Register/memory to segment register	1 0 0 0 1 1 1 0	mod 0 SR r/m	(DISP-LO)	(DISP-HI)		
Segment register to register/memory	1 0 0 0 1 1 0 0	mod 0 SR r/m	(DISP-LO)	(DISP-HI)		

PUSH = Push:

Register/memory	1 1 1 1 1 1 1 1	mod 1 1 0 r/m	(DISP-LO)	(DISP-HI)
Register	0 1 0 1 0 reg			
Segment register	0 0 0 reg 1 1 0			

POP = Pop:

Register/memory	1 0 0 0 1 1 1 1	mod 0 0 0 r/m	(DISP-LO)	(DISP-HI)
Register	0 1 0 1 1 reg			
Segment register	0 0 0 reg 1 1 1			

TABLE 15-12 (continued)

DATA TRANSFER (Cont'd.)

XCHG = Exchange:

	7 6 5 4 3 2 1 0	7 6 5 4 3 2 1 0	7 6 5 4 3 2 1 0	7 6 5 4 3 2 1 0	7 6 5 4 3 2 1 0	7 6 5 4 3 2 1 0
Register/memory with register	1 0 0 0 0 1 1 w	mod reg r/m	(DISP-LO)	(DISP-HI)		
Register with accumulator	1 0 0 1 0 reg					

IN = Input from:

Fixed port	1 1 1 0 0 1 0 w	DATA-8	
Variable port	1 1 1 0 1 1 0 w		

OUT = Output to:

Fixed port	1 1 1 0 0 1 1 w	DATA-8		
Variable port	1 1 1 0 1 1 1 w			
XLAT = Translate byte to AL	1 1 0 1 0 1 1 1			
LEA = Load EA to register	1 0 0 0 1 1 0 1	mod reg r/m	(DISP-LO)	(DISP-HI)
LDS = Load pointer to DS	1 1 0 0 0 1 0 1	mod reg r/m	(DISP-LO)	(DISP-HI)
LES = Load pointer to ES	1 1 0 0 0 1 0 0	mod reg r/m	(DISP-LO)	(DISP-HI)
LAHF = Load AH with flags	1 0 0 1 1 1 1 1			
SAHF = Store AH into flags	1 0 0 1 1 1 1 0			
PUSHF = Push flags	1 0 0 1 1 1 0 0			
POPF = Pop flags	1 0 0 1 1 1 0 1			

ARITHMETIC

ADD = Add:

Reg/memory with register to either	0 0 0 0 0 0 d w	mod reg r/m	(DISP-LO)	(DISP-HI)		
Immediate to register/memory	1 0 0 0 0 0 s w	mod 0 0 0 r/m	(DISP-LO)	(DISP-HI)	data	data if s: w=01
Immediate to accumulator	0 0 0 0 0 1 0 w	data	data if w=1			

ADC = Add with carry:

Reg/memory with register to either	0 0 0 1 0 0 d w	mod reg r/m	(DISP-LO)	(DISP-HI)		
Immediate to register/memory	1 0 0 0 0 0 s w	mod 0 1 0 r/m	(DISP-LO)	(DISP-HI)	data	data if s: w=01
Immediate to accumulator	0 0 0 1 0 1 0 w	data	data if w=1			

INC = Increment:

Register/memory	1 1 1 1 1 1 1 w	mod 0 0 0 r/m	(DISP-LO)	(DISP-HI)
Register	0 1 0 0 0 reg			
AAA = ASCII adjust for add	0 0 1 1 0 1 1 1			
DAA = Decimal adjust for add	0 0 1 0 0 1 1 1			

(continued on next page)

TABLE 15-12 (continued)

ARITHMETIC (Cont'd.)

SUB = Subtract:

	7 6 5 4 3 2 1 0	7 6 5 4 3 2 1 0	7 6 5 4 3 2 1 0	7 6 5 4 3 2 1 0	7 6 5 4 3 2 1 0	7 6 5 4 3 2 1 0
Reg/memory and register to either	0 0 1 0 1 0 d w	mod reg r/m	(DISP-LO)	(DISP-HI)		
Immediate from register/memory	1 0 0 0 0 0 s w	mod 1 0 1 r/m	(DISP-LO)	(DISP-HI)	data	data if s: w=01
Immediate from accumulator	0 0 1 0 1 1 0 w	data	data if w=1			

SBB = Subtract with borrow:

Reg/memory and register to either	0 0 0 1 1 0 d w	mod reg r/m	(DISP-LO)	(DISP-HI)		
Immediate from register/memory	1 0 0 0 0 0 s w	mod 0 1 1 r/m	(DISP-LO)	(DISP-HI)	data	data if s: w=01
Immediate from accumulator	0 0 0 1 1 1 0 w	data	data if w=1			

DEC Decrement:

Register/memory	1 1 1 1 1 1 1 w	mod 0 0 1 r/m	(DISP-LO)	(DISP-HI)
Register	0 1 0 0 1 reg			
NEG Change sign	1 1 1 1 0 1 1 w	mod 0 1 1 r/m	(DISP-LO)	(DISP-HI)

CMP = Compare:

Register/memory and register	0 0 1 1 1 0 d w	mod reg r/m	(DISP-LO)	(DISP-HI)		
Immediate with register/memory	1 0 0 0 0 0 s w	mod 1 1 1 r/m	(DISP-LO)	(DISP-HI)	data	data if s: w=1
Immediate with accumulator	0 0 1 1 1 1 0 w	data				
AAS ASCII adjust for subtract	0 0 1 1 1 1 1 1					
DAS Decimal adjust for subtract	0 0 1 0 1 1 1 1					
MUL Multiply (unsigned)	1 1 1 1 0 1 1 w	mod 1 0 0 r/m	(DISP-LO)	(DISP-HI)		
IMUL Integer multiply (signed)	1 1 1 1 0 1 1 w	mod 1 0 1 r/m	(DISP-LO)	(DISP-HI)		
AAM ASCII adjust for multiply	1 1 0 1 0 1 0 0	0 0 0 0 1 0 1 0	(DISP-LO)	(DISP-HI)		
DIV Divide (unsigned)	1 1 1 1 0 1 1 w	mod 1 1 0 r/m	(DISP-LO)	(DISP-HI)		
IDIV Integer divide (signed)	1 1 1 1 0 1 1 w	mod 1 1 1 r/m	(DISP-LO)	(DISP-HI)		
AAD ASCII adjust for divide	1 1 0 1 0 1 0 1	0 0 0 0 1 0 1 0	(DISP-LO)	(DISP-HI)		
CBW Convert byte to word	1 0 0 1 1 0 0 0					
CWD Convert word to double word	1 0 0 1 1 0 0 1					

LOGIC

NOT Invert	1 1 1 1 0 1 1 w	mod 0 1 0 r/m	(DISP-LO)	(DISP-HI)
SHL/SAL Shift logical/arithmetic left	1 1 0 1 0 0 v w	mod 1 0 0 r/m	(DISP-LO)	(DISP-HI)
SHR Shift logical right	1 1 0 1 0 0 v w	mod 1 0 1 r/m	(DISP-LO)	(DISP-HI)
SAR Shift arithmetic right	1 1 0 1 0 0 v w	mod 1 1 1 r/m	(DISP-LO)	(DISP-HI)
ROL Rotate left	1 1 0 1 0 0 v w	mod 0 0 0 r/m	(DISP-LO)	(DISP-HI)

TABLE 15-12   (continued)

LOGIC (Cont'd.)

	7 6 5 4 3 2 1 0	7 6 5 4 3 2 1 0	7 6 5 4 3 2 1 0	7 6 5 4 3 2 1 0	7 6 5 4 3 2 1 0	7 6 5 4 3 2 1 0
ROR Rotate right	1 1 0 1 0 0 v w	mod 0 0 1 r/m	(DISP-LO)	(DISP-HI)		
RCL Rotate through carry flag left	1 1 0 1 0 0 v w	mod 0 1 0 r/m	(DISP-LO)	(DISP-HI)		
RCR Rotate through carry right	1 1 0 1 0 0 v w	mod 0 1 1 r/m	(DISP-LO)	(DISP-HI)		

AND = And:

	7 6 5 4 3 2 1 0	7 6 5 4 3 2 1 0	7 6 5 4 3 2 1 0	7 6 5 4 3 2 1 0	7 6 5 4 3 2 1 0	7 6 5 4 3 2 1 0
Reg/memory with register to either	0 0 1 0 0 0 d w	mod reg r/m	(DISP-LO)	(DISP-HI)		
Immediate to register/memory	1 0 0 0 0 0 0 w	mod 1 0 0 r/m	(DISP-LO)	(DISP-HI)	data	data if w=1
Immediate to accumulator	0 0 1 0 0 1 0 w	data	data if w=1			

TEST = And function to flags no result:

	7 6 5 4 3 2 1 0	7 6 5 4 3 2 1 0	7 6 5 4 3 2 1 0	7 6 5 4 3 2 1 0	7 6 5 4 3 2 1 0	7 6 5 4 3 2 1 0
Register/memory and register	0 0 0 1 0 0 d w	mod reg r/m	(DISP-LO)	(DISP-HI)		
Immediate data and register/memory	1 1 1 1 0 1 1 w	mod 0 0 0 r/m	(DISP-LO)	(DISP-HI)	data	data if w=1
Immediate data and accumulator	1 0 1 0 1 0 0 w	data				

OR = Or:

	7 6 5 4 3 2 1 0	7 6 5 4 3 2 1 0	7 6 5 4 3 2 1 0	7 6 5 4 3 2 1 0	7 6 5 4 3 2 1 0	7 6 5 4 3 2 1 0
Reg/memory and register to either	0 0 0 0 1 0 d w	mod reg r/m	(DISP-LO)	(DISP-HI)		
Immediate to register/memory	1 0 0 0 0 0 0 w	mod 0 0 1 r/m	(DISP-LO)	(DISP-HI)	data	data if w=1
Immediate to accumulator	0 0 0 0 1 1 0 w	data	data if w=1			

XOR = Exclusive or:

	7 6 5 4 3 2 1 0	7 6 5 4 3 2 1 0	7 6 5 4 3 2 1 0	7 6 5 4 3 2 1 0	7 6 5 4 3 2 1 0	7 6 5 4 3 2 1 0
Reg/memory and register to either	0 0 1 1 0 0 d w	mod reg r/m	(DISP-LO)	(DISP-HI)		
Immediate to register/memory	0 0 1 1 0 1 0 w	data	(DISP-LO)	(DISP-HI)	data	data if w=1
Immediate to accumulator	0 0 1 1 0 1 0 w	data	data if w=1			

STRING MANIPULATION

REP = Repeat	1 1 1 1 0 0 1 z
MOVS = Move byte/word	1 0 1 0 0 1 0 w
CMPS = Compare byte/word	1 0 1 0 0 1 1 w
SCAS = Scan byte/word	1 0 1 0 1 1 1 w
LODS = Load byte/wd to AL/AX	1 0 1 0 1 1 0 w
STDS = Stor byte/wd from AL/A	1 0 1 0 1 0 1 w

(continued on next page)

TABLE 15-12 (continued)

CONTROL TRANSFER

CALL = Call:

	7 6 5 4 3 2 1 0	7 6 5 4 3 2 1 0	7 6 5 4 3 2 1 0	7 6 5 4 3 2 1 0	7 6 5 4 3 2 1 0	7 6 5 4 3 2 1 0
Direct within segment	1 1 1 0 1 0 0 0	IP-INC-LO	IP-INC-HI			
Indirect within segment	1 1 1 1 1 1 1 1	mod 0 1 0 r/m	(DISP-LO)	(DISP-HI)		
Direct intersegment	1 0 0 1 1 0 1 0	IP-lo	IP-hi			
		CS-lo	CS-hi			
Indirect intersegment	1 1 1 1 1 1 1 1	mod 0 1 1 r/m	(DISP-LO)	(DISP-HI)		

JMP = Unconditional Jump:

Direct within segment	1 1 1 0 1 0 0 1	IP-INC-LO	IP-INC-HI		
Direct within segment-short	1 1 1 0 1 0 1 1	IP-INC8			
Indirect within segment	1 1 1 1 1 1 1 1	mod 1 0 0 r/m	(DISP-LO)	(DISP-HI)	
Direct intersegment	1 1 1 0 1 0 1 0	IP-lo	IP-hi		
		CS-lo	CS-hi		
Indirect intersegment	1 1 1 1 1 1 1 1	mod 1 0 1 r/m	(DISP-LO)	(DISP-HI)	

RET = Return from CALL:

Within segment	1 1 0 0 0 0 1 1		
Within seg adding immed to SP	1 1 0 0 0 0 1 0	data-lo	data-hi
Intersegment	1 1 0 0 1 0 1 1		
Intersegment adding immediate to SP	1 1 0 0 1 0 1 0	data-lo	data-hi
JE/JZ = Jump on equal/zero	0 1 1 1 0 1 0 0	IP-INC8	
JL/JNGE = Jump on less/not greater or equal	0 1 1 1 1 1 0 0	IP-INC8	
JLE/JNG = Jump on less or equal/not greater	0 1 1 1 1 1 1 0	IP-INC8	
JB/JNAE = Jump on below/not above or equal	0 1 1 1 0 0 1 0	IP-INC8	
JBE/JNA = Jump on below or equal/not above	0 1 1 1 0 1 1 0	IP-INC8	
JP/JPE = Jump on parity/parity even	0 1 1 1 1 0 1 0	IP-INC8	
JO = Jump on overflow	0 1 1 1 0 0 0 0	IP-INC8	
JS = Jump on sign	0 1 1 1 1 0 0 0	IP-INC8	
JNE/JNZ = Jump on not equal/not zer0	0 1 1 1 0 1 0 1	IP-INC8	
JNL/JGE = Jump on not less/greater or equal	0 1 1 1 1 1 0 1	IP-INC8	
JNLE/JG = Jump on not less or equal/greater	0 1 1 1 1 1 1 1	IP-INC8	
JNB/JAE = Jump on not below/above or equal	0 1 1 1 0 0 1 1	IP-INC8	
JNBE/JA = Jump on not below or equal/above	0 1 1 1 0 1 1 1	IP-INC8	
JNP/JPO = Jump on not par/par odd	0 1 1 1 1 0 1 1	IP-INC8	
JNO = Jump on not overflow	0 1 1 1 0 0 0 1	IP-INC8	

TABLE 15-12   (continued)

CONTROL TRANSFER (Cont'd.)

		7 6 5 4 3 2 1 0	7 6 5 4 3 2 1 0	7 6 5 4 3 2 1 0	7 6 5 4 3 2 1 0	7 6 5 4 3 2 1 0	7 6 5 4 3 2 1 0
RET = Return from CALL:							
JNS = Jump on not sign		0 1 1 1 1 0 0 1	IP-INC8				
LOOP = Loop CX times		1 1 1 0 0 0 1 0	IP-INC8				
LOOPZ/LOOPE = Loop while zero/equal		1 1 1 0 0 0 0 1	IP-INC8				
LOOPNZ/LOOPNE = Loop while not zero/equal		1 1 1 0 0 0 0 0	IP-INC8				
JCXZ = Jump on CX zero		1 1 1 0 0 0 1 1	IP-INC8				

INT = Interrupt:

Type specified		1 1 0 0 1 1 0 1	DATA-8
Type 3		1 1 0 0 1 1 0 0	
INTO = Interrupt on overflow		1 1 0 0 1 1 1 0	
IRET = Interrupt return		1 1 0 0 1 1 1 1	

PROCESSOR CONTROL

CLC = Clear carry		1 1 1 1 1 0 0 0			
CMC = Complement carry		1 1 1 1 0 1 0 1			
STC = Set carry		1 1 1 1 1 0 0 1			
CLD = Clear direction		1 1 1 1 1 1 0 0			
STD = Set direction		1 1 1 1 1 1 0 1			
CLI = Clear interrupt		1 1 1 1 1 0 1 0			
STI = Set interrupt		1 1 1 1 1 0 1 1			
HLT = Halt		1 1 1 1 0 1 0 0			
WAIT = Wait		1 0 0 1 1 0 1 1			
ESC = Escape (to external device)		1 1 0 1 1 x x x	m o d y y y r / m	(DISP-LO)	(DISP-HI)
LOCK = Bus lock prefix		1 1 1 1 0 0 0 0			
SEGMENT = Override prefix		0 0 1 reg 1 1 0			

Before we look at some examples, let's define a few terms.

Definitions

d = direction
   d = 0: "from" register
   d = 1: "to" register

mod = modifier
    m   = 00: displacement = 0
    m   = 01: displacement high = 0
    mod = 10: 16-bit displacement
    mod = 11 − r/m = register field

r/m = register or memory

EXAMPLE 15-5

MOV: Memory to Accumulator (8080A = LDA mem)

$$\text{MOV AX, [2000H]}$$

SOLUTION

1010000w	addr low	addr high

= A10020

EXAMPLE 15-6

MOV: Register to Register

$$\text{MOV BX, DX}$$

SOLUTION

100010 d w	mod reg r/m

= 89DB

EXAMPLE 15-7

Increment a register.

$$\text{INC BX} \quad \text{;16-Bit increment of B}$$

1111111 w	mod 000 r/m

= FF03

    w = 1
    mod = 00
    r/m = Reg = BX

Figure 15-14 shows some examples of 8086 addressing modes and assembler syntax. The fields in a given instruction designate word (w = 1) or byte (w = 0) types of data, which registers (reg) are involved, a modifier (mod), and either a displacement (addr low) or an address (addr low, addr high).

FIGURE 15-14 Addressing modes and assembler syntax.

```
ADD   AX,BX              ;REGISTER←REGISTER
ADD   AL,5               ;REGISTER←IMMEDIATE
ADD   CX,ALPHA           ;REGISTER←MEMORY(DIRECT)
ADD   ALPHA,6            ;MEMORY(DIRECT)←IMMEDIATE
ADD   ALPHA,DX           ;MEMORY(DIRECT)←REGISTER
ADD   BL,[BX]            ;REGISTER←MEMORY(REGISTER INDIRECT)
ADD   [SI],BH            ;MEMORY(REGISTER INDIRECT)←IMMEDIATE
ADD   [PP].ALPHA,AH      ;MEMORY(BASED)←REGISTER
ADD   CX,ALPHA[SI]       ;REGISTER←MEMORY(INDEXED)
ADD   ALPHA[DI+2],10     ;MEMORY(INDEXED)←IMMEDIATE
ADD   [BX].ALPHA[SI],AL  ;MEMORY(BASED INDEXED)←REGISTER
ADD   SI,[BP+4][DI]      ;REGISTER←MEMORY(BASED INDEXED)
IN    AL,30              ;DIRECT PORT
OUT   DX,AX              ;INDIRECT PORT
```

9.2 Interrupt Modes for the 8086

Before covering the 8086 interrupt features, it is useful to discuss the RESET function. On reset, the 8086 registers are initialized as follows:

Register	State
Flags	Clear
Instruction pointer	0000 H
CS Register	FFFF H
PS Register	0000 H
SS Register	0000 H
ES Register	0000 H

Since the code segment register initializes to FFFF H and the instruction pointer is set to zero, the processor executes its first instruction at address FFFF0 H.

$$
\begin{aligned}
CS &= FFFF - \\
+IP &= \underline{0000} \\
EA &= FFFF0 \ (\text{program counter})
\end{aligned}
$$

This indicates that startup ROM must be in this range to provide a vector to an initialization routine. In addition, since all the other segment registers (DS, SS, ES), are zero, the new user need not worry about the segment registers. On a reset, the processor is in the 0–FFFF H range an 8-bit programmer is used to (except for the startup ROM vector at FFFF0).

The 8086 has three types of interrupts: a software interrupt (INT) and two hardware interrupts—nonmaskable (NMI) and maskable (INTR). The 8086 provides for up to 256 separate interrupt vectors in a table that occupies the lower 1K bytes of memory from zero to 3FF H. Figure 15-15 shows the interrupt pointer table.

FIGURE 15-15 An interrupt pointer table. (Reprinted by permission
of Intel Corporation, copyright 1982.)

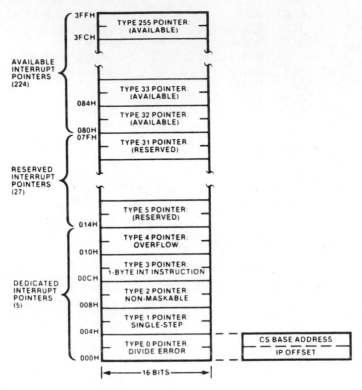

The nonmaskable interrupt (NMI) causes the MPU to vector to address 008 H.
The address is a 20-bit address contained at 008 H through 00B H. NMI is a type
2 interrupt (of 256 types).

The maskable interrupt (INRT) is enabled using the set interrupt-enable flag
instruction (STI). It is disabled by using the clear interrupt-enable flag instruction
(CLI). Use of this hardware interrupt requires an interrupt type number to be
placed on the data bus within the range of zero to FF H. This requires external
hardware, and Intel recommends the 8259A programmable interrupt controller.

The software interrupts specify a type number that cause the processor to vector
to an address within the 0–3FF H range for an interrupt service routine address.
For example, INT 33 sends the processor to address 84 H for a 20-bit address.
The format of the address is the instruction pointer offset at the first 16-bit location
and the code segment address at the second 16-bit location. If addresses 84 H
and 85 H contain 1050 H (IP) and addresses 86 H and 87 H contain 4000 H, the
interrupt service routine would be at address 41050 H.

If the trap flag (TF) is set in the flag register, the processor automatically
generates a type 1 interrupt after each instruction, and the processor can be

considered to be in single-step mode for debugging purposes. The trap flag (TF) can be altered only by pushing the flag register onto the stack (PUSHF) altering the flag (OR with 0100 H to set or AND with FEFFH to reset) and then popping the flag register off the stack (POPF).

The interrupts, in general, automatically push the flag register onto the stack, followed by the code segment register contents and the instruction pointer contents. The MPU then vectors to the appropriate service routine. The interrupt return instruction (IRET) resumes program execution.

9.3 An 8086 Program Example

Let's rewrite our "clear screen" routine in 8086 code to look at a simple 8086 program.

```
        ORG 1000H
        MOV CX, 400H      ;Set byte counter
        MOV DI, 0CC00H    ;Load destination index register
SPACE:  MOV [DI], 20H     ;Move space code to memory
                          designated by DI
        INC DI            ;Increment DI
        LOOP SPACE        ;Repeat 400H times
        RET               ;Return to calling program
        END
```

The LOOP, LOOPZ, and LOOPNZ instructions are very useful in repeating a set of program steps. The CX register is the counter when these instructions are used. Each pass through the loop automatically decrements CX by 1. The form of the LOOP instruction is as follows:

LOOP: Address

11100010	DISP

The displacement is an 8-bit signed quantity to allow forward or backward looping. When CX reaches zero, the program drops to the next instruction after LOOP. In using the displacement, the displacement is added to the contents of the instruction pointer, making this a jump within the segment that is current and within the range -128 to $+127$ bytes of the first byte of the next instruction. These are called SHORT jumps by Intel.

Intel offers two enhanced versions of the 8086. These are the 80186 (iAPX 186) and the 80286 (iAPX 286) microprocessor systems. Although based on the 8086, these units are faster and far more complex. The register architecture is similar, as is the instruction set. However, the addition of the 16-bit timers, DMA channels, interrupt controller, peripheral select logic, wait state generator, and new instruction types makes them very different.

Data sheets and instruction sets for the 80186 and 80286 are shown in Figs. 15-16 through 15-19.

FIGURE 15-16 Pin out for the 80186. (Reprinted by permission of
Intel Corporation, copyright 1983.)

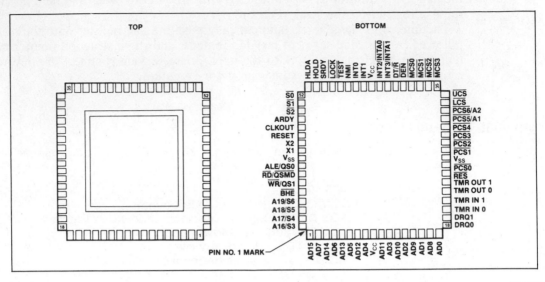

FIGURE 15-18 Pin out for the 80286. (Reprinted by permission of
Intel Corporation, copyright 1984.)

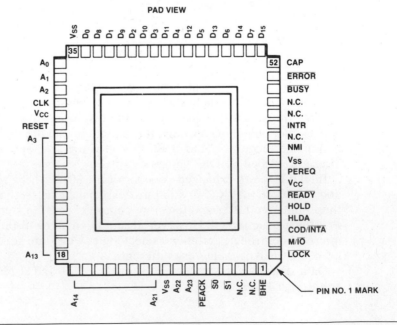

FIGURE 15-17 Data sheet for the 80186. (Reprinted by permission of Intel Corporation, copyright 1983.)

iAPX 186
HIGH INTEGRATION 16-BIT MICROPROCESSOR

- **Integrated Feature Set**
  - **—Enhanced 8086-2 CPU**
  - **—Clock Generator**
  - **—2 Independent, High-Speed DMA Channels**
  - **—Programmable Interrupt Controller**
  - **—3 Programmable 16-bit Timers**
  - **—Programmable Memory and Peripheral Chip-Select Logic**
  - **—Programmable Wait State Generator**
  - **—Local Bus Controller**
- **Available in 8 MHz (80186) and cost effective 6 MHz (80186-6) versions.**
- **High-Performance Processor**
  - **—2 Times the Performance of the Standard iAPX 86**
  - **—4 MByte/Sec Bus Bandwidth Interface**

- **Direct Addressing Capability to 1 MByte of Memory**
- **Completely Object Code Compatible with All Existing iAPX 86, 88 Software**
  - **—10 New Instruction Types**
- **Complete System Development Support**
  - **—Development Software: Assembler, PL/M, Pascal, Fortran, and System Utilities**
  - **—In-Circuit-Emulator (I²ICE™-186)**
  - **—iRMX™ 86, 88 Compatible (80130 OSF)**
- **Optional Numeric Processor Extension**
  - **—iAPX 186/20 High-Performance 80-bit Numeric Data Processor**

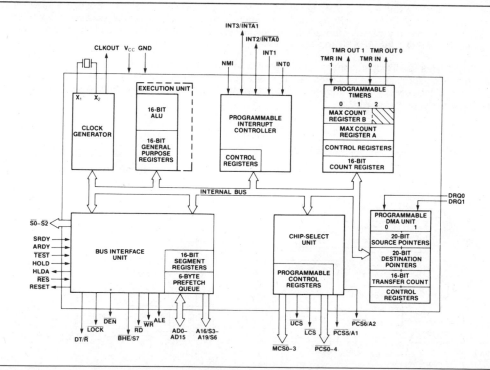

FIGURE 15-19 Data sheet for the 80286. (Reprinted by permission of Intel Corporation, copyright 1984.)

iAPX 286/10
HIGH PERFORMANCE MICROPROCESSOR
WITH MEMORY MANAGEMENT AND PROTECTION

- **High Performance 8 and 10 MHz Processor (Up to six times iAPX 86)**
- **Large Address Space:**
  - **—16 Megabytes Physical**
  - **—1 Gigabyte Virtual per Task**
- **Integrated Memory Management, Four-Level Memory Protection and Support for Virtual Memory and Operating Systems**
- **Two iAPX 86 Upward Compatible Operating Modes:**
  - **—iAPX 86 Real Address Mode**
  - **—Protected Virtual Address Mode**

- **Optional Processor Extension:**
  - **—iAPX 286/20 High Performance 80-bit Numeric Data Processor**
- **Complete System Development Support:**
  - **—Development Software: Assembler, PL/M, Pascal, FORTRAN, and System Utilities**
  - **—In-Circuit-Emulator (ICE™-286)**
- **High Bandwidth Bus Interface (8 or 10 Megabyte/Sec)**
- **Available in EXPRESS:**
  - **—Standard Temperature Range**

The iAPX 286/10 (80286 part number) is an advanced, high-performance microprocessor with specially optimized capabilities for multiple user and multi-tasking systems. The 80286 has built-in memory protection that supports operating system and task isolation as well as program and data privacy within tasks. A 10 MHz iAPX 286/10 provides up to six times greater throughput than the standard 5 MHz iAPX 86/10. The 80286 includes memory management capabilities that map up to 2^{30} bytes (one gigabyte) of virtual address space per task into 2^{24} bytes (16 megabytes) of physical memory.

The iAPX 286 is upward compatible with iAPX 86 and 88 software. Using iAPX 86 real address mode, the 80286 is object code compatible with existing iAPX 86, 88 software. In protected virtual address mode, the 80286 is source code compatible with iAPX 86, 88 software and may require upgrading to use virtual addresses supported by the 80286's integrated memory management and protection mechanism. Both modes operate at full 80286 performance and execute a superset of the iAPX 86 and 88's instructions.

The 80286 provides special operations to support the efficient implementation and execution of operating systems. For example, one instruction can end execution of one task, save its state, switch to a new task, load its state, and start execution of the new task. The 80286 also supports virtual memory systems by providing a segment-not-present exception and restartable instructions.

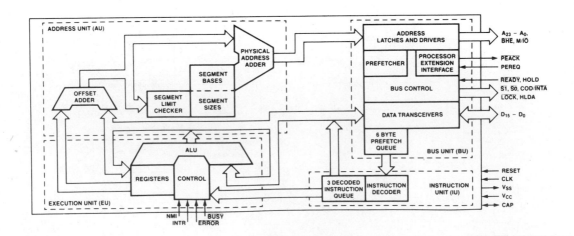

9.4 Conclusion: The 8086

The 8086 is a powerful 16-bit microprocessor, the successor to the 8-bit 8085. The 8086 is intended for applications that exceed the limits of the 8-bit MPU and require both larger memories and advanced math and interrupt capabilities. From this introduction to the 8086, it should be clear that an assembler is more than desirable for serious programming. More information on the 8086 and 8088 MPUs is available in the Intel 8086 Family User's Manual. Information on a 32 bit MPU, the 80386 is also available.

10 THE 68000 MICROPROCESSOR

The 68000 microprocessor by Motorola Semiconductors is a 16-bit microprocessor with the clear potential of being a 32-bit microprocessor, and the 68020 32-bit

FIGURE 15-20 Programming model and pin assignments for the 68000. (Courtesy of Motorola, Inc.)

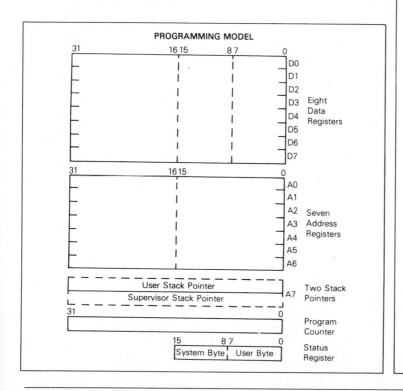

TABLE 15-13   68000 Clock Rates

CPU	Frequency (MHz)	Time (μsec)
MC68000L4	4	250–500
MC68000L6	6	167–500
MC68000L8	8	125–500
MC68000L10	10	100–500
MC68000L12	12.5	80–500

version has very recently become available. The 68000 is in a 64-pin package, thus eliminating any need for multiplexing address and data lines due to a shortage of leads on the package. The 16-bit data path (D0–D15) outside the 68000 MPU makes this a 16-bit microprocessor. However, the designers used a full 32-bit data path within the 68000 MPU to allow for a full upgrade (i.e., the 68020). The 68000 has 17 general-purpose registers, each 32 (thirty-two!) bits long. Eight of these are used for data (D0–D7), and seven are used as address registers (A0–A6). Register A7 is a stack pointer (user or supervisor), and the last 32-bit register functions as the program counter. Only 24 bits of the program counter are used (0–23), but this allows the 68000 MPU to directly address 16,777,216 bytes of memory on address lines A1–A23. There is no need for segmented memory on the 68000 due to the large program counter. Figure 15-20 shows both the 68000

TABLE 15-14   68000 Data Addressing Modes

Mode	Generation
Register Direct Addressing	
Data Register Direct	EA = Dn
Address Register Direct	EA = An
Absolute Data Addressing	
Absolute Short	EA = (Next Word)
Absolute Long	EA = (Next Two Words)
Program Counter Relative Addressing	
Relative with Offset	EA = (PC) + d_{16}
Relative with Index and Offset	EA = (PC) + (Xn) + d_8
Register Indirect Addressing	
Register Indirect	EA = (An)
Postincrement Register Indirect	EA = (An), An $\leftarrow$ An + N
Predecrement Register Indirect	An $\leftarrow$ An – N, EA = (An)
Register Indirect With Offset	EA = (An) + d_{16}
Indexed Register Indirect With Offset	EA = (An) + (Xn) + d_8
Immediate Data Addressing	
Immediate	DATA = Next Word(s)
Quick Immediate	Inherent Data
Implied Addressing	
Implied Register	EA = SR, USP, SP, PC

NOTES:

EA = Effective Address
An = Address Register
Dn = Data Register
Xn = Address or Data Register used as Index Register
SR = Status Register
PC = Program Counter

d_8 = Eight-bit Offset (displacement)
d_{16} = Sixteen-bit Offset (displacement)
N = 1 for Byte, 2 for Words and 4 for Long Words
( ) = Contents of
$\leftarrow$ = Replaces

Source: Courtesy of Motorola, Inc.

pin assignments and the register set diagram. The MC68000 is available in five different versions that run at different clock rates as shown in Table 15-13.

The 68000 requires a single 5-V d-c power supply (dissipation = 1.2 W at 8 MHz) and uses a single-phase clock signal.

In using the 32-bit registers, the data registers are used for byte (8-bit), word (16-bit), or long word (32-bit) data operations. The address registers (A0–A7) may be used as software stack pointers and base address registers and for word and long word address operations. All registers may be used as index registers. As you can see, the lack of limitations on register usage combined with the 32-bit register size makes the 68000 an extremely attractive microprocessor. Table 15-14 illustrates the 68000 data addressing modes and the manner in which the effective address is calculated.

The 68000 uses memory-mapped I/O as did the 6800, its predecessor. Memory is arranged from 8-bit bytes into bytes, words, or long words as required by the individual instructions. Figures 15-21 and 15-22 show the arrangement of 68000 memory.

10.1 The 68000 Instruction Set

The large available registers and efficient available addressing modes of the 68000 allow for an effective instruction set designed for easy programming in a wide set of applications. The instruction format (Fig. 15-23) yields resultant code that is from one to five words (2–10 bytes) long. The instruction set (Tables 15-15 and 15-16) includes instructions for signed and unsigned multiply and signed and unsigned divide.

Since the 68000 can operate on bytes, words, or long words, the length of an operand is specified as a part of the 68000 assembler syntax using B, W, or L suffixes. The MOVE instruction may appear as follows:

MOVE.B   MOVE.L   MOVE.W

depending on the size of the data to be moved. For instance, MOVE.L A0, D1 duplicates the 32-bit contents of register A0 in register D1. The format of the MOVE instruction is:

FIGURE 15-21 Word organization in the 68000's memory. (Courtesy of Motorola, Inc.)

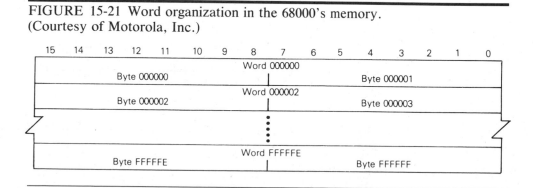

FIGURE 15-22 Data organization in memory. (Courtesy of Motorola, Inc.)

Bit Data
1 Byte = 8 Bits

7	6	5	4	3	2	1	0

Integer Data
1 Byte = 8 Bits

15	14	13	12	11	10	9	8	7	6	5	4	3	2	1	0
MSB		Byte 0					LSB				Byte 1				
		Byte 2									Byte 3				

1 Word = 16 Bits

15	14	13	12	11	10	9	8	7	6	5	4	3	2	1	0
MSB							Word 0								LSB
							Word 1								
							Word 2								

1 Long Word = 32 Bits

15	14	13	12	11	10	9	8	7	6	5	4	3	2	1	0
MSB							High Order								
— — Long Word 0 — — —						Low Order								LSB
— — Long Word 1 — — —														
— — Long Word 2 — — —														

Addresses
1 Address = 32 Bits

15	14	13	12	11	10	9	8	7	6	5	4	3	2	1	0
MSB							High Order								
— — Address 0 — — —						Low Order								LSB
— — Address 1 — — —														
— — Address 2 — — —														

MSB = Most Significant Bit
LSB = Least Significant Bit

Decimal Data
2 Binary Coded Decimal Digits = 1 Byte

15	14	13	12	11	10	9	8	7	6	5	4	3	2	1	0
MSD	BCD 0			BCD 1			LSD	BCD 2				BCD 3			
	BCD 4			BCD 5				BCD 6				BCD 7			

MSD = Most Significant Digit
LSD = Least Significant Digit

FIGURE 15-23 The 68000 instruction format. (Courtesy of Motorola, Inc.)

15	14	13	12	11	10	9	8	7	6	5	4	3	2	1	0
Operation Word (first word specifies operation and modes)																
Immediate operand (if any, one or two words)																
Source effective address extension (if any, one or two words)																
Destination effective address extension (if any, one or two words)																

TABLE 15-15 68000 Instruction Set

Mnemonic	Description
ABCD	Add Decimal with Extend
ADD	Add
AND	Logical And
ASL	Arithmetic Shift Left
ASR	Arithmetic Shift Right
B_{cc}	Branch Conditionally
BCHG	Bit Test and Change
BCLR	Bit Test and Clear
BRA	Branch Always
BSET	Bit Test and Set
BSR	Branch to Subroutine
BTST	Bit Test
CHK	Check Register Against Bounds
CLR	Clear Operand
CMP	Compare
DB_{cc}	Test Cond., Decrement and Branch
DIVS	Signed Divide
DIVU	Unsigned Divide
EOR	Exclusive Or
EXG	Exchange Registers
EXT	Sign Extend
JMP	Jump
JSR	Jump to Subroutine
LEA	Load Effective Address
LINK	Link Stack
LSL	Logical Shift Left
LSR	Logical Shift Right

Mnemonic	Description
MOVE	Move
MOVEM	Move Multiple Registers
MOVEP	Move Peripheral Data
MULS	Signed Multiply
MULU	Unsigned Multiply
NBCD	Negate Decimal with Extend
NEG	Negate
NOP	No Operation
NOT	One's Complement
OR	Logical Or
PEA	Push Effective Address
RESET	Reset External Devices
ROL	Rotate Left without Extend
ROR	Rotate Right without Extend
ROXL	Rotate Left with Extend
ROXR	Rotate Right with Extend
RTE	Return from Exception
RTR	Return and Restore
RTS	Return from Subroutine
SBCD	Subtract Decimal with Extend
S_{cc}	Set Conditional
STOP	Stop
SUB	Subtract
SWAP	Swap Data Register Halves
TAS	Test and Set Operand
TRAP	Trap
TRAPV	Trap on Overflow
TST	Test
UNLK	Unlink

Source: Courtesy of Motorola, Inc.

TABLE 15-16   68000 Variations of instruction Types

Instruction Type	Variation	Description
ADD	**ADD**	Add
	ADDA	Add Address
	ADDQ	Add Quick
	ADDI	Add Immediate
	ADDX	Add with Extend
AND	**AND**	Logical And
	ANDI	And Immediate
CMP	**CMP**	Compare
	CMPA	Compare Address
	CMPM	Compare Memory
	CMPI	Compare Immediate
EOR	**EOR**	Exclusive Or
	EORI	Exclusive Or Immediate
MOVE	**MOVE**	Move
	MOVEA	Move Address
	MOVEQ	Move Quick
	MOVE from SR	Move from Status Register
	MOVE to SR	Move to Status Register
	MOVE to CCR	Move to Condition Codes
	MOVE to USP	Move to User Stack Pointer
NEG	**NEG**	Negate
	NEGX	Negate with Extend
OR	**OR**	Logical Or
	ORI	Or Immediate
SUB	**SUB**	Subtract
	SUBA	Subtract Address
	SUBI	Subtract Immediate
	SUBQ	Subtract Quick
	SUBX	Subtract with Extend

Source: Courtesy of Motorola, Inc.

MOVE (source), (destination)

that is, MOVE S, D = move from S to D.

The 68000 assembler syntax conforms to the following conventions:

Form	Comment
$2000	Address 2000H
#48	Immediate data (base 10)
($2050)	Data at address $2050
D2	Contents of D2 register
(A2)	Contents of location specified by A2
(A2)+	Contents of address specified by A2 and increment A2 (postincrement)
−(A4)	Contents of address specified by A4 after A4 is decremented (predecrement)
d (A6)	Contents of location specified by A6 + d (displacement)

The 68000 has two absolute addressing modes: absolute short and absolute long.

Absolute short	Operand is a 16-bit address sign extended to 32 bits
Absolute long	Operand is a 32-bit address

Absolute short allows access to the lowest 32K bytes of memory (0–7FFF H) or the highest 32K bytes of memory (FF8000 H–FFFFFF H). Absolute long addressing allows access to the full 16M-byte address range. It is advantageous to use the absolute short mode when possible because one less word of instruction is required, increasing the speed of the instruction.

In immediate mode, byte or word data will be sign extended if the destination is an address register (32 bits) but will *not* be sign extended if the destination is a data register.

For example, the instruction:

$$\text{MOVE.W #\$2345, D0}$$

will load 2345 H into the lower half of D0, leaving the upper half unaffected. However, the instruction;

$$\text{MOVE.W #\$2345, A0}$$

will load FFFF2345 H into A0.

Here are some examples of 68000 assembler syntax:

Example	*Comment*
ADD.B D1, D3	Add the low byte of D1 to D3
ADD D2, D4	Add low word of D2 to D4
ADD.W D2, D4	Add low word of D2 to D4
ADD.L D3, D5	Add contents of D3 to D5
MOVE.W D1, D2	Move low word contents of D1 to D2
MOVE.W D1, (A2)	Move low word contents of D1 to address specified by A2
MOVE.L (A0), (A1)	Move contents of address (32 bits) specified by A0 to location specified by A1 (32 bits)
MOVE.W (A0)+, (A1)+	Move one data word from address in A0 to address in A1 and increment both registers
CMP (A1), D2	Compare data at address specified by A1 to D2 register

10.2 The 68000 Interrupt Modes

The reset operation of the 68000 microprocessor causes the MPU to vector to address 000000 H. At this location (4 bytes), a 32-bit address is found and loaded into the supervisor stack pointer. At the next location (000004 H), a 32-bit value is read and loaded into the program counter.

TABLE 15-17   68000 Vector Assignments

Vector Number(s)	Address			Assignment
	Dec	Hex	Space	
0	0	000	SP	Reset: Initial SSP
	4	004	SP	Reset: Initial PC
2	8	008	SD	Bus Error
3	12	00C	SD	Address Error
4	16	010	SD	Illegal Instruction
5	20	014	SD	Zero Divide
6	24	018	SD	CHK Instruction
7	28	01C	SD	TRAPV Instruction
8	32	020	SD	Privilege Violation
9	36	024	SD	Trace
10	40	028	SD	Line 1010 Emulator
11	44	02C	SD	Line 1111 Emulator
12*	48	030	SD	(Unassigned, reserved)
13*	52	034	SD	(Unassigned, reserved)
14*	56	038	SD	(Unassigned, reserved)
15*	60	03C	SD	(Unassigned, reserved)
16-23*	64	040	SD	(Unassigned, reserved)
	95	05F		—
24	96	060	SD	Spurious Interrupt
25	100	064	SD	Level 1 Interrupt Auto-Vector
26	104	068	SD	Level 2 Interrupt Auto-Vector
27	108	06C	SD	Level 3 Interrupt Auto-Vector
28	112	070	SD	Level 4 Interrupt Auto-Vector
29	116	074	SD	Level 5 Interrupt Auto-Vector
30	120	078	SD	Level 6 Interrupt Auto-Vector
31	124	07C	SD	Level 7 Interrupt Auto-Vector
32-47	128	080	SD	TRAP Instruction Vectors
	191	0BF		—
48-63*	192	0C0	SD	(Unassigned, reserved)
	255	0FF		—
64-255	256	100	SD	User Interrupt Vectors
	1023	3FF		—

*Vector numbers 12 through 23 and 48 through 63 are reserved for future enhancements by Motorola. No user peripheral devices should be assigned these numbers.

Source: Courtesy of Motorola, Inc.

The 68000 provides seven levels of vectored interrupts. These may be masked at a given level and any at a lower level are locked out. When an interrupt is acknowledged, the interrupting device must place a vector number on the data bus that selects one of 192 interrupt service routines in memory.

The 68000 also provides seven autovectors based on the priority level of the device. The 68000 has three inputs that initiate a hardware interrupt, and these are the priority level input lines ($\overline{IPL0}$, $\overline{IPL1}$, $\overline{IPL2}$). These input pins indicate the encoded priority level of the device requesting the interrupt.

Table 15-17 shows the 68000 vector assignments beginning with address 000000 H. The autovectors are the easiest to use and begin with the level 1 autovector at address 000064 H.

The interrupt control lines ($\overline{IPL0}$, $\overline{IPL1}$, $\overline{IPL2}$) establish either a priority level or an autovector request. Level 0 indicates that *no* interrupts are requested. Table 15-18 shows 68000 priority interrupt generation.

The address that the MPU vectors to (i.e., 000064 H for autovector level 1) allows the MPU to find and locate a 32-bit address to load into the program counter.

TABLE 15-18   68000 Priority Interrupt Generation

Level	$\overline{IPL2}$	$\overline{IPL1}$	$\overline{IPL0}$
0	1	1	1
1	1	1	0
2	1	0	1
3	1	0	0
4	0	1	1
5	0	1	0
6	0	0	1
7	0	0	0

10.3 A Simple 68000 Program

To be different, let's look at a program to zero 5000 H memory locations beginning at address 10000 H. You should recognize that this program is almost identical to the clear screen routine we have been using as an example.

```
        ORG $2000
        MOVE.L #$5000, D0     ;Load D0 with 5000H
        MOVE.L #$10000, A0    ;Load A0 with 10000H
LOOP:   MOVE.B #0, (A0)+      ;Move 0 to location A0 and increment A0
        SUBQ.L #1, D0         ;Subtract one from D0
        BNE LOOP             ;Repeat if not zero
        RTS                  ;Return from subroutine
```

In the 68000 instruction set, incrementing or decrementing a register is an integral part of another instruction, (A0)+, −(A1), and so on. A specific decrement might have to be done using a subtract (SUB) or quick subtract (SUBQ) instruction.

Note: The stack builds from high memory to low memory (i.e., toward zero). The 68000 does *not* use byte swapping.

10.4 Other Members of the 68000 Family

The 68000 MPU is internally a 32-bit microprocessor, but it has a 16-bit data bus externally. Since it is designed to be a higher performance MPU, it stands to reason that its users have been quick to take advantage of its excellent architecture. Other members of the 68000 family are summarized in Table 15-19.

The MC68008 is the 8-bit version of the 68000. It is available largely for designers

TABLE 15-19   68000 Family Members

MPU	Comments
MC68008	8-Bit external data bus
MC68010	Virtual memory processor
MC68020	32-Bit external data bus
MC68881	Floating-point coprocessor

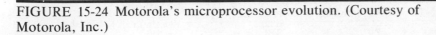

FIGURE 15-24 Motorola's microprocessor evolution. (Courtesy of Motorola, Inc.)

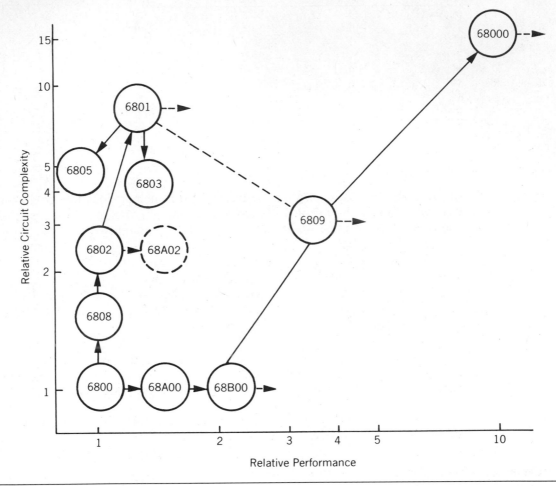

who are more familiar with such structures. The register set and instructions of the 68000 MPU are all available to the designer. Software developed for the 68000 will run on the 68008 and vice versa.

The MC68010 is very similar to the 68000 MPU. Internally it is a 32-bit machine and externally it has a 16-bit data bus. The 68010 is designed to implement a virtual memory. Virtual memory is usually a slower type of storage like a disk or diskette that is treated like RAM. When a reference is made to a virtual memory location, the data is loaded into RAM for use. In this way, an unlimited amount of RAM can effectively be made available to the user.

The MC68020 is an enhanced 68000 MPU. It is the first true 32-bit processor in that the 32-bit internal data bus is matched by an external 32-bit data bus. The

68020 is software compatible with the 68000. The 68020 has added instructions, features coprocessor operations, and has a cache memory (a special high-speed memory) to increase performance.

The MC68881 is a floating-point coprocessor designed to be used with the MC68020. The MC68881 has eight 80-bit floating-point data registers and a 67-bit ALU for arithmetic operations.

10.5 Conclusion: The 68000

The 68000 is a very impressive microprocessor; it is a 16-bit MPU, yet architecturally it is a 32-bit device. It can address 16M bytes of memory, is available to 12.5 MHz, has in excess of 100,000 transistors, operates on a single power supply, and features a large number of user-available 32-bit registers. Its nonmultiplexed address and data lines constitute a powerful feature that saves on external latching circuitry. The 68000 lends itself to mainframe architecture. Its instruction set is easy to learn and can handle data of various lengths (byte, word, long word). The use of an assembler is essential to software generation and debugging operations. Additionally, the 68000 has control signals that adapt easily to the control of 6800-type peripherals. Figure 15-24 compares members of Motorola's microprocessor family.

11 THE 8048 SINGLE-CHIP COMPUTER

The 8048 by Intel is an 8-bit MPU along with 1K bytes of ROM, 64 bytes of RAM, 27 I/O lines, and an 8-bit timer/event counter on a single chip. Included is an oscillator and clock driver to which a crystal may be attached. Figure 15-25 shows the pin configuration, logic symbol, and block diagram of the 8048, and Table 15-20 lists the members of the 8048 family. The 8048 family features a single-level interrupt line and a single-step line (see also Table 15-21).

The 8048 has an 8-bit accumulator. The first eight locations of resident data memory (64 bytes of RAM) are designated as working registers (R0–R7) and are directly addressable by several instructions. The resident program memory (1K bytes of ROM or EPROM) contains the reset vector (at zero), the external interrupt vector (at three), and the timer interrupt vector (at location seven).

TABLE 15-20   8048 Family Members

MPU	Clock (MHz)	Comments
8048	6	1K ROM, 64 bytes RAM
8748	6	EPROM version 8048
8648	6	One-time factory programmable
8035	6	8048 without ROM
8049	6	2K ROM
8022	6	Two-channel, 8-bit A/D converter

FIGURE 15-25 Pin configuration, logic symbol, and block diagram for the 8048. (Reprinted by permission of Intel Corporation, copyright 1983.)

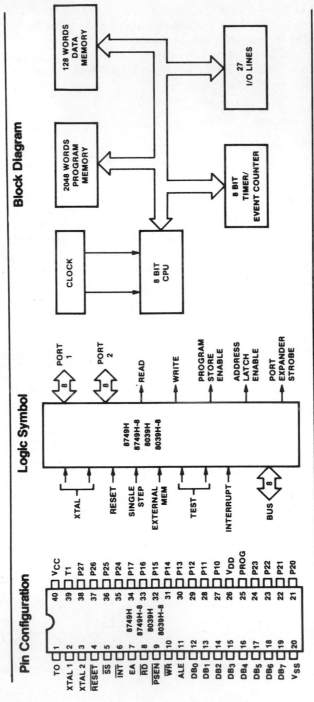

Pin Configuration

Logic Symbol

Block Diagram

TABLE 15-21   8048 Family Features

8049	8048	8021	8022	FEATURES
√	√	√	√	8-bit CPU
2K × 8	1K × 8	1K × 8	2K × 8	Program memory
128 × 8	64 × 8	64 × 8	64 × 8	Data RAM
27	27	21	28	I/O lines
√	√	√	√	Timer counter
√	√	√	√	Oscillator and clock
√	√	√	√	Reset circuit
√	√	√	√	Interrupt

Source: Reprinted by permission of Intel Corporation, copyright 1981.

11.1 The 8048/8049 Instruction Set

Compared to the 16-bit instruction sets we have had a chance to examine, an 8-bit instruction set is easy to look at. Table 15-22 shows the 8048/8049 instruction set summary. A few sample instructions illustrate how easy the 8048 is to program.

Instruction	*Comment*
MOV A, #20H	Move 20 H to register A
MOV A, R1	Move contents of R1 to A
MOV A, @R2	Move memory contents specified by R2 to A
MOV R3, A	Move A contents to R3
MOV R3, #64H	Move 64 H to R3
MOV @R3, #14H	Move 14 H to address specified in R3
INC A	Increment A
INC R2	Increment R2
INC @R2	Increment memory location contents
ADDC A, R2	Add carry and R2 contents to A

12 THE 8051 SINGLE-CHIP MICROCOMPUTER

The 8051 single-chip microcomputer by Intel uses the 8048 architecture and is enhanced with nonpaged jumps, direct addressing, four eight-register banks, stack depth to 128 bytes, and multiply, divide, subtract, and compare. Additional features of the 8051 are as follows:

4K × 8 ROM/EPROM	Full duplex serial channel
128 × 8 RAM	Boolean processor
Four 8-bit I/O ports (32 I/O lines)	Compatible with 8080A/85 peripherals
Two 16-bit timer/event counters	External memory expandable to 128K

TABLE 15-22 8048/8049 Instruction Set Summary

Accumulator

Mnemonic	Description	Bytes	Cycles
ADD A, R	Add register to A	1	1
ADD A, @R	Add data memory to A	1	1
ADD A, # data	Add immediate to A	2	2
ADDC A, R	Add register with carry	1	1
ADDC A, @R	Add data memory with carry	1	1
ADDC A, # data	Add immediate with carry	2	2
ANL A, R	And register to A	1	1
ANL A, @R	And data memory to A	1	1
ANL A, # data	And immediate to A	2	2
ORL A, R	Or register to A	1	1
ORL A @R	Or data memory to A	1	1
ORL A, # data	Or immediate to A	2	2
XRL A, R	Exclusive or register to A	1	1
XRL A, @R	Exclusive or data memory to A	1	1
XRL, A, # data	Exclusive or immediate to A	2	2
INC A	Increment A	1	1
DEC A	Decrement A	1	1
CLR A	Clear A	1	1
CPL A	Complement A	1	1
DA A	Decimal adjust A	1	1
SWAP A	Swap nibbles of A	1	1
RL A	Rotate A left	1	1
RLC A	Rotate A left through carry	1	1
RR A	Rotate A right	1	1
RRC A	Rotate A right through carry	1	1

Input/Output

Mnemonic	Description	Bytes	Cycles
IN A, P	Input port to A	1	2
OUTL P, A	Output A to port	1	2
ANL P, # data	And immediate to port	2	2
ORL P, # data	Or immediate to port	2	2
INS A, BUS	Input BUS to A	1	2
OUTL BUS, A	Output A to BUS	1	2
ANL BUS, # data	And immediate to BUS	2	2
ORL BUS, # data	Or immediate to BUS	2	2
MOVD A,P	Input expander port to A	1	2
MOVD P, A	Output A to expander port	1	2
ANLD P, A	And A to expander port	1	2
ORLD P, A	Or A to expander port	1	2

Registers

Mnemonic	Description	Bytes	Cycles
INC R	Increment register	1	1
INC @R	Increment data memory	1	1
DEC R	Decrement register	1	1

Branch

Mnemonic	Description	Bytes	Cycles
JMP addr	Jump unconditional	2	2
JMPP @A	Jump indirect	1	2
DJNZ R, addr	Decrement register and skip	2	2
JC addr	Jump on carry = 1	2	2
JNC addr	Jump on carry = 0	2	2
JZ addr	Jump on A zero	2	2
JNZ addr	Jump on A not zero	2	2
JT0 addr	Jump on TO = 1	2	2
JNT0 addr	Jump on TO = 0	2	2
JT1 addr	Jump on T1 = 1	2	2
JNT1 addr	Jump on T1 = 0	2	2
JF0 addr	Jump on F0 = 1	2	2
JF1 addr	Jump on F1 = 1	2	2
JTF addr	Jump on timer flag	2	2
JN1 addr	Jump on INT = 0	2	2
JBb addr	Jump on accumulator bit	2	2

Subroutine

Mnemonic	Description	Bytes	Cycles
CALL addr	Jump to subroutine	2	2
RETR	Return	1	2
RETR	Return and restore status	1	2

Flags

Mnemonic	Description	Bytes	Cycles
CLR C	Clear carry	1	1
CPL C	Complement carry	1	1
CLR F0	CLear flag 0	1	1
CPL F0	Complement flag 0	1	1
CLR F1	Clear flag 1	1	1
CPL F1	Complement flag 1	1	1

Data Moves

Mnemonic	Description	Bytes	Cycles
MOV A, R	Move register to A	1	1
MOV A, @R	Move data memory to A	1	1
MOV A, # data	Move immediate to A	2	2
MOV R, A	Move A to register	1	1
MOV @R, A	Move A to data memory	1	1
MOV R, # data	Move immediate to register	2	2
MOV @R, #data	Move immediate to data memory	2	2
MOV A, PSW	Move PSW to A	1	1
MOV PSW, A	Move A to PSW	1	1
XCH A, R	Exchange A and register	1	1
XCH A, @R	Exchange A and data memory	1	1
XCHD A, @R	Exchange nibble of A and register	1	1
MOVX A, @R	Move external data memory to A	1	2
MOVX @R, A	Move A to external data memory	1	2
MOVP A, @A	Move to A from current page	1	2
MOVP3 A, @	Move to A from page 3	1	2

Timer/Counter

Mnemonic	Description	Bytes	Cycles
MOV A, T	Read timer/counter	1	1
MOV T, A	Load timer/counter	1	1
STRT T	Start timer	1	1
STRT CNT	Start counter	1	1
STOP TCNT	Stop timer/counter	1	1
EN TCNT1	Enable timer/counter interrupt	1	1
DIS TCNT1	Disable timer/counter interrupt	1	1

Control

Mnemonic	Description	Bytes	Cycles
EN 1	Enable external interrupt	1	1
DIS 1	Disable external interrupt	1	1
SEL RB0	Select register bank 0	1	1
SEL RB1	Select register bank 1	1	1
SEL MB0	Select memory bank 0	1	1
SEL MB1	Select memory bank 1	1	1
ENT 0 CLK	Enable clock output on T0	1	1

Mnemonic	Description	Bytes	Cycles
NOP	No operation	1	1

Source: Reprinted by permission of Intel Corporation, copyright 1976.

With all these features, the 8051 is truly a stand-alone microcomputer on a chip; Fig. 15-26 shows its pin configuration, logic symbol, and block diagram. The 8051 family functional block diagram appears in Fig. 15-27.

FIGURE 15-26 Pin configuration, logic symbol, and block diagram for the 8051. (Reprinted by permission of Intel Corporation, copyright 1983.)

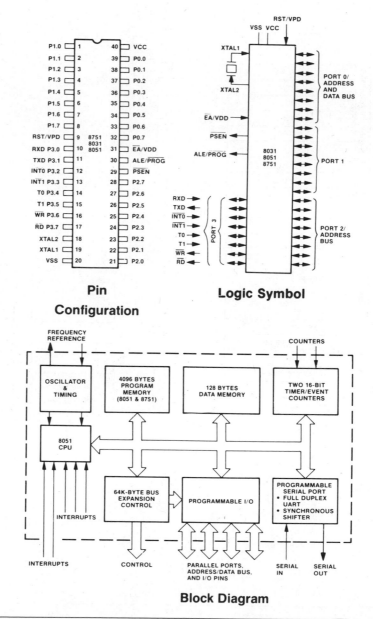

Pin Configuration

Logic Symbol

Block Diagram

FIGURE 15-27 Functional block diagram for the 8051 family.
(Reprinted by permission of Intel Corporation, copyright 1983.)

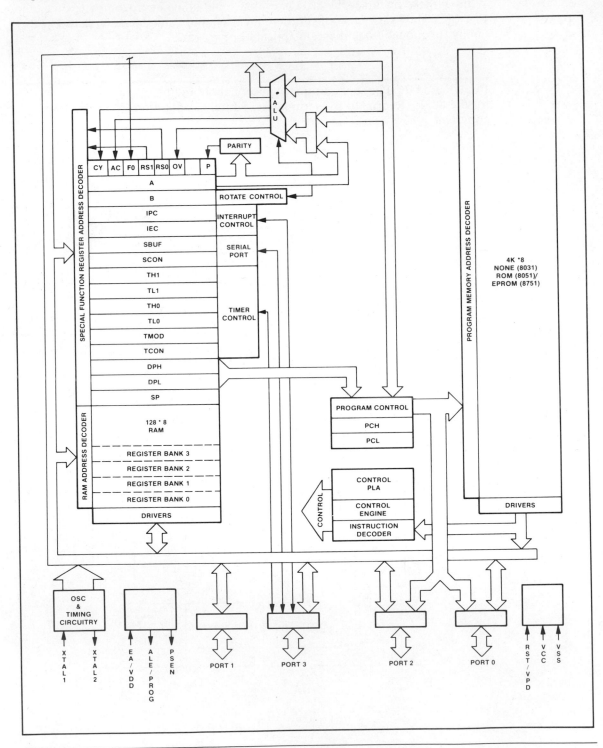

The 8051 family comes in three versions: the 8031, the 8051, and the 8751.

MPU	Comment
8031	Control-oriented MPU with RAM plus I/O
8051	An 8031 with factory-programmed ROM
8751	An 8031 with EPROM

The 8051 can operate with a 12-MHz crystal that allows 58% of the instructions to execute in 1 μsec. Instructions are 1, 2, or 3 bytes long.

The 8051 acknowledges interrupts from five sources: two hardware interrupts ($\overline{INT0}$, $\overline{INT1}$), one from each of the two internal counters, and one from the serial I/O port. Each interrupt vectors to a separate location in memory.

The 8051 has an 8-bit accumulator (A) and an 8-bit B register, which is used during multiply and divide operations.

In addition to an 8-bit stack pointer, there is a 16-bit data pointer (DPH, DPL). The PSW register contains two bits (RS0, RS1) that select one of four eight-register banks.

13   THE 8096 SIXTEEN-BIT MICROCONTROLLER

The Intel 8096/8396 16-bit microcontroller is a computer on a chip that operates at 12 MHz and is designed for high-speed control functions. The 8096 features an

FIGURE 15-28 Block diagram for the 8096/8396. (Reprinted by permission of Intel Corporation, copyright 1983.)

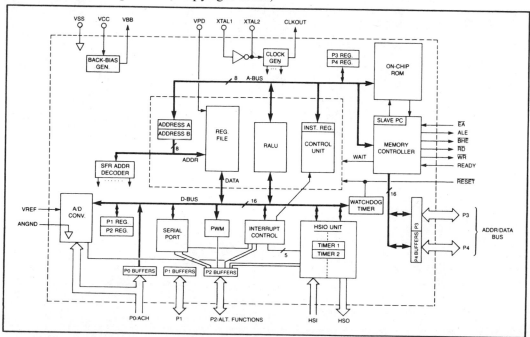

FIGURE 15-29 Packages for the 8096. (Reprinted by permission of Intel Corporation, copyright 1983.)

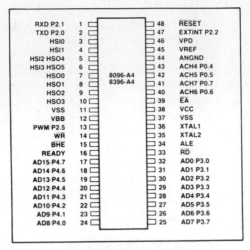

48-Pin Package

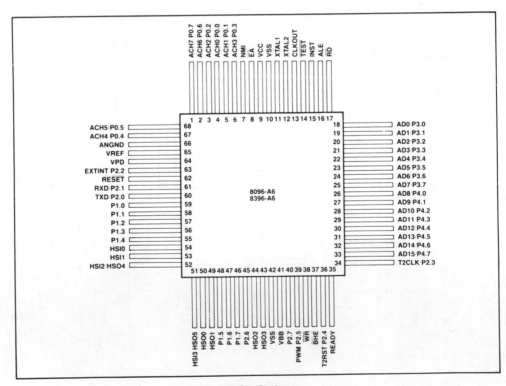

68-Pin Package

FIGURE 15-30 Memory map for the 8096. (Reprinted by permission of Intel Corporation, copyright 1983.)

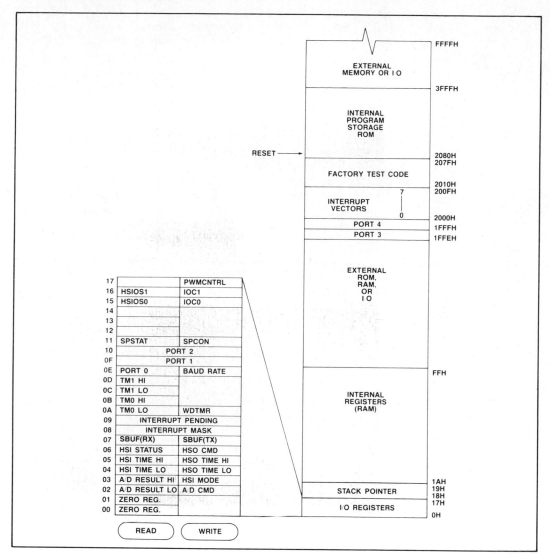

on-chip 10-bit A/D converter, five 16-bit software timers, a 232-byte register file, a serial port, five 8-bit I/O ports, and a watchdog timer. The 8396 is identical to the 8096 except that the 8396 also has 8K bytes of on-chip ROM. Each chip has an on-board clock generator. The block diagram of the 8096/8396 16-bit MCU and the two available packages for the 8096 are shown in Figs. 15-28 and 15-29, respectively.

The internal registers for the 8096 occupy address space from 1A H to FF H. These locations form the 232-byte register file. The internal registers may be used for indexing, address pointers, data pointers, and data manipulators (accumulators).

The memory map for the 8096 (Fig. 15-30) indicates that the unit supports both internal and external forms of memory. Reset causes a vector to address 2080 H. All locations are 16-bit memory locations.

The 8096 features a watchdog timer that offers a fail-safe way to reset the MPU in the event of a loss of MPU program control. The watchdog timer is a 16-bit register that is incremented every state time. At overflow, the MPU is reset. The

FIGURE 15-31 Interrupt structure for the 8096. (Reprinted by permission of Intel Corporation, copyright 1983.)

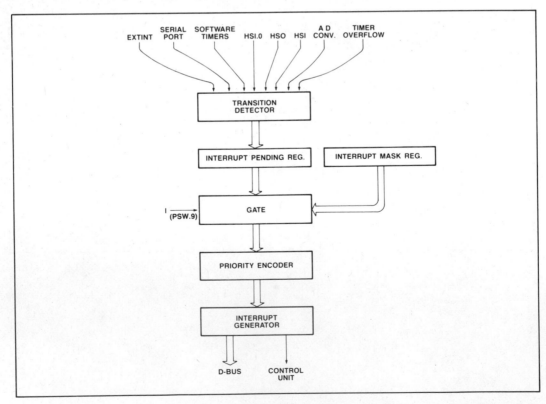

TABLE 15-23   Interrupt Vectors

Interrupt Source	Vector Location	
	High Byte	Low Byte
EXTINT	200FH	200EH
Serial port	200DH	200CH
Software timers	200BH	200AH
HSI.0	2009 H	2008 H
High-speed outputs	2007 H	2006 H
HSI data available	2005 H	2004 H
A/D conversion complete	2003 H	2002 H
Timer overflow	2001 H	2000 H

user's software must periodically reset the watchdog timer to prevent an MPU reset from occurring. The timer may be disabled by holding the reset line high.

In addition to the 10-bit A-to-D converter (which has four to six multiplexed input channels), the 8096 has a pulse width modulator output with a fixed period of 256 state times. The pulse width is programmable from zero to 255 state times. The serial port is a full duplex I/O channel that allows simultaneous transmit and receive. The baud rate is internally programmable.

The 8096's eight sources of interrupt are shown in Table 15-23, and the interrupt structure appears in Fig. 15-31.

14 THE THREE-CHIP 432 SET

The 432 is a set of three 64-pin chips. Each of these chips is a quad in-line package (QUIP), and the three-chip set has more than 200,000 devices. The QUIP is a leadless package with a socket having four staggered rows of pins on 100-mil centers. The three chips are the 43201, the 43202, and the 43203.

43201	Instruction–decode unit	100,000 devices
43202	Microexecution unit	60,000 devices
43203	Interface processor	65,000 devices

Together these three chips constitute the 432 micro mainframe. The first two chips constitute the 432's general data processor (GDP). The 432 has a segmented memory. However, the size of a segment can be up to 2^{16} bytes long. This gives rise to a new term: "structured" memory, as opposed to segmented. The 432 can address up to 2^{24} segments. The total virtual memory space is 2^{40} bytes ($2^{40} = 1,099,511,627,776$).

The 432 allows additional GDPs (43201, 43202) to be added to expand the central processing system. This can increase system performance over a wide range (from 200,000 to 2,000,000 instructions per second). In addition, a pair of GDPs can be used in parallel to perform redundancy checks on each other. Every operation is duplicated and results should agree at all times. Any difference would indicate a malfunction. The 432 will support data types to 80-bit floating-point numbers.

Figure 15-32 shows the three-chip set of the 432 micro mainframe. The 432 is two-phase clocked at 8 MHz. The GDP contains a 4K × 16-bit microprogram ROM, and the interface processor contains a 2K × 16-bit ROM. The operating system will be in microcode, making the execution time for an instruction like ''send a message'' five times the speed of a minicomputer and 20–30 times the rate of a mainframe.

At the least, the 432 is configured with a four-bus system and includes:

Address bus (24 bits)

Data bus (32 bits)

Interface processor control bus (5 bits)

Arbitration bus (8 bits)

The 432 is designed to have software ''imbedded'' in the silicon. This means that high-level user functions are already programmed into the chip itself. The new Department of Defense (DOD) language Ada is a good candidate for initial versions of the 432. Ada is a powerful language modeled after the C programming language and, to some extent, Pascal. In any event, Ada is the general replacement for FORTRAN, which has become limited in speed and performance. FORTRAN was never intended for modern programs that exceed 10,000 lines of code. Ada is intended to remedy these problems by using modular programming and streamlined programming techniques.

Overall, the 432 is being described as a major advance in the industry, in line with the steps taken from vacuum tube to transistor to microprocessor. Intel predicts that given the capabilities of the 432, its full range of applications will remain unexplored until well into the 1990s.

15 THE 2901 FOUR-BIT PROCESSOR SLICE

A different type of microprocessing unit is Advanced Micro Devices' bit slice microprocessor. These microprocessor building blocks come in 1-, 2-, or 4-bit slice units. The AMD2901 is a 4-bit processor slice. Bit slice design allows the designer to create a microcomputer of the appropriate size for the requirement of the job at hand. Since bit slice processors are made from bipolar transistors, they can operate up to five times faster than standard NMOS microprocessors. Cascading four 2901 units will produce a 16-bit microprocessor of great speed due to the true parallel nature of bit slice design.

The user of such a system must create an assembly language by designing a set of microinstructions to define each element of the assembly language. This microprogram is stored in a ROM and becomes the microcontroller for the bit slice processor. After the microprogram has been installed, a more standard assembly language program may be written for the new 16-bit microprocessor designed by cascading four 4-bit slices.

Normally, a bit slice processor would require from 100 to 1024 words of microprogram memory to define the assembly language. In microprogramming, a *new* microprocessor is designed each time. This customizing operation offers many advantages to the user, chiefly an increase of throughput or system speed over conventional microprocessors.

FIGURE 15-32 The 432 three-chip set. (Reprinted by permission of Intel Corporation, copyright 1983.)

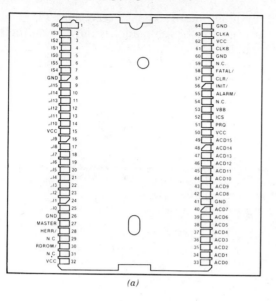

(a)

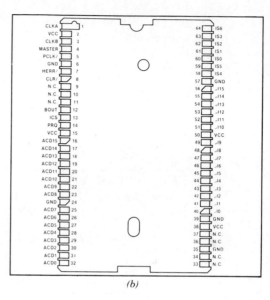

(b)

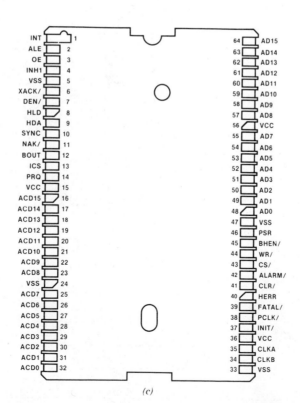

(c)

Cascading of 4-bit slices allows an arithmetic logic unit (ALU) to be created of any desired size in multiples of 4 bits, thus establishing computing power of any desired level.

The bit slice approach is generally much more expensive because the overall parts count is larger and the cost of software development is inflated during the definition stage (microprogram definition).

The AMD 2901C operates at a clock rate exceeding 16 MHz and consists of a 4-bit ALU with carry-in and C_{n+4} outputs. The 16-register RAM area (4 bit) is expanded as additional 2901C slices are cascaded together.

Figure 15-33 shows the block diagram of the 2901 4-bit microprocessor slice.

Figure 15-34 shows three 2901s in a 12-bit MPU. Each 2901 cascaded in adds

FIGURE 15-33 Block diagram for the 2901. (Copyright © 1979 Advanced Micro Devices, Inc. Reprinted with permission of copyright owner. All rights reserved.)

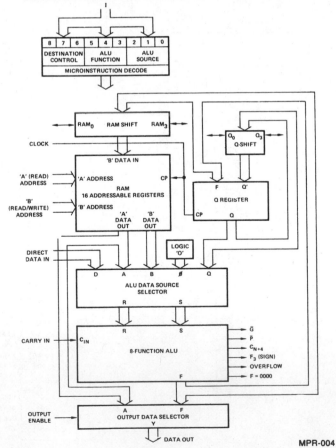

MICROPROCESSOR SLICE BLOCK DIAGRAM

MPR-004

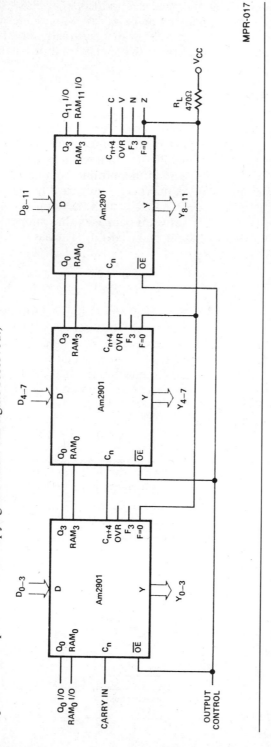

FIGURE 15-34 A 12-bit CPU using three 2901s. (Copyright © 1979 Advanced Micro Devices, Inc. Reprinted with permission of copyright owner. All rights reserved.)

MPR-017

4 bits to the external data bus and the address bus, thus expanding not only the size of data that can be handled and size of addressable memory, but also the inherent speed increase (due to the wider data paths and reduced requirements of returning to memory for operands that can be included in the basic instructions).

16 THE LSI-11 MICROCOMPUTER

Before the development of microprocessors, industry typically used minicomputers like those developed by Digital Equipment Corporation (DEC). DEC minicomputers like the PDP8 and PDP11 attained a popularity that exists heavily today. (The DEC10 is a large 36-bit mainframe computer.)

DEC has developed a 16-bit microcomputer card (printed circuit board) based on the popular PDP11 minicomputer. It uses the same architecture and the same instruction set. DEC offers software packages that support high-level languages like FORTRAN, BASIC, Pascal, and COBOL. The assembly language of the LSI-11 is called PAL-11.

16.1 THE LSI-11

The LSI-11 contains eight 16-bit registers (R0–R7). Any register or memory location can be used as an accumulator. Registers R0–R5 are general-purpose registers and can be used for addresses, data, or address pointers. Figure 15-35, which shows the LSI-11 register set, indicates that register R6 is the stack pointer and register R7 is the program counter. The instruction set is similar to that of the 68000 in that the assembler syntax is similar.

Modes of addressing are as shown in Table 15-24.

FIGURE 15-35 General registers for the LSI-11. (Copyright © 1982, Digital Equipment Corporation. All rights reserved.)

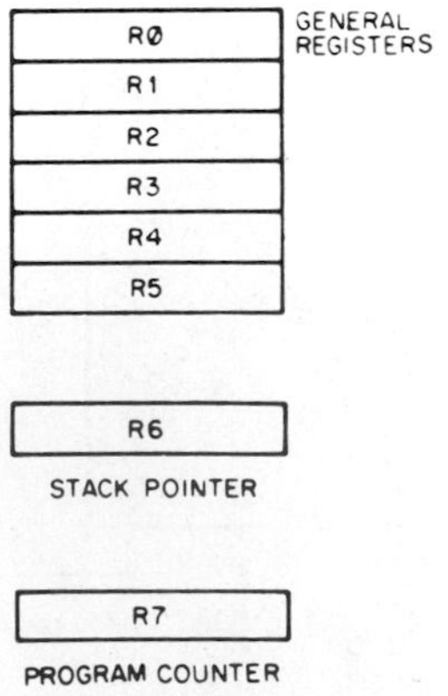

R0	GENERAL REGISTERS
R1	
R2	
R3	
R4	
R5	

R6

STACK POINTER

R7

PROGRAM COUNTER

TABLE 15-24   Modes of Addressing the LSI-11

Mode	Example	Comment
Register	MOV R1, R0	Move word from R1 to R0
	MOV R3, R1	Move word from R3 to R1
	MOVB R3, R1	Move byte from R3 to R1
Deferred	MOV (R2), R0	Move the word from memory location specified by R2 to R0
	MOV (R3), (R4)	Move memory to memory (word)
Autoincrement	MOV (R2)+, R0	Move from memory to R0, then add one to R2
	MOVB (R2)+, (R3)+	Move memory to memory, then increment both R2 and R3

Note: The @ symbol may be used in place of parentheses to indicate the deferred (indirect) mode:

MOV R4, (R3) = MOV R4, @R3

Both forms mean "move R4 to the location specified by R3."

Autoincrement deferred (register contains the address of the operand):

MOV (R3) +, R2   ;Move memory to R2, increment R3

Autodecrement:

MOV R0, −(R2)   ;Decrement R2, move R0 to
                ;memory location specified by R2
MOV −(R1), R3   ;Decrement R1, move memory to R3

Note: For PDP-11, all numbers are assumed to be *octal*.

In the *index mode*, a base address is added to an index word to produce an effective address. The base address is contained in a register; the index word is a part of the instruction.

EXAMPLE 15-8

MOV 50(R3), R1

To produce the effective address, 50 is added to R3. The word at that location is transferred to R1.

Immediate Mode:

MOV #6000, R1   ;Load R1 with 6000

Absolute Mode:

MOV R1, #6000   ;Load the contents of R1 into address 6000
MOV (R3), #2020   ;Load the contents of the address
                  ;in R3 into address 2020

FIGURE 15-36 Memory organization for the LSI-11. (Copyright ©
1982, Digital Equipment Corporation. All rights reserved.)

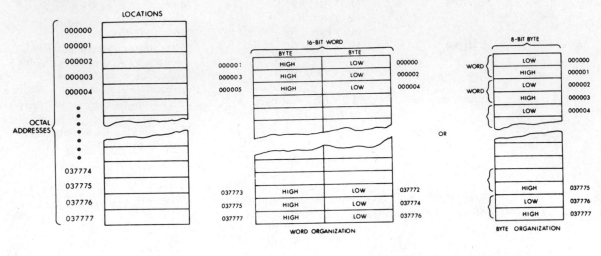

Memory Address   **Word and Byte Addresses**

FIGURE 15-37 The LSI-11 processor status word. (Copyright ©
1982, Digital Equipment Corporation. All rights reserved.)

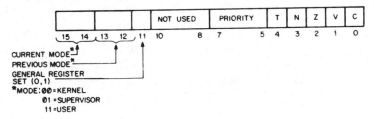

The organization of memory for the LSI-11 is the same as on the other 16-bit
processors we have studied. It is byte oriented. Since the LSI-11 has an instruction
set that is word oriented, the byte locations are always used in pairs. Hence, an
instruction or word can begin only at even addresses (0, 2, 4, etc.). Any increment
on the LSI-11 then will be by 2. Figure 15-36 shows the LSI-11 memory organization, and Fig. 15-37 gives the LSI-11 processor status word.

Here is a sample of a PDP-11 program segment.

Although the information presented here on the LSI-11 is not detailed, previous
studies in this field should allow you to begin to understand the architecture of
the LSI-11. It is not our intent to bring you "up to speed" on this processor, but
rather to expose you to the variety and styles of MPUs that are available.

PRINTOUT 15-1

```
        ; *** FIRE ***
        ;THIS ROUTINE IS USED TO PUT SUCCESSIVE LASER BLASTS INTO THE
        ;LASER DISPLAY FILE. THE PLAYER IS ONLY ALLOWED 5 AT ANY ONE
        ;TIME, SO THE ROUTINE MUST ALSO KEEP TRACK OF THIS. LASER
        ;BLASTS START AT THE CENTER OF THE SHIP AND ARE MOVED TO THE
        ;OUTER EDGE AND TURNED ON. THEN THEY PROCEED IN THE SAME
        ;DIRECTION AS THE SLOPE OF THE SHIP FOR A SET NUMBER OF
        ;CLOCK INTERRUPTS.
FIRE:   CMP     #5,@#LCNT        ;HAS THE PLAYER ALREADY FIRED 5 SHOTS
        BNE     LS
        RTS     PC               ;YES, HE CANNOT HAVE MORE YET SO RETURN
LS:     INC     @#LCNT           ;INCREASE THE LASER COUNT
        MOV     #1,@#TR5         ;SET THE INDEX COUNTER
        MOV     #LASK,R5         ;PUT THE LASER COUNT STORAGE ADDRESS INTO R5
LS3:    CMP     #0,(R5)+         ;IS THIS LASER ACTIVE
        BEQ     LS7              ;NO, LETS GO PUT ONE HERE
        INC     @#TR5            ;YES, ADD 1 TO THE INDEX COUNTER
        BR      LS3              ;AND GO ON TO THE NEXT LASER
LS7:    MOV     @#TR5,R5         ;LETS ACTIVATE A LASER, R5=TR5
        DEC     R5               ;R5=R5-1
        MOV     R5,@#LCSV        ;SAVE THIS NUMBER FOR THE FUTURE
        MOV     R5,@#TR5         ;TR5=R5
        ADD     @#TR5,R5         ;R5 NOW EQUALS 2*R5
        ADD     @#TR5,R5         ;R5 NOW EQUALS 3*R5
        ADD     @#TR5,R5         ;R5 NOW EQUALS 4*R5, BORING HUH?
        ADD     #LASER,R5        ;MAGICALLY WE NOW HAVE THE PROPER LASER
        ADD     #2,R5            ;ADD 2 TO R5 TO GET THE X POSITION
        MOV     R5,-(SP)         ;SAVE THE X POSITION
        MOV     @#SPX,(R5)+      ;START IT AT THE CENTER OF THE SHIP
        MOV     @#SPY,(R5)       ;(REMEMBER, IT'S STILL OFF)
        SUB     #4,R5            ;GET THE X POSITION AGAIN
        SUB     #LASER,R5
        ADD     #LASP,R5
        MOV     #SLOPE,R2        ;R2 POINTS TO SHIP SLOPE TABLE
        ADD     @#ROTOR2,R2      ;INDEX IT TO GET THE RIGHT ONE
        MOV     R2,R1            ;R1=R2
        MOV     R5,R2            ;R2=R5, MOVE LIKES TO USE R1 AND R2
        MOV     #5,@#TR1
LS1:    JSR     PC,MOVE          ;MOVE THE LASER A LITTLE BIT BY THE SLOPE
        DEC     @#TR1            ;DO IT UNTIL THE LASER IS OUTSIDE THE SHIP
        BNE     LS1
        MOV     (R1),R2          ;R2 EQUALS THE SLOPE CONTROL WORD
        BIC     #000777,R2       ;CLEAR UNWANTED BITS TO GET THE TIMEOUT NUMBER
        MOV     #11,@#TR1        ;SET A COUNTER AGAIN
ET:     CLC                      ;CLEAR THE CARRY JUST IN CASE
        ROR     R2               ;ROTATE R2 RIGHT
        DEC     @#TR1            ;DO THIS 11 TIMES
        BNE     ET               ;NOW WE MIGHT HAVE 100 INSTEAD OF 100000
        MOV     (SP)+,R5         ;GET THE X POSITION AGAIN
        BIS     #40000,(R5)+     ;TURN THE INTENSITY BIT ON
        BIS     #40000,(R5)      ;HERE TOO!
        MOV     @#LCSV,R5        ;REMEMBER THIS NUMBER?
        ADD     #LASK,R5         ;USE IT TO INDEX THE LASER COUNTER FILE
        ADD     @#LCSV,R5
        MOV     R2,(R5)          ;PUT THE TIMEOUT NUMBER  IN THERE
        MOV     #SLOPE,R2        ;DO THIS AGAIN,
        ADD     @#ROTOR2,R2      ;THIS TOO,
        MOV     (R2),R2          ;SO THAT R2 WILL EQUAL THE ACTUAL SLOPE WORD
        SUB     #LASK,R5         ;FIND OUT WHERE 11 GOES
        ADD     #LASR,R5         ;IN THE LASER SLOPE CONTROL FILE
        MOV     R2,(R5)          ;AND PUT IT THERE
        RTS     PC               ;CONGRATULATE YOURSELF ON A JOB WELL DONE
```

17 ARRAY PROCESSORS

The array processor is a new type of processor that has been developed to allow microprocessors to deal with math problems of types that to this point have been handled only on large mainframe computers. Many applications require very substantial number-crunching capabilities that would burden most microprocessors. The use of an array processor frees a microprocessor for other tasks because math operations go automatically to the array processor, which performs high-speed, floating-point arithmetic in parallel with a host microprocessor. The array processor is a number-crunching device for handling vast arrays of numbers using standard algorithms; its capabilities include:

- Digital filtering (a complex and time-consuming process)
- Fast Fourier transforms
- Inverse fast Fourier transform
- Power spectral density calculations
- Autocorrelation (speech and image analysis)

Applications of these parallel processors include the following:
- Seismic research
- Analysis of turbines and compressors
- Vibration and power transient analysis
- Speech recognition; pitch extraction and compression
- Nuclear medicine and tomography
- Radar and sonar signal processing
- Electronic countermeasures
- Signal sampling
- Encryption and decryption
- Simulation/automation
- Graphics
- Image enhancement

18 SUMMARY

In this chapter we have looked at three types of microprocessors:

8-bit MPUs	Z80, 6800, 6502
16-bit MPUs	8086, 68000, LSI-11
Single-chip computers	8096, 8048, 8051

Enough information has been provided to give you a clear view of the basic components of each of these popular processors, with an eye to interesting you in them and at the same time allowing you to form a few opinions on their various features. Selecting a particular processor for a specific application is not an easy job. Sometimes the price of the processor is a consideration, whereas at other times the cost of software development is the major concern.

An assembler or cross-assembler (software) alone can vary in price from less than $100 to more than $1000. And of course you need a machine to use the development software on.

In studying this chapter, you should have gotten a clear feeling of the similarity of these processors. You are encouraged to reread both this material and Chapter 9. There is plenty of literature on every processor available from the manufacturers and from commercial publishers. Deep study of this chapter will set you apart from students and engineers with superficial knowledge. The mystery of so many processors has been cracked for you.

19 GLOSSARY

Absolute jump. A jump to a specific address.

Ada. The language the Department of Defense (DOD) intends to use as a replacement for FORTRAN in government computer applications.

Array processor. A high-speed processor intended specifically to perform math operations on very large arrays of data (radar images, spectrum analysis, data arrays).

Assembler syntax. The rules according to which a line of source code must be written for recognition by an assembler.

Bit slice. A bit slice microprocessor allows an MPU of any length to be designed that can run up to five times the speed of standard MPUs.

Branch. A relative jump.

Cache memory. A special high-speed memory close to the MPU, used to enhance MPU speed.

Coprocessor. A pair of MPUs used to enhance an MPU's system operation.

Direct addressing. Often a memory reference to a location specified by a single byte.

Displacement (d). A quantity (usually 8 bits) that is added to the contents of a register, sometimes using 2's-complement arithmetic (depending on the microprocessor) to modify an address; that is, (IX + d) specifies the contents of the address in the X index register plus the displacement quantity.

Extended addressing. Often a memory reference to a location specified by a double byte.

Index register. A register used as a memory pointer.

Instruction length. The number of bytes needed to perform an instruction.

Multiplexed data bus. Address and data lines share the same pins on an MPU package.

Multiprocessor. Three or more MPUs in a system, each performing different functions.

Nonmaskable interrupt. A hardware interrupt that cannot be disabled.

Page. A 256-byte block of memory; see Zero page.

Redundant processors. Two or more MPUs that perform identical tasks. If one processor disagrees with the result of the other, an error is assumed.

Relative jump. A jump to a location *n* bytes before or after the current address.

Segmented memory.   A large memory that can be used in blocks, often 65,536 bytes at a time.

Single-chip microcomputer.   An MPU, clock, memory, and possibly I/O, D/A, and A/D on one chip.

Vectored interrupt.   An interrupt that causes the program counter to go to a specific location and continue program execution from that point.

Virtual memory.   A slower form of memory (disk) that appears to be available as RAM-type space.

Zero page.   The first 256-byte block of memory (0000H to 00FFH).

PROBLEMS

15-1   Using the chapter as a reference, fill in Table 15-25.

15-2   List the single-chip microcomputers discussed in the chapter in Table 15-26 and fill in the remaining blanks.

15-3   Write a 6800 assembly language program to place the ASCII code for the capital letters of the alphabet into memory locations beginning with $2000 (A = 41 H).

TABLE 15-25   Reference Table for Problem 15-1

MPU	Maximum Clock Rate	Data Length	Number of Registers	Maximum Memory	Reset Address
8080A					
8085					
Z80					
6800					
6502					
8086					
68000					
2900					
432					
LSI-11					

15-4 Repeat the program in Problem 15-3 using the 8086 microprocessor.

15-5 This is a research question—you must visit your computer center or consult a technical library. The study of available microprocessors leads inevitably to some consideration of large mainframe computers. Five are listed, and you are to fill in Table 15-27.

15-6 The power of a microprocessor lies in its various addressing modes. These are somewhat common and generally involve the ways in which a memory location can be used. List and define the different types of addressing (direct, inherent, indexed, etc.).

15-7 Price is a contributing factor in selecting a microprocessor for an industrial control purpose. Find out the current price for six microprocessor chips (both quantity 1 and quantity 100).

15-8 List the similarities (as they appear to you) for all microprocessors.

15-9 Why is the 6800 MPU only slightly slower than the 8080A, even though the clocks differ by a factor of 2:1?

15-10 Write to a semiconductor manufacturer, requesting the data sheets on at least one of the microprocessors studied in this chapter.

Number of Hardware Interrupts	Data Multiplexed

TABLE 15-27   Reference Table for Problem 15-5

CPU	Number of Registers	Register Sizes	Maximum Memory	Clock Rate
DEC 10				
IBM 370				
GE Sigma 7				
Honeywell 600				
VAX-11				

TABLE 15-26   Reference Table for Problem 15-2

MCU	Clock Rate	Data Length	ROM Size	RAM Size		Number of I/O Ports		D/A or A/D?
				Internal	External	Serial	Parallel	

CHAPTER 16
TROUBLESHOOTING TECHNIQUES

1 INSTRUCTIONAL OBJECTIVES

This chapter introduces the broad and elusive subject of troubleshooting. In particular, we are concerned with the problems of microprocessor-based circuitry both in the design process and in the maintenance situation. It is the objective of this chapter to start you on a logical path to troubleshooting any type of equipment including microcomputers. To be successful, however, you must add experience to this knowledge.

2 SELF-EVALUATION QUESTIONS

Keep the following questions in mind and try to answer them when you have completed the chapter:

1. What is the function of a logic analyzer?
2. In what different formats can the data acquired by a logic analyzer be displayed?
3. What is a development system?
4. What is an emulator?
5. What is a communications tester?
6. What problems arise when trying to communicate with a microcomputer using serial (RS232C) techniques?
7. In emulation mode, what is a "trace"?
8. How can a logic analyzer be triggered?
9. When servicing a microcomputer, what should be checked initially?
10. Is there a common fault of microcomputers?

3 INTRODUCTION

The high speed of microcomputers combined with the parallel transmission of multiline data makes the servicing of these devices a specialized process. A two- or four-channel real-time oscilloscope becomes an inadequate tool for observing conditions on an 8-bit data bus; they are just not suited for this technology. New equipment and techniques have been developed to answer the new needs created by microcomputers. Logic analyzers and development systems are at the top of this list, and their functions are described in this chapter, along with troubleshooting techniques for microcomputers.

4 DEVELOPMENT SYSTEMS

4.1 What Are Development Systems?

The design of a microcomputer requires both the implementation of hardware and the creation of software. It is necessary to put the required chips and components together to form a working microcomputer. To the working hardware must be added written software that will make the hardware recognize and execute commands from an external source. It is usually more than a luxury to have a computer system and a wide variety of test equipment available to aid in the design of a microcomputer. In other words, the help of a computer is needed to design a computer!

The development system is a full computer that can be used in both the hardware debug and software debug phases of designing a microcomputer. Development systems are available from many manufacturers and are categorized as either

FIGURE 16-1 The 8550 development system. (Reproduced by permission of Tektronix, Inc.)

universal or specific. For example, Intel markets a development system for Intel microprocessor development, and Motorola and Zilog offer similar machines to aid in the design of microcomputers using their respective devices. Such systems are specific (nonuniversal). Tektronix and Hewlett-Packard, on the other hand, produce development systems that can do design work for many different microprocessors and are not limited to a specific manufacturer. These are called universal systems. For example, the Tektronix 8550 development system can be used to design software for more than 40 popular microprocessors, provided you have the appropriate software. With a universal development system, the user is not locked into a specific manufacturer and thereby a specific series of microprocessor-related products. Figure 16-1 shows a Tektronix 8550 development system.

4.2 What Do Development Systems Do?

A development system can assist in both the hardware and the software design phases. In the software development stage, a development system should allow the creation, editing, and saving of assembly language programs for the target microprocessors. This capability can be satisfied by the presence of an editor and some type of disk storage unit for file retrieval at a later time. Most development systems store files (or programs) on floppy diskette, hard disk drive, or cassette tape.

In addition, a development system should be able to assemble an assembly language program into an object file (opcode form) using a particular microprocessor instruction set. Table 16-1 lists the desirable features of a development system.

In addition to these features, a development system should be able to run the emulated CPU at a speed determined by its own clock, not at the development system rate. This assures that product testing and debug occur under actual conditions (Fig. 16-3).

4.3 The Process of Emulation

There is a real difference between *simulation* and *emulation*. A simulator is a machine or a program that pretends to be the real thing. That is, it simulates a particular microprocessor but is not that machine itself. In emulation, a user microprocessor is present and a program is run on that MPU. To debug software on an actual MPU, an emulator is required. The real value of a development system occurs when a microprocessor-based system is under development (i.e., being designed and built).

Typically, the MPU chip is unplugged from the computer under development. An emulator probe from the development system plugs into the vacant socket and replaces the MPU that was removed. This gives the development system access to the vital signals of the MPU. Address and data information, control information, and clock and power supply are all monitored by the development system. With this type of access, a program developed and assembled on the development system can be tested and debugged on the computer under development. I/O ports can be tested, memory can be tested, and communication channels exercised. Since both the development system and the user computer have memory, the program can be released to the computer under test in stages.

TABLE 16-1   Desirable Features of a Development System

Feature	Description
Text editor	Creates, alters, and deletes files.
Assembler	Assembles source files into object files.
Disassembler	Disassembles the contents of a range of memory locations.
Memory dump	Dumps the contents of memory in both hexadecimal and ASCII.
File dump	Dumps the contents of a file in hexadecimal and in ASCII.
Load object file	Loads an assembled program into memory.
Save file	Saves an edited file or memory contents under a file name on disk.
List directory	Shows the names of available files stored on disk.
Emulate	Actually runs the program on the target microprocessor.
Memory map	Allocates memory as existing either on the development system or on a separate computer under development system control or a combination of both. Such memory may be defined as Read Only or Read/Write (see Fig. 16-2).
Program trace	Shows the contents of the registers on the target microprocessor for each program instruction.
Read PROM, Write PROM	Reads or burns various types of PROMs and EPROMs.
Printer spooler	Can dump/write to a slow printer without interfering with the system or slowing down the operation.
Logger capability	Records or logs every operation on the development system for future reference.
Command files	Creates files of frequently used command sequences so that they can be invoked with a single expression.
User areas	Offers separate areas or subfiles to different users, protected with an owner password.
Logic analyzer	Records high-speed data operations from remote probes connected to user's computer.
High-level language	Writes software using a language like BASIC or Pascal and has it converted to the target microprocessor code.
Real-time clock/calendar	Tags files with creation/change information for identification purposes.
Select target processor	Selects the target MPU.
Learning system	Built-in guide to learning the system's operation.

This is done by altering the memory map to show which sections of memory reside in which computer (development system or user system). In this way, even memory can be tested. In *tracing* a program, sections of program can be followed in operation on the development system. As the user computer performs its function, the contents of internal registers are printed by the development system for study and debug purposes. This provides an effective means of tracing program flow and finding sources of logical error.

FIGURE 16-2 A memory map. (Reproduced by permission of Tektronix, Inc.)

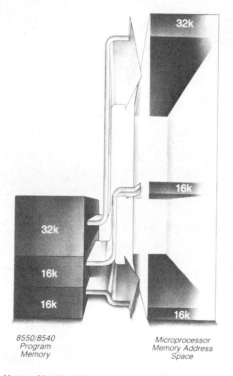

8550/8540
Program
Memory

Microprocessor
Memory Address
Space

Memory Allocation allows program memory to be assigned to different logical addresses within the microprocessor address range.

In practice, a program is developed in mnemonic form using the *editor*. The program is then *assembled*. This produces two additional files: the object file (opcodes) and the list file. The listing is a combination of the source program, object file, and addresses and includes a symbol table. The object file is *loaded* into the appropriate memory (development system or user computer). The hex code can then be *dumped* or *disassembled*. If all appears to be correct, the *emulator* will allow program execution at full speed, and during this period the details may be monitored using the *trace* function.

4.4 The Tektronix 8550

The Tektronix 8550 development system is a single-user system that has a pair of 8-in. floppy disk drives for program storage. It is capable of all the functions previously described, and attachments (including software) are available for 40 separate microprocessors. A typical command set is illustrated in Table 16-2.

FIGURE 16-3 Block diagram for the 8550 development system.
(Reproduced by permission of Tektronix, Inc.)

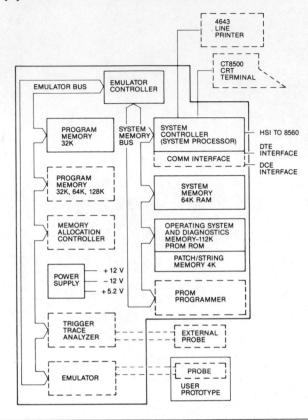

TABLE 16-2   Command Set for the 8550

Function	Comment
> DATE 1-JAN-86/12:00:00	Set date and time.
> LDIR	List disk directory.
> SEL 8080	Select 8080 software.
> LOAD TEST.OBJ	Load hex file.
> DUMP 0 0400H	Dump memory 0–400.
> FDUMP KOUNT.OBJ	Dump file contents.
> TRACE ALL	Turn on trace option.
> MAP URW 0 0FFFFH	All memory set to user read/write.
> LOG DAY.ONE	Log all subsequent operations into a file called DAY.ONE.
> G 0E800H	Jump to address E800.
> RPROM 0 2732/I 0 0FFFH	Read EPROM data to 0 from 0 to FFFH.
> WPROM 0 2716/TI 0 07FFH	Burn EPROM from 0.
> EDIT FILE.ASM	Enter editor to change existing file.

4.5 Development Systems: Conclusion

At this writing, there are more than 20 manufacturers of development systems, some universal and some specific, for various microprocessors. The general cost of a system that has the features described here is in excess of $20,000, and a pair of additional emulators can bring the total close to $30,000. The development software (assembler) for each processor is on the order of $1000–$4000, depending on manufacturer and processor popularity. Currently supported processors of Tektronix are listed in Table 16-3.

The availability of some type of development system is perhaps both a luxury and a necessity to the designer of microprocessor-based systems. If a single-board, microprocessor-based controller is under design, the software can often be tested in advance of any real hardware. Since software will eventually reside in some type of ROM/PROM memory, the development system is ideal for the debug of that software. In the hardware debug stage, sections of software can be released to the target MPU in segments, thus testing the new product in a controlled and monitored way. Although a development system is almost an essential tool in the industry, cross-assemblers for various microprocessors are also available. A cross-assembler is a program that runs on a computer other than the one being used for prototyping. Many companies have large mainframe computers that can support a cross-assembler for microprocessor-based design; these are very expensive machines, however, and the development system offers a practical means of accomplishing completion of a microprocessor-based design.

5 LOGIC ANALYZERS

5.1 Description

The analysis of non-microprocessor-based digital logic is easily carried out with traditional forms of test equipment like the logic probe or oscilloscope. Typically, two or four channels of display on an oscilloscope are enough for equipment debug or servicing if a microprocessor is not involved. When a microprocessor

TABLE 16-3   Tektronix-Supported Processors

8088/87	68000	3872
8086/87	68020	3874
8085A	6800	3876
8080A	6801	Z80A
8048	6802	Z80B
8049	6803	Z8001
8035	6808	Z8002
8039	6809	TMS9900
8021	6809E	SBP9900
8022	F8	SBP9989
8041A	3870	1802

Source: Courtesy of Tektronix.

FIGURE 16-4 A 16-channel logic analyzer. (Reproduced by permission of Tektronix, Inc.)

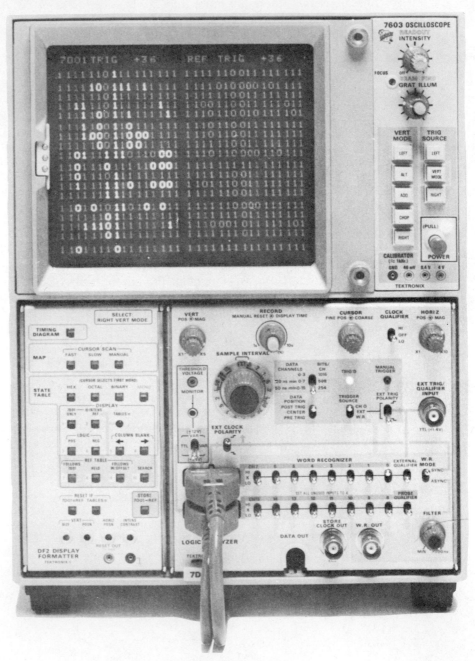

is involved, however, traditional test equipment cannot be used effectively because there are far too many signals to look at that are all time related to one another. Since a microcomputer is programmed to follow a specific sequence of activities, and since it operates at its maximum speed, the many digital signals present cannot be captured with a four-channel instrument. An 8-bit microprocessor would have 8 data lines, 16 address lines, and at least 5 lines on the control bus. Other signals including INT, INTA, Sync, and the basic clocks are also present. To observe many of these lines simultaneously requires a high-speed multichannel instrument with a large internal memory for storing data for later display and review. Such an instrument is the logic analyzer. Figure 16-4 shows a 16-channel logic analyzer.

5.2 What Do Logic Analyzers Do?

A logic analyzer generates a display on an oscilloscope-type screen that shows the state of its 16 or more inputs over a period of time. Such a display may be a timing diagram (Fig. 16-5), hex (Fig. 16-6), binary (Fig. 16-7) or octal (Fig. 16-8) format, mnemonic disassembly (Fig. 16-9), or memory map of data taken and stored in the logic analyzer's memory.

The rate at which the logic analyzer takes data is usually determined by a clock, either internal or external. The amount of data taken depends on the available memory. If a logic analyzer has a 4K-bit memory (4096 bits) and uses 16 data channels, each channel could take up to 256 data points ($16 \times 256 = 4096$). These data points would be taken 16 at a time (one per channel). This allows 256 (16-bit) samples. The state of each line at the time the sample is taken is stored in the logic analyzer's memory. Each sample (16 bits wide) may be displayed in timing diagram form, as a number system (binary, octal, or hex), or as a memory map. If it is desired to sample 8 channels instead of 16, there are still 4K bits of memory available. Twice as many samples can be ($4096 \div 8 = 512$) taken. If

FIGURE 16-5 A timing diagram display generated by a logic analyzer.

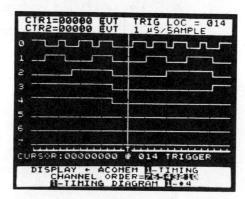

FIGURE 16-6 A hexadecimal display generated by a logic analyzer.

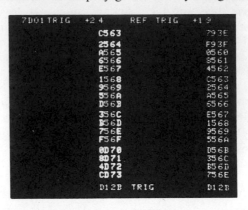

FIGURE 16-7 A binary display generated by a logic analyzer.

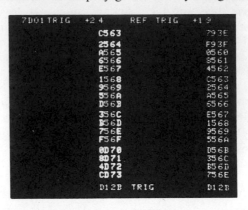

FIGURE 16-8 An octal display generated by a logic analyzer.

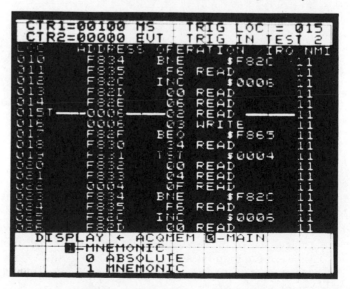

7D01 TRIG	+1 7	REF TRIG	+1 7
	15 03 06		15 03 06
	00 07 06		00 07 06
	15 57 00		15 57 00
	15 57 00		15 57 00
	16 67 01		16 67 01
	16 65 55		16 65 55
	16 65 55		16 65 55
	16 33 02		16 33 02
	16 31 56		16 31 56
	00 13 03		00 13 03
	14 53 04		14 53 04
	14 53 04		14 53 04
	15 03 05		15 03 05
	15 03 04		15 03 04
	00 07 06		00 07 06
	15 57 00		15 57 00
	15 57 00		15 57 00
	00 13 03	TRIG	00 13 03

FIGURE 16-9 A mnemonic display generated by a logic analyzer.

CTR1=00100 MS	TRIG LOC = 015
CTR2=00000 EVT	TRIG IN TEST 2

LOC	ADDRESS	OPERATION	IRQ	NMI
010	F834	BNE	$F82C	11
011	F835	F6 READ		11
012	F82C	INC	$0006	11
013	F82D	00 READ		11
014	F82E	06 READ		11
015	0006	02 READ		11
016	0006	03 WRITE		11
017	F82F	BEQ	$F865	11
018	F830	34 READ		11
019	F831	TST	$0004	11
020	F832	00 READ		11
021	F833	04 READ		11
022	0004	0F READ		11
023	F834	BNE	$F82C	11
024	F835	F6 READ		11
025	F82C	INC	$0005	11
026	F82D	00 READ		11

DISPLAY ← ACQMEM ■-MAIN
■-MNEMONIC
0 ABSOLUTE
1 MNEMONIC

FIGURE 16-10 A logic analyzer connected to the eight data lines of
a microprocessor under test. (Reproduced by permission of
Tektronix, Inc.)

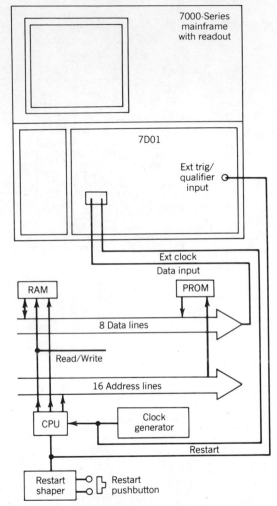

only four channels are needed, another doubling of samples can occur (4096 ÷
4 = 1024). The sample interval (i.e., the time between samples) is again determined
by a clock (internal or external). If an external clock is used, and if the rate is
equal to the rate at which data changes on an 8-bit data bus, the analyzer will
"catch" a sequence of a microcomputer's program. This use of a logic analyzer
can determine whether a microprocessor under development is functioning correctly. It can be used to isolate sections of bad memory, address, or data lines
that are hung up (stuck high or low) or to show control line malfunctions. Figure
16-10 shows a logic analyzer connected to the eight data lines of a microprocessor
under test.

5.3 Starting a Data Sample

Since the logic analyzer has a memory of fixed size, it can store only a fixed number of data points. If data is taken at 1-μsec intervals, the 16-channel, 256-sample memory is filled in 256 μsec. It is important to begin taking this data at a time chosen by the operator. For example, if the data is to be taken after hardware reset of the microprocessor, the entire startup sequence can be observed and debugged. Alternately, it may be desirable to trigger on a particular code or instruction, or perhaps on the 400th occurrence of an instruction. There are many different ways in which to trigger the logic analyzer to begin taking data. In any event, word recognition is a common means of starting the sample process. When a specific data pattern occurs, the analyzer begins taking data. Figure 16-11 shows the trigger circuitry of a typical logic analyzer.

In addition to word recognition, a separate "qualifier" input may be available to add an extra condition that must be met before the analyzer will begin its sample. This is an input that must be high or low (operator selectable) to meet the conditions of triggering. Figure 16-12 shows the block diagram of a logic analyzer.

If a logic analyzer is equipped with a personality module, it can take the acquired data and display it in the mnemonic form of the microprocessor in question. A separate personality module is required for each microprocessor, since different instruction sets are involved. This feature is particularly valuable in trouble-

FIGURE 16-11 Block diagram of the Tektronix word recognizer circuitry. (Reproduced by permission of Tektronix, Inc.)

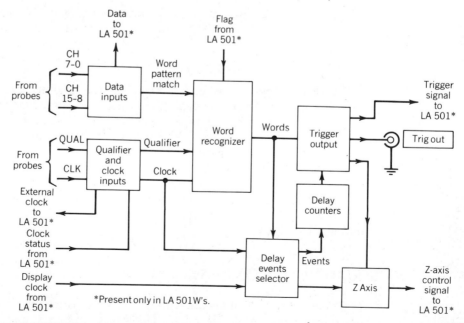

FIGURE 16-12 Block diagram of a logic analyzer. (Reproduced by permission of Tektronix, Inc.)

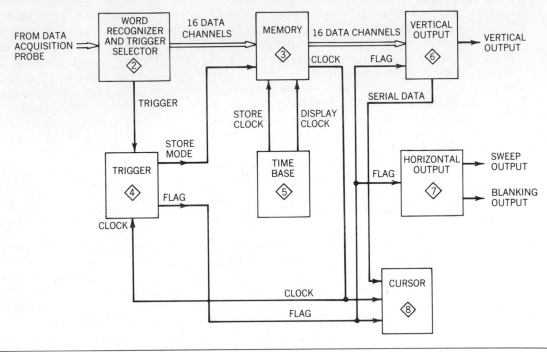

shooting both hardware and software problems. Figure 16-13 shows the display formats of the Tektronix 7D02 logic analyzer and the 7D02 itself.

5.4 Troubleshooting Using a Microprocessor

In Fig. 16-14 a logic analyzer is connected to a malfunctioning microprocessor by way of eight probes to the data bus of the errant microprocessor and the trigger of the analyzer to the reset line of the processor. This causes the analyzer to take data during the initialization sequence. In this way, the very first instruction followed by the MPU can be monitored to determine whether any data lines are stuck or inoperable. Further monitoring of the control or address lines can also help determine problems in high-speed microprocessor systems.

Since there are so many high-speed signals present in the simplest (slowest) microprocessors, it is not practical to do complete troubleshooting using a conventional oscilloscope.

6 TESTING SERIAL DATA COMMUNICATIONS

Many computers communicate with the outside world using serial data lines connected to a standard data terminal. The levels on these data lines are established

FIGURE 16-13 The 7D02 logic analyzer with personality module.
(Reproduced by permission of Tektronix, Inc.)

(continued on next page)

FIGURE 16-13 (continued)

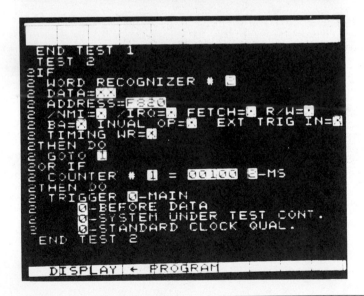

by the EIA RS232C interface standard (± 3 to ± 25 signals) and use the 11-bit transmission code. One of the main difficulties in using a new terminal or computer is found when attempting to establish contact between the two. Wiring errors, mismatched baud rates, and improper line protocol (CTS, RTS, DTR, DSR) all contribute to this problem. Often a simple voltmeter or an oscilloscope can be helpful in checking baud rates and waveforms. Since this problem arises frequently, an instrument dedicated to this application is desirable. The Tektronix 833 data communications tester (Fig. 16-15) is such an instrument.

FIGURE 16-14 Connection of a logic analyzer to a malfunctioning microprocessor. (Reproduced by permission of Tektronix, Inc.)

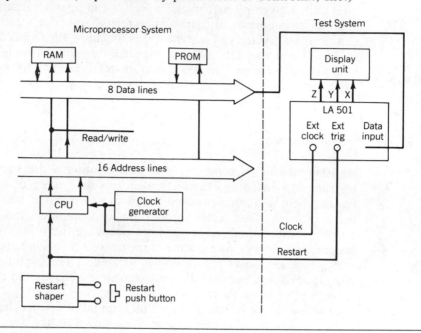

FIGURE 16-15 The 833 data communications tester. (Reproduced by permission of Tektronix., Inc.)

The 833 is initially a breakout box for all 25 pins of the standard DB25 connector. Immediate access to all the pins is provided, and the option of altering their direction is given by patching. The tester can operate at any standard baud rate and can detect errors of transmission for the operator's convenience. The tester can transmit a variety of test messages including the Quick Brown Fox Alphabet in ASCII and EBCDIC. This is especially useful in testing data terminals over long periods of time. The 833 can store acquired data for later analysis and is a useful tester for debugging RS232C lines.

7 SIGNATURE ANALYSIS

Another tool of microprocessor troubleshooting is the signature analyzer. This instrument is used to examine waveforms in a piece of errant equipment and reduce them to a number called a signature, which can be referenced on a schematic drawing. Test points within the microcomputer circuitry can be identified by the manufacturer and a signature noted on the print. When troubleshooting becomes necessary, the signature at progressive points in the circuit can be checked and a bad section of the computer located. The signature of a waveform is a hexadecimal number (usually four characters) that is the result of reducing the waveform to an easily readable form.

Looking for the problem in a circuit is a matter of checking the digital signature at key nodes in the circuit against those noted on the drawing. The signature is generated and tested when a special program is run on the computer having the problem. In this way a unique signature can be generated, which in turn becomes a test of the circuitry.

A manufacturer can take certain steps during the design phase to allow for the use of a signature analyzer later, when troubleshooting is required for the equipment. Many original equipment manufacturers (OEMs) are now taking such steps, to increase the value of their equipment to prospective buyers. Signature analysis represents yet another means of troubleshooting the microprocessor-based system.

8 TROUBLESHOOTING MICROPROCESSOR-BASED SYSTEMS

The repair or servicing of microprocessor-based equipment requires an understanding of these systems and also a knowledge of basic troubleshooting techniques. Any approach to a problem that does not incorporate a logical technique and a similar understanding is probably doomed to failure. That is not to say that a bit of luck will not shorten the time required to locate and repair a problem. It is important to keep in mind, however, that there is a reason for every system failure. It is best to locate the cause of each failure before replacing a damaged part and putting the unit back into service. To put in a new part before determining why the first one failed is to guarantee that the whole unit will fail again.

8.1 Debugging Versus Servicing

There are two types of work that are related, but require different approaches:

1. Failure in units that are new, just built, and never worked before.
2. Failure in units that once worked but failed after long use.

The first type is the worst, and its remedies fall into the "debug" category. That is, we have a product that we desire to make functional. The second type is easier to repair because the unit was working and had already proved both the original design and the construction. Each category has its own types of problem.

1. Never worked
   - Miswired (calls for laborious tracing or wiring).
   - Design flaw—can never work.
   - Chips in the wrong places.
   - Other components inserted wrong.
   - Electrolytic capacitors inserted backward.
   - Diodes inserted backward.
   - Transistors inserted incorrectly.
   - Chips installed backward.
   - Power supply problems.
2. Worked but failed:
   - Component failure due to age or overstress.
   - Damaged by accident (e.g., dropped a screwdriver into it, misconnected the power supply, plugged a chip in backward possibly with power on, pulled a card from an edge connector with power on).

8.2 A Few Warnings

First of all, don't attempt any extensive repair without a schematic of the unit. It is also a good idea to have the manufacturer's service manual and suggestions.

Second, find out whether you are the first person to look at the unit. If a technically illiterate do-gooder has already touched the project, your job will be ten times as difficult because, for example, you will have to determine what additional damage has been done by nonlogical techniques. (You probably won't be able to talk to the previous "repairman.")

Third, ascertain the symptoms of the problem. How did the unit fail? Talk to any witnesses. Try to get a good description of the unit's problem before touching the job. Symptoms frequently lead the trained troubleshooter to a correct diagnosis. Did the unit fail at power-up? Did the unit fail after extensive use? Was the area in which it was operating excessively hot or cold? Was something spilled into the unit? Was something dropped onto the circuit board? Was a thunderstorm in progress when the unit failed (indicating a powerline surge or damaged I/O lines)?

Before continuing with troubleshooting techniques, we present an important idea. The microprocessor represents a significant advance from traditional circuitry. It is high speed, data oriented, and heavily programmed. Efficient repair of this type of equipment usually requires specialized instruments like logic analyzers and development systems.

8.3 Troubleshooting Begins with Observation

The repair of micro-based equipment can be a highly organized enterprise. If you develop a system of locating problems, the unit can frequently be repaired quickly. Be forewarned that hit-and-miss techniques are the result of desperation and take a long time to succeed. As you proceed, write down the areas you have worked on, the chips you have changed, and so on.

8.4 Obvious Physical Damage

Look the unit over carefully. Check the fuses. If one is blown, you may decide to change it and see if that solves the problem. Occasionally, application of a-c power occurs at the peak of the a-c wave. This creates a large start-up current, and it is not unusual for a fuse to blow. In that case, the problem was not worth mentioning. However, if the fuse blows again, there is a heavy load on the supply and there is work ahead to find the problem.

Other physical damage includes burned or damaged components or printed circuit lands. Resistors may be discolored, indicating excessive heating, or capacitors may have material oozing out. An integrated circuit may be incinerated. Physical damage is a welcome sight to the repairman. The problem area has been located! When there is no physical damage, you must become a technical Sherlock Holmes. Study the symptoms; observe; deduce. Don't overlook something simple like a broken wire. Wires leading from edge connectors to switches or panels may have been broken as cabinet doors were opened, and so on. These are welcome problems and easy to repair.

Look for solder bridges, be suspicious of rewiring and component changes, and check the location of all parts. An IC inserted backwards or into the wrong socket can easily complicate the problem.

A simple but often overlooked problem is the corrosion of edge connectors. These components can be effectively cleaned (polished) with a pencil eraser. Never, NEVER apply solder to gold contacts in an attempt to improve the conductivity of edge connectors.

8.5 Power Supply Problems

The power supply is a likely suspect in any problem. Micro-based systems require low voltage ($+5$, $+12$, -12 V, etc.) but at substantial current levels. Systems that have a video display also have a high-voltage section to supply the anode of the CRT. The two basic types of power supply in use today are the standard supply and the switching supply. The switching power supply weighs less because it has a smaller power transformer due to the use of a higher frequency. In the switcher, the a-c line is rectified to produce 170 V d-c. This drives a multivibrator, which drives a small step-down transformer. The low-voltage alternating current produced is rectified and regulated. Since the multivibrator operates at a frequency

(> 20 kHz) greater than the a-c line, a smaller transformer is required. These supplies are also more compact.

Power supplies contain filter capacitors (electrolytics) that through age may deteriorate, and the usual symptom is internal electrical leakage. The ideal high resistance between the plates decreases, causing heating and then chemical leakage. If a component of the main circuit fails, a heavy load may be placed on the supply. A look at areas of the main circuit that require heavy current may reveal the fault. Other high-current areas include vertical and horizontal sweep circuitry in video displays. Audio sections require more current than the average IC, so it is wise to check these also.

8.6 Clock Circuit Problems

All microprocessor-based systems require a precision crystal-controlled clock. Obviously, if the clock fails, the micro cannot run. An oscilloscope should be used to check the clock. If it is a two-phase nonoverlapping clock, the two phases must not overlap.

Other common problems include crystals that quit, clock chips that quit, and related components that also quit.

8.7 Intermittent Problems

Locating intermittent problems can be quite annoying. The unit works for a while and then develops some peculiarity, usually a heating problem related to component aging. Frequently such problems can be located with the aid of freeze mist sprayed slowly and judiciously on the circuit board. If a cold IC resolves the problem, you have probably found the culprit. However, if a timing problem exists, changing the temperature of an IC may only correct the difficulty synthetically. Thus timing problems are much more difficult to locate.

Occasionally a cracked land (almost microscopic) or a poorly plated-through hole shows up. These two problems are very difficult to locate, and fortunately they do not occur too often. Careful troubleshooting (described later) and the use of an ohmmeter can help.

8.8 Components Change Value with Time

As a piece of equipment ages, its reliability generally decreases. This is due more than anything else to aging of components caused by heating. When each component is originally selected with a sufficient margin from maximum values, the equipment can run nearly forever. However, a product is likely to fail either at the beginning of its career or well down the road of a useful life. A bathtub curve (Fig. 16-16) demonstrates the typical reliability of equipment containing electronic components. This curve is subject to a certain amount of variation and depends on the selection of components during the design and construction phase, but it gives an idea of when to expect failures due to aging.

Components undergo certain changes as they age or "burn in," tending to settle into specific values. Resistors, particularly the composition variety, can open up

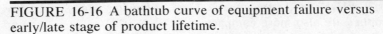

FIGURE 16-16 A bathtub curve of equipment failure versus
early/late stage of product lifetime.

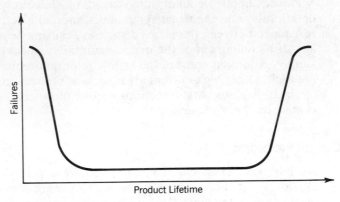

with use and not show any sign of physical change. Capacitors can open up,
although electrolytic (power supply) capacitors tend to short or to draw excessive
current.

Resistors that fail are usually in high-current areas: Whenever a component is
operated at a level close to its rating, the probability of failure increases greatly.
For instance, a 0.5-W resistor operated at 0.5 W has a significantly shorter life
than a 0.5-W resistor used at 0.25 W. Capacitors, however, are best used near
their rated value. The selection of components in the design phase is important
to the life of the product. Often, the price of a component figures into the decision.
Such decisions become economic tradeoffs by the original designer. When you
replace a component, be sure to use at least the original rating.

Integrated circuits fail for a number of reasons. If an IC is forced to drive
currents close to its maximum rating, it will probably fail at some time. ICs
manufactured under nonclean conditions contain contaminants that can eat the
interconnect wires inside the package. There is nothing you can do about improper
production methods, but that is why there are data codes on each IC manufactured.
Quality assurance personnel in many companies open a sample of ICs purchased
to determine the conditions under which they were manufactured. You would be
quite surprised at the things found inside. Such contaminants are the cause of
failures after a year or more. Not the kind of problem that could be tolerated on
a Voyager spacecraft passing Pluto!

8.9 Static Problems

The high voltages associated with static electricity can have a very adverse effect
on electronic circuitry. Be sure to check the environment in which a unit that
failed is used. Is there a carpet on the floor? Are people allowed to generate static
near the machines by taking off their heavy winter nylon coats in the area?

Although all ICs can suffer from static problems, CMOS devices (most microprocessor chips and support chips) suffer the worst damage. Static that enters

from a person's fingers through a keyboard can destroy the CMOS keyboard encoder chip commonly used. Static can find its way into the main circuit board and, for this reason, any CMOS chips should be carefully handled and individuals using the equipment should take proper precautions to ensure that they are "static-free." They are also suspect in any equipment failure.

Note that if a printed circuit board is to be moved, it should be carried in a static bag, placed on conductive foam, or wrapped in aluminum foil. If you receive a board for repair in a paper bag, we suggest you send it back untouched. This is the work of a technical illiterate. Well, perhaps you could look at the problem, but be sure to educate the "carrier."

8.10 A Logical Approach to Troubleshooting

It is absolutely essential to have available a set of correct schematics for the unit to be repaired. Certain surface troubleshooting is possible without them, but for problems greater than a blown fuse, the drawings are needed. A suggested sequence of events for troubleshooting a nonworking printed circuit board is summarized in Section 8.17. If you choose a hit-and-miss method, you are probably desperate. Learn to use a logical approach to the problem.

8.11 Desirable Equipment

We list a few items in order of need. The first two are a minimum and can handle 75% of the problems encountered in troubleshooting microprocessor-based equipment.

1. A d-c multimeter, consisting of a voltmeter to monitor supply voltages, an ohmmeter to check resistances, and an ammeter to monitor supply current.

2. A laboratory oscilloscope (10 MHz or better) to observe digital waveforms and measure pulse widths and periods.

3. A logic probe to catch high-speed glitches (spikes) that may occur, a logic clip to monitor individual ICs, and a logic pen to insert pulses.

4. A logic analyzer to observe the high-speed waveforms on both data and address busses.

5. A development system with an emulator to monitor program execution and determine program errors possibly leading to bad memory or timing problems.

8.12 Getting Started

The first steps for troubleshooting a microprocessor-based system have already been mentioned: look for obvious damage, and check the power supply. Do the chips actually have power when the supply is loaded down?

Next, check the clock for the microprocessor to be sure it is functioning. Look at areas of high current. In a video circuit, the sweep circuitry requires high currents; in audio sections, the power output circuitry will use larger currents. Is the power supply loaded down? Is it supplying so much current that its output voltage has dropped? If so, you have a shorted IC, a bad capacitor, or damage to another component. Is the MPU running? Check the MPU power and clock signals, and then use the oscilloscope or logic probe to see whether there is bus activity. Are there missing signals? Do RAM/ROM areas have enable signals? Are the chip enables getting used on occasion?

8.13 Finding the Problem with Power Applied

If there is no physical damage, you will have to apply power and observe the operation of the circuitry. First, apply the ohmmeter to the power supply lands on the board to determine whether any shorts exist. Use the R × 1 scale and observe correct polarity: 5–10 Ω for a dozen or more ICs is not unusual. Look for a direct short circuit. When you apply power, don't waste a lot of time after power is on. Get to it. Measure the voltages immediately, and be observant. Look for heating/smoking effects. Use your fingers to find hot spots. If you have an ammeter and can monitor supply current, do so.

If everything seems normal in a unit that nevertheless will not function correctly, note the symptoms. What is it doing that is not right? Think it through. What areas must be causing a problem? In a microprocessor-based system, static electricity damage may mean a memory problem. Stored-memory (ROM, PROM, EPROM) damage means that the micro does not have a valid program to run and may be running garbage. Check activity at each ROM location. It may be necessary to dump the contents of the ROMs to determine whether they are damaged. You need a development system to check the contents of the ROMs. Is a bit stuck high or low? If so, you've found it!

8.14 Doing the Repair

Use caution when replacing burned components. First find out why the component burned up. What else went that may have caused the failure? Never replace components with power applied to the board.

If you must remove a component that is soldered in, use a good solder sucker to remove the part. If you use too much heat in detaching the part, the lands will become damaged and you will have created both a mess and a nightmare. Use the right tool for the job.

When selecting replacement components, the best practice is to use a perfect match. Frequently, the original part can be found. Swapping TTL parts or LS parts can be risky. They are not all the same and may have different characteristics (e.g., 7400 vs. 74LS00).

Desperate? Have a working board from which to swap parts. Change only suspect parts, and write down any swaps so you can go back to at least one working unit.

8.15 Power Supply Problems

When a power supply problem is evident, use an ohmmeter to check the rectifiers. A conventional power supply may experience a rectifier failure due to high current, whereas the switcher type may incur such a failure because of high voltage. In either case, rectifiers will fail under heavy load conditions. Are the filter capacitors shorted? Was the supply itself shorted? Check filter capacitors for bulging (or oozing) and overheating. A capacitor should never run warm. If aged, it can develop leakage and become warm due to power losses.

8.16 Repairing the Power Supply

When replacing parts in the power supply, use components rated higher than original parts if possible. Apply power to the supply without the microprocessor-based circuitry connected. In other words, run the supply with no load first. If all seems to be in order, connect a resistive load and draw-rated current for awhile before reconnecting the main circuitry to the supply. The main circuitry should be checked with an ohmmeter first anyway to rule out any short-circuit problems.

If the supply shows no damage but is not producing one or more voltages as required, check the raw d-c source, then the individual regulators. You are bound to find a problem.

8.17 Conclusion

The repair of any electronic circuit requires a certain amount of expertise, and ability improves with practice. Record problems for ready reference in the future. Keep a notebook or card catalog of all repair jobs; the information will be useful when similar problems come along. In addition:

1. Have repair equipment on hand to help you.
2. Have the schematics handy for ready reference.
3. Look for obvious physical damage first.
4. Check the power supply, then the microprocessor clock.
5. Compare the symptoms with the circuitry to determine the problem area.
6. Use a logical approach and necessary equipment to find the problem.
7. Don't get angry. Be patient, and use a logical approach. It takes time and experience to become a good troubleshooter in any area. The special nature of the microprocessor makes patience and a logical approach even more important. Study the ideas of this section and work toward proficiency at servicing microcomputer-related products.

9 SUMMARY

In this chapter we presented the tools of troubleshooting microprocessor-based equipment and also discussed troubleshooting in general from a pragmatic view-point. Study of these techniques and types of equipment will give you a good start when confronted with an errant microcomputer. Real expertise can come only from experience, and that clearly means that you need to spend time using both the logic analyzer and the development system to really appreciate their power. The more experience you have with these tools, the more valuable they will become to you.

10 GLOSSARY

Cross-assembler. An assembler that runs on a computer whose language is different from that of the program being assembled.

Development system. A computer used to write and debug software for a microprocessor. This system may also be used to debug hardware under development.

Emulator. A system that can act in place of a microprocessor. When a development system is in emulation mode, it replaces a particular MPU for monitoring purposes.

List file (.LST). A display of source file, object file, addresses and comments and a symbol table.

Logic analyzer. A multichannel instrument that can sample at least eight and sometimes more than 100 channels of data. Each channel can store many samples in a memory for later observation.

Memory map. A display of how different sections of memory are used (e.g., where ROM or RAM is located, and which ranges of memory are empty).

Object file (.OBJ or .HEX). The machine language file derived from a memory map or produced by an assembler.

Qualifier. A condition necessary to cause a logic analyzer to start taking data. Used as an AND with the trigger word.

Sample interval. The time between samples (reciprocal of sample rate).

Sample rate. The rate at which data is taken.

Signature analysis. An electronic method of checking patterns in digital circuits.

Simulator. Usually clever software pretending to be hardware.

Source file (.ASM). The assembly language file containing mnemonics.

Target processor. The microprocessor the software is being written for.

Text editor. Development system software to aid in creating assembly language source files.

Trace. A printout of a program in progress displaying the contents of MPU registers as the program runs.

Trigger word. A pattern that is used to begin data acquisition.

PROBLEMS

16-1 A logic analyzer has a 4096-bit memory. If eight channels of data are used, how many data samples can be stored per channel?

16-2 If the logic analyzer of Problem 16-1 uses 16 channels, how many data samples can be stored per channel?

16-3 If the sample interval is 100 nsec, how long can a 4096-bit, 16-channel analyzer take data?

16-4 What is the value of a logic analyzer?

16-5 What is the value of a development system?

16-6 What is the function of a communication tester?

16-7 Use your own words to define emulation as opposed to simulation.

16-8 Why is memory mapping desirable on a development system?

16-9 In your own words, describe the overall design process that is invoked when you begin to build a microprocessor-based control system. Discuss the use and need for an assembler and the desirability of equipment covered in this chapter. How could you design a project without such equipment?

16-10 List some simple things to observe as you begin to look for problems in an errant computer. Where should you begin? Be sure to discuss the basics that every electronic/digital system requires.

16-11 Discuss the use of a signature analyzer in the repair of a troublesome microprocessor-based system.

16-12 Get manufacturers' literature on a logic analyzer and a development system. Look up both subjects in the trade magazines in your technical library, and find articles and advertisements for these test instruments.

16-13 If you have these items of test and development equipment available to you, read and study their equipment manuals. The equipment is very sturdy and there is little you can do to hurt it. With this in mind, learn to use the equipment.

16-14 Suppose that in the construction of a microprocessor-based system a pair of data lines was accidentally shorted together, perhaps during soldering or wire wrapping. With so many wires to look at, finding such a problem is no small task. How would this problem show up on a logic analyzer? How would the logic analzyer show a data or address wire that was stuck either high or low? How would you see this?

16-15 If a pair of wires was shorted together on either a data bus or an address bus, why would an ordinary oscilloscope be difficult to use in finding the problem? Why would a logic analyzer be a much better instrument?

16-16 You can learn a great deal about a microprocessor system if you already have an idea of what "normal" signals look like on address, data, and control lines. Take an ordinary oscilloscope, ground it correctly to the microcomputer, and observe the signals on all the lines on an operating microprocessor chip. Let the microprocessor run a simple program and carefully look at these typical signals. The shapes and amplitudes may surprise you; sketch them for future reference. Also note the relative times involved.

16-17 Observe the effect of using the RESET line in a microcomputer. What happens to all the signals when you use the RESET button or line? Do they all go high, low, or perhaps float? Consult the manual on the microprocessor involved to answer this question and then observe some of them in a working microprocessor.

16-18 Study the schematics of several microcomputers. Identify the various parts and determine their relationship to the block diagram of a computer. How are the various memory chips decoded to specific memory areas? How are the I/O areas established? Are the address and data busses buffered? How are status signals obtained?

16-19 Effective troubleshooting is based on general knowledge, experience, and continued practice. What experiences have you had in repairing electronic and microprocessor-based equipment? What can you do to get more experience in this area? (One answer is to study and monitor working systems.)

16-20 A simple but major problem in getting a computer working is the basic communication between the terminal and the computer. If RS232C is used, use an oscilloscope to observe the transmit and receive lines of the system and also the control lines (DTR, RTS, CTS, etc). Be sure to note normal levels and the data rates involved. Where are the control signals supposed to be in a working system? (Is DTR high or low in normal operation? RTS? CTS?)

APPENDIX A

HEXADECIMAL/DECIMAL CONVERSION; POWERS OF 2 AND 16

Hexadecimal Columns

___6__		___5__		___4__		___3__		___2__		___1__	
HEX	DEC	HEX	DEC	HEX	DEC	HEX	DEC	HEX	DEC	HEX	DEC
0	0	0	0	0	0	0	0	0	0	0	0
1	1,048,576	1	65,536	1	4,096	1	256	1	16	1	1
2	2,097,152	2	131,072	2	8,192	2	512	2	32	2	2
3	3,145,728	3	196,608	3	12,288	3	768	3	48	3	3
4	4,194,304	4	262,144	4	16,384	4	1,024	4	64	4	4
5	5,242,880	5	327,680	5	20,480	5	1,280	5	80	5	5
6	6,291,456	6	393,216	6	24,576	6	1,536	6	96	6	6
7	7,340,032	7	458,752	7	28,672	7	1,792	7	112	7	7
8	8,388,608	8	524,288	8	32,768	8	2,048	8	128	8	8
9	9,437,184	9	589,824	9	36,864	9	2,304	9	144	9	9
A	10,485,760	A	655,360	A	40,960	A	2,560	A	160	A	10
B	11,534,336	B	720,896	B	45,056	B	2,816	B	176	B	11
C	12,582,912	C	786,432	C	49,152	C	3,072	C	192	C	12
D	13,631,488	D	851,968	D	53,248	D	3,328	D	208	D	13
E	14,680,064	E	917,504	E	57,344	E	3,584	E	224	E	14
F	15,728,640	F	983,040	F	61,440	F	3,840	F	240	F	15

Powers of 2

2^n	n		
256	8	2^0	$= 16^0$
512	9	2^4	$= 16^1$
1 024	10	2^8	$= 16^2$
2 048	11	2^{12}	$= 16^3$
4 096	12	2^{16}	$= 16^4$
8 192	13	2^{20}	$= 16^5$
16 384	14	2^{24}	$= 16^6$
32 768	15	2^{28}	$= 16^7$
65 536	16	2^{32}	$= 16^8$
131 072	17	2^{36}	$= 16^9$
262 144	18	2^{40}	$= 16^{10}$
524 288	19	2^{44}	$= 16^{11}$
1 048 576	20	2^{48}	$= 16^{12}$
2 097 152	21	2^{52}	$= 16^{13}$
4 194 304	22	2^{56}	$= 16^{14}$
8 388 608	23	2^{60}	$= 16^{15}$
16 777 216	24		

Powers of 16

16^n	n
1	0
16	1
256	2
4 096	3
65 536	4
1 048 576	5
16 777 216	6
268 435 456	7
4 294 967 296	8
68 719 476 736	9
1 099 511 627 776	10
17 592 186 044 416	11
281 474 976 710 656	12
4 503 599 627 370 496	13
72 057 594 037 927 936	14
1 152 921 504 606 846 976	15

APPENDIX B

8085 MICROCOMPUTER PROGRAMS FOR LABORATORY USE

SAMPLE ASSEMBLY LANGUAGE SOFTWARE

These programs were designed to be used with a single-board computer having an 8085 microprocessor and a display terminal. Each program uses the keyboard input subroutine (CIN) and the display output subroutine (COUT) that are resident in the ROM-type memory on the computer card. It is assumed that to run this software you will have available a means of entering the machine language, modifying and displaying this code, and running a program. A minimal monitor resident in ROM is typically active at power-up.

Each program uses a CALL instruction to do the I/O with the display terminal. You will need to know the addresses of your particular CIN and COUT routines to try these programs. The standard used here assumes the input routine to be located at address 01DA H. Characters entered on the keyboard appear in the A register of the 8085. We also assume the character output routine to be located at address 020E H. Characters are output to the display from the C register of the 8085. It should not be difficult to modify these example programs to work on your own computer.

DETAILS: CIN: 01DAH   to Register A
                 COUT: 020EH   from Register C
                 RAM begins at 2000 H.
                 STACK begins at 2800 H and uses decreasing memory locations.
                 The monitor in ROM begins at 0 H.

Typical monitor commands are as follows:

D   Display contents of memory
I   Insert hex into memory
X   Examine registers
S   Examine memory with the option to substitute data
G   Go and run a program
M   Move a block of memory

This software was run on a Pro-Log PLS858 8085 computer card. Each program shows address, machine language, assembly language, and comments. The programs cover the following areas:

1. Electric typewriter
2. Printing an ASCII string to the terminal
3. Recognizing a particular ASCII character
4. Producing delay using a loop
5. Printing a counter on the screen
6. Printing your name using ASCII codes
7. Printing a message using a subroutine
8. Adding two numbers
9. Storing a number in a block of memory
10. BCD addition of numbers

Each of the program listings was created on a computer using an 8085 assembler. It is suggested that you do this on your own computer using your own assembler. Change the program ORG line to match the beginning of RAM on your 8085 system. Change the location of CIN and COUT to match your own 8085 system.

A PRO-LOG PRIMER

The Pro-Log is a single-board computer that can use a variety of microprocessors. All our cards use the 8085. The 8085 is capable of addressing 65,536 memory locations. The Pro-Log CPU card does not have this many locations on it, however; it has RAM located from 2000 H to 27FF H. The MONITOR program is in an EPROM that is located from address 0 to over 400 H.

These 10 exercises will familiarize you with the seven single-letter commands of the monitor and give you a feeling for the way programs written in assembly language are entered into memory and executed.

Load and run each program and note its effects. Study the information given as certain programming concepts are also presented here. It is anticipated that you will study and run all 10 programs. You are also expected to change the programs for different effects.

1. Get the information on your monitor, which details the six commands: D, I, S, X, G, M. Try each command and be sure that you understand its use and operation.

2. Learn to zero a block of memory in the following way:

```
.S2000 4A-00
.M2000,2300,2001
```

3. Run each program several times.

Note: The input and output routines of the monitor are located as follows:

CIN: 01DAH   COUT: 020EH
ROM: 0 to 1FFFH
RAM: 2000H to 27FFH
STACK: Usually set to 2800H

```
                           1        ;PROLOG DEMONSTRATION PROGRAM ONE
                           2        ;
                           3        ;** THE ELECTRIC TYPEWRITER **
                           4        ;
                           5        ;THIS PROGRAM CAUSES THE TERMINAL TO BEHAVE AS AN
                           6        ;ELECTRIC TYPEWRITER INSTEAD OF A COMPUTER.
                           7        ;CHARACTERS THAT ARE ENTERED ARE ECHOED ONTO THE
                           8        ;TERMINAL.  THE ONLY WAY OUT OF THIS PROGRAM IS
                           9        ;TO PRESS THE RESET BUTTON AND REENTER THE MONITOR.
                          10        ;
                          11        ;THIS PROGRAM USES THE MONITOR I/O ROUTINES CALLED
                          12        ;CIN AND COUT.  CIN RECEIVES A CHARACTER FROM THE
                          13        ;KEYBOARD AND PLACES IT INTO THE ACCUMULATOR.
                          14        ;COUT SENDS A CHARACTER FROM THE C REGISTER.
                          15        ;
                          16        ORG 2000H            ;BEGINNING OF USER RAM
2000   310028             17        LXI SP, 2800H        ;USER STACK AREA
2003   CDDA01             18  NEXT: CALL CIN             ;GET A CHARACTER FROM THE KEYBOARD
2006   4F                 19        MOV C, A             ;TRANSFER IT TO THE C REGISTER
2007   CD0E02             20        CALL COUT            ;ECHO IT TO THE DISPLAY
200A   C30320             21        JMP NEXT             ;REPEAT THE PROCESS
                          22  CIN:  SET 01DAH            ;LOCATION OF INPUT ROUTINE
                          23  COUT: SET 020EH            ;LOCATION OF OUTPUT ROUTINE
                          24        END                  ;END OF PROGRAM


                           1        ;PROLOG DEMONSTRATION PROGRAM TWO
                           2        ;
                           3        ;** ASCII CHARACTER STRINGS **
                           4        ;
                           5        ;THIS PROGRAM PRINTS CONTINUOUS STRINGS OF ASCII
                           6        ;CHARACTERS ON THE VIDEO DISPLAY TERMINAL. A CARRIAGE
                           7        ;RETURN-LINE FEED (CRLF) IS SENT AFTER EACH LINE.
                           8        ;
                           9        ;THIS PROGRAM ALSO USES THE MONITOR I/O ROUTINES
                          10        ;TO SEND CHARACTERS TO THE TERMINAL
                          11        ;
                          12        ORG 2100H            ;ORIGIN OF THIS PROGRAM
2100   310028             13        LXI SP, 2800H        ;USER STACK AREA
2103   114000             14  START: LXI D, 64D          ;COUNTER FOR 64 CHARACTERS
2106   0E30               15        MVI C, 30H           ;CODE FOR AN ASCII ZERO
2108   CD0E02             16  TOP:  CALL COUT            ;SEND A CHARACTER TO DISPLAY
210B   1B                 17        DCX D                ;DECREMENT THE CHARACTER COUNTER
210C   7A                 18        MOV A, D             ;TEST TO SEE IF DE IS ZERO YET
210D   B3                 19        ORA E                ;
210E   CA1521             20        JZ CRLF              ;IF ZERO, SEND A CRLF
2111   0C                 21        INR C                ;NEXT SEQUENTIAL ASCII CHARACTER
2112   C30821             22        JMP TOP              ;CONTINUE
2115   0E0D               23  CRLF: MVI C, 0DH           ;ASCII CARRIAGE RETURN
2117   CD0E02             24        CALL COUT            ;SEND IT
211A   0E0A               25        MVI C, 0AH           ;ASCII LINE FEED
211C   CD0E02             26        CALL COUT            ;SEND IT
211F   C30321             27        JMP START            ;START WITH A NEW LINE
                          28  COUT: SET 020EH            ;LOCATION OF MONITOR OUTPUT ROUTINE
                          29        END
```

```
                           1              ;PROLOG DEMONSTRATION PROGRAM THREE
                           2              ;
                           3              ;** RECOGNIZE A CHARACTER FROM THE KEYBOARD **
                           4              ;
                           5              ;THIS PROGRAM IS ESSENTIALLY THE ELECTRIC TYPEWRITER
                           6              ;PROGRAM EXCEPT THAT WHEN AN EXCLAMATION
                           7              ;POINT (!) IS TYPED, A STORED MESAGE IS PRINTED. AFTER
                           8              ;THE PRINTING OF THE MESSAGE, WE RETURN TO THE ELECTRIC
                           9              ;TYPEWRITER PROGRAM
                          10              ORG 2200H             ;ORIGIN OF THIS PROGRAM
2200    310028            11              LXI SP, 2800H         ;USER STACK AREA
2203    212522            12    START:    LXI H, MESAG          ;ADDRESS OF STORED MESSAGE
2206    7E                13    MORE:     MOV A, M              ;TRANSFER CHARACTER TO ACCUMULATOR
2207    FE00              14              CPI 0                 ;END OF MESSAGE IF A ZERO
2209    CA1422            15              JZ DONE               ;SO BACK TO TYPEWRITER
220C    4F                16              MOV C, A              ;OTHERWISE PRINT ANOTHER CHARACTER
220D    CD0E02            17              CALL COUT             ;SEND IT
2210    23                18              INX H                 ;POINT TO NEXT CHARACTER
2211    C30622            19              JMP MORE              ;AND CONTINUE PRINTING
2214    CDDA01            20    DONE:     CALL CIN              ;GET A CHARACTER FROM KEYBOARD
2217    E67F              21              ANI 7FH               ;SET PARITY BIT=0
2219    FE21              22              CPI 21H               ;IS IT AN EXCLAMATION POINT?
221B    CA0322            23              JZ START              ;IF SO, PRINT MESSAGE
221E    4F                24              MOV C, A              ;OTHERWISE, TYPEWRITER
221F    CD0E02            25              CALL COUT             ;SEND IT
2222    C31422            26              JMP DONE              ;CONTINUE
2225    535543            27    MESAG:    DB 'SUCCESS *** ON TO PROGRAM FOUR!',0DH,0AH,0
2228    43455353202A2A2A204F4E20544F2050524F4752414D20464F5552210D0A00
                          28    CIN:      SET 01DAH             ;LOCATION OF INPUT ROUTINE
                          29    COUT:     SET 020EH             ;LOCATION OF OUTPUT ROUTINE
                          30              END
```

```
                           1              ;PROLOG DEMONSTRATION PROGRAM FOUR
                           2              ;
                           3              ;** A DELAY PROGRAM **
                           4              ;
                           5              ;THE PURPOSE OF THIS PROGRAM IS TO SHOW HOW TO
                           6              ;PRODUCE A REAL TIME DELAY USING A 16 BIT REGISTER.
                           7              ;EVEN THOUGH THE MICROPROCESSOR IS VERY FAST, TIME
                           8              ;IS TAKEN UP WITH EACH OPERATION. IN THIS EXAMPLE
                           9              ;A 16 BIT REGISTER (DE) IS DECREMENTED F000H TIMES OR
                          10              ;61440D TIMES. THE RESULT IS A CONSIDERABLE DELAY
                          11              ;NOTED BY PRESSING A KEY ON THE KEYBOARD AND OBSERVING
                          12              ;THE TIME IT TAKES FOR THE CHARACTER TO BE ECHOED
                          13              ;TO THE TERMINAL.
                          14              ORG 2000H
2000    310028            15              LXI SP, 2800H         ;USER STACK AREA
2003    CDDA01            16    NEXT1:    CALL CIN              ;GET A CHARACTER
2006    4F                17              MOV C, A              ;MUST BE IN C FOR OUTPUT SUBROUTINE
2007    1100F0            18              LXI D, 0F000H         ;DE REGISTER PAIR IS FOR DELAY COUNTER
200A    1B                19    DEC:      DCX D                 ;DECREMENT DE REGISTER PAIR
200B    7A                20              MOV A, D              ;TRANSFER A COPY OF D TO A
200C    B3                21              ORA E                 ;ARE D AND E EMPTY?
200D    C20A20            22              JNZ DEC               ;IF NOT, CONTINUE DECREMENTING
2010    CD0E02            23              CALL COUT             ;IF SO, OUTPUT CHARACTER IN C
2013    C30320            24              JMP NEXT1             ;CONTINUE THE PROGRAM
                          25    CIN:      SET 01DAH             ;MONITOR INPUT SUBROUTINE
                          26    COUT:     SET 020EH             ;MONITOR OUTPUT SUBROUTINE
                          27              END
```

```
                         1                    ;PROLOG DEMONSTRATION PROGRAM FIVE
                         2                    ;
                         3                    ;** A COUNTER ON THE SCREEN **
                         4                    ;
                         5                    ;THIS PROGRAM PRODUCES A COUNT ON THE SCREEN.
                         6                    ;THE SCREEN IS FIRST CLEARED BY SENDING A FORM
                         7                    ;FEED (ASCII 12=0CH). THEN A COUNT IS PRINTED.
                         8                    ;AFTER A DELAY, THE NEXT NUMBER IS PRINTED.
                         9                    ORG 2700H            ;ORIGIN OF THIS PROGRAM
2700    310028           10         LXI SP, 2800H        ;USER STACK AREA
2703    212F27           11    TOP:  LXI H, DATA           ;LOCATION OF WORD DATA
2706    0605             12         MVI B, 5             ;NUMBER OF DIGITS TO BE PRINTED
2708    0E0C             13         MVI C, 12            ;FORM FEED CODE
270A    CD0E02           14         CALL COUT            ;SEND THE FORM FEED
270D    7E               15    NOW:  MOV A, M             ;GET A CHARACTER FROM THE TABLE
270E    FE00             16         CPI 0               ;ZERO MARKS THE END OF THE WORD
2710    CA1B27           17         JZ DELAY            ;IF A ZERO, LET'S WAIT A BIT
2713    4F               18         MOV C, A            ;OTHERWISE, OUTPUT IT
2714    CD0E02           19         CALL COUT           ;USE THE OUTPUT SUBROUTINE
2717    23               20         INX H               ;POINT TO THE NEXT CHARACTER
2718    C30D27           21         JMP NOW             ;PRINT ANOTHER LETTER
271B    110080           22    DELAY: LXI D, 8000H        ;LET'S WASTE SOME TIME
271E    1B               23    DEL:  DCX D               ;USE DE REGISTERS FOR DELAY
271F    AF               24         XRA A               ;ZERO THE A REGISTER
2720    B2               25         ORA D               ;IS D EMPTY YET?
2721    CA2727           26         JZ NEXT             ;IF SO, DELAY IS DONE
2724    C31E27           27         JMP DEL             ;OTHERWISE, WAIT SOME MORE
2727    05               28    NEXT:  DCR B             ;ONLY WANT TO DO 5 NUMBERS
2728    CA0327           29         JZ TOP              ;START OVER
272B    23               30         INX H               ;IF NOT DONE, KEEP GOING
272C    C30D27           31         JMP NOW             ;WITH NEXT DIGIT
                         32    COUT:  SET 020EH          ;MONITOR OUTPUT SUBROUTINE
272F    4F4E45           33    DATA:  DB 'ONE',0DH,0AH,0,'TWO',0DH,0AH,0,'THREE',0DH,0AH,0
2732    0D0A0054574F0D0A0054485245450D0A00
2743    464F55           34         DB 'FOUR',0DH,0AH,0,'FIVE',0DH,0AH,0
2746    520D0A00464956450D0A00
                         35         END
```

```
                         1                    ;PROLOG DEMONSTRATION PROGRAM SIX
                         2                    ;
                         3                    ;** PRINT YOUR OWN NAME **
                         4                    ;
                         5                    ;THIS PROGRAM REQUIRES YOU TO LOOK UP THE ASCII CODES
                         6                    ;FOR THE LETTERS OF YOUR NAME OR OTHER MESSAGE YOU MAY
                         7                    ;CARE TO PRINT.  THESE ARE FOUND ON THE BACK OF YOUR
                         8                    ;8085 PROGRAMMING CARD IN HEXADECIMAL. THESE ARE ENTERED
                         9                    ;INTO SEQUENTIAL MEMORY LOCATIONS BEGINNING WITH
                         10                   ;ADDRESS 2600H. ENTER THE PROGRAM AT ADDRESS 2500H AND
                         11                   ;THEN PUT YOUR DATA INTO MEMORY STARTING AT 2600H.
                         12                   ;MAKE SURE IT FINISHES WITH A CRLF AND A ZERO.
                         13                   ;(0DH,0AH,0)
                         14                   ORG 2500H            ;BEGINNING OF PROGRAM
2500    310028           15         LXI SP, 2800H        ;USER STACK AREA
2503    210026           16    START: LXI H, 2600H        ;START OF DATA TABLE
2506    7E               17    DOIT:  MOV A, M             ;GET THE FIRST LETTER
2507    FE00             18         CPI 0               ;IF ZERO WE ARE DONE
2509    CA0325           19         JZ START            ;SO LET'S START OVER
250C    4E               20         MOV C, M            ;PUT THE CHARACTER IN C
250D    CD0E02           21         CALL COUT           ;AND OUTPUT IT
2510    23               22         INX H               ;POINT TO THE NEXT LETTER
2511    C30625           23         JMP DOIT            ;START ALL OVER.
                         24
                         25                   ORG 2600H            ;NEW ORG FOR DATA TABLE
2600    414C41           26    WORDS: DB 'ALAN WAS HERE BEFORE YOU',0DH,0AH,0
2603    4E2057415320484552452042454F524520594F550D0A00
                         27    COUT:  SET 020EH          ;CANNED OUTPUT ROUTINE
                         28         END
```

```
                        1              ;PROLOG DEMONSTRATION PROGRAM SEVEN
                        2              ;
                        3              ;** THIS MESSAGE PROGRAM USES SUBROUTINES **
                        4              ;
                        5              ;THIS PROGRAM PRINTS A MESSAGE AND THEN CALLS
                        6              ;FOR SOME DELAY. IT THEN ISSUES A CONTROL G (BEL)
                        7              ;SIGNAL. A SHORTER DELAY IS PRODUCED AND THE
                        8              ;PROGRAM STARTS OVER AGAIN.
                        9              ORG 2600H          ;START OF PROGRAM
2600    310028          10             LXI SP,2800H       ;USER STACK AREA
2603    213026          11   AGAIN:    LXI H, MSG         ;MESSAGE DATA TABLE
2606    CD1D26          12             CALL SEND          ;CALL THE SEND SUBROUTINE
2609    110080          13             LXI D, 8000H       ;SET UP FOR A LONG DELAY
260C    CD2926          14             CALL DELAY         ;CALL FOR THE DELAY
260F    0E07            15             MVI C, 7           ;READY FOR A BEL SIGNAL
2611    CD0E02          16             CALL COUT          ;SEND IT
2614    110040          17             LXI D, 4000H       ;SET UP FOR A SHORTER DELAY
2617    CD2926          18             CALL DELAY         ;CALL FOR THE DELAY
261A    C30326          19             JMP AGAIN          ;START THE PROGRAM OVER
261D    7E              20   SEND:     MOV A, M           ;GET A CHARACTER
261E    FE00            21             CPI 0              ;IS IT THE LAST CHARACTER?
2620    C8              22             RZ                 ;IF SO RETURN TO MAIN PROGRAM
2621    4F              23             MOV C, A           ;IF NOT, SEND THE CHARACTER
2622    CD0E02          24             CALL COUT          ;USING THE MONITOR OUTPUT ROUTINE
2625    23              25             INX H              ;POINT TO THE NEXT CHARACTER
2626    C31D26          26             JMP SEND           ;SEND MORE CHARACTERS
2629    1B              27   DELAY:    DCX D              ;BEGINNING OF DELAY SUBROUTINE
262A    7A              28             MOV A, D           ;CHECK TO SEE IF D=E=0
262B    B3              29             ORA E              ;USING THIS TRICK
262C    C8              30             RZ                 ;DONE IF ZERO
262D    C32926          31             JMP DELAY          ;OTHERWISE NOT DONE YET
                        32   COUT:     SET 020EH          ;MONITOR I/O ROUTINE
2630    544849          33   MSG:      DB 'THIS IS AN 8085 MICROPROCESSOR',0DH,0AH,00
2633    5320495320414E2038303835204D4943524F50524F434553534F520D0A00
                        34             END


                        1              ;PROLOG DEMONSTRATION PROGRAM EIGHT
                        2              ;
                        3              ;** ADD TWO NUMBERS TOGETHER AND STORE THE RESULT **
                        4              ;
                        5              ;THIS PROGRAM REQUIRES YOU TO PLACE TWO NUMBERS IN
                        6              ;SEQUENTIAL MEMORY LOCATIONS. PUT THE FIRST NUMBER
                        7              ;INTO 2300H AND THE SECOND ONE INTO 2301H.
                        8              ;THE PROGRAM PUTS THE SUM OF THE NUMBERS INTO ADDRESS
                        9              ;2302H. IF THERE IS AN OVERFLOW, ADDRESS 2303H
                        10             ;IS SET TO A ONE, OTHERWISE IT IS ZERO.
                        11             ;AFTER EXECUTION, THE PROGRAM JUMPS TO THE MONITOR
                        12             ;SO THAT YOU CAN CHECK THE CONTENTS OF THE LOCATIONS.
                        13             ORG 2250H          ;START OF PROGRAM
2250    310028          14             LXI SP, 2800H      ;USER STACK AREA
2253    210023          15             LXI H, 2300H       ;FIRST NUMBER LOCATION
2256    7E              16             MOV A, M           ;GET THE FIRST NUMBER INTO A
2257    23              17             INX H              ;POINT TO THE SECOND NUMBER
2258    86              18             ADD M              ;ADD THEM TOGETHER
2259    23              19             INX H              ;POINT TO THE RESULT LOCATION
225A    77              20             MOV M, A           ;PUT THE ANSWER IN 2302H
225B    DA6422          21             JC ONE             ;IF A CARRY OCCURS, STORE A ONE
225E    AF              22             XRA A              ;ZERO THE ACCUMULATOR
225F    23              23             INX H              ;POINT TO THE CARRY LOCATION
2260    77              24             MOV M, A           ;PUT A ZERO IN THERE
2261    C30000          25             JMP 0              ;RETURN TO MONITOR
2264    3E01            26   ONE:      MVI A, 1           ;READY TO STORE A ONE
2266    23              27             INX H              ;POINT TO CARRY LOCATION
2267    77              28             MOV M, A           ;STORE A 1 IN 2303H
2268    C30000          29             JMP 0              ;RETURN TO MONITOR
                        30             END
```

```
                           1              ;PROLOG DEMONSTRATION PROGRAM NINE
                           2              ;
                           3              ;** STORE A NUMBER IN A BLOCK OF MEMORY **
                           4              ;
                           5              ;THIS PROGRAM ACCEPTS A DIGIT (0 TO 9) FROM
                           6              ;THE KEYBOARD, STRIPS OFF THE ASCII BIAS (30H)
                           7              ;AND STORES THE NUMBER IN 100H MEMORY LOCATIONS
                           8              ;STARTING AT 2000H. YOU THEN RETURN TO THE MONITOR.
                           9              ;AT THIS TIME, DUMP THE MEMORY LOCATIONS (2000H - 20FFH)
                          10              ;AND NOTE THAT THE NUMBER WAS STORED.
                          11              ORG 2400H              ;START OF PROGRAM
2400   310028            12              LXI SP, 2800H         ;USER STACK LOCATION
2403   CDDA01            13              CALL CIN              ;GET THE DIGIT
2406   E67F              14              ANI 7FH               ;MASK OFF THE PARITY BIT
2408   D630              15              SUI 30H               ;STRIP OFF ASCII BIAS
240A   210020            16              LXI H, 2000H          ;SET UP MEMORY POINTER (HL=M)
240D   110001            17              LXI D, 100H           ;GOING TO FILL 100H LOCATIONS
2410   77                18      GO:     MOV M, A              ;PUT DATA IN A LOCATION
2411   23                19              INX H                 ;POINT TO NEXT LOCATION
2412   F5                20              PUSH A                ;SAVE A REGISTER
2413   1B                21              DCX D                 ;DECREMENT THE COUNTER
2414   7A                22              MOV A, D              ;TEST TO SEE IF BOTH D AND E
2415   B3                23              ORA E                 ;ARE AT ZERO
2416   C22124            24              JNZ MORE              ;IF NOT, KEEP GOING
2419   0E07              25      DONE:   MVI C, 7              ;LET'S RING THE BEL
241B   CD0E02            26              CALL COUT             ;SEND A CONTROL G
241E   C30000            27              JMP 0                 ;RETURN TO THE MONITOR
2421   F1                28      MORE:   POP A                 ;RESTORE A
2422   C31024            29              JMP GO                ;KEEP GOING
                          30      CIN:    SET 01DAH            ;MONITOR INPUT ROUTINE
                          31      COUT:   SET 020EH            ;MONITOR OUTPUT ROUTINE
                          32              END


                           1      ;THIS IS A DEMONSTRATION OF BCD ADDITION
                           2      ;AND ACCEPTS A PAIR OF 2 DIGIT NUMBERS.
                           3      ;THE SOFTWARE CHECKS FOR VALID DIGITS
                           4      ;WITHIN THE RANGE OF 0 - 9.  THE SUM
                           5      ;OF THE TWO NUMBERS IS PRINTED ON THE
                           6      ;TERMINAL.  THE ADDRESSES USED FOR CIN
                           7      ;AND COUT ARE FOR THE PROLOG SBC. IF YOU
                           8      ;ARE UP FOR IT, MODIFY THE PROGRAM TO
                           9      ;ADD LARGER NUMBERS.
                          10      ;SEE IF YOU CAN ALTER THIS PROGRAM TO ECHO
                          11      ;NUMBERS AS THEY ARE ENTERED.
                          12              ORG 2000H
2000   315E21            13      RSTRT:  LXI SP, STAK          ;SET UP STACK AREA
                          14      ;MAIN PROGRAM
2003   21A720            15      BEGIN:  LXI H, MSG0           ;SIGNON MESSAGE
2006   CD3020            16              CALL SEND             ;SEND IT TO CRT
2009   21CC20            17              LXI H, MSG1           ;FIRST PROMPT MESSAGE
200C   CD3020            18              CALL SEND             ;SEND IT TO CRT
200F   CD3C20            19              CALL GETCH            ;GET FIRST BCD PAIR
2012   323A21            20              STA NUM1              ;STORE AT TEMP LOC 1
2015   21E620            21              LXI H, MSG2           ;SECOND PROMPT MESSAGE
2018   CD3020            22              CALL SEND             ;SEND IT TO CRT
201B   CD3C20            23              CALL GETCH            ;GET SECOND BCD PAIR
201E   323B21            24              STA NUM2              ;STORE AT TEMP LOC 2
2021   CD6720            25              CALL ADDR             ;ADD THE NUMBERS TOGETHER
2024   210121            26              LXI H, MSG3           ;ANSWER PROMPT MESSAGE
2027   CD3020            27              CALL SEND
202A   CD7E20            28              CALL ANSWR            ;PRINT THE RESULTS
202D   C30320            29              JMP BEGIN             ;DO ANOTHER PROBLEM
                          30      ;
                          31      ;
                          32.     ;SEND SUBROUTINE
```

(continued on next page)

```
2030   7E        33   SEND:    MOV A, M           ;GET CHARACTER
2031   FE2A      34            CPI '*'            ;LAST CHARACTER YET?
2033   C8        35            RZ                 ;DONE IF END CHARACTER (*)
2034   4F        36            MOV C, A           ;MOVE TO B REGISTER
2035   CD0E02    37            CALL COUT          ;SEND IT
2038   23        38            INX H              ;POINT TO NEXT CHARACTER
2039   C33020    39            JMP SEND           ;SEND ANOTHER
                 40   ;
                 41   ;GET A PAIR OF NUMBERS SUBROUTINE
203C   CDDA01    42   GETCH:   CALL CIN           ;GET AN ASCII CHARACTER FROM TTY
203F   CD5320    43            CALL VALDG         ;CHECK TO SEE IF 0 - 9
2042   E60F      44            ANI 0FH            ;MASKS OFF LS NIBBLE
2044   07        45            RLC                ;MOVE BCD NUMBER TO MS BYTE
2045   07        46            RLC
2046   07        47            RLC
2047   07        48            RLC
2048   57        49            MOV D, A           ;TEMPORARY STORAGE
2049   CDDA01    50            CALL CIN           ;GET SECOND CHARACTER
204C   CD5320    51            CALL VALDG
204F   E60F      52            ANI 0FH            ;MASKS OFF LS NIBBLE
2051   B2        53            ORA D              ;COMBINE THE TWO BCD NUMBERS
2052   C9        54            RET
                 55   ;
                 56   ;VALID DIGIT SUBROUTINE
2053   FE30      57   VALDG:   CPI 30H            ;IS IT LESS THAN ZERO?
2055   FA5E20    58            JM ERROR
2058   FE3A      59            CPI 3AH            ;IS IT MORE THAN NINE?
205A   F25E20    60            JP ERROR
205D   C9        61            RET                ;DIGIT OK. RETURN.
                 62   ;
205E   211D21    63   ERROR:   LXI H, MSG4        ;ERROR MESSAGE
2061   CD3020    64            CALL SEND          ;SEND IT
2064   C30020    65            JMP RSTRT
                 66   ;
                 67   ;ADDER SUBROUTINE
2067   37        68   ADDR:    STC                ;MUST START WITH CARRY = 0
2068   3F        69            CMC
2069   113A21    70            LXI D, NUM1        ;DE POINTS TO FIRST PAIR
206C   213B21    71            LXI H, NUM2        ;HL POINTS TO SECOND PAIR
206F   7E        72            MOV A, M           ;GET SECOND NUMBER
2070   EB        73            XCHG               ;POINT TO FIRST NUMBER
2071   8E        74            ADC M              ;GET HEX SUM OF PAIR
2072   27        75            DAA                ;CONVERT TO BCD
2073   323C21    76            STA NUM3           ;PUT ANSWER IN TEMP LOC
2076   3E00      77            MVI A, 0           ;GUESS WHAT
2078   CE00      78            ACI 0              ;ADD WITH CARRY IMMEDIATE
207A   323D21    79            STA NUM4           ;SAVE CARRY INFORMATION
207D   C9        80            RET                ;RESULT IS IN ACCUMULATOR
                 81   ;
                 82   ;CONVERT TO ASCII AND PRINT SUBROUTINE
207E   3A3D21    83   ANSWR:   LDA NUM4           ;GET CARRY INFORMATION
2081   E60F      84            ANI 0FH            ;MASK OFF MSBYTE
2083   CA8C20    85            JZ SKIP            ;IF ZERO, DON'T PRINT IT
2086   C630      86            ADI 30H            ;ADD ASCII BIAS
2088   4F        87            MOV C, A           ;
2089   CD0E02    88            CALL COUT          ;SENT IT
208C   3A3C21    89   SKIP:    LDA NUM3           ;GET BCD PAIR (ANSWER)
208F   E6F0      90            ANI 0F0H           ;WANT MS NIBBLE FIRST
2091   0F        91            RRC                ;GET IT INTO LS NIBBLE
2092   0F        92            RRC
2093   0F        93            RRC
2094   0F        94            RRC
2095   C630      95            ADI 30H            ;ADD ASCII BIAS
2097   4F        96            MOV C, A           ;READY TO OUTPUT
2098   CD0E02    97            CALL COUT
209B   3A3C21    98            LDA NUM3           ;GET BCD PAIR (ANSWER) BACK
```

```
209E   E60F       99            ANI  OFH          ;WANT LS NIBBLE
20A0   C630      100            ADI  30H              ;ADD ASCII BIAS
20A2   4F        101            MOV  C, A             ;READY TO OUTPUT
20A3   CDOE02    102            CALL COUT
20A6   C9        103            RET
                 104   ;
                 105   ;CONSTANTS
                 106   ;
                 107   CIN:      SET  01DAH
                 108   COUT:     SET  020EH
20A7   0D0A57    109   MSG0:     DB 0DH,0AH,'WELCOME TO THE BCD ADDER PROGRAM',0DH,0AH,'*'
20AA   454C434F4D4520544F2054484520424344204144444552205052524F4752414D0D0A2A
20CC   454E54    110   MSG1:     DB 'ENTER FIRST TWO NUMBERS', 0DH, 0AH, '*'
20CF   4552204649525354205457F204E554D424552530D0A2A
20E6   454E54    111   MSG2:     DB 'ENTER SECOND TWO NUMBERS', 0DH, 0AH, '*'
20E9   4552205345434F4E442054574F204E554D424552530D0A2A
2101   544845    112   MSG3:     DB 'THE SUM OF THE NUMBERS IS', 0DH, 0AH, '*'
2104   2053554D204F4620544845204E554D42455253204953 0D0A2A
211D   484559    113   MSG4:     DB 'HEY!!!, ENTER TWO DIGITS.',07H, 0DH, 0AH, '*'
2120   2121212C20454E5445522054574F204449474954532E070D0A2A
213A   00        114   NUM1:     DS 1
213B   00        115   NUM2:     DS 1
213C   00        116   NUM3:     DS 1
213D   00        117   NUM4:     DS 1
213E   000000    118            DS 20H                     ;STACK AREA
2141   000000000000000000000000000000000000000000000000000000000000
215E   00        119   STAK:     DS 1
                 120            END
```

APPENDIX C

555 TIMER
DETAILS

Signetics

LINEAR INTEGRATED CIRCUITS

DESCRIPTION

The NE/SE 555 monolithic timing circuit is a highly stable controller capable of producing accurate time delays, or oscillation. Additional terminals are provided for triggering or resetting if desired. In the time delay mode of operation, the time is precisely controlled by one external resistor and capacitor. For a stable operation as an oscillator, the free running frequency and the duty cycle are both accurately controlled with two external resistors and one capacitor. The circuit may be triggered and reset on falling waveforms, and the output structure can source or sink up to 200mA or drive TTL circuits.

FEATURES

- TIMING FROM MICROSECONDS THROUGH HOURS
- OPERATES IN BOTH ASTABLE AND MONOSTABLE MODES
- ADJUSTABLE DUTY CYCLE
- HIGH CURRENT OUTPUT CAN SOURCE OR SINK 200mA
- OUTPUT CAN DRIVE TTL
- TEMPERATURE STABILITY OF 0.005% PER $^\circ$C
- NORMALLY ON AND NORMALLY OFF OUTPUT

APPLICATIONS

PRECISION TIMING
PULSE GENERATION
SEQUENTIAL TIMING
TIME DELAY GENERATION
PULSE WIDTH MODULATION
PULSE POSITION MODULATION
MISSING PULSE DETECTOR

PIN CONFIGURATIONS (Top View)

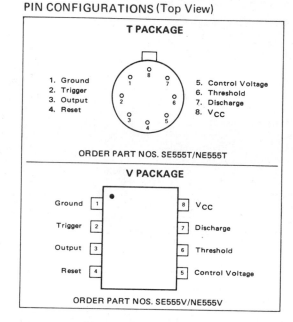

T PACKAGE

1. Ground
2. Trigger
3. Output
4. Reset

5. Control Voltage
6. Threshold
7. Discharge
8. V_{CC}

ORDER PART NOS. SE555T/NE555T

V PACKAGE

Ground 1   8 V_{CC}
Trigger 2   7 Discharge
Output 3   6 Threshold
Reset 4   5 Control Voltage

ORDER PART NOS. SE555V/NE555V

ABSOLUTE MAXIMUM RATINGS

Supply Voltage	+18V
Power Dissipation	600 mW
Operating Temperature Range	
NE555	0°C to +70°C
SE555	−55°C to +125°C
Storage Temperature Range	−65°C to +150°C
Lead Temperature (Soldering, 60 seconds)	+300°C

BLOCK DIAGRAM

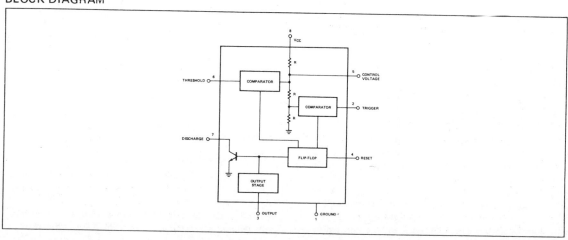

APPLICATIONS INFORMATION
MONOSTABLE OPERATION

In this mode of operation, the timer functions as a one-shot. Referring to Figure 1a the external capacitor is initially held discharged by a transistor inside the timer.

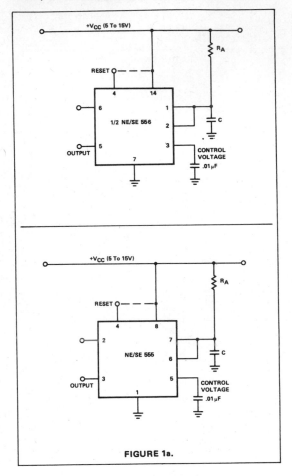

FIGURE 1a.

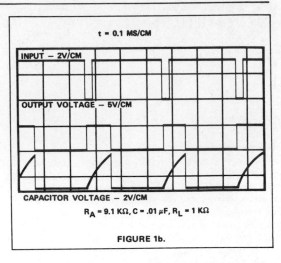

$$R_A = 9.1 \text{ K}\Omega, C = .01 \, \mu F, R_L = 1 \text{ K}\Omega$$

FIGURE 1b.

interval is independent of supply. Applying a negative pulse simultaneously to the reset terminal (pin 4) and the trigger terminal (pin 2) during the timing cycle discharges the external capacitor and causes the cycle to start over again. The timing cycle will now commence on the positive edge of the reset pulse. During the time the reset pulse is applied, the output is driven to its low state.

When the reset function is not in use, it is recommended that it be connected to V_{CC} to avoid any possibility of false triggering.

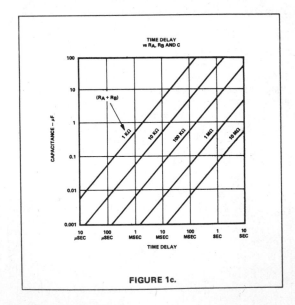

FIGURE 1c.

ASTABLE OPERATION

If the circuit is connected as shown in Figure 2a (pins 2 and 6 connected) it will trigger itself and free run as a multivibrator. The external capacitor charges through R_A and R_B and discharges through R_B only. Thus the duty cycle may be precisely set by the ratio of these two resistors.

Upon application of a negative trigger pulse to pin 2, the flip-flop is set which releases the short circuit across the external capacitor and drives the output high. The voltage across the capacitor, now, increases exponentially with the time constant $\tau = R_A C$. When the voltage across the capacitor equals 2/3 V_{CC}, the comparator resets the flip-flop which in turn discharges the capacitor rapidly and drives the output to its low state. Figure 1b shows the actual waveforms generated in this mode of operation.

The circuit triggers on a negative going input signal when the level reaches 1/3 V_{CC}. Once triggered, the circuit will remain in this state until the set time is elapsed, even if it is triggered again during this interval. The time that the output is in the high state is given by $t = 1.1 \, R_A C$ and can easily be determined by Figure 1c. Notice that since the charge rate, and the threshold level of the comparator are both directly proportional to supply voltage, the timing

APPLICATIONS INFORMATION (Cont'd)

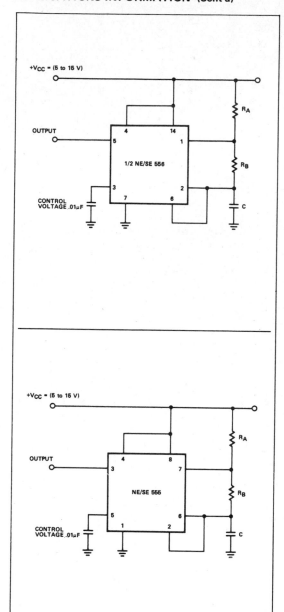

FIGURE 2a.

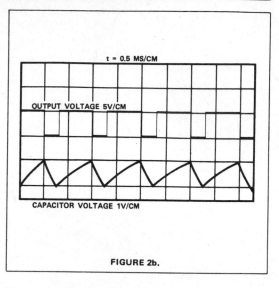

FIGURE 2b.

The charge time (output high) is given by:

$$t_1 = 0.693 \, (R_A + R_B) \, C$$

and the discharge time (output low) by:

$$t_2 = 0.693 \, (R_B) \, C$$

Thus the total period is given by:

$$T = t_1 + t_2 = 0.693 \, (R_A + 2R_B) \, C$$

The frequency of oscillation is then:

$$f = \frac{1}{T} = \frac{1.44}{(R_A + 2R_B) \, C}$$

and may be easily found by Figure 2c.

The duty cycle is given by:

$$D = \frac{R_B}{R_A + 2R_B}$$

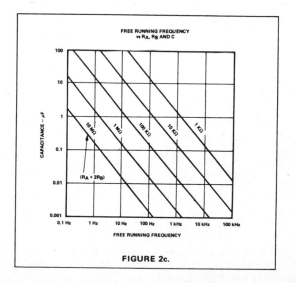

FIGURE 2c.

In this mode of operation, the capacitor charges and discharges between 1/3 V_{CC} and 2/3 V_{CC}. As in the triggered mode, the charge and discharge times, and therefore the frequency are independent of the supply voltage.

Figure 2b shows actual waveforms generated in this mode of operation.

Source: Reprinted material © 1984 Signetics Corp.

APPENDIX D

DIGITAL AND MICROPROCESSOR INTEGRATED CIRCUITS

Part No.	Pins	Function
74LS00	14	Quad 2 NAND Gate
74LS01	14	Quad 2-Input NAND Gate
74LS02	14	Quad 2 NOR Gate
74LS03	14	Quad 2 NAND Gate
74LS04	14	Hex Inverter
74LS05	14	Hex Inverter (Open Collector)
74LS08	14	Quad 2-Input AND Gate
74LS09	14	Quad 2-Input AND Gate (Open Collector)
74LS10	14	Triple 3 NAND Gate
74LS11	14	Triple 3-Input AND Gate
74LS12	14	3-Input NAND Gate (Open Collector)
74LS13	14	Dual Schmitt Trigger
74LS14	14	Hex Schmitt Trigger
74LS15	14	Triple 3-Input AND Gate
74LS20	14	Dual 4 NAND Gate
74LS21	14	Dual 4-Input AND Gate
74LS22	14	Dual 4-Input NAND Gate (Open Collector)
74LS26	14	Quad 2-Input Interface Positive NAND Gate
74LS27	14	Triple 3 NOR Gate
74LS28	14	Quad 2-Input Positive NOR Buffer
74LS30	14	8-Input NAND Gate
74LS32	14	Quad 2-Input Positive OR Gate
74LS33	14	Quad 2-Input NOR Buffer (Open Collector)
74LS37	14	Quad 2-Input NAND Buffer
74LS38	14	Quad 2-Input NAND Buffer (Open Collector)
74LS40	14	Dual 4 NAND Buffer
74LS42	16	BCD-to-Decimal Decoder
74LS47	16	7-Segment Decoder/Driver
74LS48	16	BCD to 7-Segment Decoder/Driver
74LS49	14	BCD to 7-Segment Decoder/Driver
74LS51	14	Dual 2-Input AND/OR Invert Gate
74LS54	14	Quad 2-Input AND/OR Invert Gate
74LS55	14	Dual 4-Input AND/OR Invert Gate
74LS73	14	Dual JK Flip Flop with Clear
74LS74	14	Dual D Flip Flop
74LS75	16	Quad Latch
74LS76	16	Dual JK Flip Flop with Preset and Clear
74LS78	14	Dual JK Flip Flop w/Preset Comm. Clk & Clr
74LS83	16	4-Bit Full Adder
74LS85	16	4-Bit Magnitude Comparator
74LS86	14	Quad Exclusive OR Gate 2-Input
74LS90	14	Decade Counter
74LS91	14	8-Bit Shift Register
74LS92	14	Divide-by-Twelve Counter
74LS93	14	4-Bit Binary Counter
74LS95	14	4-Bit Parallel-Access Shift Register
74LS96	16	5-Bit Shift Register
74LS107	14	Dual JK Master-Slave Flip Flop
74LS109	16	Dual JK Positive Edge Flip Flop
74LS112	16	Dual JK Negative Edge Flip Flop
74LS113	14	Dual JK Negative Edge Flip Flop
74LS114	14	Dual JK Negative Edge Flip Flop
74LS122	14	Retriggerable Monostable Multivibrator
74LS123	16	Monostable Multivibrator with Clear
74LS125	14	Tri-State Quad Buffer
74LS126	14	Quad Buffer (Tri-State)
74LS132	14	Quad Schmitt Trigger
74LS133	16	13-Input NAND Gate
74LS136	14	Quad Exclusive OR Gate
74LS138	16	Expandable 3/8 Decoder
74LS139	16	Expandable Dual 2/4 Decoder
74LS145	16	BCD-to-Decimal Decoder/Driver
74LS147	16	10-Line to 4-Line Priority Encoder
74LS148	16	8-Line to 3-Line Priority Encoder
74LS151	16	8-Input Multiplexer
74LS153	16	Dual 4-Input Multiplexer
74LS154	24	Single 4-16 Decoder
74LS155	16	Dual 2/4 Demultiplexer
74LS156	16	Dual 2/4 Demultiplexer (Open Collector)
74LS157	16	Quad 2/1 Multiplexer
74LS158	16	Quad 2/1 Multiplexer (Inv. Out)
74LS160	16	Presettable Decade Counter
74LS161	16	Presettable Binary Counter
74LS162	16	Presettable Decade Counter with Clear
74LS163	16	Presettable Binary Counter with Clear
74LS164	14	8-Bit Shift Register
74LS165	16	Parallel Load 8-Bit Serial Shift Register
74LS166	16	8-Bit Shift Register

Part No.	Pins	Function
74LS168	16	Synch. Decade Up/Down Counter (74LS668)
74LS169	16	Synch. Binary Up/Down Counter (74LS669)
74LS170	16	4x4 Register File
74LS173	16	Quad D Register (Tri-State)
74LS174	16	Hex D Flip Flop with Clear
74LS175	16	Quad D Flip Flop
74LS181	24	Arithmetic Logic Unit
74LS189	16	64-Bit RAM Tri-State
74LS190	16	Up/Down Decade Counter
74LS191	16	Up/Down Binary Counter
74LS192	16	4-Bit Up/Down BCD Counter
74LS193	16	4-Bit Up/Down Binary Counter
74LS194	16	4-Bit Bi-Directional Universal Shift Register
74LS195	16	4-Bit Parallel-Access Shift Register
74LS196	14	Presettable Decade Counter
74LS197	14	Preset Binary Counter
74LS221	16	Dual One-Shot
74LS240	20	Octal Inverting Bus/Line Driver
74LS241	20	Octal Bus/Line Driver
74LS242	14	Quad Bus Transceiver Inverting
74LS243	14	Tri-State Quad Transceiver
74LS244	20	Octal Driver Non-Inverting Tri-State
74LS245	20	Octal Bus Transceiver Non-Inverting
74LS247	16	BCD to 7-Segment Decoder/Driver
74LS248	16	BCD to 7-Segment Decoder/Driver
74LS249	16	BCD to 7-Segment Decoder/Driver
74LS251	16	Tri-State 8-Channel Multiplexer
74LS253	16	Dual 4-Input Multiplexer Tri-State
74LS257	16	Quad 2-Input Multiplexer Tri-State
74LS258	16	Quad 2/1 Multiplexer
74LS259	16	8-Bit Addressable Latch
74LS260	14	Dual 5-Input NOR Gate
74LS266	14	Quad Ex-NOR Gate
74LS273	20	8-Bit D Type Register
74LS279	16	Quad S-R Latches
74LS280	14	9-Bit Odd/Even Parity Generator/Checker
74LS283	16	4-Bit Full Adder
74LS289	16	64-Bit RAM Open Collector
74LS293	14	4-Bit Binary Counter
74LS298	16	Quad 2-Input Multiplexer with Storage
74LS299	20	8-Bit Universal Shift/Storage Register
74LS322	20	8-Bit Serial/Parallel Reg. (25LS22)
74LS323	20	8-Bit Univ. Shift/Storage Reg. (25LS23)
74LS347	16	BCD to 7-Segment Decoder
74LS352	16	Dual 4-Bit Multiplexer (Inv. Out)
74LS353	16	Dual 4-Bit Multiplexer (Inv. Out)
74LS364	20	Octal D-Type Transparent Latch
74LS365	16	Hex Buffer (Tri-State)
74LS366	16	Hex Inverter (Tri-State)
74LS367	16	Hex Buffer (Tri-State)
74LS368	16	Hex Inverter (Tri-State)
74LS373	20	Octal Transparent Latch
74LS374	20	Octal Dual Flip Flop (Tri-State)
74LS375	16	Quad Latch
74LS377	20	Octal D Register, Common Enable (25LS07)
74LS378	16	Hex D Register, Common Enable (25LS08)
74LS379	16	4-Bit Register, Common Enable
74LS386	14	Quad Ex-OR Gate
74LS388	16	Quad D Reg., Std & 3-State Out. (25LS2518)
74LS390	16	Dual 4-Bit Binary Counter
74LS393	14	Dual 4-Bit Binary Counter
74LS399	16	Quad 2-Mux with Q & Q Outputs (25LS09)
74LS490	16	Dual 4-Bit Decade Counter
74LS533	20	Octal Transparent Latch (Tri-State)
74LS534	20	Octal D-Type Flip Flop (Tri-State)
74LS540	20	Octal Buffer/Line Driver (Tri-State)
74LS541	20	Octal Buffer/Line Driver (Tri-State)
74LS640	20	Octal Bus Transceiver (Inverting) Tri-State
74LS641	20	Octal Bus Transceiver (True) O.C.
74LS644	20	Octal Bus Transceiver (True/Inverting) O.C.
74LS645	20	Octal Bus Transceiver (True) Tri-State
74LS670	16	4x4 Register File (Tri-State)
74LS688	20	8-Bit Magnitude Comparator (25LS2521)
81LS95	20	Tri-State Octal Buffer - True (74LS465)
81LS96	20	Tri-State Octal Buffer - Inverting (74LS466)
81LS97	20	Tri-State Octal Buffer - True (74LS467)
81LS98	20	Tri-State Octal Buffer - Inverting (74LS468)

Source: Reprinted courtesy of Jameco Electronics.

MICROPROCESSOR COMPONENTS

Z80, Z80A, Z80B, Z8000 SERIES

Part No.	Pins	Function
Z80–2MHz		
Z80	40	CPU (MK3880N)(780C) 2MHz
Z80-CTC	28	Counter Timer Circuit (MK3882)
Z80-DART	40	Dual Asynchronous Receiver/Transmitter
Z80-DMA	40	Direct Memory Access Circuit (MK3883)
Z80-PIO	40	Parallel I/O Interface Controller (MK3881)
Z80-SIO/0	40	Serial I/O (TxCB & RxCB Bonded) (MK3884)
Z80-SIO/1	40	Serial Input/Output (Lacks DTRB)
Z80-SIO/2	40	Serial Input/Output (Lacks SYNCB)
Z80-SIO/9	40	Serial Input/Output
Z80A–4MHz		
Z80A	40	CPU (MK3880N-4)(780C-1) 4MHz
Z80A-CTC	28	Counter Timer Circuit (MK3882-4)
Z80A-DART	40	Dual Asynchronous Receiver/Transmitter
Z80A-DMA	40	Direct Memory Access Circuit (MK3883-4)
Z80A-PIO	40	Parallel I/O Interface Controller (MK3881-4)
Z80A-SIO/0	40	Serial I/O (TxCB & RxCB Bonded) (MK3884-4)
Z80A-SIO/1	40	Serial I/O (Lacks DTRB)
Z80A-SIO/2	40	Serial I/O (Lacks SYNCB)
Z80A-SIO/9	40	Serial Input/Output
Z80B–6MHz		
Z80B	40	CPU (MK3880N-6) 6MHz
Z80B-CTC	28	Counter Timer Circuit (MK3882-6)
Z80B-DART	40	Dual Asynchronous Receiver/Transmitter
Z80B-PIO	40	Parallel I/O Interface Controller (MK3881-6)
Z8000		
Z8001B	48	CPU Segmented (10MHz)
Z8002B	40	CPU Non-Segmented (10MHz)
Z8030A	40	Serial Communications Controller (6MHz)
Z8036A	40	Counter/Timer and Parallel I/O Unit (6MHz)

6500 SERIES

Part No.	Pins	Function
6502	40	MPU with Clock (1MHz)
6502A	40	MPU with Clock (2MHz)
6502B	40	MPU with Clock (3MHz)
6520	40	Peripheral Inter. Adapter
6522	40	Versatile Inter. Adapter
6545	40	CRT Controller (CRTC)
6551	28	Asynchronous Comm. Interface Adapter

6800, 68B00, 68000 SERIES

Part No.	Pins	Function
6800–1MHz		
6800	40	MPU–8-Bit
6802	40	MPU–8-Bit with Clock and RAM
6809	40	CPU–8-Bit (On-Chip Oscillator)
6809E	40	CPU–8-Bit (External Clocking)
6810	24	128x8 Static RAM
6821	40	Peripheral Inter. Adapt (MC6820)
6830L8	24	1024x8-Bit ROM Mik/Min Bug Program
6845	40	CRT Controller (CRTC)
6850	24	Asynchronous Comm. Adapter (ACIA)
6860	24	0-600 bps Digital MODEM
6880A	16	Quad 3-State Bus Trans. (N8T26)
68B00–2MHz		
68B00	40	MPU–8-Bit
68B02	40	MPU–8-Bit with Clock and RAM
68B09	40	CPU–8-Bit (On-Chip Oscillator)
68B09E	40	CPU–8-Bit (External Clocking)
68B10	24	128x8 Static RAM
68B21	40	Peripheral Interface Adapter
68B45	40	CRT Controller (CRTC)
68B50	24	Asynchronous Comm. Int. Adapter (ACIA)
68000 SERIES		
MC68000L8	64	MPU 16-Bit (8MHz)
MC68488P	40	General Purpose Interface Adapter
MC68652P2	40	Multi-Protocol Comm. Cont. 2MHz(2652)
MC68661PB	28	Enhanced Prog. Comm. Interface (2661-2)
MC68764C	24	8192x8-Bit UV EPROM (21V) 450ns

8000, 80000 SERIES

Part No.	Pins	Function
8000		
8031	40	CPU with RAM and I/O
8035	40	MPU–8-Bit
8039	40	CPU–Single Chip 8-Bit (128 Bytes RAM)
8040	40	CPU–8-Bit (256 Bytes RAM)
8048**	40	MPU–8-Bit (Can use for 8035)
8049**	40	CPU–Single 8-Bit (Can use for 8039)
8060	40	CPU–8-Bit NMOS
8070	40	CPU–64 Bytes RAM
8073	40	CPU–with Basic Micro Interpreter
8080A	40	CPU
8085	40	CPU–8-Bit N-Channel
8085A-2	40	CPU–8-Bit N-Channel (5MHz)
8086-2	40	CPU–16-Bit (8MHz)
8087	40	Arithmetic Processor
8088	40	CPU–16-Bit (8-Bit Data Bus)
8100		
8154	40	1K (128x8) RAM 16-Bit I/O
8155	40	HMOS RAM I/O Port and Timer
8156	40	RAM with I/O Port and Timer
8185-2	18	1024x8 Bit Static RAM for MCS-85
8200		
8205	16	High Speed 1 out of 8 Binary Decoder (3205)
8212	24	8-Bit Input/Output (74S412)
8214	24	Priority Interrupt Control
8216	16	Bi-Directional Bus Driver
8224	16	Clock Generator/Driver
8226	16	Bus Driver
8228	28	System Controller/Bus Driver (74S428)
8237	40	Programmable DMA Controller
8237-5	40	Programmable DMA Controller
8238	28	System Controller (74S438)
8243	24	I/O Expander for 48 Series
8245	18	16-Key Keyboard Encoder (74C922)
8246	20	20-Key Keyboard Encoder (74C923)
8247	28	Display Controller (74C911)
8248	28	Display Controller (74C912)
8250	40	Asynchronous Comm. Element
8251	28	Prog. Comm. I/O (USART)
8251A	28	Prog. Comm. Interface (USART)
8253	24	Prog. Interval Timer
8253-5	24	Prog. Interval Timer
8255	40	Prog. Peripheral I/O (PPI)
8255A-5	40	Prog. Peripheral I/O (PPI)
8257	40	Prog. DMA Controller
8257-5	40	Prog. DMA Controller
8259	28	Prog. Interrupt Controller
8259-5	28	Prog. Interrupt Controller
8272	40	Sgle/Dble Density Floppy Disk Cont. (D765A)
8274	40	Multi-Protocol Serial Controller (7201)
8275	40	Prog. CRT Controller
8279	40	Prog. Keyboard/Display Interface
8279-5	40	Prog. Keyboard/Display Interface
8282	20	Octal Latch
8284	18	Clock Generator and Driver
8286	20	Octal Bus Transceiver
8287	20	Octal Bus Transceiver (Inverted)
8288	20	Bus Controller
8300		
8303	20	8-Bit Tri-State Bi-Directional Transceiver
8304	20	8-Bit Bi-Directional Receiver
8307	20	8-Bit Bi-Directional Receiver
8308	20	8-Bit Bi-Directional Receiver
8310	20	Octal Latched Peripheral Driver
8311	20	Octal Latched Peripheral Driver
8700		
8741	40	8-Bit Universal Peripheral Interface
8748	40	MPU–HMOS EPROM
8749	40	MPU–8-Bit (EPROM Version of 8049)
8755	40	16K EPROM with I/O
80000		
80186-6	68*	High Integration 16-Bit MPU
80188	68*	High Integration 16-Bit MPU (8-Bit Data Bus)

**8048 MPU AND 8049 CPU

To convert an 8048 to an 8035, or an 8049 to an 8039,
just tie Pin 7 high to by-pass the Internal ROM.

*REQUIRES SPECIAL 68-PIN SOCKET

Source: Reprinted courtesy of Jameco Electronics.

APPENDIX E

DIGITAL LABORATORY EXERCISES FOR THE 7400 SERIES

These exercises deal with the 7400 family of TTL integrated circuits. They are intended to familiarize you with the basic logic gates and their uses. They also support the various concepts from digital theory covered in this text. Each exercise presents a general concept of what you are to do. You will find the details relating to the subject in the chapters of this book.

To do these exercises, you will need some basic equipment, including a power supply, an oscilloscope, a pulse generator, a logic probe, and a prototype board in which to hold the 7400 family of integrated circuits. It is very convenient to have available a prototype system similar to the E & L Instruments Digi-Designer shown in Fig. E-1. This designer module contains a 5-V/1-A power supply, bounceless switches, a clock, indicator lamps, and a prototyping socket suitable for dual in-line packages (DIPs). It is also important to have a TTL data manual for the 7400 family so that you can learn to correctly look up the various outline drawings and electrical specifications for the DIPs you will need for each exercise.

You must learn to wire the circuitry very carefully. Place the wires in the correct holes and develop a neat-looking layout of the wires to aid in troubleshooting. Always remember to connect power (ground and +5 V) to the DIPs according to the TTL data manual. These connections are frequently not shown on the drawings, but it is assumed that you will be able to properly connect power and ground to the ICs. Before applying power to any IC, however, be certain of the location of pin 1 of the DIP and of your wiring in general.

As you work on the exercises, you should be developing your own troubleshooting skills in using these integrated circuits. Do not be too quick to look for help from someone else. It is your own skills that need to be developed, and hands-on ability does not always come easily. If you are having a problem getting a circuit working properly, puzzle it out. Find the wiring error or the missing ground or the missing power connection. It is helpful to label the drawing with the pin numbers of the various logic gates obtained from the TTL data manual. This will also aid in troubleshooting the circuit if problems develop.

You will get the most from these exercises if you let your own curiosity take over. Thoroughly explore the capabilities of the DIP involved. Why are the pins

FIGURE E-1 A Digi designer (courtesy of E&L Instruments).

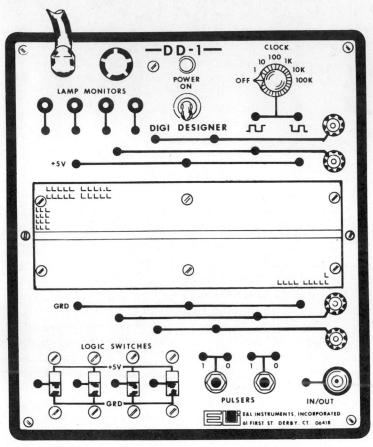

you did not have to use there? What other things can this particular IC be made to do for you? Where is it used? Which pins need to be grounded for the chip to function correctly?

The few exercises that do not use 7400 family ICs are included because of their interest value and their appropriateness. The D/A and A/D converter exercises, for example, use an MC1408L8 DAC and a 741 op amp. The last lab presented uses a microcomputer to produce waveforms from a DAC. Take your time with all the exercises and try to get the most out of each one.

List of Experiments

1	Basic Logic Gates	7400, 7402, 7404, 7408, 7432, 7486
2	Ring Oscillator	7404
3	Bounceless Switch	7400

FIGURE E-1 (continued)

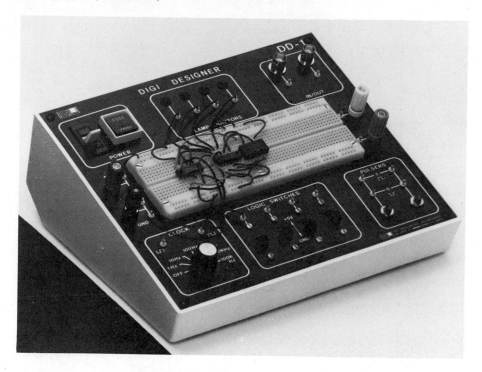

4	JK Flip-Flop	7473 or 7476
5	Four-Bit Counter	7473
6	Binary and Decade Counters	7493, 7490
7	MAN 1 LED Display	7490, 7447, MAN 1
8	Dual LED Display	7490, 7447, MAN 1
9	Modulo N Counter	7493, 7400, 7404
10	7445 BCD to decimal Decoder	7490, 7445
11	Four-Bit Shift Register	7473 or 7476, 7404
12	Eight-Bit Shift Register	7491
13	Four-Bit PISO Register	7494
14	One-Shots	74121

15	Four-Bit Digital-to-Analog Converter	7490, 741 op amp
16	Eight-Bit Digital-to-Analog Converter	7493, MC1408L8 DAC or DAC 808
17	Four-Bit Analog-to-Digital Converter	7490, 7447, MAN 1, 741 op amp
18	Burst Clock	7400, 7493
19	Sixteen-Bit Scratchpad RAM	7481
20	Verification of DeMorgan's Theorem	7493, 7400
21	ASCII "A" Generator (RS232C)	7400, 7410, 7493
22	Waveform Generation with an Eight-Bit Digital-to-Analog Converter and a Computer	MC1408L8 or DAC 808, Microcomputer

GETTING STARTED

Figure E-2 shows a prototype socket and the wiring necessary to connect a 4-digit BCD counter. This gives you an idea of transferring a schematic drawing over to the appropriate wiring diagram for a prototype socket. It is suggested that you make a diagram like this and transfer it to actual circuitry on the prototype socket.

Laboratory: #1 Basic Logic Gates
DIPs Needed (approximate): 7400, 7402, 7404, 7408, 7432, 7486
Chapter Reference: Chapter 3
Object: To examine the following basic logic gates: AND, OR, INVERT, NAND, NOR, XOR. To learn basic wiring of the prototype board. To verify the truth tables for the gates.

INTRODUCTION
Connect the basic gates one at a time. There are several gates in each package. Sketch each logic gate and its corresponding truth table. Check each gate to be sure that it follows the truth table for the appropriate gate.

THINGS TO OBSERVE
The DigiDesigner has a set of slide switches and a set of indicator lights that will help you test the basic logic gates. Use a light to check and observe each gate output. Be sure to try all the entries on the truth table. Look up the logic gate in the Electrical Specifications section of the manual and find the following items: temperature specifications, range of values for a "1" and a "0," gate propagation delay time, and input and output current under various conditions.

FIGURE E-2 Prototype wiring example (copyright © 1984 Signetics Corp.).

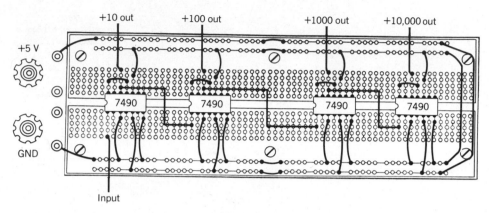

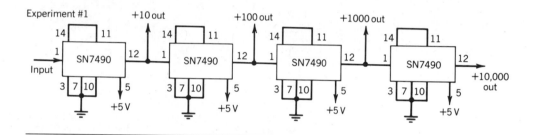

DRAWING
See Fig. E-3.

Laboratory: #2 Ring Oscillator
DIPs Needed (approximate): 7404
Chapter Reference: Chapter 3
Object: To use three TTL inverters in an unstable circuit that will produce a train of pulses. The frequency of the pulses is governed by the size of a capacitor.

INTRODUCTION
The ring oscillator is a quick way to produce a string of pulses for triggering circuitry like flip-flops and shift registers. The frequency of this clock is not particularly stable and varies with temperature. But it does produce a waveform easily and quickly.

THINGS TO OBSERVE
The circuit will oscillate at a very high frequency without any capacitor present. You will need an oscilloscope that can show frequencies above 30 MHz to see the output at full amplitude. A 10-MHz oscilloscope will work but will show the

FIGURE E-3 Lab 1: basic logic gates (copyright © 1984 Signetics Corp.).

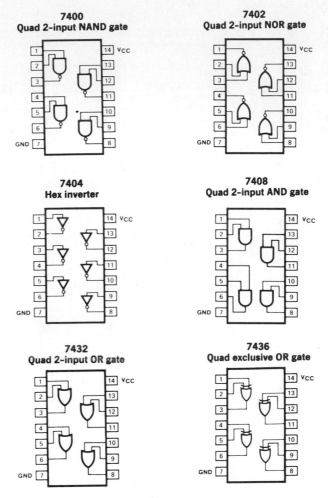

normal 5-V pulse amplitude as a much smaller amplitude. Correct operation of this circuit requires that the capacitor be very close to the IC. To get around this problem, place a small (0.01 μF) capacitor very close to the IC. Then use longer leads to try larger values of C. Try 10 values of C from 0.1 to 1 μF and take data of C versus frequency. Plot these on graph paper. See if you can determine the gate delay time from data taken with NO capacitor present. See if you can find a capacitor that is too LARGE for operation.

DRAWING
See Fig. E-4.

FIGURE E-4 Lab 2: ring oscillator (copyright © 1984 Signetics Corp.).

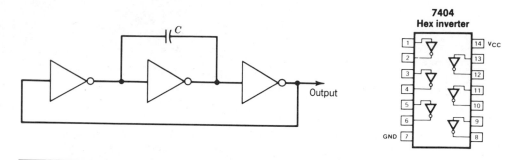

Laboratory: #3 Bounceless Switch
DIPs Needed (approximate): 7400
Chapter Reference: Chapter 3
Object: To learn the operation of an electronic switch used to debounce the normally noisy contacts of a mechanical switch. The bounceless switch circuit produces a clean edge.

INTRODUCTION
The two NAND gates form a cross-coupled latch that switches the instant a contact closure to ground is achieved. After this switching action, the contact can bounce as much as it likes; the output is already established.

THINGS TO OBSERVE
The circuit has two outputs, and each one is the complement of the other. Actuate the switch as fast as you can and observe the resulting clean edges at the "1" and "0" outputs. In contrast, observe the dirty signal at the mechanical switch contacts themselves. Also, notice that the two outputs are never at the same state. Observe this on an oscilloscope and then also on the indicator lights until you are sure of the circuit's correct operation.

DRAWING
See Fig. E-5.

Laboratory: #4 JK Flip-Flop
DIPs Needed (approximate): 7473 or 7476
Chapter Reference: Chapter 5
Object: To learn about the JK flip-flop, to investigate the different inputs to the device, and to learn about the complementary outputs.

INTRODUCTION
The JK flip-flop you will be looking at has both synchronous and asynchronous inputs. The $\overline{\text{Clear}}$ and $\overline{\text{Clock}}$ inputs are asynchronous and need no other signal to accomplish their function. They are both active low inputs. The J and K inputs are both synchronous inputs and are used with the clock input.

FIGURE E-5 Lab 3: bounceless switch (copyright © 1984 Signetics Corp.).

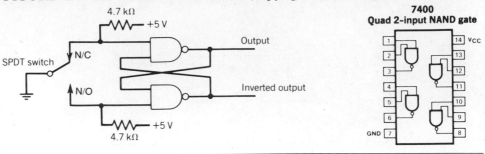

THINGS TO OBSERVE

Start by pulsing the clock input (J and K are tied high, $\overline{\text{CLEAR}}$ is also tied high) and observing the Q and $\overline{\text{Q}}$ outputs. The outputs should toggle each time the $\overline{\text{CLOCK}}$ receives a negative edge. If you leave this part of the circuitry functioning, you can test the J and K inputs. If you ground J and alternately K, the data on J and K should transfer to Q and $\overline{\text{Q}}$ on the negative edge of the clock. Reread Chapter 5 and verify that the flip-flop is completely functional. If you are using a 7476 instead of a 7473, you can also test the $\overline{\text{PRESET}}$ input. $\overline{\text{PRESET}}$ is asynchronous and causes the Q output to go high.

DRAWING
See Fig. E-6.

Laboratory: #5 Four-Bit Counter
DIPs needed (approximate): 7473 (2)
Chapter Reference: Chapter 5
Object: To use four flip-flops connected as an up counter and to observe the Q outputs of the flip-flops both on indicator lights and on a four-channel oscilloscope.

INTRODUCTION
The basic counter connection is shown in the drawing for this lab. The J and K inputs should be tied high, since we intend to toggle the flip-flop. Also tie the $\overline{\text{CLEAR}}$ (and $\overline{\text{PRESET}}$ if a 7476) lines high.

THINGS TO OBSERVE
The counter will sequence through a count of 0000 to 1111. Observe this in two ways. Run the counter manually using a bounceless switch and watch the Q outputs on four indicator lights. Run the counter using a fast clock and watch the Q outputs on a four-channel oscilloscope. Sketch the timing diagram from the oscilloscope. What happens to the counter if you momentarily ground all the $\overline{\text{CLEAR}}$ lines?

FIGURE E-6 Lab 4: the JK flip-flop (copyright © 1984 Signetics Corp.).

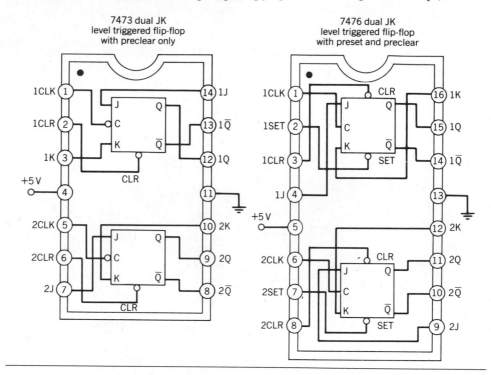

DRAWING
See Fig. E-7.

Laboratory: #6 Binary and Decade Counters
DIPs Needed (approximate): 7493, 7490
Chapter Reference: Chapter 5
Object: To observe two prepackaged 4-bit counters and learn the proper connections to allow them to count. The 7493 is a binary counter; the 7490 is a decade counter.

INTRODUCTION
Each of these counters has four flip-flops in one DIP. Look up the 7493 and the 7490 in a TTL data manual and sketch the circuitry from the electrical specs page. You must connect the A flip-flop output to the B flip-flop input (pin 12 to 1) for correct operation.

THINGS TO OBSERVE
You must also ground the reset lines (R_{01}, R_{02}, R_{91}, R_{92}) or the counter will not function. Observe the four outputs both on the indicator lamps and on the four-channel oscilloscope. Sketch the timing diagrams very carefully. Notice that the 7490 decade counter counts from 0000 to 1001 and then starts over. Which output is the MSB of the count and which is the LSB?

FIGURE E-7 Lab 5: the four-bit counter (copyright © 1984
Signetics Corp.).

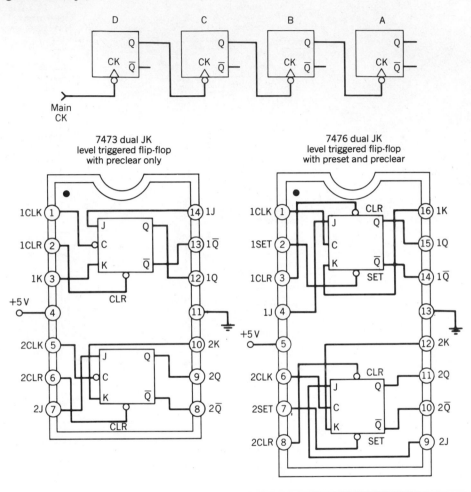

DRAWING
See Fig. E-8.

Laboratory: #7 MAN 1 LED Display
DIPs Needed (approximate): 7490, 7447, MAN 1
Chapter Reference: Chapter 4
Object: To drive a BCD to 7-segment decoder (7447) with a decade counter (7490)
and observe the resulting digits on a 7-segment LED readout (MAN 1).

INTRODUCTION
Refer to Lab #6 to be sure of the connections on the 7490 counter. The 7447 is
a convenient method of converting BCD data back to 7-segment data for a 7-

FIGURE E-8 Lab 6: binary and decade counters (copyright © 1984 Signetics Corp.).

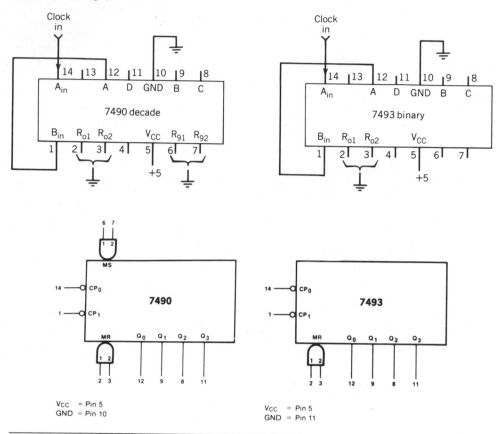

segment display. The MAN 1 is a common anode display, and the segments are lit by grounding the cathodes.

THINGS TO OBSERVE

As the output of the decade counter progresses from 0000 to 1001, the MAN 1 display should show the digits 0 to 9. If you ground the $\overline{RBI}$ input as the display is counting, you will notice that the "0" digit is blanked out. At this time, the $\overline{RBO}$ output should also go low. The $\overline{LAMPTEST}$ input can be grounded to see whether all the segments are functional. Try lifting the R_0 and then the R_9 counter inputs to see the counter reset to 0 and to 9.

DRAWING
See Fig. E-9.

Laboratory: #8 Dual LED Display
DIPs Needed (approximate): 7447 (2), 7490 (2), MAN 1 (2)

FIGURE E-9 Lab 7: MAN 1 LED display (copyright © 1984 Signetics Corp.).

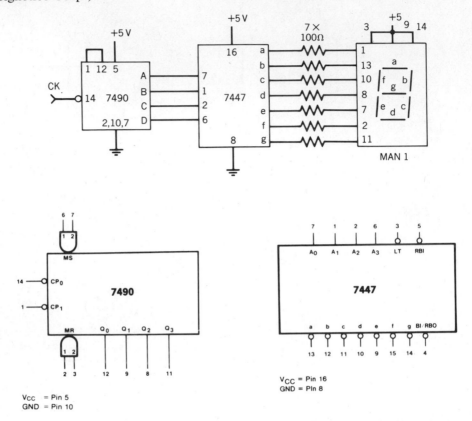

Chapter Reference: Chapter 4

Object: To set up a pair of LED readout devices and build a counter that counts from 00 to 99. To "multiplex" the digits at a rate greater than 20 Hz.

INTRODUCTION

A two-digit counter is relatively simple and we will make it more interesting by turning the devices on and off at a rate greater than the persistence of the human eye. You will need a clock set at a frequency greater than 20 Hz to alternate the displays and a bounceless switch to run the counters.

THINGS TO OBSERVE

Vary the rate of the clock and determine the frequency at which you can no longer notice any flickering. Use a bounceless switch to advance the counter from 00 to 99. What effect does the 50% duty cycle of the clock have on the apparent intensity of the displays?

DRAWING

See Fig. E-10.

FIGURE E-10 Lab 8: dual LED display (copyright © 1984 Signetics Corp.).

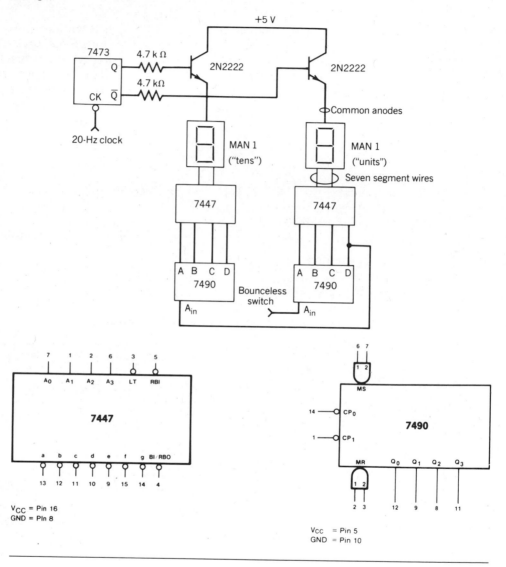

Laboratory: #9 Modulo *N* Counter
DIPs Needed (approximate): 7493, 7400, 7404
Chapter Reference: Chapter 5
Object: To build a counter that can count to different counts before resetting. To build a circuit that will reset a counter at any desired count.

INTRODUCTION
We will use the reset lines on a 7493 counter to reset the counter when we desire. Do not ground the reset inputs during this exercise; the circuit we design will do

this. You are already familiar with the 7490 decade counter. Now we will build a different counter.

THINGS TO OBSERVE

The drawing shows a counter that has five states. It will count from 0 to 4 (0000 to 0100) and then reset as the count attempts to reach 5 (0101). The AND gate causes the counter to reset itself when the desired count is reached. You can build a four-input AND gate from two-input AND gates or you can get a 7421 dual four-input AND gate. Don't expect to see a signal at the output from the AND gate—it is too fast. Just watch the counter as it counts from 0 to 4 and starts over. Design a couple of circuits to reset the counter at other counts.

DRAWING

See Fig. E-11.

Laboratory: #10 7445 BCD to decimal Decoder
DIPs Needed (approximate): 7490, 7445
Chapter Reference: Chapter 5
Object: To observe a BCD to decimal decoder (selector) circuit in operation.

INTRODUCTION

The 7445 is a data selector that has 10 outputs that can be used to provide a ground to one of 10 indicator lamps. The BCD inputs determine which output will be connected to ground. To observe the outputs on an oscilloscope, you will have to use pull up resistors, since each output is an open-collector output.

FIGURE E-11 Lab 9: modulo N counters (copyright © 1984 Signetics Corp.).

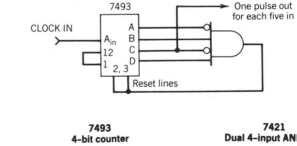

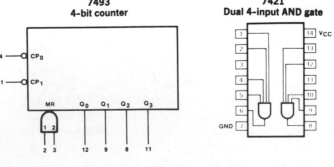

THINGS TO OBSERVE

Trigger the oscilloscope from the MSB on the counter. You should notice that each of the 10 outputs of the 7445 goes low for one clock interval. The time position of each increases with each sequential output (0 to 9). Sketch the relative positions of the outputs on a timing diagram for the 7490 counter. Use a 7404 inverter to turn the output pulses over so that they are upgoing pulses.

Note: The outputs are open-collector outputs and you must use pullup resistors on them. (Try 4.7 kΩ.)

DRAWING
See Fig. E-12.

FIGURE E-12 Lab 10: 7445 1 of 10 decoder (copyright © 1984 Signetics Corp.).

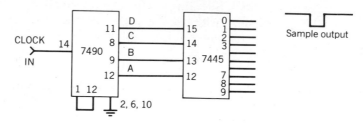

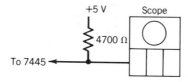

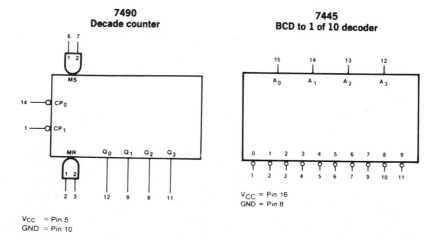

Laboratory: #11 Four-Bit Shift Register
DIPs Needed (approximate): 7473 or 7476 (2 each), 7404
Chapter Reference: Chapter 5
Object: To build a simple 4-bit shift register and store a 4-bit number in it. To shift the number out of the register. To circulate the 4 bits in the register.

INTRODUCTION

The shift register is used in many operations. A binary number can be entered serially or in parallel. The number can be read either in serial or in parallel. To build this circuit you will be using the J and K inputs of a JK flip-flop. Be certain that the flip-flops are operational before you begin.

THINGS TO OBSERVE

Use a bounceless switch for the clock on the shift register. Be sure that you do not connect a logic level to the serial data input at the same time that you connect the circulate line. Test the shift register by initially setting the data input low and pulsing the clock line four times. The register should then be filled with zeros. Do the same thing with ones. If you can store all zeros and all ones, your register is functional. Now enter a binary number (try 0111). You should be able to circulate the data now.

FIGURE E-13 Lab 11: four-bit shift register (copyright © 1984 Signetics Corp.).

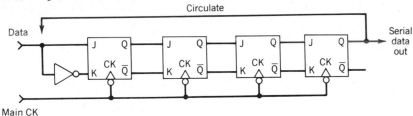

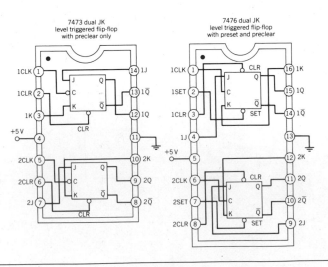

DRAWING
See Fig. E-13.

Laboratory: #12 Eight-Bit Shift Register
DIPs Needed (approximate): 7491
Chapter Reference: Chapter 5
Object: To use a larger shift register contained in a single DIP package. To store
an 8-bit quantity and then circulate the data.

INTRODUCTION
The 7491 is an 8-bit serial-in/serial-out shift register. The serial data output is the
Q output of the last flip-flop in the chain. This is the only way to observe the
contents of the shift register. By using the clock line, you can see all the bits in
the register, one at a time.

FIGURE E-14 Lab 12: eight-bit shift register (copyright © 1984
Signetics Corp.).

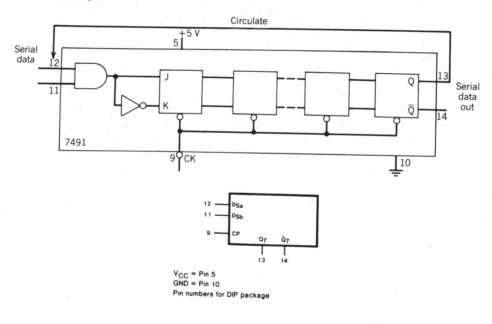

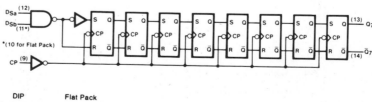

THINGS TO OBSERVE

Use one of the serial inputs and let the other float. Use a bounceless switch to drive the clock input. Observe the serial data output on an indicator lamp. Be careful not to connect the serial data input to a source and the circulate line at the same time. Check the register by first loading it with all zeros and then all ones.

Enter a binary pattern (e.g., 01100101) one bit at a time in conjunction with the clock line. Use the circulate line to refresh the data as you clock (pulse) it around the loop. If you have time, use a second 7491 8-bit shift register to make a 16-bit register.

DRAWING
See Fig. E-14.

FIGURE E-15 Lab 13: four-bit PISO register (copyright © 1984 Signetics Corp.).

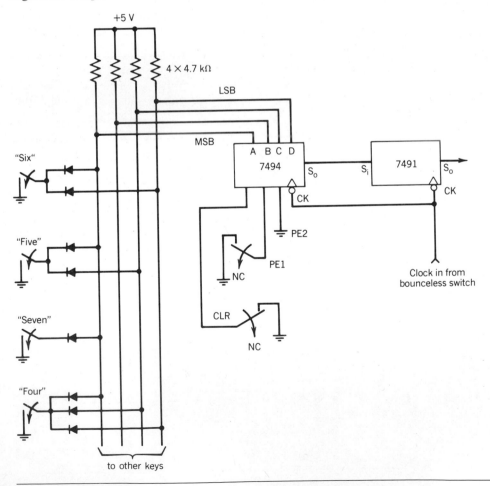

Laboratory: #13 Four-Bit PISO Register
DIPs Needed (approximate): 7494
Chapter Reference: Chapter 5
Object: To use a 5-bit parallel-input/serial-output (PISO) register to build a simple keyboard encoder for a few keys. To convert a single keystroke into serial BCD data.

INTRODUCTION
When you press a key on a hand-held calculator, a single contact switch causes a 4-bit code (in BCD) to be generated. The 7494 can be used to accept this data and convert it to serial data so that it can be entered into a larger register. *Note:* The diodes in the drawing are used to decouple one key from another.

FIGURE E-15 (continued)

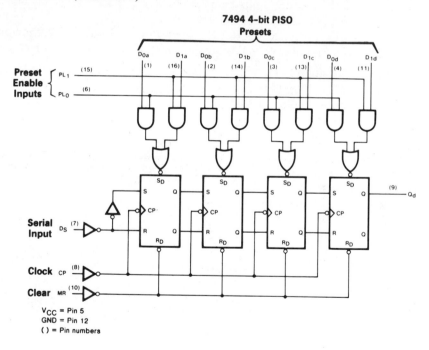

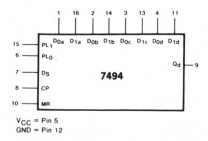

THINGS TO OBSERVE

For correct operation, CLR and PE1 and PE2 are grounded. You must first clear or zero the four flip-flops using the CLR input. Momentarily lift it high and then reground the CLR line. Place the 4-bit data at the ABCD inputs by pressing a keyboard button. Drop the data into the register by momentarily lifting the PE1 line and regrounding it. You should now be able to clock the data out of the register using a bounceless switch on the clock input and observing the serial data output. You can connect a 7491 8-bit register as shown in the drawing to store two "digits" for future use. The 7494 has a second parallel input path but we are not using it. You may wish to try it.

DRAWING

See Fig. E-15.

Laboratory: #14 One-Shots
DIPs Needed (approximate): 74121
Chapter Reference: Chapter 11
Object: To learn about the operation of a pulse stretcher or one-shot. To select component values to change the resulting pulse duration.

INTRODUCTION

The one-shot or pulse stretcher is used when a very narrow pulse or edge must be made wider. The 74121 is a TTL integrated circuit that uses an external resistor

FIGURE E-16 Lab 14: single shots (copyright © 1984 Signetics Corp.).

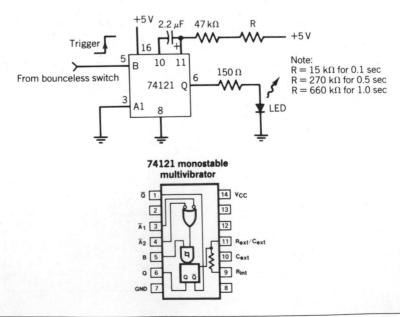

74121 monostable multivibrator

and capacitor to set the width of the output pulse. The resulting output pulse can be observed on an oscilloscope or an LED.

THINGS TO OBSERVE

As you change the size of the timing resistor, notice the change in the duration of the output pulse. This can also be done using the timing capacitor. Look for the relationship between pulse width and R and C. Determine an expression from your data for pulse width versus R and C. To measure the pulse width, trigger the one-shot circuit from a continuous clock.

DRAWING
See Fig. E-16.

Laboratory: #15 Four-Bit Digital-to-Analog Converter (DAC)
DIPs Needed (approximate): 7490, 741 op amp
Chapter Reference: Chapter 12
Object: To build a simple 4-bit D-to-A converter. To convert a binary count ranging from 0000 to 1001 to an analog voltage ranging from 0 to −9 V in steps of 1 V.

INTRODUCTION

The D-to-A converter is the link between the world of computers and the analog world in which we live. An operational amplifier connected as a summing amplifier is used to convert the digital signal to an analog level. The gain of each input is selected to give the proper weight to each bit of the input signal. For instance, the MSB has more weight than the LSB.

THINGS TO OBSERVE

You should see a negative-going staircase waveform at the output of the 741 op amp. If you have selected the resistor values carefully, each step should be exactly 1 V in amplitude. You can adjust the overall amplitude of the staircase waveform by adjusting the size of R_F.

Step the counter manually using a bounceless switch to be certain that you have the idea of the operation of the DAC. What is the relation between the output of the counter and the output of the op amp? What resolution is possible with this 4-bit DAC? What would be the resolution of an 8-bit DAC? Where might a DAC be used?

DRAWING
See Fig. E-17.

Laboratory: #16 Eight-Bit Digital-to-Analog Converter
DIPs Needed (approximate): 7493 (2), MC1408L8 DAC or DAC0808
Chapter Reference: Chapter 12
Object: To further investigate the conversion of digital data to an analog level. In this case an 8-bit commercial D/A converter is used. The resistor values are all internal within the DIP.

INTRODUCTION

An 8-bit D/A converter will have $2^8 = 256$ possible output levels. If driven with an 8-bit binary counter, the resulting staircase waveform will have 256 steps with about 20 mV/step if a 5-V reference is used.

FIGURE E-17 Lab 15: four-bit digital-to-analog converter (DAC)
(copyright © 1984 Signetics Corp.).

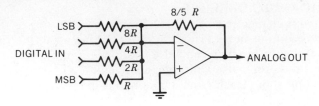

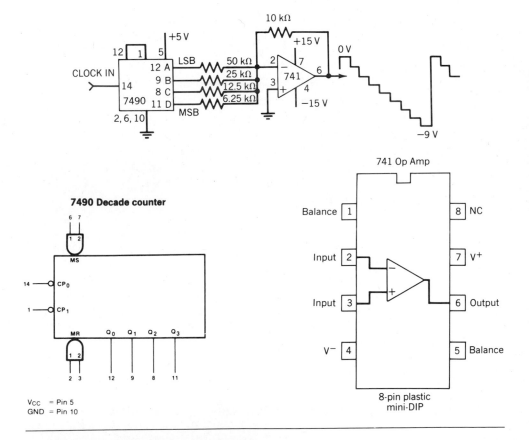

THINGS TO OBSERVE

Verify that you have a negative-going staircase waveform with 256 steps. Determine the d-c level of the first and last steps. Are all the steps uniform in size? What is the size of each step? Is this the resolution of the DAC? If you double the size of the load resistor, what happens to the amplitude of the staircase waveform?

Step the counters manually and watch the resulting d-c output level.

Try the alternate output circuit to obtain a positive-going output waveform.

FIGURE E-18 Lab 16: eight-bit D/A converter.

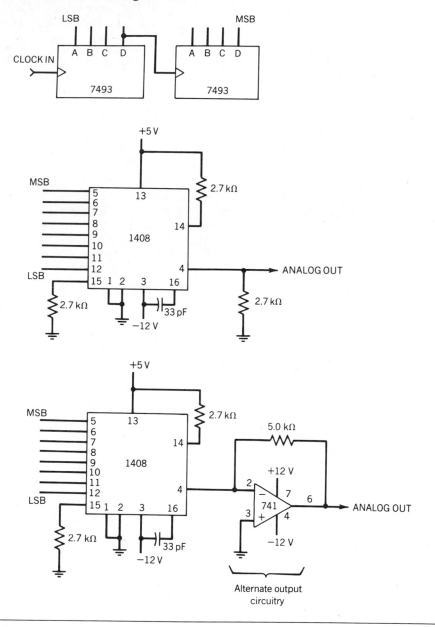

DRAWING
See Fig. E-18.

Laboratory: #17 Four-Bit Analog-to-Digital Converter
DIPs Needed (approximate): 7490, 7447, MAN 1, 741 op amp

Chapter Reference: Chapter 12
Object: To build a simple 4-bit A-to-D converter and use it to measure a negative voltage from 0 to -9 V. To learn the principles of A-to-D conversion.

INTRODUCTION
This A/D converter starts with a working D/A converter. The 4-bit D-to-A converter is constructed as before, and this time the output of the 7490 counter is monitored with an LED 7-segment display. The circuit becomes an A/D when an op amp used as a comparator is added to the circuit.

THINGS TO OBSERVE
The potentiometer is used to produce an unknown analog voltage to measure. Assume that it is set to -7 V d-c. With the 7490 reset, the output of the DAC should be 0. The comparator has 0 V on its $+$ input and -7 V on its $-$ input. The output of the comparator is close to $+15$ V. As we step the counter, the DAC output begins to go negative. When the polarity of the voltage between the comparator inputs changes, the output of the comparator will go negative. This is your signal to look at the MAN 1 and read the unknown voltage. It may be 8. This shows that with a 4-bit DAC the resolution isn't too good! But it is close (within 1 V). This is the principle that many voltmeters use to measure voltage. With an 8- or 10-bit DAC, you can get much closer to the actual voltage.

FIGURE E-19 Lab 17: four-bit A/D converter.

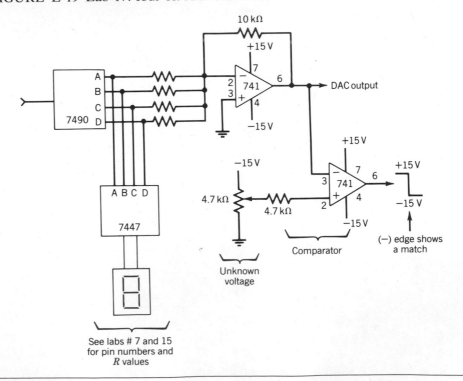

DRAWING
See Fig. E-19.

Laboratory: #18 Burst Clock
DIPs Needed (approximate): 7400, 7493
Chapter Reference: Chapter 5
Object: To build a circuit that produces a specific number of pulses when started.
To investigate a circuit that produces a burst of exactly four pulses when started.

INTRODUCTION

This type of circuit is used to circulate or load 4 bits of serial data from a 4-bit
shift register. It can be used with the circuit of Lab #13 to enter parallel data and
then shift out the BCD number into a larger serial register. A start pulse will cause
exactly 4 pulses to be produced.

THINGS TO OBSERVE

The start pulse causes the circuit to begin operation. Notice that the rate of the
output pulses is half the rate of the input clock.
Modify the circuit to produce exactly 8 output pulses.

DRAWING
See Fig. E-20.

Laboratory: #19 Sixteen-Bit Scratchpad RAM
DIPs Needed (approximate): 7481
Chapter Reference: Chapter 7
Object: To become familiar with the operation of a simple memory circuit capable
of storing 16 bits of binary information. To learn how a location is addressed and
the information read.

FIGURE E-20 Lab 18: burst clock.

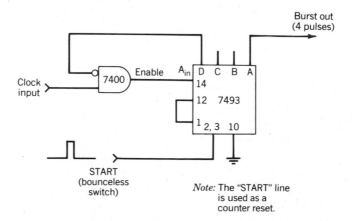

INTRODUCTION

The 16 bits are addressed one bit at a time by lifting one row (x) line and one column (y) line. To address the stored bit at the X_1Y_1 location, raise address lines X_1 and Y_1 to a high level with the remainder of the address lines at a zero level. This procedure is used to address any bit.

THINGS TO OBSERVE

You can work with only one bit at a time. The sense outputs are open-collector outputs and must have pullup resistors. After you address (get to) a location, you can identify the bit in that location by looking at the sense outputs. The sense output that is low (zero) is the contents of the addressed location. (If S_0 is low, there is a "0" in the addressed location.) To write a "0" into the addressed location, momentarily lift W_0. To write a "1" into the addressed location, momentarily lift W_1.

Write 16 bits into the memory using the X and Y wires. Then read out the locations and confirm that the data was stored. The additional circuitry is shown as an example of the necessary decoding circuitry for a RAM.

DRAWING

See Fig. E-21.

Laboratory: #20 Verification of DeMorgan's Theorem
DIPs Needed: (approximate): 7493, 7400
Chapter Reference: Chapter 4
Object: To demonstrate the validity of DeMorgan's theorem when used to convert logic to all-NAND logic. To use Karnaugh mapping to simplify a circuit design.

INTRODUCTION

A 4-input digital black box is constructed and driven with a 4-bit binary counter. This provides a dynamic test of the DBB circuitry and introduces you to the generation of sequential pulse patterns. The counter outputs must be correctly matched to the DBB circuitry. To do this, the MSB of the counter must match the MSB of the digital black box.

THINGS TO OBSERVE

As you connect the DBB to the counter, the counter MSB (D out) must be connected to the digital black box MSB (A input). Then the other outputs must be correctly connected until the LSB of the counter (A out) is connected to the digital black box LSB (D input). Don't get confused with the counter output labels (A B C D) and the DBB labels (D C B A); they don't match. It is important to deal with these connections in terms of LSBs or MSBs.

Trigger the oscilloscope from the slowest counter waveform available (MSB = D out). The waveform resulting from the DBB should match the f column in the truth table. If the oscilloscope is triggered from D out ($-$ edge) the "0" bit should be at the left.

DRAWING

See Fig. E-22.

Figure E-21 Lab 19: sixteen-bit scratch pad memory.

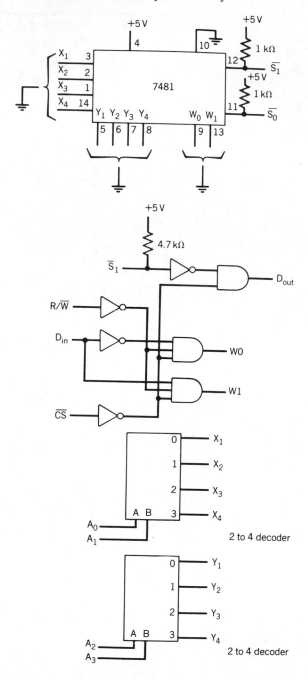

Circuitry for address decoding (optional). "Data out" is always Ø unless $\overline{CS}$ is active (re: low); then D_{out} will follow D_{in} (write) or the data saved in the addressed location (read).

FIGURE E-22 Lab 20: DeMorgan's theorem.

A	B	C	D	f
0	0	0	0	0
0	0	0	1	1
0	0	1	0	0
0	0	1	1	0
0	1	0	0	0
0	1	0	1	1
0	1	1	0	0
0	1	1	1	0
1	0	0	0	0
1	0	0	1	1
1	0	1	0	0
1	0	1	1	0
1	1	0	0	1
1	1	0	1	1
1	1	1	0	1
1	1	1	1	1

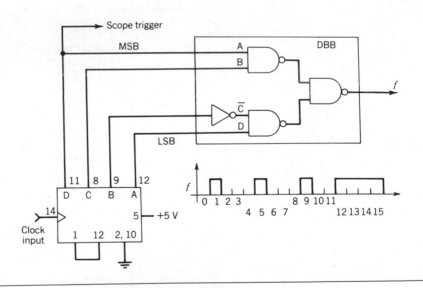

Build a digital black box such that
$$f = M\ (1, 5, 9, 12, 13, 14, 15)$$

The Karnaugh map simplifies to:
$$f = AB + \overline{C}D$$
DeMorgan's theorem converts to NAND
logic producing the equivalent:
$$f = \overline{\overline{AB} \cdot \overline{\overline{C}D}}$$

Laboratory: #21 ASCII "A" Generator (RS232C)
DIPs Needed (approximate): 7400, 7410
Chapter Reference: Chapter 4,8
Object: To design a digital black box that will generate a 300-baud RS232C waveform for an ASCII "A" and have it print over and over on a computer terminal.

INTRODUCTION
We can use the idea from Lab #20 to produce a waveform that will cause a computer terminal (CRT) to receive an ASCII letter "A." The clock must be set to exactly 300 Hz and the terminal must be set to exactly 300 baud. You will need a good-quality pulse generator set to 300 pulses per second and + 5 V to drive the 7493 counter.

THINGS TO OBSERVE
The output of the DBB is an uninverted 11-bit transmission code waveform. You must invert it to get the RS232C format that the terminal will expect. Then $\bar{f}$ is what is sent to the terminal. The terminal will have a DB25 connector on it. Pin 7 is the signal ground and pin 2 is the receive data input for the terminal. Design your own DBB for a different letter of the alphabet and get it working.

DRAWING
See Fig. E-23.

Laboratory: #22 Waveform Generation with an Eight-Bit Digital-to-Analog Converter and a Computer
DIPs Needed (approximate): MC1408L8 or DAC808; microcomputer
Chapter Reference: Chapter 12
Object: To generate three waveforms using a microcomputer and an 8-bit digital-to-analog converter. To study the synthesis of waveforms using a computer and a DAC.

INTRODUCTION
You will need an 8-bit output port available on an 8085-based microcomputer. (Port 01 is assumed here.) The 8-bit output wires are connected to the MC1408L8 DAC. The three programs shown will produce a sawtooth, a triangle, and a sine wave. You have to install sine wave data at address 1000 H.

THINGS TO OBSERVE
Predict the frequency of the waveforms from the time that each program takes to execute (see the 8080A timing sheet in Chapter 10 and adjust for the 8085). How many data points does each program output to produce a complete cycle of each wave? How could you generate the sine wave data that is supplied on the accompanying sheet? Can you produce a data table to create some other type of waveform? Consider the possibility of producing data that would sound like speech. For the sine wave, give the data for sin 0°, sin 90°, and sin 180°.

DRAWING
See Fig. E-24.

FIGURE E-23 Lab 21: ASCII "A" generator.

$$b_1 \cdots\cdots \text{Data} \cdots b_7$$

1 1 1 1 0 1 0 0 0 0 0 1 0 1 1 1 1

SB   ASCII "A"   PB   Stop bits

We will need a digital black box to produce this pattern of bits at the correct rate to transmit an "A" to the computer or receiver unit. The design begins with a truth table and is simplified with a Karnaugh map.

A	B	C	D	f
0	0	0	0	1
0	0	0	1	0
0	0	1	0	1
0	0	1	1	0
0	1	0	0	0
0	1	0	1	0
0	1	1	0	0
0	1	1	1	0
1	0	0	0	1
1	0	0	1	0
1	0	1	0	1
1	0	1	1	1
1	1	0	0	1
1	1	0	1	1
1	1	1	0	1
1	1	1	1	1

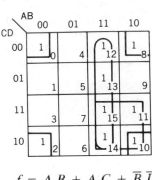

$$f = A B + A C + \overline{B}\,\overline{D}$$

In NAND logic:

$$f = \overline{\overline{A B} \cdot \overline{A C} \cdot \overline{\overline{B}\,\overline{D}}}$$

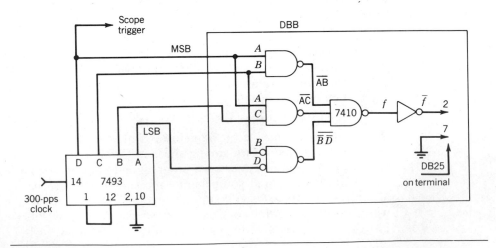

FIGURE E-24 Lab 22: generating waveforms using computer and DAC.

1. Sawtooth: TOP: INR A
                            OUT 01
                            JMP TOP

2. Triangle: TOP: INR A
                            OUT 01
                            JNZ TOP
              MID: DCR A
                            OUT 01
                            JNZ MID
                            JMP TOP

3. Sine: SINE: SET 1000H
              TOP: LXI H, SINE
                            LXI D, 100H
                            MOV A, M
              NEXT: OUT 01
                            INX H
                            DCX D
                            MOV A, D
                            ORA E
                            JNZ NEXT
                            JMP TOP

	0	1	2	3	4	5	6	7	8	9	A	B	C	D	E	F
0000	82	85	88	8B	8E	91	94	97	9B	9E	A1	A4	A7	AA	AD	AF
0010	B2	B5	B8	BB	BE	C0	C3	C6	C8	CB	CD	D0	D2	D4	D7	D9
0020	DB	DD	DF	E1	E3	E5	E7	E9	EB	EC	EE	EF	F1	F2	F4	F5
0030	F6	F7	F8	F9	FA	FB	FB	FC	FD	FD	FE	FE	FE	FE	FE	FF
0040	FE	FE	FE	FE	FE	FD	FD	FC	FB	FB	FA	F9	F8	F7	F6	F5
0050	F4	F2	F1	EF	EE	EC	EB	E9	E7	E5	E3	E1	DF	DD	DB	D9
0060	D7	D4	D2	D0	CD	CB	C8	C6	C3	C0	BE	BB	B8	B5	B2	AF
0070	AD	AA	A7	A4	A1	9E	9B	97	94	91	8E	8B	88	85	82	7E
0080	7B	78	75	72	6F	6C	69	66	62	5F	5C	59	56	53	50	4E
0090	4B	48	45	42	3F	3D	3A	37	35	32	30	2D	2B	29	26	24
00A0	22	20	1E	1C	1A	18	16	14	12	11	0F	0E	0C	0B	09	08
00B0	07	06	05	04	03	02	02	01	00	00	00	00	00	00	00	00
00C0	00	00	00	00	00	00	00	01	02	02	03	04	05	06	07	08
00D0	09	0B	0C	0E	0F	11	12	14	16	18	1A	1C	1E	20	22	24
00E0	26	29	2B	2D	30	32	35	37	3A	3D	3F	42	45	48	4B	4E
00F0	50	53	56	59	5C	5F	62	66	69	6C	6F	72	75	78	7B	7F

ANSWERS TO SELECTED PROBLEMS

Chapter 1

1-1 (a) 13.625   (b) 55.75   (c) 890.75   (d) 0.114257812
    (e) 3539   (f) 183.875   (g) 255.15625   (h) 879.578125

1-2 (a) 22.40625   (b) 0.00352478   (c) 19921.546875
    (d) 2874.787109375

1-3 (a) 1708.29296875   (b) .276072502   (c) 19706.6953125
    (d) 9777.2578125

1-4 (a) 09A2 H   (b) FDE7 H   (c) 0F9F H   (d) 1FFE H

1-5 (a) 8193rd   (b) 513th   (c) 8206th   (d) 52225th

1-6 (a) 4B H = 1001011B = 113 Q
    (b) 121 H = 100100001B = 441 Q
    (c) FE7 H = 111111100111B = 7747 Q
    (d) 6F0F H = 110111100001111B = 67417 Q

1-7 (a) .333$\overline{3}$ H = .0011001100$\overline{11}$B = .1463$\overline{1463}$ Q
    (b) .199$\overline{9}$ H = .0001100110011$\overline{1001}$B = .063$\overline{1463}$ Q
    (c) .0BCD H = .0000101111001101B = .02746 Q

1-8 (a) 11111110   (c) 01010101   (e) 11000011   (g) 10010100

1-9 21 = 00100001
    00 = 00000000
    40 = 01000000
    DB = 11011011
    40 = 01000000
    FE = 11111110
    02 = 00000010
    C2 = 11000010
    03 = 00000011
    40 = 01000000
    76 = 01110110

1-12 $2^N - 1 = 2^{20} - 1 = 1,048,575 =$ FFFFF H

1-14 (b) 001000001010.010011001101 B

1-15 (a) 14011.59375   (b) 18239.297363281   (c) 219.609375

1-16 (a) 1243.20 Q   (b) 14232.77 Q   (c) 6624.0020304 Q

Chapter 2

2-1   (a) 66 D = 42 H = 01000010 B   (b) 169 D = A9 H = 10101001B
      (c) 44.8 D = 2C.CC$\overline{C}$H = 101100.1100110$\overline{1100}$ B
      (d) 85.3 D = 55.4C$\overline{C}$H = 1010101.010011001$\overline{100}$ B

2-2   (a) 62 D = 3E = 111110 B   (b) −44 D = −2C H = −101100B
      (c) 3.9 D = 3.E6$\overline{6}$H = 11.11100110$\overline{0110}$ B

2-3   (a) 19 D = 13 H = 10011 B   (b) −26 D = −1A H = −11010 B
      (c) 8.14D = 8.23D H = 1000.001000111101 B

2-4   (a) 36 D = 24 H = 100100 B   (b) 116 D = 74 H = 1110100 B
      (c) −58 D = 3A H = 111010 B

2-5   (a) Accumulator entries:

```
        206.4          11001110.011001100110
      ×    36          × 100100.
        7430.4                        ?

            1100111001.1001100110
            1100111001100.1100110011
            1110100000110.0110011001
                  (7430.399414062)
```

      (b) Accumulator entries:

```
        141.3          10001101.010011001100
      ×     .6          ×        .100110011001
        84.78                       ?
                       1000110.1010011001100
                          1000.1101010011001
                           100.0110101001100
                             .1000110101001
                             .0100011010100
                       1010100.1011100101110
                            (84.724365234)
```

2-6   (a)   0010 0110
             + 0100 1001
               0110 1111
             + 0000 0110
               0111 0101
             (75 Answer)

      (c)   1001 0011.0010
               + 0100 0110.1000
                 1101 1001.1010
               + 0110 0000.0110
               1 0011 1010.0000
           + 0000 0000 0110.0000
             0001 0100 0000.0000
                (140.0 Answer)

2-7 (a)   0010 0111   9A
            + 1000 0111   − 13
          ————————   ————————
              1010 1110   87 Calculation of complement
            + 0110 0110
          ————————
            1 0001 0100
              (14 Answer)

2-8 (b) 101011.0001 B
2-9 (b)   28 H
            − 79 H   − 79 H = 10000111 B
          ————————
            − 51 H

              0010 1000
            + 1000 0111
          ————————
            0 1010 1111
            (−)0101 0000
          ————————
                  + 1
          ————————
            (−)0101 0001   Answer

2-10 (b) Accumulator entries:

$$
\begin{aligned}
&101111.11\\
&1011111.10\\
&10111111.00\\
&1011.1111\\
&\underline{\qquad 10.111111}\\
&101011101.001011\\
&\quad(349.171875 \qquad \text{Answer})
\end{aligned}
$$

2-13 19C6.74FD H = 1100111000110.0111010011111101 B = 14706.35176 Q

Chapter 3

3-1 (b) f is a 1 when A is a 1 or B is a 0 or C is a 1
    (d) f is a 1 when you do not have (not($A = 1$ and $B = 1$) or ($A = 0$ and $B = 1$))
3-4 (a) Good   (b) Bad   (c) Bad   (d) Good
3-5 (a) $f = \overline{A\,B} \cdot \overline{C}\,D$
    (c) $f = \overline{\overline{A\,B} \cdot \overline{\overline{A}\,\overline{B}} \cdot \overline{A}\,B\,C}$
3-6 (a) $f = \overline{A + \overline{B} + \overline{\overline{B} + \overline{C}}}$
    (c) $f = \overline{A + \overline{B} + \overline{\overline{B} + \overline{C}}}$
3-7 $X = \overline{A\,B}$ $Y = \overline{B + C}$
3-8 $f = C\,D$; Yes.

Chapter 4

4-1 (a) $f = A + BC + \overline{B}\,\overline{C}$
    (c) $f = \overline{B}\,\overline{C} + \overline{A}\,B$
4-2 (a) $f = (A + \overline{A})\,\overline{B}\,\overline{C} = \overline{B}\,\overline{C}$ (3 gates vs 6 gates)
    (c) $f = A\,\overline{B}$ (7 gates vs 2 gates; fewer inputs)

4-3 (a) S/P $f = \overline{A}\,C + A\,\overline{C}$ P/S $f = \overline{\overline{A}\,C + A\,C}$
   (c) S/P $f = \overline{A}\,B + A\,\overline{B}\,C$ P/S $f = \overline{\overline{A}\,B + A\,B + A\,\overline{C}}$
4-4 (a) S/P $f = \overline{\overline{A}\,\overline{C} \cdot A\,\overline{C}}$ (NAND)
           $f = \overline{\overline{A + \overline{C}} + \overline{\overline{A} + C}}$ (NOR)
       P/S $f = \overline{\overline{A}\,C \cdot A\,C}$ (NAND)
           $f = \overline{\overline{A + C} + \overline{\overline{A} + \overline{C}}}$ (NOR)
4-5 S/P $f = \overline{A}\,\overline{B} + A\,B + A\,\overline{C}$ (or $\overline{B}\,\overline{C}$)
   P/S $f = \overline{A\,B + \overline{A}\,\overline{B}\,C}$
4-6 (a) S/P $f = \overline{A}\,\overline{C} + \overline{A}\,B + A\,\overline{B}\,C\,\overline{D}$
       P/S $f = \overline{A}\,\overline{C} + \overline{A}\,D + \overline{A}\,\overline{B}\,C + A\,B$
   (c) S/P $f = A\,\overline{C} + A\,B\,\overline{D} + \overline{A}\,\overline{B}\,C$
       P/S $f = \overline{A}\,C + \overline{A}\,B + A\,C\,D + A\,\overline{B}\,\overline{C}$
4-7 (a) $f = B\,\overline{C} + A\,\overline{B}\,C$ $\overline{f} = B\,C + \overline{A}\,\overline{B} + \overline{B}\,\overline{C}$
   (b) $f = \overline{A}\,B + A\,\overline{B} + A\,C$ $\overline{f} = \overline{A}\,\overline{B} + A\,B\overline{C}$
   (c) $f = \overline{B}$ $\overline{f} = B$
4-8 (a) $f = \overline{A}\,\overline{B}\,\overline{C}\,\overline{D} + A\,B\,\overline{C} + \overline{A}\,B\,C + B\,C\,\overline{D} + A\,\overline{B}\,C\,D$
   (c) $f = \overline{A}\,\overline{C} + A\,C$ $\overline{f} = A\,\overline{C} + \overline{A}\,C$

Chapter 5

5-2 The flip-flop requires a falling edge to cause the clock operation.

5-3 Asynchronous inputs produce immediate effects regardless of the state of the clock. Synchronous inputs act only with a valid clock input.

5-7 Counts backward (from 15 to 0).

Chapter 6

6-1 $2^5 = 32$ $2^8 = 256$

6-2 $Y = \overline{\overline{A\,\overline{S}} \cdot \overline{B\,S}}$
      Case 1: $S = 0$, then $Y = A$
      Case 2: $S = 1$, then $Y = B$

6-3 The circuit in Fig. 6-32b is preferred because it uses one chip instead of three.

6-4 (a) LEDs 5 and 11   (b) LEDs 2 and 15
   (c) LEDs 3 and 12   (d) LEDs 6 and 11

Chapter 7

7-1 (a) $2^{15} = 32{,}768$
    (b) $32{,}768 \times 8 = 262{,}144$

7-2 The chip enable is used to select individual RAM chips in large memory systems.
    (a) Yes

7-3 (b) No

7-4 Read and write do not occur simultaneously. A single control line prevents this from happening. A high on the read/write line will cause a read operation. A low on this line will cause a write operation.

7-5 (a) 1024 bits
    (b) 4096 bits

7-6 35 tracks $\times$ 10 sectors/track $\times$ 256 bytes/sector = 89,600

7-7 2^{12} = 4096 (Twelve is the answer.)

7-8 Thirteen.

7-9 4096

7-10 2^{14} = 16384

7-11 1K = 1024 = 400 H

4K = 4096 = 1000 H

8K = 8192 = 2000 H

Therefore, if memory starts at 2000 H, an 8K memory will end at 3FFF H.

7-12 16K = 16384 = 4000 H

If memory starts at 5000 H, it ends at 8FFF H

7-13 (a) 2^{14} = 16384

2^{13} = 8192

For a 12K memory, you need 14 address lines.

(b) 12K = 4K + 8K = 4096 + 8192 = 12288 = 3000 H

Starting at 0, a 12K memory ends at 2FFF H

(c) At 5000 H, memory ends at 7FFF H

7-14 (a) 16: 2^{16} = 65,536

(b) 20: 2^{20} = 1,048,576

(c) 24: 2^{24} = 16,777,216

(d) 28: 2^{28} = 268,435,456

(e) 32: 2^{32} = 4,294,967,300

7-15 Word processing   Multiuser systems

Image processing   Large data bases

Very large information systems   Pattern recognition

Speech analysis/recognition

Chapter 8

8-1 (a) 11000110   (c) 01001101   (e) 01100101

(g) 01010110   (i) 00010100

8-2 (a)

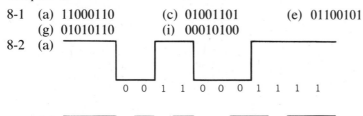

0 0 1 1 0 0 0 1 1 1 1

(e)

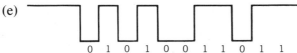

0 1 0 1 0 0 1 1 0 1 1

8-3 1800/11 = 163.63$\overline{63}$ characters per second

8-7 (a) DC1 = ↑ Q = 00010001

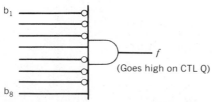

(Goes high on CTL Q)

(d) b_6 = 0, b_7 = 0 → f = $\overline{b_6} \cdot \overline{b_7}$ = $\overline{b_6 + b_7}$

8-8   * = 2AH or AAH (even parity)
      $f = M(2,4,6,8,9,10,11,12,13,14,15)$
      $f = A + C \overline{D} + B \overline{C} \overline{D}$ *(answer)*
8-15 S/P: $f = A + \overline{B}C + D$
      P/S: $f = \overline{\overline{A} \overline{C} \overline{D} + \overline{A} B \overline{D}}$
8-16 (a) CR, LF, DEL
      (b) 8085
      (c) IBM

Chapter 9

9-3 $c = 300 \times 10^6$ m/sec $\times$ 39.370 in./m = 1.1811×10^{10} in./sec
      delay = $0.2/(1.1811 \times 10^{10} \times \frac{2}{3})$ = 25.4 psec
9-4 $c = 300 \times 10^6$ m/sec $\frac{2}{3}c = 200 \times 10^6$ m/sec

      delay = $\dfrac{1.5}{200 \times 10^6}$ = 7 ns

9-20 (a) 0A H   (c) 20 H
9-21 (a) CC H   (c) 7F H
9-22 (a) 6B H   (c) FBH
9-23 Bit 1 = ANI 01
      Bit 2 = ANI 02
      Bit 3 = ANI 04
      Bit 4 = ANI 08
      Bit 5 = ANI 10H
      Bit 6 = ANI 20H
      Bit 7 = ANI 40H
      Bit 8 = ANI 80H

Chapter 10

10-2   3E 20 06 30 05 C2 04 20 76
10-3   370.5 μsec
10-5   00000
10-6   00011
10-8   2167 μsec
10-9   80 H
10-11 1000 D = 3E8 H, therefore 1000 H $-$ 3E8 H = C18 H

Chapter 13

13-1   500 nS (nanoseconds)
       7(500) = 3.5 μS (micro-seconds)
13-4   The 8224 uses an internal $\div$ 9 counter to generate the two clock phases.
13-8

13-13 15 RAM ghosts (+1 original RAM area)
   7 ROM ghosts (+1 original ROM area)

Chapter 16

16-1 512
16-2 256
16-3 25.6 μsec
16-4 The logic analyzer freezes high-speed data on many channels, thus allowing the user to study the data at leisure.
16-8 Software under test can be initially tested in memory that is known to be good. Then gradually the software can be released to prototype memory for testing. This allows testing of the prototype memory as well as the software.
16-10 Check the power supply for correct d-c level, then a-c ripple. Check to be sure that the clock is running correctly. See if there is activity on the address and data lines. Check the level of the ready, hold, and interrupt lines. Look for obvious physical damage (burned parts or lands). Check the I/O devices to be sure they are operational.
16-14 By examining data on data lines, it would be easy to find two bits that were always the same, indicating that they are shorted together. It would also be easy to find a data line that is always tied low or always tied high. Observing the address lines may turn up the same type of problem.
16-15 It is not easy holding even two probes for an oscilloscope in place to determine if signals appear to be the same. And which two should be tested? The logic analyzer, with its many inputs, allows simultaneous monitoring of many lines.

INDEX

Access time, 195
Accumulator, 27
ACIA, 247
A-to-D conversion, 4
A-to-D converter, 474, 480
Addition, binary, 26
Address, memory, 21
Address bus, 261
Addressing, RAM, ROM, 524
Addressing modes, 270
Add/subtract instruction, 267
Alarm circuit, 82
Analog signals, 4
AND gate, 65
Arithmetic logic unit (ALU), 27
Arithmetic processor, 567
Array processors, 662
ASCII code, 230
Assembler, 265, 400
Assembler syntax, 323
Assembly language, 264
Asynchronous clock, 134
Asynchronous data transmission, 236
Auxiliary carry bit, 44

Babbage, Charles, 2
Bar Code Reader Program (WAND), 459
Base conversions, software, 388
Base (radix), 5
Baudot, J. E. (Emile), 237
Baud rates, 237
BCD addition, 43
BCD arithmetic, 42
  advanced program, 442
  program, 392

BCD-to-decimal decoder, 173
BCD subtraction, 45
Bidirectional transceiver, 176
Binary adder, 85
Binary counter, 146
Binary numbers, 5
Bistable flip-flop, 154
Bit, 5, 7, 8
Bit flipping, 348
Bit slice processor, 2901, 654
Black box design (DBB), 98
Block move, Z80, 592
Block search, Z80, 592
Boolean algebra, 60
Boolean identities, 63
Boolean reduction, 99, 107
Bounceless switch, 81, 130
Branch instruction, 269
Bubble memory, 221
Buffer, 64
Burglar alarm, 161
Burning, EPROM, 213
Bus drivers, 176
Byte, 7

Cable length, recommended, 240
Carry bit, 26, 33
Caterpillar program, 347
Central processing unit (CPU), 260
  8085, 529
Character generator, 496
Character generator ROM, 207
Chip enable (CE) line, 197
Chip select (CS) line, 196
CIN routine, 355

Clock, 259
Clock pulse, 36
CMOS logic, 52, 57, 61
421 Code, 18
8421 Code, 18
Compare instruction, 269
Complement, 28
Computer memory, 21
Conditioning interrupt pulses, 416
Control bus, 262
Control characters, 357
   (ASCII), 233
Control word, 37
Core memory, 214
Cost, of design, 102
Counter(s), 129, 139
   8-bit binary, 8085, 326
Counting in different bases, 12
COUT routine, 355
Cross assembler, 692
Current loop transmission, 242

D-to-A conversion, 4
D-to-A converter, 474, 480
Data bus, 261
   8-bit, 8080A, 522
Data control devices, 157
Data selector, 82
Data terminal, 230
Data transmission, 229
Data conversion, 386
DB25 connector, 239
DeMorgan's theorem, 73
Demultiplexers, 157, 168
Decade counter, 140
Decimal numbers, 5
Decoder(s), 82, 157, 168
Delay, double resister, 336
Delay circuit, 151
Deposit, 21
Destructive read out memory, 217
Development systems, 668
Digital signals, 4
Diode matrix ROM, 210
Diode resistor logic, 52
Direct memory access (DMA), 272
Disassembler, 265
Disassembler program:
   6502, 437
   8085, 424
Disco light show, 395
Disk memory, 223
Displays, 118
Display system, multiplexed, 485
Division, 35

Dot matrix, 208
Downline loading, 497
D-type flip-flop, 132
Dual in-line package (DIP), 53
Dynamic RAM, 163
Dynamic RAM controller, 561

EBCDIC code, 248
Edge trigger, 132
Editor, 670
EEPLA, 124
Electronic Industries Association, 238
Eleven-bit transmission code, 235
Emitter coupled logic, 52, 61
Emulation, 669
Enabling, pulse, 71, 139
End around carry, 29
ENIAC, 2
EPROM:
   2708, 211
   2716, 213
   EAPROM, PROM, 206
Exclusive OR gate (XOR), 66

Fall time (t_f), 81
Fanout, gate, 65
Fetch, decode, execute cycle, 263
Field programmable array (FPLA), 122
Flags, Z,S,P,C,AC, 288
Flip-flops, 129, 130
   testing, 371
Floating bus, 183
Floating point arithmetic, 447
Floating point processor, 563
Floppy disk controller, 572
Four-line to sixteen-line decoder, 175
Fractional part of number, 13
Fusible link, 122

Gate array logic, 121
Gate propagation delay time, 53
Gate substitutions, 73
General purpose computer, 258

Hardware, 8080A, 513
Hexadecimal numbers, 9
Hex dump, program, 419
Hex pairs, 20

Inhibit operation, 220
Input/output ports, 321
Input/output section (I/O), 262, 270
Instructions, 264
   logical group instructions, 8080A/8085:
      ANA r, 298

ANA m, 298
ANI data8, 299
XRA r, 299
XRA m, 300
XRI data8, 301
ORA r, 301
ORA m, 302
ORI data8, 303
CMP r, 303
CMP m, 304
CPI data8, 304
RLC, RRC, 305
RAL, RAR, 306
CMA, 306
CMC, 307
STC, 307
branch group instructions, 8080A/8085:
JMP, 308
RST n, 311
Jcc, 308
PCHL, 312
CALL addr, 310
data transfer instructions, 8080A/8085:
MOV R1, R2, 281
MOV R, M, 281
MOV M, R, 282
MVI R, data8, 282
MVI M, data8, 283
LXI RP, data16, 283
LDA addr, 284
STA addr, 284
LXLD addr, 285
SHLD addr, 285
LDAX RP, 286
STAX RP, 287
XCHG, 287
arithmetic group instructions, 8080A/8085:
ADD r, 288
ADC r, 289
ADI data8, 290
ACI data8, 291
SUB r, 291
SBB r, 293
SUI data8, 293
SBI data8, 294
INR r, 294
DCR r, 295
INX rp, 295
DCX rp, 296
DAD rp, 296
stack I/O machine control, 8080A/8085:
PUSH RP, 313
PUSH PSW, 313
POP RP, 314
POP PSW, 314
SPHL, 315
XTHL, 314
IN port, 316
OUT port, 316
EI, DI, 317
HLT, 317
NOP, 317
RIM, SIM (8085 only), 317
Integer part of number, 13
Integrated circuit test program, 366
Integrator, 218
Interface circuits, 176, 177
Intermittent circuits, 687
Internal organization, 8080A, 514
Interrupt(s), 272
60Hz, 415
software, 379
Interrupt handling routine, 379
Interrupt priority, 381
Inverter, 53, 59
I/O ports for 80804, 528
I/O routines, serial, 350

J-K flip-flop, 137
Jump instruction, 269

Karnaugh mapping, 108
Keypad scanner program, 374

Label, 324
Lamp drivers, 79
Large-scale integration (LSI), 57
Largest numbers, 8
Latch, d-c, 130
Least significant bit, 6
Leibniz, 2
Life expectancy (IC's), 688
Load, see Deposit
Location, see Address, memory
Logic, positive/negative, 57
Logical operations, 267
Logic analyzers, 673
Logic circuit design, 97
Logic families, 52
Logic gates, 51, 59
Logic levels, positive/negative, 57
Long word, 7
Lookup tables, 361

Machine cycle, 8085, 553
Machine cycle states, 8080A, 516
Machine language, 265
Maximum data rates, 237
Memories, 189
Memory, composition, 260

Memory map, 200
Memory mapped video display, 382
Message center program, 359
Microcomputers, comparisons, 258
Microprocessors:
  history, 278
  types of:
    8080A/8085, 585
    Z80, 585
    6800, 597
    6502, 609
    8086/8088, 617
    80186, 630
    80286, 632
    68000, 633
    8048/8049, 643
    8051, 645
    8096, 649
    432, 653
    2901, 654
    LSI-11, 658
Minimum system, 8085, 536
Minterms/maxterms, 98
Mnemonic, 324
Modems, 249
Most significant bit, 6
MOVE instruction, 266
M terms, 98
Multiplexed display, 181
Multiplexer(s), 84, 120, 157
Multiplication, 35
  by decimal, 38
  by integer, 37

NAND gate, 67
Nibble, 7
Nine's complement, 45
Nixie display, 174
Noise margin, 58
NOR gate, 68
Number conversion, 12
Number systems, 1

Octal D-type flip-flop, 178
Octal numbers, 8
One's complement, 28
Open collector gates, 60
Operand, 324
Organization of computers, 257
OR gate, 66
ORG statement, 323
OR-tying, 161
Oscillators, 78

Parallel adder, 86

Parallel data transmission, 234
Parallel input, 149
Parallel I/O, 364
Parity bit, 233
Partitioning memory, 526
Pascal, Blaise, 2
Peripheral interface adapter, 365
Physical damage, 686
Polling, 411
Power dissipation per gate, 53
Powers of two, 6, 7
Power supply problems, 686, 690
Processor cycle, 8080A, 514
Processor registers, 264
Product of sums (P/S), 103
Programmable array logic (PAL), 121
Programmable CRT controller, 574
Programmable interval timer, 571
Programmable keyboard/display interface, 575
Propagation delay time, 76
Protocol, data transmission, 239
Pseudo-opcodes, 324
Pseudo-random generator, 193
Pulses, software generated, 370
Pulse width (t_{pw}), 81

Race conditions, 153
Radix (base), 11, 22
Random access memory (RAM), 195
Read cycle, 197
Reading memory, 196
Read only memory (ROM), 195
Receiver ready bit (RRDY), 351
Receiver-transmitter chips, 246
Recirculate line, 191
Recognizing commands, 357
Redundant terms, 106
Refresh of dynamic memories, 195
Register organization, 8080A, 280
Registers, 21
Relay driver, 398
Remainder, 13
Representation of numbers, 11
Resistor transistor logic, 52, 59, 61
Restart addresses (RST n), 378
Ripple blanking, 117
Ripple carry, 147
Rise time (t_r), 81
RS flip-flop, 130
RS232C standard, 238
RS422 standard, 252

Saturated logic, 59
Saving registers on stack, 344
Schmitt trigger gate, 63

Schottky TTL, 53
Screen memory, 167
Scrolling, video display, 384
Sectors, on disk, 224
Security system, 411
Segmented memory, 617
Selectors, 157
Sense amplifier, 218
Serial data communication tester, 680
Serial data transmission, 234
Serial input, 148
Serial I/O, 350, 354
Seven-segment readout, 116
Shift register(s), 36, 129, 148
Shift register memories, 190
Shortcut between bases, 17
Signal definitions, 182
Signature analysis, 684
Signed 2's complement arithmetic, 31
Simplex, half and full duplex, 241
Simulation, 669
Single shots/one shots, 397
Software clock (real time), 412
Software flip-flop, 348
Software logic gates, 87
Software UART program, 406
Sorting numbers, 504
Stack use, 343
Start bit, 235
Static electricity problems, 688
Static vs. dynamic, 194
Status lines, 8085, 532
Status word, 8080A, 517
Stop bits, 235
Strobe pulse, 119
Structured memory, 653
Subroutines, 345
Subtraction, binary, 28
Sum of Products (S/P), 103
Symbol table, 401
Synchronous clock, 133
Synchronous data transmission, 236
System bus, 262

Tape memory, 222
Telephone dialer program, 440
Temperature measurement, 399
Temperature range, 53

Ten's complement, 45
Time delay, 183, 338
  for 8085, 531
  microprocessor, 320
Timing diagram, 141
Timing diagram fundamentals, 180
Toggle, 155
Tone generator, 172
Tracing, program, 670
Tracks, on disk, 224
Traffic signal control program, 372
Transceivers and latches, 176
Transistor transistor logic, 52, 53
Transmitter ready bit (TRDY), 351
Trigger word, 692
Tri-state logic, 79
Tri-state outputs, 162
Troubleshooting, 68
Troubleshooting microprocessors, 684
Troubleshooting techniques, 667
Turing, Alan M., 2
TV typewriter, 191
Two-phase clock, 516
Two's complement arithmetic, 29
Typewriter programs, 355, 358

UART, 351
UART, USART, 247
Universal peripheral interface, 558
Unsigned 2's complement arithmetic, 33
USART–8251A, 351

Valid hex digits (VALDIG), 388
Variable baud rate, 410
Very-large-scale integration, 57
Voltage controlled oscillator, 171
Voltmeter (digital) program, 474

Waveform generation, 477
Weight, 5
White noise source, 163
Wired OR function, 78
Word, 7
Write cycle, 198
Writing memory, 196

Zero crossing detection, 485
Zeroing, accumulator, 326